Prealgebra

Prealgebra

FIFTH EDITION

Richard N. Aufmann

Palomar College, California

Vernon C. Barker

Palomar College, California

Joanne S. Lockwood

New Hampshire Community Technical College, New Hampshire

HOUGHTON MIFFLIN COMPANY
Boston   New York

Publisher: Richard Stratton
Executive Editor: Mary Finch
Senior Marketing Manager: Katherine Greig
Assistant Editor: Janine Tangney
Senior Project Editor: Kerry Falvey
Art and Design Manager: Jill Haber
Cover Design Manager: Anne S. Katzeff
Senior Photo Editor: Jennifer Meyer Dare
Senior Composition Buyer: Chuck Dutton
New Title Project Manager: James Lonergan
Editorial Assistant: Nicole Catavolos
Marketing Assistant: Erin Timm
Editorial Assistant: Emily Meyer

Cover Photograph © iStock International, Inc.

Photo Credits   p. xxv Stockxpert; p. xxviii Stockbyte/Getty Images; p. xxix Stockxpert; p. xxxi–xxxv Stockxpert; p. 1 Blend Images/Getty Images; p. 8 AP/Wide World Photos; p. 17 © CORBIS; p. 20 Vince Streano/CORBIS; p. 38 Dave Bartuff/CORBIS; p. 65 Wally McNamee/CORBIS; p. 66 AFP/CORBIS; p. 77 Blend Images/Getty Images; p. 84 AP/Wide World Photos; p. 87 Spencer Grant/PhotoEdit, Inc.; p. 88 AP/Wide World Photos; p. 100 Tannen Maury/The Image Works; p. 115 DPA/The Image Works; p. 127 AP/Wide World Photos; p. 148 Bettmann/CORBIS; p. 149 © Dynamic Graphics Group/IT Stock Free/Alamy; p. 150 Getty Images/Hola Images; p. 157 Getty Images; p. 170 Torc, from the Snettisham Hoard, Iron Age, c.75 B.C. (gold) by Celtic (1st century B.C.) © British Museum, London, UK/Photo © Boltin Picture Library/The Bridgeman Art Library; p. 171 Bettmann/CORBIS; p. 188 (top) Dennis MacDonald/PhotoEdit, Inc.; p. 188 (bottom) Superstudio/Getty Images; p. 189 (top) AP/Wide World Photos; p. 189 (bottom) Stephen Frank/CORBIS; p. 204 John Giustina/Getty Images; p. 205 AP/Wide World Photos; p. 206 Bob Daemmrich/The Image Works; p. 213 © Mark E. Gibson/CORBIS; p. 224 © Roger Hagadone/SuperStock; p. 230 AFP/CORBIS; p. 235 AP/Wide World Photos; p. 242 Bettmann/CORBIS; p. 245 Frank Siteman/PhotoEdit, Inc.; p. 276 AP/Wide World Photos; p. 282 © mediacolor's/Alamy; p. 293 NASA; p. 303 Paul Seheult/Eye Ubiquitous; p. 304 Bill Aron/PhotoEdit, Inc.; p. 315 Jay Freis/Getty Images; p. 316 Daisuke Morita/Photodisc/Getty Images; p. 338 Javier Pierini/Digital Vision/Getty Images; p. 343 Richard Hamilton Smith/CORBIS; p. 360 Reuters NewMedia, Inc./CORBIS; p. 368 (top) CORBIS; p . 368 (bottom) Tom Prettyman/PhotoEdit, Inc.; p. 369 Reproduced by permission of The State Hermitage Museum, St. Petersburg, Russia; p. 374 Photodisc/Getty Images; p. 376 Adam Crowley/Photodisc/Getty Images; p. 379 © Reuters/CORBIS; p. 380 Tony Freeman/PhotoEdit, Inc.; p. 408 © Najlah Feanny/CORBIS/SABA; p. 409 (top) Frank Gaglione/Getty Images; p. 409 (bottom) © J. A. Giordano/CORBIS/SABA; p. 439 Rich Pilling/Getty Images; p. 447 Franck Fife/AFP/Getty Images; p. 448 Justin Sullivan/Getty Images; p. 456 Andy Lyons/Getty Images; p. 461 Photodisc/Getty Images; p. 469 Brad Wilson/Getty Images; p. 470 (top) Stephen Shugerman/Getty Images; p. 470 Reza Estakhrian/Getty Images; p. 482 Jeff Gross/Getty Images; p. 491 © Stockdisc/age fotostock; p. 492 Roy Morsch/CORBIS; p. 507 Bruce Miller/Alamy; p. 509 Robert Brenner/PhotoEdit, Inc.; p. 510 © Bob Krist/CORBIS; p. 511 © Robert Essel NYC/CORBIS; p. 518 Paul Conklin/PhotoEdit, Inc.; p. 521 Photodisc/CORBIS; p. 526 Felicia Martinez/PhotoEdit, Inc.; p. 530 Dwayne Newton/PhotoEdit, Inc.; p. 534 Johannes Simon/AFP/Getty Images; p. 535 © Andrew Holt/Alamy; p. 539 © Getty Images; p. 553 AP/Wide World Photos; p. 568 Royalty-Free/CORBIS; p. 573 Patrick Ward/CORBIS; p. 589 © dk/Alamy; p. 590 © Kim Karpeles/Alamy; p. 592 Thinkstock Images/Jupiter Images; p. 605 John Zich/AFP/Getty Images; p. 607 © StockShot/Alamy; p. 611 David Young-Wolff/PhotoEdit, Inc.; p. 619 © Tony Freeman/PhotoEdit, Inc.; p. 622 Syracuse Newspapers/The Image Works; p. 623 Steve Chenn/CORBIS; p. 626 Cat Gwyn/CORBIS; p. 634 Michael Newman/PhotoEdit, Inc.; p. 634 Visuals Unlimited; p. 635 L. Clarke/CORBIS; p. 646 © Michael Newman/PhotoEdit, Inc.

Printed in the U.S.A.

Library of Congress Control Number: 2007926194

ISBNs
Instructor's Annotated Edition:
ISBN-13: 978-0-618-96681-3
ISBN-10: 0-618-96681-1

For orders, use student text ISBNs:
ISBN-13: 978-0-618-95688-3
ISBN-10: 0-618-95688-3

2 3 4 5 6 7 8 9—WEB—11 10 09 08

Contents

3 Fractions 149

4 Decimals and Real Numbers 235

5 Variable Expressions   315

6 First-Degree Equations   379

Applications

7 Measurement and Proportion   **439**

8 Percent 491

9 Geometry 535

10 Statistics and Probability 605

Preface

Welcome to *Prealgebra!*

Prealgebra, Fifth Edition, builds on the strong pedagogical features of the previous edition. Always with an eye toward supporting student success, we have increased our emphasis on conceptual understanding, quantitative reasoning, and applications. You will find in this edition new features that support student success and understanding of the concepts presented. We hope you enjoy using this new edition of the text!

Aufmann Interactive Method (AIM)

By incorporating many interactive learning techniques, including the key features outlined below, *Prealgebra* helps students to understand concepts, to work independently, and to obtain greater mathematical proficiency.

- *AIM for Success Student Preface* This updated student preface encourages the student to interact with the textbook. Students are asked to fill in the blanks, answer questions, prepare a weekly time schedule, find information within the text and on the class syllabus, write explanations, and provide solutions. This approach will ensure their understanding of how the textbook works and what they need to do to succeed in this course.

  - **The AIM for Success PowerPoint Slide Show is available on the instructor website!** This PowerPoint presentation offers a lesson plan for the AIM for Success Student Preface in the text. Visit **college.hmco.com/pic/aufmannPA5e** to access this content and more.

- *You Try It Exercises* Each objective of *Prealgebra* contains numbered Examples. To the right of each Example is the You Try It feature, which encourages students to test their understanding by working an exercise similar to the Example. Page references are provided beneath the You Try Its, directing students to check their solutions using the fully–worked-out solutions in the Appendix. This interaction among the Examples, You Try Its, and Solutions to the You Try Its serves as a checkpoint for students as they read the text and study a section.

- **NEW!** *Interactive Exercises in the Exercise Sets* These exercises provide students with guided practice on core concepts in each objective. We are confident that these Interactive Exercises will lead to greater student success in mastering the essential skills. Read more about the new Interactive Exercises below.

Changes for the Fifth Edition

Content Enhancements to this Edition

- **NEW!** *Interactive Exercises* These exercises test students' understanding of the basic concepts presented in a lesson. Included when appropriate, these exercises:
  - Generally appear at the beginning of an objective's exercise set.
  - Provide students with guided practice on some of the objective's underlying principles.

- Test students' knowledge of the terms associated with a topic OR provide fill-in-the-blank exercises in which students are given part of a solution to a problem and are asked to complete the missing portions.

- Act as stepping stones to the remaining exercises for the objective.

- **NEW!** *Think About It Exercises* These exercises are conceptual in nature. They generally appear near the end of an objective's exercise set and ask the students to think about the objective's concepts, make generalizations, and apply them to more abstract problems. The focus is on mental mathematics, not calculation or computation, and these exercises are designed to help students synthesize concepts.

- **Revised!** *Exercise Sets* We have thoroughly reviewed each exercise set. In addition to updating and adding contemporary applications, we have focused our revisions on providing a smooth progression from routine exercises to exercises that are more challenging.

- **Revised!** *Important Points* We have highlighted more of the important points within the body of the text. This will help students to locate and focus on major concepts when reading the text and when studying for an exam.

- **Revised!** *Annotated Examples* In many of the numbered Examples in the text, annotations have been added to the steps within a solution. These annotations assist the student in moving from step to step and help explain the solution.

- **Revised!** *Application Problems* Throughout the text, data problems have been updated to reflect current data and trends. These applications require students to use problem-solving strategies and newly learned skills to solve practical problems that demonstrate the value of mathematics. Where appropriate, application exercises are accompanied by a diagram that helps students visualize the mathematics of the application. We have included many new application exercises from a wide range of disciplines, including:

  - Astronomy
  - Sports
  - Travel
  - Finance
  - And much more! For a full listing, see the Index of Applications on the front inside cover of this text.

Organizational Changes to this Edition

- **Revised!** *Chapter 3* In response to user requests, the presentation of operations on fractions has been changed from presenting addition and subtraction of fractions first, followed by multiplication and division. The chapter now covers multiplication and division of fractions first, followed by addition and subtraction.

- **Revised!** *Chapter 4* In the last edition, Section 2 included all four operations on decimals. Users and reviewers considered that to be too much material for one section. Therefore, in this edition, these concepts are covered in two separate sections:

  - Section 2 now covers addition and subtraction of decimals and includes an objective on applications that require addition and subtraction of decimals for their solutions.

  - Section 3 covers multiplication and division of decimals, converting between fractions and decimals, order relations between fractions and decimals, and applications that require multiplication and division of decimals for their solutions.

We are confident students will find this organization fosters their learning the concepts.

■ **Revised!** *Chapter 9* This chapter has been shortened to more closely reflect the geometry topics generally covered in a prealgebra course. The material on composite figures has been deleted.

NEW! Updates to the Instructional Media Package

■ **HM Testing**™ (powered by Diploma®) offers all the tools needed to create, deliver, and customize multiple types of tests—including authoring and editing algorithmic questions. In addition to producing an unlimited number of tests for each chapter, including cumulative tests and final exams, HM Testing also offers instructors the ability to deliver tests online, or by paper and pencil.

■ **HM MathSPACE**™ encompasses the interactive online products and services integrated with Houghton Mifflin mathematics programs. Students and instructors can access HM MathSPACE content through text-specific Student and Instructor Websites and via online learning platforms. Use HM MathSPACE to access course management tools, which allow you to assign, collect, grade, and record homework assignments via the Web.

■ **Online Multimedia eBook** integrates numerous assets such as Video Explanations and Interactive Lessons to expand upon and reinforce concepts as they appear in the text.

ACKNOWLEDGMENTS

We would especially like to thank users of the previous edition for their helpful suggestions on improving the text. Also, we sincerely appreciate the time, effort, and suggestions of the reviewers of this edition.

Becky Bradshaw, *Lake Superior College, MN*

Judith Carter, *North Shore Community College, MA*

James Jenkins, *Tidewater Community College—Virginia Beach, VA*

Maryann Justinger, *Erie Community College, NY*

Steve Meidinger, *Merced College, CA*

Sherry Steele, *Lamar State College—Port Arthur, TX*

Gowribalan Vamadeva, *University of Cincinnati, OH*

Cynthia Williams, *Pulaski Technical College, AR*

Thank you also to our panel of student reviewers, who provided valuable feedback on the AIM for Success preface:

Matthew F. Berg, *Vanderbilt University, TN*

Gregory Fulchino, *Middlebury College, VT*

Emma Goehring, *Trinity College, CT*

Gili Malinsky, *Boston University, MA*

Julia Ong, *Boston University, MA*

Anjali Parasnis-Samar, *Mount Holyoke College, MA*

Teresa Reilly, *University of Massachusetts—Amherst, MA*

Special thanks to Christi Verity for her diligent preparation of the solutions manuals and for her contribution to the accuracy of the textbooks.

Student Success:
Aufmann Interactive Method

Prealgebra uses an interactive style that engages students in trying new skills and reinforcing learning through structured exercises.

Updated

AIM for Success Student Preface

This preface helps students develop the study skills necessary to achieve success in college mathematics.

It also provides students with an explanation of how to effectively use the features of the text.

AIM for Success can be used as a lesson on the first day of class or as a student project.

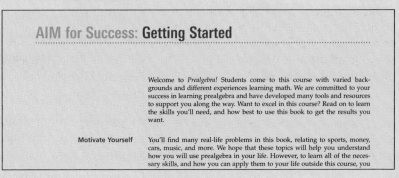

AIM for Success: **Getting Started**

Welcome to *Prealgebra!* Students come to this course with varied backgrounds and different experiences learning math. We are committed to your success in learning prealgebra and have developed many tools and resources to support you along the way. Want to excel in this course? Read on to learn the skills you'll need, and how best to use this book to get the results you want.

Motivate Yourself   You'll find many real-life problems in this book, relating to sports, money, cars, music, and more. We hope that these topics will help you understand how you will use prealgebra in your life. However, to learn all of the necessary skills, and how you can apply them to your life outside this course, you

page xxv

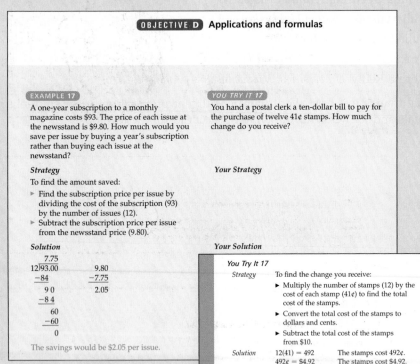

page 268

page S12

Interactive Approach

Each section of the text is divided into objectives, and every objective contains one or more sets of matched-pair examples. The first example in each set is worked out.

The second example, called "You Try It," is for the student to work. By solving this problem, the student actively practices concepts as they are presented in the text.

Complete worked-out solutions to these examples appear in the Appendix for students to check their work.

1.2 Exercises

OBJECTIVE A Addition of whole numbers

1. Find the sum for the addition problem shown at the right.
   a. The first step is to add the numbers in the ones column. The result is _____.
      Write _____ below the ones column and carry _____ to the tens column.
   b. To find the sum of the tens column, add three numbers:
      _____ + _____ + _____ = _____.
   c. The sum of 23 and 69 is _____.

$$\begin{array}{r} 23 \\ + 69 \end{array}$$

2. To estimate the sum of 5,789 + 78,230, begin by rounding 5,789 to the nearest _____ and rounding 78,230 to the nearest _____.

page 33

NEW! **Interactive Exercises**

Placed at the beginning of an objective's exercise set (when appropriate), these exercises provide guided practice and test students' understanding of the underlying concepts in a lesson. They also act as stepping stones to the remaining exercises for the objective.

Student Success:
Objective-Based Approach

Prealgebra is designed to foster student success through an integrated text and media program.

Each chapter's objectives are listed on the chapter opener page, which serves as a guide to student learning. All lessons, exercise sets, tests, and supplements are organized around this carefully constructed hierarchy of objectives.

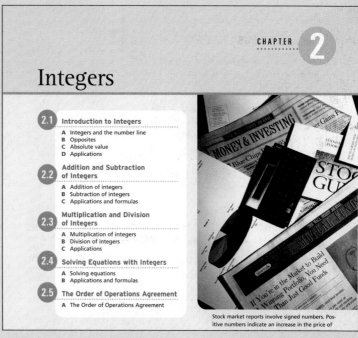

page 87

Objectives describe the topic of each lesson.

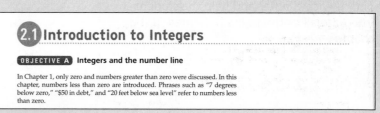

page 89

All exercise sets correspond directly to objectives.

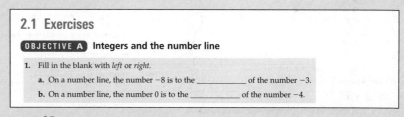

page 95

Answers to the Prep Tests, Chapter Review Exercises, Chapter Tests, and Cumulative Review Exercises refer students back to the original objectives for further study.

Answers to Chapter 2 *Exercises*

Prep Test *(page 88)*

1. 54 > 45 [1.1A]   **2.** 4 units [1.1A]   **3.** 15,847 [1.2A]   **4.** 3,779 [1.2B]   **5.** 26,432 [1.3A]   **6.** 6 [1.3B]   **7.** 13 [1.4A]
8. 5 [1.4A]   **9.** $172 [1.2C]   **10.** 31 [1.5A]

2.1 Exercises *(pages 95–100)*

1a. left   **b.** right   **3.** ⟵┼─●┼┼┼┼┼┼┼┼┼⟶   **5.** ⟵●┼┼┼┼┼┼┼┼┼┼┼⟶
                                    -6 -5 -4 -3 -2 -1 0 1 2 3 4 5 6   -6 -5 -4 -3 -2 -1 0 1 2 3 4 5 6
7. ⟵┼┼┼┼┼┼┼┼┼┼●┼⟶   **9.** ⟵┼┼┼┼┼●┼┼┼┼┼┼┼⟶   **11.** 1   **13.** −1   **15.** 3
        -6 -5 -4 -3 -2 -1 0 1 2 3 4 5 6   -6 -5 -4 -3 -2 -1 0 1 2 3 4 5 6
17. A is −4. C is −2.   **19.** A is −7. D is −4.   **21.** −2 > −5   **23.** 3 > −7   **25.** −42 < 27   **27.** 53 > −46
29. −51 < −20   **31.** −131 < 101   **33.** −7, −2, 0, 3   **35.** −5, −3, 1, 4   **37.** −4, 0, 5, 9   **39.** −10, −7, −5, 4, 12
41a. never true   **b.** sometimes true   **c.** sometimes true   **d.** always true   **43.** minus; negative   **45.** −45   **47.** 88
49. −n   **51.** d   **53.** the opposite of negative thirteen   **55.** the opposite of negative p   **57.** five plus negative ten
59. negative fourteen minus negative three   **61.** negative thirteen minus eight   **63.** m plus negative n   **65.** 7
67. −46   **69.** 73   **71.** z   **73.** −p   **75.** negative   **77a.** −6   **b.** 6   **c.** 6   **79.** 4   **81.** 9   **83.** 11

page A2

Student Success:
Assessment and Review

Prealgebra appeals to students' different study styles with a variety of review methods.

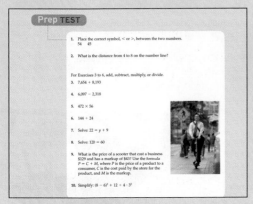

page 88

Prep Tests assess students' mastery of prerequisite skills for the upcoming chapter.

Chapter Summaries include *Key Words* and *Essential Rules and Procedures* covered in the chapter. Each concept references the objective and page number from the lesson where the concept is introduced.

Chapter 2 Summary

Key Words	Examples
A number n is a **positive number** if $n > 0$. A number n is a **negative number** if $n < 0$. [2.1A, p. 89]	Positive numbers are numbers greater than zero. 9, 87, and 603 are positive numbers. Negative numbers are numbers less than zero. -5, -41, and -729 are negative numbers.
The **integers** are . . . $-4, -3, -2, -1, 0, 1, 2, 3, 4, ...$ The integers can be defined as the whole numbers and their opposites. **Positive integers** are to the right of zero on the number line. **Negative integers** are to the left of zero on the number line. [2.1A, p. 89]	-729, -41, -5, 9, 87, and 603 are integers. 0 is an integer, but it is neither a positive nor a negative integer.
Opposite numbers are two numbers that are the same distance from zero on the number line but on opposite sides of zero. The opposite of a number is called its **additive inverse**. [2.1B, p. 91; 2.2A, p. 103]	8 is the opposite, or additive inverse, of -8. -2 is the opposite, or additive inverse, of 2.
The **absolute value** of a number is the distance from zero to the number on the number line. The absolute value of a number is a positive number or zero. The symbol for absolute value is "| |". [2.1C, p. 92]	$\|9\| = 9$ $\|-9\| = 9$ $-\|9\| = -9$

Essential Rules and Procedures

To add integers with the same sign, add the absolute values of the numbers. Then attach the sign of the addends. [2.2A, p. 102]	$6 + 4 = 10$ $-6 + (-4) = -10$

page 141

Chapter Review Exercises are found at the end of each chapter and help the student integrate all of the topics presented in the chapter.

Chapter 2 Review Exercises

1. Write the expression $8 - (-1)$ in words.

2. Evaluate $-|-36|$.

3. Find the product of -40 and -5.

4. Evaluate $-a \div b$ for $a = -27$ and $b = -3$.

page 143

Chapter 2 Test

1. Write the expression $-3 + (-5)$ in words.

2. Evaluate $-|-34|$.

3. What is 3 minus -15?

4. Evaluate $a + b$ for $a = -11$ and $b = -9$.

page 145

Chapter Tests are designed to simulate a possible test of the material in the chapter.

Cumulative Review Exercises

1. Find the difference between -27 and -32.

2. Estimate the product of 439 and 28.

3. Divide: $19,254 \div 6$

4. Simplify: $16 \div (3 + 5) \cdot 9 - 2^4$

page 147

Cumulative Review Exercises, which appear at the end of each chapter beginning with Chapter 2, help students maintain skills learned in previous chapters.

Student Website
Need help? For online student resources, visit college.hmco.com/pic/aufmannPA5e.

page 149

Fully Integrated Online Resources provide students with the opportunity to review and test the skills they have learned. The ⊙ and ⟁ icons on each chapter opener remind the student of the many and varied additional resources available for each chapter.

Student Success:
Conceptual Understanding

Prealgebra helps students understand the course concepts through the textbook exposition and feature set.

Determine whether each statement is always true, sometimes true, or never true. Assume a and b are integers.

70. If $a > 0$ and $b > 0$, then $a + b > 0$.

71. If $a > 0$ and $b < 0$, then $a + b > 0$.

72. If $a = -b$, then $a + b < 0$.

73. If $a < 0$ and $b < 0$, then $a + b < 0$.

page 113

NEW! **Think About It Exercises** are conceptual in nature and help develop students' critical thinking skills.

38. For the equation $0.375x = 0.6$, a student offered the solution shown at the right. Is this a correct method of solving the equation? Explain your answer.

39. Consider the equation $12 = \frac{x}{a}$, where a is any positive number. Explain how increasing values of a affect the solution, x, of the equation.

page 282

Writing Exercises require students to verbalize concepts.

NEW! Important points are now highlighted, to help students recognize what is most important and to study more effectively.

Note in this last example that we are adding a number and its opposite (-8 and 8), and the sum is 0. The opposite of a number is called its **additive inverse.** The opposite or additive inverse of -8 is 8, and the opposite or additive inverse of 8 is -8. The sum of a number and its additive inverse is always zero. This is known as the Inverse Property of Addition.

page 103

Examples indicated by vertical dots use explanatory comments to describe what is happening in key steps of the complete, worked-out solutions.

Add: $321 + 6{,}472$

Add the digits in each column.

page 19

Numbers greater than zero are called **positive numbers.** Numbers less than zero are called **negative numbers.**

> **Positive and Negative Numbers**
>
> A number n is positive if $n > 0$.
> A number n is negative if $n < 0$.

page 89

Key words, in bold, emphasize important terms. **Key concepts** are presented in green boxes to highlight these important concepts and to provide for easy reference.

Students can now reference the new **Glossary** to find the definitions of key words.

Take Note

$\frac{3}{4} \div 6 = \frac{1}{8}$ means that if $\frac{3}{4}$ is divided into 6 equal parts, each equal part is $\frac{1}{8}$. For example, if 6 people share $\frac{3}{4}$ of a pizza, each person eats $\frac{1}{8}$ of the pizza.

The **Take Note** feature amplifies the concept under discussion or alerts students to points requiring special attention.

page 178

Student Success:
Problem Solving

Prealgebra emphasizes applications, problem solving, and critical thinking.

Integrated Real-Life Applications

Wherever appropriate, the last objective of a section presents applications requiring students to use problem-solving skills and strategies to solve practical problems.

Applications are taken from many disciplines, and a complete list of the topics can be found in the Index of Applications.

page 242

Real Data

Real data examples and exercises,

identified by 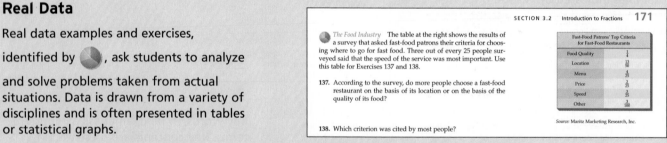, ask students to analyze

and solve problems taken from actual situations. Data is drawn from a variety of disciplines and is often presented in tables or statistical graphs.

page 171

Problem-Solving Strategies

A carefully developed approach to problem solving emphasizes the importance of strategy when solving problems.

- Students are encouraged to develop their own strategies as part of their solutions to problems.
- Model strategies are always presented as guides for students to follow as they attempt the parallel You Try Its.

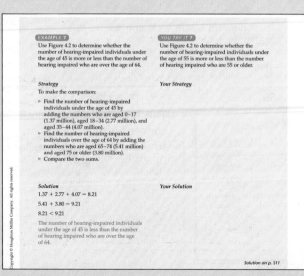

page 251

Student Success:
Problem Solving

Prealgebra emphasizes applications, problem solving, and critical thinking.

Focus on Problem Solving

From Concrete to Abstract

As you progress in your study of algebra, you will find that the problems become less concrete and more abstract. Problems that are concrete provide information pertaining to a specific instance. Abstract problems are theoretical; they are stated without reference to a specific instance. Let's look at an example of an abstract problem.

How many cents are in d dollars?

How can you solve this problem? Are you able to solve the same problem if the information given is concrete?

How many cents are in 5 dollars?

You know that there are 100 cents in 1 dollar. To find the number of cents in 5 dollars, multiply 5 by 100.

$100 \cdot 5 = 500$ There are 500 cents in 5 dollars.

Use the same procedure to find the number of cents in d dollars: multiply d by 100.

$100 \cdot d = 100d$ There are $100d$ cents in d dollars.

page 305

Focus on Problem Solving

Students are introduced to various successful problem-solving strategies within the end-of-chapter material.

- Drawing a diagram
- Applying solutions to other problems
- Looking for a pattern
- Making a table
- Trial and error

CRITICAL THINKING

65. Classify each number as a whole number, an integer, a positive integer, a negative integer, a rational number, an irrational number, and/or a real number.

   a. -2 **b.** 18 **c.** $-\dfrac{9}{37}$ **d.** -6.606 **e.** $4.5\overline{6}$ **f.** $3.050050005\ldots$

66. Using the variable x, write an inequality to represent the graph.

   a. $\xleftarrow{\ \ \ \ }|\ \ |\ \ |\ \ |\ \ |\ \ |\ \ |\ \ |\ \ |\ \xrightarrow{\ \ \ \ }$
       $-4\ -3\ -2\ -1\ \ 0\ \ 1\ \ 2\ \ 3\ \ 4$

   b. $\xleftarrow{\ \ \ \ }|\ \ |\ \ |\ \ |\ \ |\ \ |\ \ |\ \ |\ \ |\ \xrightarrow{\ \ \ \ }$
       $-4\ -3\ -2\ -1\ \ 0\ \ 1\ \ 2\ \ 3\ \ 4$

67. For the given inequality, which of the numbers in parentheses make the inequality true?

   a. $|x| < 9$ $(-2.5, 0, 9, 15.8)$ **b.** $|x| > -3$ $(-6.3, -3, 0, 6.7)$
   c. $|x| \geq 4$ $(-1.5, 0, 4, 13.6)$ **d.** $|x| \leq 5$ $(-4.9, 0, 2.1, 5)$

68. Given that a, b, c, and d are real numbers, which will ensure that $a + c < b + d$?

   a. $a < b$ and $c < d$ **b.** $a > b$ and $c > d$
   c. $a < b$ and $c > d$ **d.** $a > b$ and $c < d$

69. Determine whether the statement is always true, sometimes true, or never true.

   a. Given that $a > 0$ and $b > 0$, then $ab > 0$.
   b. Given that $a < 0$, then $a^2 > 0$.
   c. Given that $a > 0$ and $b > 0$, then $a^2 > b$.

page 304

Critical Thinking

Included in each exercise set are Critical Thinking exercises that present extensions of topics, require analysis, or offer challenge problems.

Projects & Group Activities

Customer Billing

Chris works at B & W Garage as an auto mechanic and has just completed an engine overhaul for a customer. To determine the cost of the repair job, Chris keeps a list of times worked and parts used. A price list and a list of parts used and times worked are shown below. Use these tables, and the fact that the charge for labor is $46.75 per hour, to determine the total cost for parts and labor.

Parts Used		Time Spent	
Item	Quantity	Day	Hours
Gasket set	1	Monday	7.0
Ring set	1	Tuesday	7.5
Valves	8	Wednesday	6.5
Wrist pins	8	Thursday	8.5
Valve springs	16	Friday	9.0
Rod bearings	8		
Main bearings	5		
Valve seals	16		
Timing chain	1		

Price List		
Item Number	Description	Unit Price
27345	Valve spring	$9.25
41257	Main bearing	$17.49
54678	Valve	$16.99
29753	Ring set	$169.99
45837	Gasket set	$174.90
23751	Timing chain	$50.49
23765	Fuel pump	$229.99
28632	Wrist pin	$23.55
34922	Rod bearing	$13.69
2871	Valve seal	$1.69

page 306

Projects & Group Activities

The Projects & Group Activities feature at the end of each chapter can be used as extra credit or for cooperative learning activities.

Additional Resources—Get More from Your Textbook!

Instructor Resources

Instructor's Annotated Edition (IAE)
This edition contains a replica of the student text and additional items just for the instructor. Answers to all exercises are provided.

Instructor Website
This website offers instructors a variety of resources, including instructor's solutions, digital art and figures, course sequences, a printed test bank, and more.

HM Testing™ (Powered by Diploma®)
"Testing the way you want it"
HM Testing provides instructors with a wide array of new algorithmic exercises, along with improved functionality and ease of use. Instructors can create, author and edit algorithmic questions, customize, and deliver multiple types of tests.

Student Resources

Student Solutions Manual
Contains complete solutions to all odd-numbered exercises, and all of the solutions to the end-of-chapter material.

Math Study Skills Workbook, Third Edition, by Paul Nolting
Helps students identify their strengths, weaknesses, and personal learning styles in math. Nolting offers study tips, proven test-taking strategies, a homework system, and recommendations for reducing anxiety and improving grades.

Student Website
This free website gives students access to ACE practice tests, glossary flashcards, chapter summaries, and other study resources.

Instructional DVDs

DVD Hosted by Dana Mosely, these text-specific DVDs cover all sections of the text and provide explanations of key concepts, examples, exercises, and applications in a lecture-based format. DVDs are now close-captioned for the hearing-impaired.

HM MathSPACE™

HM MathSPACE encompasses the interactive online products and services integrated with Houghton Mifflin mathematics programs. Students and instructors can access HM MathSPACE content through text-specific Student and Instructor Websites and via online learning platforms. Use HM MathSPACE to access course management tools, which allow you to assign, collect, grade, and record homework assignments via the Web.

Online Course Management Content for Blackboard®, WebCT®, and eCollege®

Deliver program- or text-specific Houghton Mifflin content online using your institution's local course management system. Houghton Mifflin offers tutorials, videos, and other resources via Blackboard, WebCT, eCollege, and other course management systems. Add to an existing online course or create a new one by selecting from a wide range of powerful learning and instructional materials.

For more information, visit **college.hmco.com/pic/aufmannPA5e** or contact your local Houghton Mifflin sales representative.

AIM for Success: Getting Started

Welcome to *Prealgebra!* Students come to this course with varied backgrounds and different experiences learning math. We are committed to your success in learning prealgebra and have developed many tools and resources to support you along the way. Want to excel in this course? Read on to learn the skills you'll need, and how best to use this book to get the results you want.

Motivate Yourself

You'll find many real-life problems in this book, relating to sports, money, cars, music, and more. We hope that these topics will help you understand how you will use prealgebra in your life. However, to learn all of the necessary skills, and how you can apply them to your life outside this course, you need to stay motivated.

THINK ABOUT WHY YOU WANT TO SUCCEED IN THIS COURSE. LIST THE REASONS HERE (NOT IN YOUR HEAD . . . ON THE PAPER!)

We also know that this course may be a requirement for you to graduate or complete your major. That's OK. If you have a goal for the future, such as becoming a nurse or a teacher, you will need to succeed in prealgebra first. Picture yourself where you want to be, and use this image to stay on track.

Make the Commitment

Stay committed to success! With practice, you will improve your prealgebra skills. Skeptical? Think about when you first learned to ride a bike or drive a car. You probably felt self-conscious and worried that you might fail. But with time and practice, it became second nature to you.

 You will also need to put in the time and practice to do well in prealgebra. Think of us as your "driving" instructors. We'll lead you along the path to success, but we need you to stay focused and energized along the way.

LIST A SITUATION IN WHICH YOU ACCOMPLISHED YOUR GOAL BY SPENDING TIME PRACTICING AND PERFECTING YOUR SKILLS (SUCH AS LEARNING TO PLAY THE PIANO, PLAYING BASKETBALL, ETC.).

If you spend time learning and practicing the skills in this book, you will also succeed in prealgebra.

Think You Can't Do Math? Think Again!

You can do math! When you first learned the skills listed above, you may have not done them well. With practice, you got better. With practice, you will be better at math. Stay focused, motivated, and committed to success.

It is difficult for us to emphasize how important it is to overcome the "I Can't Do Math Syndrome." If you listen to interviews of very successful athletes after a particularly bad performance, you will note that they focus on the positive aspect of what they did, not the negative. Sports psychologists encourage athletes to always be positive and to have a "Can Do" attitude. Develop this attitude toward math and you will succeed.

Skills for Success

Get the Big Picture If this were an English class, we wouldn't encourage you to look ahead in the book. But this is prealgebra—go right ahead! Take a few minutes to read the Table of Contents. Then, look through the entire book. Move quickly: scan titles, look at pictures, notice diagrams.

Getting this big-picture view will help you see where this course is going. To reach your goal, it's important to get an idea of the steps you will need to take along the way.

As you look through the book, find topics that interest you. What's your preference? Horse racing? Sailing? TV? Amusement parks? Find the Index of Applications on the front inside cover and pull out three subjects that interest you. Then, flip to the pages in the book where the topics are featured, and read the exercises or problems where they appear. Write these topics here:

WRITE THE TOPIC HERE	WRITE THE CORRESPONDING EXERCISE/PROBLEM HERE

You'll find it's easier to work at learning the material if you are interested in how it can be used in your everyday life.

Use the following activities to think about more ways you might use prealgebra in your daily life. Flip open your book to the exercises below to answer the questions.

■ (see p. 84, #31) I've hired a contractor to work on my house. I need to use prealgebra to. . . _____

_____

_____

- (see p. 188, #162) I just started a new job and will be paid hourly, but my hours change every week. I need to use prealgebra to. . . _____

_____

_____

- (see p. 368, #71) I have to buy a new suit, but it's very expensive! Luckily, I see it's marked down. I need prealgebra to. . . _____

_____

_____

You know that the activities you just completed are from daily life, but do you notice anything else they have in common? That's right—they are **word problems.** Try not to be intimidated by word problems. You just need a strategy. It's true that word problems can be challenging because we need to use multiple steps to solve them:

- Read the problem.
- Determine the quantity we must find.
- Think of a method to find it.
- Solve the problem.
- Check the answer.

In short, we must come up with a strategy and then use that strategy to find the *solution*.

We'll teach you about strategies for tackling word problems that will make you feel more confident solving these problems from daily life. After all, even though no one will ever come up to you on the street and ask you to solve a multiplication problem, you will need to use math every day to balance your checkbook, evaluate credit card offers, etc.

Take a look at the example below. You'll see that solving a word problem includes finding a *strategy* and using that strategy to find a *solution*. If you find yourself struggling with a word problem, try writing down the information you know about the problem. Be as specific as you can. Write out a phrase or a sentence that states what you are trying to find. Ask yourself whether there is a formula that expresses the known and unknown quantities. Then, try again!

EXAMPLE 10

The daily low temperatures during one week were: $-10°$, $2°$, $-1°$, $-9°$, $1°$, $0°$, and $3°$. Find the average daily low temperature for the week.

Strategy

To find the average daily low temperature:

▶ Add the seven temperature readings.
▶ Divide by 7.

Solution

$-10 + 2 + (-1) + (-9) + 1 + 0 + 3 = -14$

$-14 \div 7 = -2$

The average daily low temperature was $-2°$.

YOU TRY IT 10

The daily high temperatures during one week were: $-7°$, $-8°$, $0°$, $-1°$, $-6°$, $-11°$, and $-2°$. Find the average daily high temperature for the week.

Your Strategy

Your Solution

Solution on p. S5

---

page 122

Take Note

Take a look at your syllabus to see if your instructor has an **attendance policy** that is part of your overall grade in the course.

The attendance policy will tell you:

- How many classes you can miss without a penalty
- What to do if you miss an exam or quiz
- If you can get the lecture notes from the professor when you miss a class

Get the Basics On the first day of class, your instructor will hand out a **syllabus** listing the requirements of your course. Think of this syllabus as your personal roadmap to success. It shows you the *destinations* (topics you need to learn) and the *dates* you need to arrive at those destinations (when you need to learn the topics by). Learning prealgebra is a journey. But, to get the most out of this course, you'll need to

know what the important stops are and what skills you'll need to learn for your arrival at those stops.

You've quickly scanned the Table of Contents, but now we want you to take a closer look. Flip open to the Table of Contents and look at it next to your syllabus. Identify when your major exams are and what material you'll need to learn by those dates. For example, if you know you have an exam in the second month of the semester, how many chapters of this text will you need to learn by then? What homework do you have to do during this time? Write it down using the chart below:

CHART YOUR PROGRESS. . . STAY ON TRACK FOR SUCCESS!

Write the dates of your exams in this column	Write the chapters you'll need to learn by each exam	Write the homework you need to do by each exam	Write the grades you receive on your exams here

Managing these important dates will help keep you on track for success.

Manage Your Time We know how busy you are outside of school. Do you have a full-time or a part-time job? Do you have children? Visit your family often? Play basketball or write for the school newspaper? It can be stressful to balance all of the important activities and responsibilities in your life. Making a **time management plan** will help you create a schedule that gives you enough time for everything you need to do.

Let's get started! Use the grid on the next page to fill in your weekly schedule.

First, fill in all of your responsibilities that take up certain set hours during the week. Be sure to include:

- Each class you are taking

- Time you spend at work

- Any other commitments (child care, tutoring, volunteering, etc.)

Then, fill in all of your responsibilities that are more flexible. Remember to make time for:

- Studying—You'll need to study to succeed, but luckily you get to choose what times work best for you. Keep in mind:

  - Most instructors ask students to spend twice as much time studying as they do in class. (3 hours of class each week = 6 hours of study each week)

  - Try studying in chunks. We've found it works better to study an hour each day, rather than studying for 6 hours on one day.

  - Studying can be even more helpful if you're able to do it right after your class meets, when the material is fresh in your mind.

- **Meals**—Eating well gives you energy and stamina for attending classes and studying.

- **Entertainment**—It's impossible to stay focused on your responsibilities 100% of the time. Giving yourself a break for entertainment will reduce your stress and help keep you on track.

- **Exercise**—Exercise contributes to overall health. You'll find you're at your most productive when you have both a healthy mind **and** a healthy body.

Here is a sample of what part of your schedule might look like:

	8–9	9–10	10–11	11–12	12–1	1–2	2–3	3–4	4–5	5–6
Monday	History class Jenkins Hall 8–9:15	Eat 9:15–10	Study/Homework for History 10–12 AM		Lunch & Nap 12–1:30 PM		Work 2–6 PM			
Tuesday	Sleep!	Math Class Douglas Hall 9–9:45 AM	Study/Homework for Math 10–12 AM		Eat 12–1 PM	English Class Scott Hall 1–1:45 PM	Study/Homework for English 2–4 PM		Hang out with Alli & Mike 4 until whenever	

XXX

	MORNING				AFTERNOON							EVENING			
	8-9	9-10	10-11	11-12	12-1	1-2	2-3	3-4	4-5	5-6	6-7	7-8	8-9	9-10	10-11
Monday															
Tuesday															
Wednesday															
Thursday															
Friday															
Saturday															
Sunday															

Tear along the dotted line to remove

Features for Success in This Text

Organization   Let's look again at the Table of Contents. There are 10 chapters in this book. You'll see that every chapter is divided into **sections,** and each section contains a number of **learning objectives.** Each learning objective is labeled with a letter from A to E. Knowing how this book is organized will help you locate important topics and concepts as you're studying.

Preparation   Ready to start a new chapter? Take a few minutes to be sure you're ready, using some of the tools in this book.

- **Cumulative Review Exercises:** You'll find these exercises after every chapter, starting with Chapter 2. The questions in the Cumulative Review Exercises are taken from the previous chapters. For example, the Cumulative Review for Chapter 3 will test all of the skills you have learned in Chapters 1, 2, and 3. Use this review to test what you know before a big exam.

Here's an example of how to use the Cumulative Review:

- Turn to page 313 and look at the questions for the Chapter 4 Cumulative Review, which are taken from the current chapter and the previous chapters.
- We have the answers to all of the Cumulative Review exercises in the back of the book. Flip to page A8 to see the answers for this chapter.
- Got the answer wrong? We can tell you where to go in the book for help! For example, scroll down page A8 to find the answer for the first exercise, which is 0.03879. You'll see that after this answer, there is an **objective reference** [4.3B]. This means that the question was taken from Chapter 4, Section 3, Objective B. Go there to re-study the objective.

- **Prep Tests:** These tests are found at the beginning of every chapter, and they will help you see if you've mastered all of the skills needed for the new chapter.

Here's an example of how to use the Prep Test:

- Turn to page 236 and look at the Prep Test for Chapter 4.
- All of the answers to the Prep Tests are in the back of the book. You'll find them in the first set of answers in each answer section for a chapter. Turn to page A6 to see the answers for this Prep Test.
- Re-study the objectives if you need some extra help.

- Before you start a new section, take a few minutes to read the **Objective Statement** for that section. Then, browse through the objective material. Especially note the words or phrases in bold type—these are important concepts that you'll need as you're moving along in the course.
- As you start moving through the chapter, pay special attention to the **rule boxes.** These rules give you the reasons certain types of problems are solved the way they are. When you see a rule, try to rewrite the rule in your own words.

> ### Rule for Adding Two Integers
>
> **To add two integers with the same sign,** add the absolute values of the numbers. Then attach the sign of the addends.
>
> **To add two integers with different signs,** find the absolute values of the numbers. Subtract the smaller absolute value from the larger absolute value. Then attach the sign of the addend with the larger absolute value.

Knowing what to pay attention to as you move through a chapter will help you study and prepare.

Interaction We want you to be actively involved in learning prealgebra, and have given you many ways to get hands-on with this book.

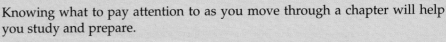

- **Annotated Examples** Take a look at page 104 below. See the dotted blue lines next to the example? These lines show you that the example includes explanations for steps in the solution.

Evaluate $-x + y$ for $x = -15$ and $y = -5$.	$-x + y$
Replace x with -15 and y with -5.	$-(-15) + (-5)$
Simplify $-(-15)$.	$= 15 + (-5)$
Add.	$= 10$

page 104

Grab a paper and pencil and work along as you're reading through each example. When you're done, get a clean sheet of paper. Write down the problem and try to complete the solution without looking at your notes or at the book. When you're done, check your answer. If you got it right, you're ready to move on.

- **Example/You Try It Pairs** You'll need hands-on practice to succeed in prealgebra. When we show you an Example, work it out beside our solution. Use the Example/You Try It Pairs to get the practice you need.

EXAMPLE 6

Subtract: $-\dfrac{5}{6} - \left(-\dfrac{3}{8}\right)$

Solution

$$-\frac{5}{6} - \left(-\frac{3}{8}\right) = -\frac{5}{6} + \frac{3}{8} = \frac{-20}{24} + \frac{9}{24}$$

$$= \frac{-20 + 9}{24}$$

$$= \frac{-11}{24} = -\frac{11}{24}$$

YOU TRY IT 6

Subtract: $-\dfrac{5}{6} - \dfrac{7}{9}$

Your Solution

page 197

You'll see that each Example is fully worked out. Study this Example carefully by working through each step. Then, try your hand at it by completing the You Try It. If you get stuck, the solutions to the You Try Its are provided in the back of the book. There is a page number following the You Try It, which shows you where you can find the completely worked-out solution. Use the solution to get a hint for the step on which you are stuck. Then, try again!

When you've finished the solution, check your work against the solution in the back of the book. Turn to page S8 to see the solution for You Try It 6 above.

Remember that sometimes there can be more than one way to solve a problem. But your answer should always match the answers we've given in the back of the book. If you have any questions about whether your method will always work, check with your instructor.

Review We have provided many opportunities for you to practice and review the skills you have learned in each chapter.

- **Section Exercises** After you're done studying a section, flip to the end of the section and complete the exercises. If you immediately practice what you've learned, you'll find it easier to master the core skills. Want to know if you answered the questions correctly? The answers to the odd-numbered exercises are given in the back of the book.

- **Chapter Summary** Once you've completed a chapter, look at the Chapter Summary. This is divided into two sections: *Key Words* and *Essential Rules and Procedures*. Flip to page 141 to see the Chapter Summary for Chapter 2. This summary shows all of the important topics covered in the chapter. See the objective reference and page number following each topic? This shows you the page in the text where you can find more information on the concept.

- **Chapter Review Exercises** You'll find the Chapter Review Exercises after the Chapter Summary. Flip to page 143 to see the Chapter Review Exercises for Chapter 2. When you do the review exercises, you're giving yourself an important opportunity to test your understanding of the chapter. The answer to each review exercise is given at the back of the book, along with the objective the question relates to. When you're done with the Chapter Review Exercises, check your answers. If you had trouble with any of the questions, you can re-study the objectives from which they are taken and re-try some of the exercises in those objectives for extra help.

- **Chapter Tests** The Chapter Tests can be found after the Chapter Review Exercises, and can be used to prepare for your exams. Think of these tests as a "practice run" for your in-class tests. Take the test in a quiet place and try to work through it in the same amount of time you will be allowed for your exam.

Here are some strategies for success when you're taking your exams:

- Scan the entire test to get a feel for the questions (get the big picture).
- Read the directions carefully.
- Work the problems that are easiest for you first.
- Stay calm, and remember that you will have lots of opportunities for success in this class!

Additional Tools for Success

Finished with a chapter? Want to check your work and refresh on key concepts? We have provided a number of resources to help you as you're studying:

- The **Student Solutions Manual** contains complete solutions to all odd-numbered exercises, and all of the solutions to the end-of-chapter material.

- The *Prealgebra* **Student Website** contains a wide variety of resources to help you study, including:
  - **Glossary Flashcards** will help test your knowledge of key terms.
  - **Online Multimedia eBook** includes Video Explanations and Interactive Lessons to help reinforce concepts you've learned in the text.
  - **ACE Quizzes** will help improve your understanding of the course concepts. These quiz questions include step-by-step tutorial help and are based on problems from your textbook. Use these quizzes for virtually unlimited practice!

Get Involved

Have a question? Ask! Your professor and your classmates are there to help. Here are some tips to help you jump in to the action:

■ See something you don't understand? There are a few ways to get help:

  ■ Raise your hand in class.

  ■ Your instructor may have a website where students can write in with questions, or your professor may ask you to email or call him/her directly. Take advantage of these ways to get your questions answered.

  ■ Visit a **math center.** Ask your instructor for more information about the math center services available on your campus.

  ■ Your instructor will have **office hours** where he/she will be available to help you. Take note of where and when your instructor holds office hours. Use this time for one-on-one help, if you need it. Write down your instructor's office hours here:

Office Hours: _____
Office Location: _____

■ Form a **study group** with students from your class. This is a great way to prepare for tests, catch up on topics you may have missed, or get extra help on problems you're struggling with. Here are a few suggestions to make the most of your study group:

  ■ **Test each other by asking questions.** Have each person bring a few sample questions when you get together.

  ■ **Practice teaching each other.** We've found that you can learn a lot about what you know when you have to explain it to someone else.

  ■ **Compare class notes.** Couldn't understand the last five minutes of class? Missed class because you were sick? Chances are someone in your group has the notes for the topics you missed.

  ■ **Brainstorm test questions.**

  ■ **Make a plan for your meeting.** Agree on what topics you'll talk about, and how long you'll be meeting for. When you make a plan, you'll be sure that you make the most of your meeting.

Ready, Set, Succeed!

It takes hard work and commitment to succeed, but we know you can do it! Doing well in prealgebra is just one step you'll take along the path to success.

We want you to check the box below, once you have accomplished your goals for this course.

☐   I succeeded in prealgebra!

We are confident that if you follow our suggestions, you will succeed. Good luck!

Whole Numbers

Cost of living is based on the costs for groceries, housing, clothes, transportation, medical care, recreation, and education. The cost of living varies depending on where people live. Knowing the cost of living in a new city helps people figure out what salary they need to earn there in order to maintain the standard of living they are enjoying in their present location. The **Project on page 79** shows you how to calculate the amount of money you would need to earn in another city in order to maintain your current standard of living.

DVD   SSM

Student Website

Need help? For online student resources,
***visit* college.hmco.com/pic/aufmannPA5e.**

1. Name the number of ◆s shown below.

   ◆ ◆ ◆ ◆ ◆ ◆ ◆ ◆

2. Write the numbers from 1 to 10.

   1 __ __ __ __ __ __ __ __ 10

3. Match the number with its word form.

   a. 4   A. five
   b. 2   B. one
   c. 5   C. zero
   d. 1   D. four
   e. 3   E. two
   f. 0   F. three

4. How many American flags contain the color green?

5. Write the number of states in the United States of America as a word, not a number.

GO Figure

Five adults and two children want to cross a river in a rowboat. The boat can hold one adult or two children or one child. Everyone is able to row the boat. What is the minimum number of trips that will be necessary for everyone to get to the other side?

1.1 Introduction to Whole Numbers

OBJECTIVE A   Order relations between whole numbers

The **natural numbers** are 1, 2, 3, 4, 5, 6, 7, 8, 9, 10, 11, ...

The three dots mean that the list continues on and on and there is no largest natural number. The natural numbers are also called the **counting numbers.**

The **whole numbers** are 0, 1, 2, 3, 4, 5, 6, 7, 8, 9, 10, 11, ... Note that the whole numbers include the natural numbers and zero.

Just as distances are associated with markings on the edge of a ruler, the whole numbers can be associated with points on a line. This line is called the **number line** and is shown below.

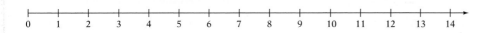

The arrowhead at the right indicates that the number line continues to the right.

The **graph** of a whole number is shown by placing a heavy dot on the number line directly above the number. Shown below is the graph of 6 on the number line.

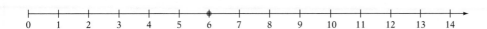

On the number line, the numbers get larger as we move from left to right. The numbers get smaller as we move from right to left. Therefore, the number line can be used to visualize the order relation between two whole numbers.

A number that appears to the right of a given number is **greater than** the given number. The symbol for *is greater than* is >.

8 is to the right of 3.
8 is greater than 3.
8 > 3

A number that appears to the left of a given number is **less than** the given number. The symbol for *is less than* is <.

5 is to the left of 12.
5 is less than 12.
5 < 12

An **inequality** expresses the relative order of two mathematical expressions. 8 > 3 and 5 < 12 are inequalities.

Point of Interest

Among the slang words for zero are *zilch*, *zip*, and *goose egg*. The word *love* for zero in scoring a tennis game comes from the French for "the egg": *l'oeuf.*

Take Note

An inequality symbol, < or >, points to the smaller number. The symbol opens toward the larger number.

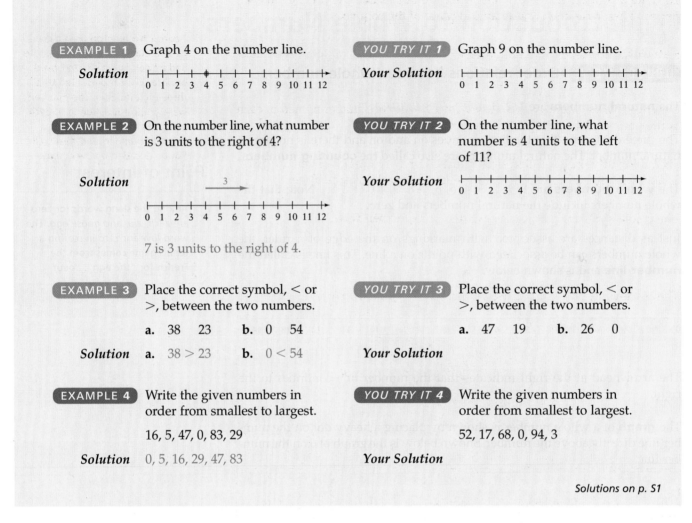

EXAMPLE 1 Graph 4 on the number line.

Solution
0 1 2 3 4 5 6 7 8 9 10 11 12

YOU TRY IT 1 Graph 9 on the number line.

Your Solution
0 1 2 3 4 5 6 7 8 9 10 11 12

EXAMPLE 2 On the number line, what number is 3 units to the right of 4?

Solution

3

0 1 2 3 4 5 6 7 8 9 10 11 12

7 is 3 units to the right of 4.

YOU TRY IT 2 On the number line, what number is 4 units to the left of 11?

Your Solution
0 1 2 3 4 5 6 7 8 9 10 11 12

EXAMPLE 3 Place the correct symbol, < or >, between the two numbers.

a. 38 23 **b.** 0 54

Solution **a.** 38 > 23 **b.** 0 < 54

YOU TRY IT 3 Place the correct symbol, < or >, between the two numbers.

a. 47 19 **b.** 26 0

Your Solution

EXAMPLE 4 Write the given numbers in order from smallest to largest.

16, 5, 47, 0, 83, 29

Solution 0, 5, 16, 29, 47, 83

YOU TRY IT 4 Write the given numbers in order from smallest to largest.

52, 17, 68, 0, 94, 3

Your Solution

Solutions on p. S1

OBJECTIVE B   **Place value**

When a whole number is written using the digits 0, 1, 2, 3, 4, 5, 6, 7, 8, and 9, it is said to be in **standard form.** The position of each digit in the number determines the digit's **place value.** The diagram below shows a **place-value chart** naming the first twelve place values. The number 64,273 is in standard form and has been entered in the chart.

In the number 64,273, the position of the digit 6 determines that its place value is ten-thousands.

When a number is written in standard form, each group of digits separated by a comma is called a **period.** The number 5,316,709,842 has four periods. The period names are shown in red in the place-value chart above.

Point of Interest

The Romans represented numbers using M for 1,000, D for 500, C for 100, L for 50, X for 10, V for 5, and I for 1. For example, MMDCCCLXXVI represented 2,876. The Romans could represent any number up to the largest they would need for their everyday life, except zero.

To write a number in words, start from the left. Name the number in each period. Then write the period name in place of the comma.

5,316,709,842 is read "five billion three hundred sixteen million seven hundred nine thousand eight hundred forty-two."

To write a whole number in standard form, write the number named in each period, and replace each period name with a comma.

Six million fifty-one thousand eight hundred seventy-four is written 6,051,874. The zero is used as a place holder for the hundred-thousands place.

The whole number 37,286 can be written in **expanded form** as

$$30,000 + 7,000 + 200 + 80 + 6$$

The place-value chart can be used to find the expanded form of a number.

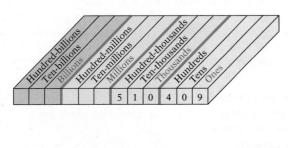

3		7		2		8		6
Ten-thousands	+	Thousands	+	Hundreds	+	Tens	+	Ones
30,000	+	7,000	+	200	+	80	+	6

Write the number 510,409 in expanded form.

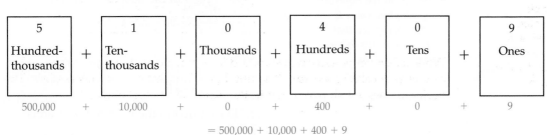

5		1		0		4		0		9
Hundred-thousands	+	Ten-thousands	+	Thousands	+	Hundreds	+	Tens	+	Ones
500,000	+	10,000	+	0	+	400	+	0	+	9

$$= 500,000 + 10,000 + 400 + 9$$

Point of Interest

George Washington used a code to communicate with his men. He had a book in which each word or phrase was represented by a three-digit number. The numbers were arbitrarily assigned to each entry. Messages appeared as a string of numbers and thus could not be decoded by the enemy.

EXAMPLE 5

Write 82,593,071 in words.

Solution

eighty-two million five hundred ninety-three thousand seventy-one

YOU TRY IT 5

Write 46,032,715 in words.

Your Solution

EXAMPLE 6

Write four hundred six thousand nine in standard form.

Solution

406,009

YOU TRY IT 6

Write nine hundred twenty thousand eight in standard form.

Your Solution

EXAMPLE 7

Write 32,598 in expanded form.

Solution

30,000 + 2,000 + 500 + 90 + 8

YOU TRY IT 7

Write 76,245 in expanded form.

Your Solution

Solutions on p. S1

OBJECTIVE C **Rounding**

When the distance to the sun is given as 93,000,000 mi, the number represents an approximation to the true distance. Giving an approximate value for an exact number is called **rounding.** A number is rounded to a given place value.

48 is closer to 50 than it is to 40. 48 rounded to the nearest ten is 50.

4,872 rounded to the nearest ten is 4,870.

4,872 rounded to the nearest hundred is 4,900.

A number is rounded to a given place value without using the number line by looking at the first digit to the right of the given place value.

If the digit to the right of the given place value is less than 5, replace that digit and all digits to the right of it by zeros.

Round 12,743 to the nearest hundred.

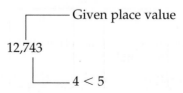

Given place value

12,743

4 < 5

12,743 rounded to the nearest hundred is 12,700.

If the digit to the right of the given place value is greater than or equal to 5, increase the digit in the given place value by 1, and replace all other digits to the right by zeros.

Round 46,738 to the nearest thousand.

```
                ┌──────── Given place value
                │
         46,738
                └──────
                        └─── 7 > 5
```

46,738 rounded to the nearest thousand is 47,000.

Round 29,873 to the nearest thousand.

```
         ┌──────── Given place value
         │
  29,873
         └──────
                └─── 8 > 5    Round up by adding 1 to the 9 (9 + 1 = 10).
                              Carry the 1 to the ten-thousands place (2 + 1 = 3).
```

29,873 rounded to the nearest thousand is 30,000.

EXAMPLE 8

Round 435,278 to the nearest ten-thousand.

Solution

```
             ┌──────── Given place value
             │
    435,278
             └──────
                     └─── 5 = 5
```

435,278 rounded to the nearest ten-thousand is 440,000.

YOU TRY IT 8

Round 529,374 to the nearest ten-thousand.

Your Solution

EXAMPLE 9

Round 1,967 to the nearest hundred.

Solution

```
           ┌──────── Given place value
           │
    1,967
           └──────
                   └─── 6 > 5
```

1,967 rounded to the nearest hundred is 2,000.

YOU TRY IT 9

Round 7,985 to the nearest hundred.

Your Solution

Solutions on p. S1

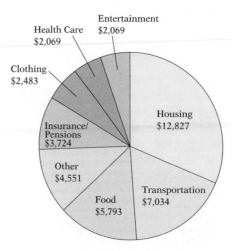

Bill Gates

OBJECTIVE D **Applications and statistical graphs**

Graphs are displays that provide a pictorial representation of data. The advantage of graphs is that they present information in a way that is easily read.

A **pictograph** uses symbols to represent information. The symbol chosen usually has a connection to the data it represents.

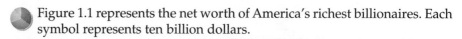 Figure 1.1 represents the net worth of America's richest billionaires. Each symbol represents ten billion dollars.

	Net Worth (in tens of billions of dollars)
Bill Gates	$ $ $ $ $
Warren Buffett	$ $ $ $ $
Sheldon Adelson	$ $
Larry Ellison	$ $
Paul Allen	$ $

Figure 1.1 Net Worth of America's Richest Billionaires
Source: **www.Forbes.com**

From the pictograph, we can see that Bill Gates and Warren Buffett have the greatest net worth. Warren Buffett's net worth is $30 billion more than Sheldon Adelson's net worth.

A typical household in the United States has an average after-tax income of $40,550. The **circle graph** in Figure 1.2 represents how this annual income is spent. The complete circle represents the total amount, $40,550. Each sector of the circle represents the amount spent on a particular expense.

Health Care $2,069
Entertainment $2,069
Clothing $2,483
Insurance/ Pensions $3,724
Other $4,551
Food $5,793
Housing $12,827
Transportation $7,034

From the circle graph, we can see that the largest amount is spent on housing. We can see that the amount spent on food ($5,793) is less than the amount spent on transportation ($7,034).

Figure 1.2 Average Annual Expenses in a U.S. Household
Source: American Demographics

The **bar graph** in Figure 1.3 shows the expected U.S. population aged 100 and over for various years.

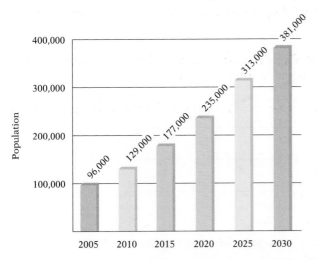

Figure 1.3   Expected U.S. Population Aged 100 and Over
Source: Census Bureau

In this bar graph, the horizontal axis is labeled with the years (2005, 2010, 2015, etc.) and the vertical axis is labeled with the numbers for the population. For each year, the height of the bar indicates the population for that year. For example, we can see that the expected population of those aged 100 and over in the year 2015 is 177,000. The graph indicates that the population of people aged 100 and over keeps increasing.

A **double-bar graph** is used to display data for the purposes of comparison.

The double-bar graph in Figure 1.4 shows the fuel efficiency of four vehicles, as rated by the Environmental Protection Agency. These are among the most fuel-efficent 2006 model-year cars for city and highway mileage.

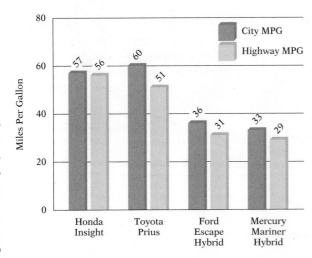

Figure 1.4

From the graph, we can see that the fuel efficiency of the Honda Insight is less on the highway (56 mpg) than it is for city driving (57 mpg).

The **broken-line graph** in Figure 1.5 shows the effect of inflation on the value of a $100,000 life insurance policy. (An inflation rate of 5 percent is used here.)

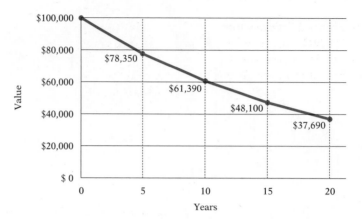

Figure 1.5   Effect of Inflation on the Value of a $100,000 Life
Insurance Policy

According to the line graph, after five years the purchasing power of the $100,000 has decreased to $78,350. We can see that the value of the $100,000 keeps decreasing over the 20-year period.

Two broken-line graphs can be used to compare data. Figure 1.6 shows the populations of California and Texas. The figures are those of the U.S. Census for the years 1900, 1925, 1950, 1975, and 2000. The numbers are rounded to the nearest thousand.

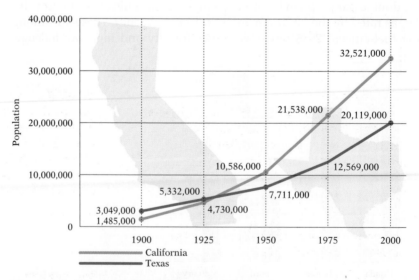

Figure 1.6   Populations of California and Texas

From the graph, we can see that the population was greater in Texas in 1900 and 1925, while the population was greater in California in 1950, 1975, and 2000.

To solve an application problem, first read the problem carefully. The **Strategy** involves identifying the quantity to be found and planning the steps that are necessary to find that quantity. The **Solution** involves performing each operation stated in the Strategy and writing the answer.

The circle graph in Figure 1.7 shows the result of a survey of 300 people who were asked to name their favorite sport. Use this graph for Example 10 and You Try It 10.

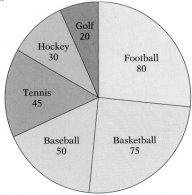

Figure 1.7   Distribution of Responses in a Survey

EXAMPLE 10

According to Figure 1.7, which sport was named by the least number of people?

Strategy

To find the sport named by the least number of people, find the smallest number given in the circle graph.

Solution

The smallest number given in the graph is 20.

The sport named by the least number of people was golf.

YOU TRY IT 10

According to Figure 1.7, which sport was named by the greatest number of people?

Your Strategy

Your Solution

EXAMPLE 11

The distance between St. Louis, Missouri, and Portland, Oregon, is 2,057 mi. The distance between St. Louis, Missouri, and Seattle, Washington, is 2,135 mi. Which distance is greater, St. Louis to Portland or St. Louis to Seattle?

Strategy

To find the greater distance, compare the numbers 2,057 and 2,135.

Solution

$2,135 > 2,057$

The greater distance is from St. Louis to Seattle.

YOU TRY IT 11

The distance between Los Angeles, California, and San Jose, California, is 347 mi. The distance between Los Angeles, California, and San Francisco, California, is 387 mi. Which distance is shorter, Los Angeles to San Jose or Los Angeles to San Francisco?

Your Strategy

Your Solution

Solutions on p. S1

The bar graph in Figure 1.8 shows the states with the most sanctioned league bowlers. Use this graph for Example 12 and You Try It 12.

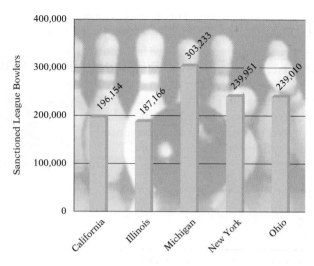

Figure 1.8 States with the Most Sanctioned League Bowlers
Sources: American Bowling Congress, Women's International Bowling Congress, Young American Bowling Alliance

---

EXAMPLE 12

According to Figure 1.8, which state has the most sanctioned league bowlers?

Strategy

To determine which state has the most sanctioned league bowlers, locate the state that corresponds to the highest bar.

Solution

The highest bar corresponds to Michigan.

Michigan is the state with the most sanctioned league bowlers.

YOU TRY IT 12

According to Figure 1.8, which state has fewer sanctioned league bowlers, New York or Ohio?

Your Strategy

Your Solution

EXAMPLE 13

The land area of the United States is 3,539,341 mi². What is the land area of the United States to the nearest ten-thousand square miles?

Strategy

To find the land area to the nearest ten-thousand square miles, round 3,539,341 to the nearest ten-thousand.

Solution

3,539,341 rounded to the nearest ten-thousand is 3,540,000.

To the nearest ten-thousand square miles, the land area of the United States is 3,540,000 mi².

YOU TRY IT 13

The land area of Canada is 3,851,809 mi². What is the land area of Canada to the nearest thousand square miles?

Your Strategy

Your Solution

Solutions on p. S1

1.1 Exercises

OBJECTIVE A   **Order relations between whole numbers**

1.   The inequality 7 > 4 is read "seven _____ four."

2.   Fill in the blank with < or >: On the number line, 2 is to the left of 8, so 2 _____ 8.

Graph the number on the number line.

3.   2

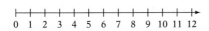

4.   7

5.   10

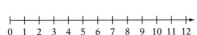

6.   1

7.   5

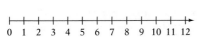

8.   11

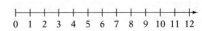

On the number line, which number is:

9.   4 units to the left of 9

10.   5 units to the left of 8

11.   3 units to the right of 2

12.   4 units to the right of 6

13.   7 units to the left of 7

14.   8 units to the left of 11

Place the correct symbol, < or >, between the two numbers.

15.   27   39

16.   68   41

17.   0   52

18.   61   0

19.   273   194

20.   419   502

21.   2,761   3,857

22.   3,827   6,915

23.   4,610   4,061

24.   5,600   56,000

25.   8,005   8,050

26.   92,010   92,001

27.   Do the inequalities 15 > 12 and 12 < 15 express the same order relation?

28.   Use the inequality symbol < to rewrite the order relation expressed by the inequality 23 > 10.

Write the given numbers in order from smallest to largest.

29.   21, 14, 32, 16, 11

30.   18, 60, 35, 71, 27

31.   72, 48, 84, 93, 13

32. 54, 45, 63, 28, 109   **33.** 26, 49, 106, 90, 77   **34.** 505, 496, 155, 358, 271

35. 736, 662, 204, 981, 399   **36.** 440, 404, 400, 444, 4,000   **37.** 377, 370, 307, 3,700, 3,077

OBJECTIVE B   Place value

Write the number in words.

38. To write the number 72,405 in words, first write *seventy-two*. Next, replace the comma with the word _____. Then write the words
_____.

39. To write the number eight hundred twenty-two thousand in standard form, write "822." Then replace the word *thousand* with a _____ followed by _____ zeros.

40. 704   **41.** 508   **42.** 374

43. 635   **44.** 2,861   **45.** 4,790

46. 48,297   **47.** 53,614   **48.** 563,078

49. 246,053   **50.** 6,379,482   **51.** 3,842,905

Write the number in standard form.

52. seventy-five   **53.** four hundred ninety-six

54. two thousand eight hundred fifty-one   **55.** fifty-three thousand three hundred forty

56. one hundred thirty thousand two hundred twelve   **57.** five hundred two thousand one hundred forty

58. eight thousand seventy-three   **59.** nine thousand seven hundred six

60. six hundred three thousand one hundred thirty-two   **61.** five million twelve thousand nine hundred seven

62. three million four thousand eight   **63.** eight million five thousand ten

Write the number in expanded form.

64. 6,398   **65.** 7,245   **66.** 46,182

67. 532,791   **68.** 328,476   **69.** 5,064

70. 90,834   **71.** 20,397   **72.** 400,635

73. 402,708   **74.** 504,603   **75.** 8,000,316

76. What is the place value of the leftmost number in a five-digit number?

77. What is the place value of the third number from the left in a four-digit number?

OBJECTIVE C Rounding

78. The number 921 rounded to the nearest ten is 920 because the ones digit in 921 is _____ than 5.

79. The number 927 rounded to the nearest ten is 930 because the ones digit in 927 is _____ than 5.

Round the number to the given place value.

80. 3,049; tens   **81.** 7,108; tens   **82.** 1,638; hundreds

83. 4,962; hundreds   **84.** 17,639; hundreds   **85.** 28,551; hundreds

86. 5,326; thousands   **87.** 6,809; thousands   **88.** 84,608; thousands

89. 93,825; thousands   **90.** 389,702; thousands   **91.** 629,513; thousands

92. 746,898; ten-thousands   **93.** 352,876; ten-thousands   **94.** 36,702,599; millions

Determine whether each statement is sometimes true, never true, or always true.

95. A six-digit number rounded to the nearest thousand is greater than the same number rounded to the nearest ten-thousand.

96. If a number rounded to the nearest ten is equal to itself, then the ones digit of the number is 0.

97. If the ones digit of a number is greater than 5, then the number rounded to the nearest ten is less than the original number.

98. Use the circle graph in Figure 1.7 on page 11. To decide whether baseball or football is the more popular sport, compare the numbers _____ and _____.

99. Use the double-line graph in Figure 1.6 on page 10. To determine the population of Texas in 1950, follow the vertical line above 1950 up to the _____ line.

OBJECTIVE D **Applications and statistical graphs**

100. *Sports* During his baseball career, Eddie Collins had a record of 743 stolen bases. Max Carey had a record of 738 stolen bases during his baseball career. Who had more stolen bases, Eddie Collins or Max Carey?

101. *Sports* During his baseball career, Ty Cobb had a record of 892 stolen bases. Billy Hamilton had a record of 937 stolen bases during his baseball career. Who had more stolen bases, Ty Cobb or Billy Hamilton?

102. *Nutrition* The figure at the right shows the annual per capita turkey consumption in different countries. **a.** What is the annual per capita turkey consumption in the United States? **b.** In which country is the annual per capita turkey consumption the highest?

Britain	🦃🦃🦃🦃
Canada	🦃🦃🦃🦃🦃
France	🦃🦃🦃🦃🦃🦃
Ireland	🦃🦃🦃🦃
Israel	🦃🦃🦃🦃🦃🦃🦃🦃🦃🦃
Italy	🦃🦃🦃🦃🦃
U.S.	🦃🦃🦃🦃🦃🦃🦃🦃🦃

Each 🦃 represents 2 lb.

Per Capita Turkey Consumption
Source: National Turkey Federation

103. *The Arts*   The play *Hello Dolly* was performed 2,844 times on Broadway. The play *Fiddler on the Roof* was performed 3,242 times on Broadway. Which play had the greater number of performances, *Hello Dolly* or *Fiddler on the Roof*?

104. *The Arts*   The play *Annie* was performed 2,377 times on Broadway. The play *My Fair Lady* was performed 2,717 times on Broadway. Which play had the greater number of performances, *Annie* or *My Fair Lady*?

105. *Nutrition*   Two tablespoons of peanut butter contain 190 calories. Two tablespoons of grape jelly contain 114 calories. Which contains more calories, two tablespoons of peanut butter or two tablespoons of grape jelly?

106. *History*   In 1892, the diesel engine was patented. In 1844, Samuel F. B. Morse patented the telegraph. Which was patented first, the diesel engine or the telegraph?

Samuel F. B. Morse

107. *Geography*   The distance between St. Louis, Missouri and Reno, Nevada is 1,892 mi. The distance between St. Louis, Missouri and San Diego, California is 1,833 mi. Which is the shorter distance, St. Louis to Reno or St. Louis to San Diego?

108. *Consumerism*   The circle graph at the right shows the result of a survey of 150 people who were asked, "What bothers you most about movie theaters?" **a.** Among the respondents, what was the most often mentioned complaint? **b.** What was the least often mentioned complaint?

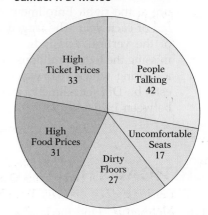

Distribution of Responses in a Survey

109. *Astronomy*   As measured at the equator, the diameter of the planet Uranus is 32,200 mi and the diameter of the planet Neptune is 30,800 mi. Which planet is smaller, Uranus or Neptune?

110. *Aviation*   The cruising speed of a Boeing 747 is 589 mph. What is the cruising speed of a Boeing 747 to the nearest ten miles per hour?

111. *Physics*   Light travels at a speed of 299,800 km/s. What is the speed of light to the nearest thousand kilometers per second?

112. *Geography*   The land area of Alaska is 570,833 mi^2. What is the land area of Alaska to the nearest thousand square miles?

Alaska

113. *Geography*   The acreage of the Appalachian Trail is 161,546. What is the acreage of the Appalachian Trail to the nearest ten-thousand acres?

114. Travel The figure below shows the number of crashes on U.S. roadways during each of the last six months of a recent year. Also shown is the number of vehicles involved in those crashes. **a.** Which was greater, the number of crashes in July or in October? **b.** Were there fewer vehicles involved in crashes in July or in December?

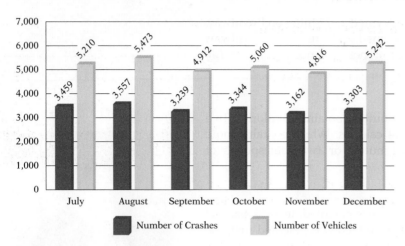

Accidents on U.S. Roadways
Source: National Highway Traffic Safety Administration

115. Education Actual and projected student enrollment in elementary and secondary schools in the United States is shown in the figure at the right. Enrollment figures are for the fall of each year. The jagged line at the bottom of the vertical axis indicates that this scale is missing the tens of millions from 0 to 30,000,000. **a.** During which year was enrollment the lowest? **b.** Did enrollment increase or decrease between 1975 and 1980?

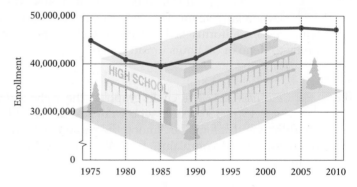

Enrollment in Elementary and Secondary Schools
Source: National Center for Education Statistics

CRITICAL THINKING

116. Geography Find the land areas of the seven continents. List the continents in order from largest to smallest.

117. Mathematics What is the largest three-digit number? What is the smallest five-digit number?

118. What is the total enrollment of your school? To what place value would it be reasonable to round this number? Why? To what place value is the population of your town or city rounded? Why? To what place value is the population of your state rounded? To what place value is the population of the United States rounded?

1.2 Addition and Subtraction of Whole Numbers

OBJECTIVE A   **Addition of whole numbers**

Addition is the process of finding the total of two or more numbers.

On Arbor Day, a community group planted 3 trees along one street and 5 trees along another street. By counting, we can see that a total of 8 trees were planted.

3   +   5   =   8

The 3 and 5 are called **addends.** The **sum** is 8.

The basic addition facts for adding one digit to one digit should be memorized. Addition of larger numbers requires the repeated use of the basic addition facts.

To add large numbers, begin by arranging the numbers vertically, keeping the digits of the same place value in the same column.

Add: 321 + 6,472

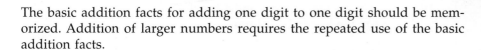

Add the digits in each column.

$$\begin{array}{r} 3\ 2\ 1 \\ +\ 6\ 4\ 7\ 2 \\ \hline 6\ 7\ 9\ 3 \end{array}$$

Find the sum of 211, 45, 23, and 410.

Remember that a *sum* is the answer to an addition problem.
Arrange the numbers vertically, keeping digits of the same place value in the same column.
Add the numbers in each column.

$$\begin{array}{r} 211 \\ 45 \\ 23 \\ +\ 410 \\ \hline 689 \end{array}$$

The phrase *the sum of* was used in the example above to indicate the operation of addition. All of the phrases listed below indicate addition. An example of each is shown to the right of each phrase.

added to	6 added to 9	9 + 6
more than	3 more than 8	8 + 3
the sum of	the sum of 7 and 4	7 + 4
increased by	2 increased by 5	2 + 5
the total of	the total of 1 and 6	1 + 6
plus	8 plus 10	8 + 10

When the sum of the numbers in a column exceeds 9, addition involves "carrying."

Add: 359 + 478

Add the ones column.
9 + 8 = 17 (1 ten + 7 ones).
Write the 7 in the ones column and carry the 1 ten to the tens column.

Hundreds Tens Ones
```
     1
  3 5 9
+ 4 7 8
      7
```

Add the tens column.
1 + 5 + 7 = 13 (1 hundred + 3 tens).
Write the 3 in the tens column and carry the 1 hundred to the hundreds column.

```
  1 1
  359
+ 478
   37
```

Add the hundreds column.
1 + 3 + 4 = 8 (8 hundreds).
Write the 8 in the hundreds column.

```
 1 1
  359
+ 478
  837
```

The bar graph in Figure 1.9 shows the seating capacity in 2006 of the five largest National Football League stadiums. What is the total seating capacity of these five stadiums? *Note:* The jagged line below 70,000 on the vertical axis indicates that this scale is missing the numbers less than 70,000.

Arrowhead Stadium, Kansas City

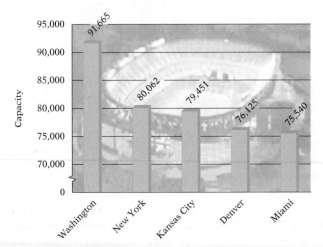

Figure 1.9 Seating Capacity of the Five Largest NFL Stadiums

```
  91,665
  80,062
  79,451
  76,125
+ 75,540
 402,843
```

The total capacity of the five stadiums is 402,843 people.

An important skill in mathematics is the ability to determine whether an answer to a problem is reasonable. One method of determining whether an answer is reasonable is to use estimation. An **estimate** is an approximation.

Estimation is especially valuable when using a calculator. Suppose that you are adding 1,497 and 2,568 on a calculator. You enter the number 1,497 correctly, but you inadvertently enter 256 instead of 2,568 for the second addend. The sum reads 1,753. If you quickly make an estimate of the answer, you can determine that the sum 1,753 is not reasonable and that an error has been made.

$$\begin{array}{r} 1,497 \\ + 2,568 \\ \hline 4,065 \end{array} \qquad \begin{array}{r} 1,497 \\ + 256 \\ \hline 1,753 \end{array}$$

Calculator Note

Here is an example of how estimation is important when using a calculator.

To estimate the answer to a calculation, round each number to the highest place value of the number; the first digit of each number will be nonzero and all other digits will be zero. Perform the calculation using the rounded numbers.

$$\begin{array}{rcl} 1,497 & \longrightarrow & 1,000 \\ 2,568 & \longrightarrow & + 3,000 \\ & & \hline 4,000 \end{array}$$

As shown above, the sum 4,000 is an estimate of the sum of 1,497 and 2,568; it is very close to the actual sum, 4,065. 4,000 is not close to the incorrectly calculated sum, 1,753.

Estimate the sum of 35,498, 17,264, and 81,093.

Round each number to the nearest ten-thousand.

$$\begin{array}{rcl} 35,498 & \longrightarrow & 40,000 \\ 17,264 & \longrightarrow & 20,000 \\ 81,093 & \longrightarrow & + 80,000 \\ & & \hline 140,000 \end{array}$$

Add the rounded numbers.

Note that 140,000 is close to the actual sum, 133,855.

Just as the word *it* is used in language to stand for an object, a letter of the alphabet can be used in mathematics to stand for a number. Such a letter is called a **variable.**

A mathematical expression that contains one or more variables is a **variable expression.** Replacing the variables in a variable expression with numbers and then simplifying the numerical expression is called **evaluating the variable expression.**

Evaluate $a + b$ for $a = 678$ and $b = 294$.

$$a + b$$

Replace a with 678 and b with 294.

$$678 + 294$$

Arrange the numbers vertically.

$$\begin{array}{r} \overset{1\ 1}{678} \\ + 294 \\ \hline 972 \end{array}$$

Add.

Variables are often used in algebra to describe mathematical relationships. Variables are used below to describe three properties, or rules, of addition. An example of each property is shown at the right.

The Addition Property of Zero

$a + 0 = a$ or $0 + a = a$

$5 + 0 = 5$

The Addition Property of Zero states that the sum of a number and zero is the number. The variable a is used here to represent any whole number. It can even represent the number zero because $0 + 0 = 0$.

The Commutative Property of Addition

$a + b = b + a$

$5 + 7 = 7 + 5$
$12 = 12$

The Commutative Property of Addition states that two numbers can be added in either order; the sum will be the same. Here the variables a and b represent any whole numbers. Therefore, if you know that the sum of 5 and 7 is 12, then you also know that the sum of 7 and 5 is 12, because $5 + 7 = 7 + 5$.

The Associative Property of Addition

$(a + b) + c = a + (b + c)$

$(2 + 3) + 4 = 2 + (3 + 4)$
$5 + 4 = 2 + 7$
$9 = 9$

The Associative Property of Addition states that when adding three or more numbers, we can group the numbers in any order; the sum will be the same. Note in the example at the right above that we can add the sum of 2 and 3 to 4, or we can add 2 to the sum of 3 and 4. In either case, the sum of the three numbers is 9.

Rewrite the expression by using the Associative Property of Addition.

$(3 + x) + y$

The Associative Property of Addition states that addends can be grouped in any order.

$(3 + x) + y = 3 + (x + y)$

An **equation** expresses the equality of two numerical or variable expressions. In the preceding example, $(3 + x) + y$ is an expression; it does not contain an equals sign. $(3 + x) + y = 3 + (x + y)$ is an equation; it contains an equals sign.

Here is another example of an equation. The **left side** of the equation is the variable expression $n + 4$. The **right side** of the equation is the number 9.

$$n + 4 = 9$$

Point of Interest

The equals sign (=) is generally credited to Robert Recorde. In his 1557 treatise on algebra, *The Whetstone of Witte,* he wrote, "No two things could be more equal (than two parallel lines)." His equals sign gained popular usage, even though continental mathematicians preferred a dash.

Just as a statement in English can be true or false, an equation may be true or false. The equation shown above is *true* if the variable is replaced by 5.

$$n + 4 = 9$$
$$5 + 4 = 9 \quad \text{True}$$

The equation is *false* if the variable is replaced by 8.

$$8 + 4 = 9 \quad \text{False}$$

A **solution** of an equation is a number that, when substituted for the variable, produces a true equation. The solution of the equation $n + 4 = 9$ is 5 because replacing n by 5 results in a true equation. When 8 is substituted for n, the result is a false equation; therefore, 8 is not a solution of the equation.

10 is a solution of $x + 5 = 15$ because $10 + 5 = 15$ is a true equation.

20 is not a solution of $x + 5 = 15$ because $20 + 5 = 15$ is a false equation.

Is 9 a solution of the equation $11 = 2 + x$?

Replace x by 9.

Simplify the right side of the equation. Compare the results. If the results are equal, the given number is a solution of the equation. If the results are not equal, the given number is not a solution.

$$11 = 2 + x$$
$$\frac{11 \mid 2 + 9}{}$$
$$11 = 11$$

Yes, 9 is a solution of the equation.

EXAMPLE 1   Estimate the sum of 379, 842, 693, and 518.

Solution

379 $\longrightarrow$	400
842 $\longrightarrow$	800
693 $\longrightarrow$	700
518 $\longrightarrow$	+ 500
	2,400

YOU TRY IT 1   Estimate the total of 6,285, 3,972, and 5,140.

Your Solution

EXAMPLE 2   Identify the property that justifies the statement.

$$7 + 2 = 2 + 7$$

Solution   The Commutative Property of Addition

YOU TRY IT 2   Identify the property that justifies the statement.

$$33 + 0 = 33$$

Your Solution

Solutions on p. S1

The topic of the circle graph in Figure 1.10 is the eggs produced in the United States in a recent year. The graph shows where the eggs that were produced went or how they were used. Use this graph for Example 3 and You Try It 3.

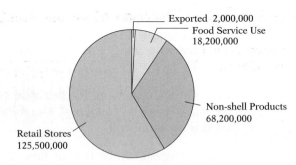

Exported 2,000,000
Food Service Use 18,200,000
Non-shell Products 68,200,000
Retail Stores 125,500,000

Figure 1.10 Eggs Produced in the United States (in cases)
Source: U.S. Department of Agriculture.

EXAMPLE 3   Use Figure 1.10 to determine the sum of the number of cases of eggs sold by retail stores or used for non-shell products.

Solution   125,500,000 cases of eggs were sold by retail stores. 68,200,000 cases of eggs were used for non-shell products.

125,500,000
+ 68,200,000
———————
193,700,000

193,700,000 cases of eggs were sold by retail stores or used for non-shell products.

YOU TRY IT 3   Use Figure 1.10 to determine the total number of cases of eggs produced during the given year.

Your Solution

EXAMPLE 4   Evaluate $x + y + z$ for $x = 8{,}427$, $y = 3{,}659$, and $z = 6{,}281$.

Solution $x + y + z$
$8{,}427 + 3{,}659 + 6{,}281$

  ¹ ¹¹
  8,427
  3,659
+ 6,281
————
 18,367

YOU TRY IT 4   Evaluate $x + y + z$ for $x = 1{,}692$, $y = 4{,}783$, and $z = 5{,}046$.

Your Solution

EXAMPLE 5   Is 6 a solution of the equation $9 + y = 14$?

Solution $9 + y = 14$
$\overline{9 + 6 \mid 14}$
$15 \neq 14$ ▶ The symbol $\neq$ is read "is not equal to."

No, 6 is not a solution of the equation $9 + y = 14$.

YOU TRY IT 5   Is 7 a solution of the equation $13 = b + 6$?

Your Solution

Solutions on p. S1

OBJECTIVE B   Subtraction of whole numbers

Subtraction is the process of finding the difference between two numbers.

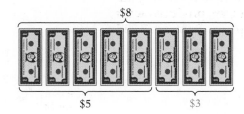

By counting, we see that the difference between $8 and $5 is $3.

$8	−	$5	=	$3

Minuend − Subtrahend = Difference

Note that addition and subtraction are related.

Subtrahend	5
+ Difference	+ 3
= Minuend	8

The fact that the sum of the subtrahend and the difference equals the minuend can be used to check subtraction.

To subtract large numbers, begin by arranging the numbers vertically, keeping the digits of the same place value in the same column. Then subtract the numbers in each column.

Find the difference between 8,955 and 2,432.

A *difference* is the answer to a subtraction problem.

```
    Thousands Hundreds Tens Ones
      8    9    5    5
    − 2    4    3    2
      6    5    2    3
```

Check:
Subtrahend	2,432
+ Difference	+ 6,523
= Minuend	8,955

In the subtraction example above, the lower digit in each place value is smaller than the upper digit. When the lower digit is larger than the upper digit, subtraction involves "borrowing."

Subtract: 692 − 378

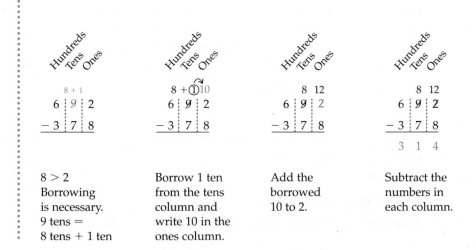

8 > 2 Borrowing is necessary. 9 tens = 8 tens + 1 ten	Borrow 1 ten from the tens column and write 10 in the ones column.	Add the borrowed 10 to 2.	Subtract the numbers in each column.

Subtraction may involve repeated borrowing.

Subtract: $7{,}325 - 4{,}698$

$$
\begin{array}{r}
{}^{1}\ {}^{15} \\
7{,}3\ \cancel{2}\ \cancel{5} \\
-\,4{,}6\ 9\ 8 \\
\hline
7
\end{array}
\qquad
\begin{array}{r}
{}^{11} \\
{}^{2}\ \cancel{1}\ {}^{15} \\
7{,}\cancel{3}\ \cancel{2}\ \cancel{5} \\
-\,4{,}6\ 9\ 8 \\
\hline
2\ 7
\end{array}
\qquad
\begin{array}{r}
{}^{12}\ {}^{11} \\
{}^{6}\ {}^{2}\ \cancel{1}\ {}^{15} \\
\cancel{7}{,}\ \cancel{3}\ \cancel{2}\ \cancel{5} \\
-\,4{,}\ 6\ 9\ 8 \\
\hline
2{,}\ 6\ 2\ 7
\end{array}
$$

Borrow 1 ten (10 ones) from the tens column and add 10 to the 5 in the ones column. Subtract $15 - 8$.	Borrow 1 hundred (10 tens) from the hundreds column and add 10 to the 1 in the tens column. Subtract $11 - 9$.	Borrow 1 thousand (10 hundreds) from the thousands column and add 10 to the 2 in the hundreds column. Subtract $12 - 6$ and $6 - 4$.

When there is a zero in the minuend, subtraction involves repeated borrowing.

Subtract: $3{,}904 - 1{,}775$

$$
\begin{array}{r}
{}^{8}\ {}^{10} \\
3{,}\cancel{9}\ \cancel{0}\ 4 \\
-\,1{,}7\ 7\ 5 \\
\hline
\end{array}
\qquad
\begin{array}{r}
{}^{9} \\
{}^{8}\ \cancel{10}\ {}^{14} \\
3{,}\cancel{9}\ \cancel{0}\ \cancel{4} \\
-\,1{,}7\ 7\ 5 \\
\hline
\end{array}
\qquad
\begin{array}{r}
{}^{9} \\
{}^{8}\ \cancel{10}\ {}^{14} \\
3{,}\cancel{9}\ \cancel{0}\ \cancel{4} \\
-\,1{,}7\ 7\ 5 \\
\hline
2{,}\ 1\ 2\ 9
\end{array}
$$

There is a 0 in the tens column. Borrow 1 hundred (10 tens) from the hundreds column and write 10 in the tens column.	Borrow 1 ten from the tens column and add 10 to the 4 in the ones column.	Subtract the numbers in each column.

Note that, for the preceding example, the borrowing could be performed as shown below.

Borrow 1 from 90. ($90 - 1 = 89$. The 8 is in the hundreds column. The 9 is in the tens column.) Add 10 to the 4 in the ones column. Then subtract the numbers in each column.

$$
\begin{array}{r}
{}^{8}\ {}^{9}\ {}^{14} \\
3{,}9\cancel{0}\cancel{4} \\
-\,1{,}7\,7\,5 \\
\hline
2{,}1\,2\,9
\end{array}
$$

Estimate the difference between 49,601 and 35,872.

Round each number to the nearest ten-thousand.

$$
\begin{array}{r}
49{,}601 \longrightarrow 50{,}000 \\
35{,}872 \longrightarrow -\,40{,}000 \\
\hline
10{,}000
\end{array}
$$

Subtract the rounded numbers.

Note that 10,000 is close to the actual difference, 13,729.

The phrase *the difference between* was used in the preceding example to indicate the operation of subtraction. All of the phrases listed below indicate subtraction. An example of each is shown to the right of each phrase.

minus	10 minus 3	$10 - 3$
less	8 less 4	$8 - 4$
less than	2 less than 9	$9 - 2$
the difference between	the difference between 6 and 1	$6 - 1$
decreased by	7 decreased by 5	$7 - 5$
subtract . . . from	subtract 11 from 20	$20 - 11$

> ### Take Note
> Note the order in which the numbers are subtracted when the phrase *less than* is used. Suppose that you have $10 and I have $6 *less than* you do; then I have $6 *less than* $10, or $10 − $6 = $4.

Evaluate $c - d$ for $c = 6{,}183$ and $d = 2{,}759$.

Replace c with 6,183 and d with 2,759.

$$c - d$$
$$6{,}183 - 2{,}759$$

Arrange the numbers vertically and then subtract.

$$\begin{array}{r} {}^{5\ \ 11\ 7\ 13} \\ 6{,}183 \\ -\,2{,}759 \\ \hline 3{,}424 \end{array}$$

Is 23 a solution of the equation $41 - n = 17$?

Replace n by 23.
Simplify the left side of the equation.
The results are not equal.

$$41 - n = 17$$
$$41 - 23 \mid 17$$
$$18 \ne 17$$

No, 23 is not a solution of the equation.

Point of Interest

Someone who is our equal is our peer. Two make a pair. Both of the words *peer* and *pair* come from the Latin *par, paris,* meaning "equal."

EXAMPLE 6 Subtract and check:
57,004 − 26,189

Solution
$$\begin{array}{r} {}^{6\ \ \ \ 9\ 9\ 14} \\ 5\cancel{7{,}00}4 \\ -\,26{,}189 \\ \hline 30{,}815 \end{array}$$

Check: $\begin{array}{r} 26{,}189 \\ +\,30{,}815 \\ \hline 57{,}004 \end{array}$

YOU TRY IT 6 Subtract and check:
49,002 − 31,865

Your Solution

EXAMPLE 7 Estimate the difference between 7,261 and 4,315. Then find the exact answer.

Solution
$$\begin{array}{rccc} 7{,}261 & \longrightarrow & 7{,}000 & 7{,}261 \\ 4{,}315 & \longrightarrow & -\,4{,}000 & -\,4{,}315 \\ & & \overline{3{,}000} & \overline{2{,}946} \end{array}$$

YOU TRY IT 7 Estimate the difference between 8,544 and 3,621. Then find the exact answer.

Your Solution

Solutions on p. S1

The graph in Figure 1.11 shows the actual and projected world energy consumption in quadrillion British thermal units (Btu). Use this graph for Example 8 and You Try It 8.

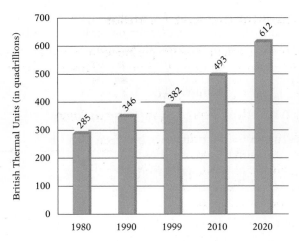

Figure 1.11   World Energy Consumption (in quadrillion British thermal units)

Sources: Energy Information Administration; Office of Energy Markets and End Use; *International Statistics Database and International Energy Annual*; World Energy Projection System

---

EXAMPLE 8   Use Figure 1.11 to find the difference between the world energy consumption in 1980 and that projected for 2010.

Solution   2010: 493 quadrillion Btu
1980: 285 quadrillion Btu

$$\begin{array}{r} 493 \\ -\ 285 \\ \hline 208 \end{array}$$

The difference between the world energy consumption in 1980 and that projected for 2010 is 208 quadrillion Btu.

YOU TRY IT 8   Use Figure 1.11 to find the difference between the world energy consumption in 1990 and that projected for 2020.

Your Solution

EXAMPLE 9   Evaluate $x - y$ for $x = 3{,}506$ and $y = 2{,}477$.

Solution $x - y$
$3{,}506 - 2{,}477$

$$\begin{array}{r} \overset{4\ \ 9\ \ 16}{3{,}5\cancel{0}\cancel{6}} \\ -\ 2{,}477 \\ \hline 1{,}029 \end{array}$$

YOU TRY IT 9   Evaluate $x - y$ for $x = 7{,}061$ and $y = 3{,}229$.

Your Solution

EXAMPLE 10   Is 39 a solution of the equation $24 = m - 15$?

Solution $24 = m - 15$

$$\begin{array}{c|c} 24 & 39 - 15 \\ \end{array}$$
$24 = 24$

Yes, 39 is a solution of the equation.

YOU TRY IT 10   Is 11 a solution of the equation $46 = 58 - p$?

Your Solution

Solutions on pp. S1–S2

OBJECTIVE C   Applications and formulas

One application of addition is calculating the perimeter of a figure. However, before defining perimeter, we will introduce some terms from geometry.

Two basic concepts in the study of geometry are point and line.

A **point** is symbolized by drawing a dot. A **line** is determined by two distinct points and extends indefinitely in both directions, as the arrows on the line shown at the right indicate. This line contains points *A* and *B*.

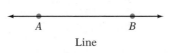
Line

A **ray** starts at a point and extends indefinitely in *one* direction. The point at which a ray starts is called the **endpoint** of the ray. Point *A* is the endpoint of the ray shown at the right.

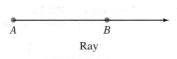
Ray

A **line segment** is part of a line and has two endpoints. The line segment shown at the right has endpoints *A* and *B*.

Line Segment

An **angle** is formed by two rays with the same endpoint. An angle is measured in **degrees.** The symbol for degrees is a small raised circle, °. A **right angle** is an angle whose measure is 90°.

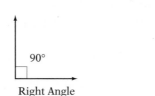
Right Angle

> ### *Take Note*
> The corner of a page of this book is a good model of a right angle.

A **plane** is a flat surface and can be pictured as a floor or a wall. Figures that lie in a plane are called **plane figures.**

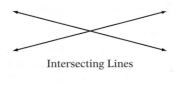
Intersecting Lines

Lines in a plane can be intersecting or parallel. **Intersecting lines** cross at a point in the plane. **Parallel lines** never meet. The distance between them is always the same.

Parallel Lines

A **polygon** is a closed figure determined by three or more line segments that lie in a plane. The line segments that form the polygon are called its **sides.** The figures below are examples of polygons.

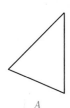

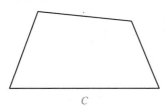

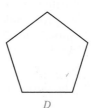

A   *B*   *C*   *D*   *E*

Rectangle

The name of a polygon is based on the number of its sides. A polygon with three sides is a **triangle.** Figure A on the previous page is a triangle. A polygon with four sides is a **quadrilateral.** Figures B and C are quadrilaterals.

Quadrilaterals are one of the most common types of polygons. Quadrilaterals are distinguished by their sides and angles. For example, a **rectangle** is a quadrilateral in which opposite sides are parallel, opposite sides are equal in length, and all four angles measure 90°.

The **perimeter** of a plane geometric figure is a measure of the distance around the figure.

The perimeter of a triangle is the sum of the lengths of the three sides.

Perimeter of a Triangle

The formula for the perimeter of a triangle is $P = a + b + c$, where P is the perimeter of the triangle and a, b, and c are the lengths of the sides of the triangle.

4 in.   5 in.
8 in.

Find the perimeter of the triangle shown at the left.

Use the formula for the perimeter of a triangle.
It does not matter which side you label a, b, or c.
Add.

$P = a + b + c$
$P = 4 + 5 + 8$
$P = 17$

The perimeter of the triangle is 17 in.

The perimeter of a quadrilateral is the sum of the lengths of its four sides.

In a rectangle, opposite sides are equal in length. Usually the length, L, of a rectangle refers to the length of one of the longer sides of the rectangle, and the width, W, refers to the length of one of the shorter sides. The perimeter can then be represented as $P = L + W + L + W$.

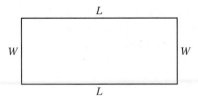

Use the formula $P = L + W + L + W$ to find the perimeter of the rectangle shown at the left.

32 ft
16 ft

Write the given formula for the perimeter of a rectangle.
Substitute 32 for L and 16 for W.
Add.

$P = L + W + L + W$
$P = 32 + 16 + 32 + 16$
$P = 96$

The perimeter of the rectangle is 96 ft.

In this section, some of the phrases used to indicate the operations of addition and subtraction were presented. In solving application problems, you might also look for the types of questions listed below.

Addition	Subtraction
How many . . . altogether?	How many more (or fewer) . . . ?
How many . . . in all?	How much is left?
How many . . . and . . . ?	How much larger (or smaller) . . . ?

The bar graph in Figure 1.12 shows the number of fatal accidents on amusement rides in the United States each year during the 1990s. Use this graph for Example 11 and You Try It 11.

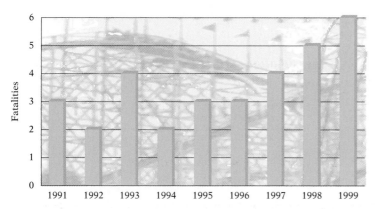

Figure 1.12   Number of Fatal Accidents on Amusement Rides
Source: USA Today, April 7, 2000

EXAMPLE 11 Use Figure 1.12 to determine how many more fatal accidents occurred during the years 1995 through 1998 than occurred during the years 1991 through 1994.

Strategy To find how many more occurred in 1995 through 1998 than occurred in 1991 through 1994:

▶ Find the total number of fatalities that occurred from 1995 to 1998 and the total number that occurred from 1991 to 1994.

▶ Subtract the smaller number from the larger.

Solution Fatalities during 1995–1998: 15
Fatalities during 1991–1994: 11
$15 - 11 = 4$
4 more fatalities occurred from 1995 to 1998 than occurred from 1991 to 1994.

YOU TRY IT 11 Use Figure 1.12 to find the total number of fatal accidents on amusement rides during 1991 through 1999.

Your Strategy

Your Solution

Solution on p. S2

EXAMPLE 12

What is the price of a pair of skates that cost a business $109 and has a markup of $49? Use the formula $P = C + M$, where P is the price of a product to the consumer, C is the cost paid by the store for the product, and M is the markup.

Strategy

To find the price, replace C by 109 and M by 49 in the given formula and solve for P.

Solution

$P = C + M$

$P = 109 + 49$

$P = 158$

The price of the skates is $158.

YOU TRY IT 12

What is the price of a leather jacket that cost a business $148 and has a markup of $74? Use the formula $P = C + M$, where P is the price of a product to the consumer, C is the cost paid by the store for the product, and M is the markup.

Your Strategy

Your Solution

EXAMPLE 13

Find the length of decorative molding needed to edge the top of the walls in a rectangular room that is 12 ft long and 8 ft wide.

Strategy
Draw a diagram.

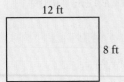

12 ft

8 ft

To find the length of molding needed, use the formula for the perimeter of a rectangle, $P = L + W + L + W$. $L = 12$ and $W = 8$.

Solution

$P = L + W + L + W$

$P = 12 + 8 + 12 + 8$

$P = 40$

40 ft of decorative molding are needed.

YOU TRY IT 13

Find the length of fencing needed to surround a rectangular corral that measures 60 ft on each side.

Your Strategy

Your Solution

Solutions on p. S2

128. *Finances*   The repair bill on your car includes $358 for parts, $156 for labor, and a sales tax of $30. What is the total amount owed?

129. *Finances*   The computer system you would like to purchase includes an operating system priced at $830, a monitor that costs $245, an extended keyboard priced at $175, and a printer that sells for $395. What is the total cost of the computer system?

130.  *Geography*   The area of Lake Superior is 81,000 mi²; the area of Lake Michigan is 67,900 mi²; the area of Lake Huron is 74,000 mi²; the area of Lake Erie is 32,630 mi²; and the area of Lake Ontario is 34,850 mi². Estimate the total area of the five Great Lakes.

The Great Lakes

131. *Consumerism*   The odometer on your car read 58,376 this time last year. It now reads 77,912. Estimate the number of miles your car has been driven during the past year.

The figure at the right shows the number of cars sold by a dealership for the first four months of 2005 and 2006. Use this graph for Exercises 132 to 134.

132. *Business*   Between which two months did car sales decrease the most in 2006? What was the amount of decrease?

133. *Business*   Between which two months did car sales increase the most in 2005? What was the amount of increase?

134. *Business*   In which year were more cars sold during the four months shown?

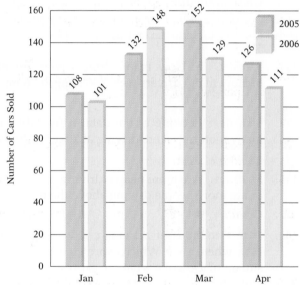

Car Sales at a Dealership

135. *Finances*   Use the formula $A = P + I$, where A is the value of an investment, P is the original investment, and I is the interest earned, to find the value of an investment that earned $775 in interest on an original investment of $12,500.

136. *Finances*   Use the formula $A = P + I$, where A is the value of an investment, P is the original investment, and I is the interest earned, to find the value of an investment that earned $484 in interest on an original investment of $8,800.

137. *Finances*   What is the mortgage loan amount on a home that sells for $290,000 with a down payment of $29,000? Use the formula $M = S - D$, where M is the mortgage loan amount, S is the selling price, and D is the down payment.

138. *Finances* What is the mortgage loan amount on a home that sells for $236,000 with a down payment of $47,200? Use the formula $M = S - D$, where M is the mortgage loan amount, S is the selling price, and D is the down payment.

139. *Physics* What is the ground speed of an airplane traveling into a 25 mph headwind with an air speed of 375 mph? Use the formula $g = a - h$, where g is the ground speed, a is the air speed, and h is the speed of the headwind.

140. *Physics* Find the ground speed of an airplane traveling into a 15 mph headwind with an air speed of 425 mph. Use the formula $g = a - h$, where g is the ground speed, a is the air speed, and h is the speed of the headwind.

In some states, the speed limit on certain sections of highway is 70 mph. To test drivers' compliance with the speed limit, the highway patrol conducted a one-week study during which it recorded the speeds of motorists on one of these sections of highway. The results are recorded in the table at the right. Use this table for Exercises 141 to 143.

Speed	Number of Cars
> 80	1,708
76 – 80	2,503
71 – 75	3,651
66 – 70	3,717
61 – 65	2,984
< 61	2,870

141. *Statistics* **a.** How many drivers were traveling at 70 mph or less? **b.** How many drivers were traveling at 76 mph or more?

142. *Statistics* Looking at the data in the table, is it possible to tell how many motorists were driving at 70 mph? Explain your answer.

143. *Statistics* Are more people driving at or below the posted speed limit, or are more people driving above the posted speed limit?

144. Two sides of a triangle have lengths of a inches and b inches, where $a < b$. Which expression, $a - b$ or $b - a$, has meaning in this situation? Describe what the expression represents.

CRITICAL THINKING

145. *Dice* If you roll two ordinary six-sided dice and add the two numbers that appear on top, how many different sums are possible?

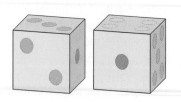

146. *Mathematics* How many two-digit numbers are there? How many three-digit numbers are there?

147. Determine whether the statement is always true, sometimes true, or never true.
a. If a is any whole number, then $a - 0 = a$.
b. If a is any whole number, then $a - a = 0$.

148. What estimate is given for the expected population of your state by the year 2025? What is the expected growth in the population of your state between now and 2025?

Multiply: 3(20)(10)(4)

Multiply the first two numbers.	$3(20)(10)4 = 60(10)(4)$
Multiply the product by the third number.	$= (600)(4)$
Continue multiplying until all the numbers have been multiplied.	$= 2,400$

Figure 1.13 shows the average weekly earnings of full-time workers in the United States. Using these figures, calculate the earnings of a female full-time worker, age 27, for working for 4 weeks.

Multiply the number of weeks (4) times the amount earned for one week ($573).

$$4(573) = 2,292$$

The average earnings of a 27-year-old, female, full-time worker for working for 4 weeks are $2,292.

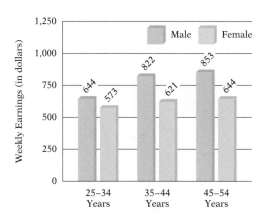

Figure 1.13   Average Weekly Earnings of Full-Time Workers
Source: Bureau of Labor Statistics

Estimate the product of 345 and 92.

Round each number to its highest place value.	345 ⟶ 300
	92 ⟶ 90
Multiply the rounded numbers.	$300 \cdot 90 = 27,000$

27,000 is an estimate of the product of 345 and 92.

The phrase *the product of* was used in the example above to indicate the operation of multiplication. All of the phrases below indicate multiplication. An example of each is shown to the right of each phrase.

times	8 times 4	$8 \cdot 4$
the product of	the product of 9 and 5	$9 \cdot 5$
multiplied by	7 multiplied by 3	$3 \cdot 7$
twice	twice 6	$2 \cdot 6$

Evaluate xyz for $x = 50$, $y = 2$, and $z = 7$.

xyz means $x \cdot y \cdot z$.	xyz
Replace each variable by its value.	$50 \cdot 2 \cdot 7$
Multiply the first two numbers.	$= 100 \cdot 7$
Multiply the product by the next number.	$= 700$

As for addition, there are properties of multiplication.

> ### The Multiplication Property of Zero
> ....
> $a \cdot 0 = 0$ or $0 \cdot a = 0$

$8 \cdot 0 = 0$

The Multiplication Property of Zero states that the product of a number and zero is zero. The variable a is used here to represent any whole number. It can even represent the number zero because $0 \cdot 0 = 0$.

> ### The Multiplication Property of One
> ....
> $a \cdot 1 = a$ or $1 \cdot a = a$

$1 \cdot 9 = 9$

The Multiplication Property of One states that the product of a number and 1 is the number. Multiplying a number by 1 does not change the number.

> ### The Commutative Property of Multiplication
> ....
> $a \cdot b = b \cdot a$

$4 \cdot 9 = 9 \cdot 4$
$36 = 36$

The Commutative Property of Multiplication states that two numbers can be multiplied in either order; the product will be the same. Here the variables a and b represent any whole numbers. Therefore, for example, if you know that the product of 4 and 9 is 36, then you also know that the product of 9 and 4 is 36 because $4 \cdot 9 = 9 \cdot 4$.

> ### The Associative Property of Multiplication
> ....
> $(a \cdot b) \cdot c = a \cdot (b \cdot c)$

$(2 \cdot 3) \cdot 4 = 2 \cdot (3 \cdot 4)$
$6 \cdot 4 = 2 \cdot 12$
$24 = 24$

The Associative Property of Multiplication states that when multiplying three numbers, the numbers can be grouped in any order; the product will be the same. Note in the example at the right above that we can multiply the product of 2 and 3 by 4, or we can multiply 2 by the product of 3 and 4. In either case, the product of the three numbers is 24.

What is the solution of the equation $5x = 5$?

By the Multiplication Property of One, the product of a number and 1 is the number.

$$\begin{array}{c|c} 5x = 5 \\ \hline 5(1) & 5 \\ 5 = 5 \end{array}$$

The solution is 1.

The check is shown at the right.

Is 7 a solution of the equation $3m = 21$?

Replace m by 7.

Simplify the left side of the equation.

The results are equal.

$$\begin{array}{c|c} 3m = 21 \\ \hline 3(7) & 21 \\ 21 = 21 \end{array}$$

Yes, 7 is a solution of the equation.

 Figure 1.14 shows the average monthly savings of individuals in seven different countries. Use this graph for Example 1 and You Try It 1.

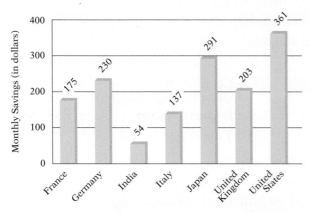

Figure 1.14   Average Monthly Savings
Source: Taylor Nelson - Sofres for American Express

EXAMPLE 1   Use Figure 1.14 to determine the average annual savings of individuals in Japan.

Solution   The average monthly savings in Japan is $291. The number of months in one year is 12.

$$\begin{array}{r} 291 \\ \times\ 12 \\ \hline 582 \\ 291 \\ \hline 3,492 \end{array}$$

The average annual savings of individuals in Japan is $3,492.

YOU TRY IT 1   According to Figure 1.14, what is the average annual savings of individuals in France?

Your Solution

Solution on p. *S2*

EXAMPLE 2 Estimate the product of 2,871 and 49.

Solution
$$2,871 \longrightarrow 3,000$$
$$49 \longrightarrow 50$$

$$3,000 \cdot 50 = 150,000$$

YOU TRY IT 2 Estimate the product of 8,704 and 93.

Your Solution

EXAMPLE 3 Evaluate $3ab$ for $a = 10$ and $b = 40$.

Solution
$$3ab$$
$$3(10)(40) = 30(40)$$
$$= 1,200$$

YOU TRY IT 3 Evaluate $5xy$ for $x = 20$ and $y = 60$.

Your Solution

EXAMPLE 4 What is 800 times 300?

Solution $800 \cdot 300 = 240,000$

YOU TRY IT 4 What is 90 multiplied by 7,000?

Your Solution

EXAMPLE 5 Complete the statement by using the Associative Property of Multiplication.

$$(7 \cdot 8) \cdot 5 = 7 \cdot (? \cdot 5)$$

Solution $(7 \cdot 8) \cdot 5 = 7 \cdot (8 \cdot 5)$

YOU TRY IT 5 Complete the statement by using the Multiplication Property of Zero.

$$? \cdot 10 = 0$$

Your Solution

EXAMPLE 6 Is 9 a solution of the equation $82 = 9q$?

Solution
$$82 = 9q$$
$$\begin{array}{c|c} 82 & 9(9) \end{array}$$
$$82 \neq 81$$

No, 9 is not a solution of the equation.

YOU TRY IT 6 Is 11 a solution of the equation $7a = 77$?

Your Solution

Solutions on p. S2

Point of Interest

Lao-tzu, founder of Taoism, wrote: Counting gave birth to Addition, Addition gave birth to Multiplication, Multiplication gave birth to Exponentiation, Exponentiation gave birth to all the myriad operations.

OBJECTIVE B **Exponents**

Repeated multiplication of the same factor can be written in two ways:

$$4 \cdot 4 \cdot 4 \cdot 4 \cdot 4 \quad \text{or} \quad 4^5 \longleftarrow \text{exponent}$$
$$\uparrow\text{_____ base}$$

The expression 4^5 is in **exponential form.** The **exponent,** 5, indicates how many times the **base,** 4, occurs as a factor in the multiplication.

It is important to be able to read numbers written in exponential form.

$$2 = 2^1$$ Read "two to the first power" or just "two."
Usually the 1 is not written.

$$2 \cdot 2 = 2^2$$ Read "two squared" or "two to the second power."

$$2 \cdot 2 \cdot 2 = 2^3$$ Read "two cubed" or "two to the third power."

$$2 \cdot 2 \cdot 2 \cdot 2 = 2^4$$ Read "two to the fourth power."

$$2 \cdot 2 \cdot 2 \cdot 2 \cdot 2 = 2^5$$ Read "two to the fifth power."

Variable expressions can contain exponents.

$$x^1 = x$$ x to the first power is usually written simply as x.

$$x^2 = x \cdot x$$ x^2 means x times x.

$$x^3 = x \cdot x \cdot x$$ x^3 means x occurs as a factor 3 times.

$$x^4 = x \cdot x \cdot x \cdot x$$ x^4 means x occurs as a factor 4 times.

Each place value in the place-value chart can be expressed as a power of 10.

$$\begin{aligned}
\text{Ten} = \quad 10 &= \quad\quad\quad 10 &&= 10^1 \\
\text{Hundred} = \quad 100 &= \quad\quad 10 \cdot 10 &&= 10^2 \\
\text{Thousand} = \quad 1{,}000 &= \quad 10 \cdot 10 \cdot 10 &&= 10^3 \\
\text{Ten-thousand} = \quad 10{,}000 &= \quad 10 \cdot 10 \cdot 10 \cdot 10 &&= 10^4 \\
\text{Hundred-thousand} = \quad 100{,}000 &= 10 \cdot 10 \cdot 10 \cdot 10 \cdot 10 &&= 10^5 \\
\text{Million} = 1{,}000{,}000 &= 10 \cdot 10 \cdot 10 \cdot 10 \cdot 10 \cdot 10 &&= 10^6
\end{aligned}$$

Note that the exponent on 10 when the number is written in exponential form is the same as the number of zeros in the number written in standard form. For example, $10^5 = 100{,}000$; the exponent on 10 is 5, and the number 100,000 has 5 zeros.

To evaluate a numerical expression containing exponents, write each factor as many times as indicated by the exponent and then multiply.

$$5^3 = 5 \cdot 5 \cdot 5 = 25 \cdot 5 = 125$$

$$2^3 \cdot 6^2 = (2 \cdot 2 \cdot 2) \cdot (6 \cdot 6) = 8 \cdot 36 = 288$$

Evaluate the variable expression c^3 for $c = 4$.

$$c^3 = c \cdot c \cdot c$$
$$4^3 = 4 \cdot 4 \cdot 4$$
$$= 16 \cdot 4 = 64$$

Replace c with 4 and then evaluate the exponential expression.

Point of Interest

One billion is too large a number for most of us to comprehend. If a computer were to start counting from 1 to 1 billion, writing to the screen one number every second of every day, it would take over 31 years for the computer to complete the task.

And if a billion is a large number, consider a googol. A googol is 1 with 100 zeros after it, or 10^{100}. Edward Kasner is the mathematician credited with thinking up this number, and his nine-year-old nephew is said to have thought up the name. The two then coined the word googolplex, which is 10^{googol}.

Calculator Note

A calculator can be used to evaluate an exponential expression. The y^x key (or on some calculators an x^y key or $\wedge$ key) is used to enter the exponent. For instance, for the example at the left, enter 4 y^x 3 $=$. The display reads 64.

EXAMPLE 7   Write $7 \cdot 7 \cdot 7 \cdot 4 \cdot 4$ in exponential form.

Solution $7 \cdot 7 \cdot 7 \cdot 4 \cdot 4 = 7^3 \cdot 4^2$

YOU TRY IT 7   Write $2 \cdot 2 \cdot 2 \cdot 3 \cdot 3 \cdot 3 \cdot 3$ in exponential form.

Your Solution

Solution on p. S2

EXAMPLE 8   Evaluate 8^3.

Solution $8^3 = 8 \cdot 8 \cdot 8 = 64 \cdot 8 = 512$

YOU TRY IT 8   Evaluate 6^4.

Your Solution

EXAMPLE 9   Evaluate 10^7.

Solution $10^7 = 10,000,000$

(The exponent on 10 is 7. There are 7 zeros in 10,000,000.)

YOU TRY IT 9   Evaluate 10^8.

Your Solution

EXAMPLE 10   Evaluate $3^3 \cdot 5^2$.

Solution $3^3 \cdot 5^2 = (3 \cdot 3 \cdot 3) \cdot (5 \cdot 5)$
$= 27 \cdot 25 = 675$

YOU TRY IT 10   Evaluate $2^4 \cdot 3^2$.

Your Solution

EXAMPLE 11   Evaluate $x^2 y^3$ for $x = 4$ and $y = 2$.

Solution $x^2 y^3$ ($x^2 y^3$ means x^2 times y^3.)

$4^2 \cdot 2^3 = (4 \cdot 4) \cdot (2 \cdot 2 \cdot 2)$
$= 16 \cdot 8$
$= 128$

YOU TRY IT 11   Evaluate $x^4 y^2$ for $x = 1$ and $y = 3$.

Your Solution

Solutions on p. S2

OBJECTIVE C   **Division of whole numbers**

Division is used to separate objects into equal groups.

A grocer wants to distribute 24 new products equally on 4 shelves. From the diagram, we see that the grocer would place 6 products on each shelf.

The grocer's problem could be written

Number on each shelf
Quotient

Number of shelves ⟶ $4\overline{)24}$ ⟵ Number of objects
Divisor   **Dividend**

Note that the quotient multiplied by the divisor equals the dividend.

$4\overline{)24}^{\,6}$ because $\boxed{\begin{array}{c}6\\ \text{Quotient}\end{array}} \times \boxed{\begin{array}{c}4\\ \text{Divisor}\end{array}} = \boxed{\begin{array}{c}24\\ \text{Dividend}\end{array}}$

Point of Interest

The Chinese divided a day into 100 k'o, which was a unit equal to a little less than 15 min. Sundials were used to measure time during the daylight hours, and by A.D. 500, candles, water clocks, and incense sticks were used to measure time at night.

Division is also represented by the symbol ÷ or by a fraction bar. Both are read "divided by."

$$9\overline{)54}^{\,6} \qquad 54 \div 9 = 6 \qquad \frac{54}{9} = 6$$

The fact that the quotient times the divisor equals the dividend can be used to illustrate properties of division.

$0 \div 4 = 0$ because $0 \cdot 4 = 0$.

$4 \div 4 = 1$ because $1 \cdot 4 = 4$.

$4 \div 1 = 4$ because $4 \cdot 1 = 4$.

$4 \div 0 = ?$ What number can be multiplied by 0 to get 4? $? \cdot 0 = 4$
There is no number whose product with 0 is 4
because the product of a number and zero is 0.
Division by zero is undefined.

Calculator Note

Enter 4 ÷ 0 = . An error message is displayed because division by zero is undefined.

The properties of division are stated below. In these statements, the symbol ≠ is read "is not equal to."

Division Properties of Zero and One

If $a \neq 0$, $0 \div a = 0$.   Zero divided by any number other than zero is zero.

If $a \neq 0$, $a \div a = 1$.   Any number other than zero divided by itself is one.

$a \div 1 = a$ A number divided by one is the number.

$a \div 0$ is undefined.   Division by zero is undefined.

Take Note

Recall that the variable *a* represents any whole number. Therefore, for the first two properties, we must state that $a \neq 0$ in order to ensure that we are not dividing by zero.

The example below illustrates division of a larger whole number by a one-digit number.

Divide and check: $3,192 \div 4$

$$
\begin{array}{r}
7 \\
4\overline{)3,192} \\
-2\,8 \\
\hline
39
\end{array}
$$
Think $31 \div 4$.
Subtract 7×4.
Bring down the 9.

$$
\begin{array}{r}
79 \\
4\overline{)3,192} \\
-2\,8 \\
\hline
39 \\
-36 \\
\hline
32
\end{array}
$$
Think $39 \div 4$.
Subtract 9×4.
Bring down the 2.

$$
\begin{array}{r}
798 \\
4\overline{)3,192} \\
-2\,8 \\
\hline
39 \\
-36 \\
\hline
32 \\
-32 \\
\hline
0
\end{array}
$$
Think $32 \div 4$.
Subtract 8×4.

Check:
$$
\begin{array}{r}
798 \\
\times\ \ 4 \\
\hline
3,192
\end{array}
$$

The place-value chart is used to show why this method works.

$$\begin{array}{r} \text{Hundreds Tens Ones} \\ 7\,9\,8 \\ 4\overline{)3,1\,9\,2} \\ -2\,8\,0\,0 \qquad \text{7 hundreds} \times 4 \\ \hline 3\,9\,2 \\ -3\,6\,0 \qquad \text{9 tens} \times 4 \\ \hline 3\,2 \\ -3\,2 \qquad \text{8 ones} \times 4 \\ \hline 0 \end{array}$$

Sometimes it is not possible to separate objects into a whole number of equal groups.

A packer at a bakery has 14 muffins to pack into 3 boxes. Each box will hold 4 muffins. From the diagram, we see that after the packer places 4 muffins in each box, there are 2 muffins left over. The 2 is called the **remainder.**

The packer's division problem could be written

$$\begin{array}{c} \text{Number in each box} \\ 4 \leftarrow \quad \textbf{Quotient} \\ \text{Number of boxes} \longrightarrow 3\overline{)14} \longleftarrow \text{Total number of muffins} \qquad \text{or} \qquad 4\,\text{r}2 \\ \textbf{Divisor} \qquad -12 \qquad \qquad \textbf{Dividend} \qquad \qquad \qquad 3\overline{)14} \\ \hline 2 \longleftarrow \text{Number left over} \\ \textbf{Remainder} \end{array}$$

For any division problem, **(quotient · divisor) + remainder = dividend.** This result can be used to check a division problem.

Find the quotient of 389 and 24.

$$\begin{array}{r} 16\,\text{r}5 \\ 24\overline{)389} \\ -24 \\ \hline 149 \\ -144 \\ \hline 5 \end{array}$$

Check: $(16 \cdot 24) + 5 = 384 + 5 = 389$

The phrase *the quotient of* was used in the example above to indicate the operation of division. The phrase *divided by* also indicates division.

the quotient of	the quotient of 8 and 4	$8 \div 4$
divided by	9 divided by 3	$9 \div 3$

Estimate the result when 56,497 is divided by 28.

Round each number to its highest place value.   56,497 $\longrightarrow$ 60,000

28   $\longrightarrow$ 30

Divide the rounded numbers. $60,000 \div 30 = 2,000$

2,000 is an estimate of $56,497 \div 28$.

Evaluate $\dfrac{x}{y}$ for $x = 4,284$ and $y = 18$. $\dfrac{x}{y}$

Replace x with 4,284 and y with 18. $\dfrac{4,284}{18} = 238$

$\dfrac{4,284}{18}$ means $4,284 \div 18$.

Is 42 a solution of the equation $\dfrac{x}{6} = 7$? $\dfrac{x}{6} = 7$

Replace x by 42. $\dfrac{42}{6}\ \Big|\ 7$

Simplify the left side of the equation.

The results are equal. $7 = 7$

42 is a solution of the equation.

EXAMPLE 12   What is the quotient of 8,856 and 42?

Solution
$$\begin{array}{r} 210 \text{ r36} \\ 42\overline{)8,856} \\ -\ 84 \\ \hline 45 \\ -\ 42 \\ \hline 36 \\ -\ \ 0 \\ \hline 36 \end{array}$$

Think $42\overline{)36}$.
Subtract $0 \cdot 42$.

Check: $(210 \cdot 42) + 36$
$= 8,820 + 36 = 8,856$

YOU TRY IT 12   What is 7,694 divided by 24?

Your Solution

Solution on p. S2

Figure 1.15 shows a household's annual expenses of $44,000. Use this graph for Example 13 and You Try It 13.

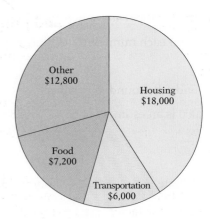

Figure 1.15 Annual Household Expenses

| EXAMPLE 13 | Use Figure 1.15 to find the household's monthly expense for housing. | YOU TRY IT 13 | Use Figure 1.15 to find the household's monthly expense for food. |

Solution   The annual expense for housing is $18,000.

$18,000 \div 12 = 1,500$

The monthly expense is $1,500.

Your Solution

| EXAMPLE 14 | Estimate the quotient of 55,272 and 392. | YOU TRY IT 14 | Estimate the quotient of 216,936 and 207. |

Solution $55,272 \longrightarrow 60,000$
$392 \longrightarrow 400$

$60,000 \div 400 = 150$

Your Solution

| EXAMPLE 15 | Evaluate $\frac{x}{y}$ for $x = 342$ and $y = 9$. | YOU TRY IT 15 | Evaluate $\frac{x}{y}$ for $x = 672$ and $y = 8$. |

Solution $\frac{x}{y}$

$\frac{342}{9} = 38$

Your Solution

| EXAMPLE 16 | Is 28 a solution of the equation $\frac{x}{7} = 4$? | YOU TRY IT 16 | Is 12 a solution of the equation $\frac{60}{y} = 2$? |

Solution $\frac{x}{7} = 4$

$\frac{28}{7}\ \bigg|\ 4$

$4 = 4$

Yes, 28 is a solution of the equation.

Your Solution

Solutions on p. S2

OBJECTIVE D   Factors and prime factorization

Natural number factors of a number divide that number evenly (there is no remainder).

1, 2, 3, and 6 are natural number factors of 6 because they divide 6 evenly.

Note that both the divisor and the quotient are factors of the dividend.

$$\overset{6}{1)6} \quad \overset{3}{2)6} \quad \overset{2}{3)6} \quad \overset{1}{6)6}$$

To find the factors of a number, try dividing the number by 1, 2, 3, 4, 5, ... Those numbers that divide the number evenly are its factors. Continue this process until the factors start to repeat.

Find all the factors of 42.

$42 \div 1 = 42$	1 and 42 are factors of 42.
$42 \div 2 = 21$	2 and 21 are factors of 42.
$42 \div 3 = 14$	3 and 14 are factors of 42.
$42 \div 4$	4 will not divide 42 evenly.
$42 \div 5$	5 will not divide 42 evenly.
$42 \div 6 = 7$	6 and 7 are factors of 42.
$42 \div 7 = 6$	7 and 6 are factors of 42.

The factors are repeating.
All the factors of 42 have been found.

The factors of 42 are 1, 2, 3, 6, 7, 14, 21, and 42.

The following rules are helpful in finding the factors of a number.

2 is a factor of a number if the digit in the ones' place of the number is 0, 2, 4, 6, or 8.

436 ends in 6.
Therefore, 2 is a factor of 436
($436 \div 2 = 218$).

3 is a factor of a number if the sum of the digits of the number is divisible by 3.

The sum of the digits of 489 is $4 + 8 + 9 = 21$.
21 is divisible by 3.
Therefore, 3 is a factor of 489
($489 \div 3 = 163$).

4 is a factor of a number if the last two digits of the number are divisible by 4.

556 ends in 56.
56 is divisible by 4 ($56 \div 4 = 14$).
Therefore, 4 is a factor of 556
($556 \div 4 = 139$).

5 is a factor of a number if the ones' digit of the number is 0 or 5.

520 ends in 0.
Therefore, 5 is a factor of 520
($520 \div 5 = 104$).

A **prime number** is a natural number greater than 1 that has exactly two natural number factors, 1 and the number itself. 7 is prime because its only factors are 1 and 7. If a number is not prime, it is a **composite number.** Because 6 has factors of 2 and 3, 6 is a composite number. The prime numbers less than 50 are

2, 3, 5, 7, 11, 13, 17, 19, 23, 29, 31, 37, 41, 43, 47

Point of Interest

Twelve is the smallest *abundant number,* or number whose proper divisors add up to more than the number itself. The proper divisors of a number are all of its factors except the number itself. The proper divisors of 12 are 1, 2, 3, 4, and 6, which add up to 16, which is greater than 12. There are 246 abundant numbers between 1 and 1,000.

A *perfect number* is one whose proper divisors add up to exactly that number. For example, the proper divisors of 6 are 1, 2, and 3, which add up to 6. There are only three perfect numbers less than 1,000: 6, 28, and 496.

The **prime factorization** of a number is the expression of the number as a product of its prime factors. To find the prime factors of 90, begin with the smallest prime number as a trial divisor and continue with prime numbers as trial divisors until the final quotient is prime.

Find the prime factorization of 90.

$$\frac{45}{2)90}$$

$$\frac{15}{3)45}$$
$$\overline{2)90}$$

$$\frac{5}{3)15}$$
$$\overline{3)45}$$
$$\overline{2)90}$$

Divide 90 by 2.

45 is not divisible by 2. Divide 45 by 3.

Divide 15 by 3. 5 is prime.

The prime factorization of 90 is $2 \cdot 3 \cdot 3 \cdot 5$, or $2 \cdot 3^2 \cdot 5$.

Finding the prime factorization of larger numbers can be more difficult. Try each prime number as a trial divisor. Stop when the square of the trial divisor is greater than the number being factored.

Find the prime factorization of 201.

$$\frac{67}{3)201}$$

67 cannot be divided evenly by 2, 3, 5, 7, or 11. Prime numbers greater than 11 need not be tried because $11^2 = 121$ and $121 > 67$.

The prime factorization of 201 is $3 \cdot 67$.

EXAMPLE 17 Find all the factors of 40.

Solution
$40 \div 1 = 40$
$40 \div 2 = 20$
$40 \div 3$ Does not divide evenly.
$40 \div 4 = 10$
$40 \div 5 = 8$
$40 \div 6$ Does not divide evenly.
$40 \div 7$ Does not divide evenly.
$40 \div 8 = 5$ The factors are repeating.

The factors of 40 are 1, 2, 4, 5, 8, 10, 20, and 40.

YOU TRY IT 17 Find all the factors of 30.

Your Solution

EXAMPLE 18 Find the prime factorization of 84.

Solution
$$\frac{7}{3)21}$$
$$\overline{2)42}$$
$$\overline{2)84}$$

$84 = 2 \cdot 2 \cdot 3 \cdot 7 = 2^2 \cdot 3 \cdot 7$

YOU TRY IT 18 Find the prime factorization of 88.

Your Solution

Solutions on p. S3

EXAMPLE 19	Find the prime factorization of 141.	**YOU TRY IT 19**	Find the prime factorization of 295.

Solution

$$\begin{array}{r} 47 \\ 3\overline{)141} \end{array}$$ ▶ Try only 2, 3, 5, and 7 because $7^2 = 49$ and $49 > 47$.

$141 = 3 \cdot 47$

Your Solution

Solution on p. S3

OBJECTIVE E   **Applications and formulas**

In Section 1.2, we defined perimeter as the distance around a plane figure. The perimeter of a rectangle was given as $P = L + W + L + W$. This formula is commonly written as $P = 2L + 2W$.

Perimeter of a Rectangle

The formula for the perimeter of a rectangle is $P = 2L + 2W$, where P is the perimeter of the rectangle, L is the length, and W is the width.

> **Take Note**
>
> Remember that $2L$ means 2 times L, and $2W$ means 2 times W.

Find the perimeter of the rectangle shown at the right.

Use the formula for the perimeter of a rectangle.	$P = 2L + 2W$
Substitute 32 for L and 16 for W.	$P = 2(32) + 2(16)$
Find the product of 2 and 32 and the product of 2 and 16.	$P = 64 + 32$
Add.	$P = 96$

The perimeter of the rectangle is 96 ft.

32 ft

16 ft

A **square** is a rectangle in which each side has the same length. Letting s represent the length of each side of a square, the perimeter of a square can be represented as $P = s + s + s + s$. Note that we are adding *four* s's. We can write the addition as multiplication: $P = 4s$.

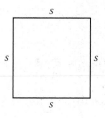

$P = s + s + s + s$
$P = 4s$

Perimeter of a Square

The formula for the perimeter of a square is $P = 4s$, where P is the perimeter and s is the length of a side of a square.

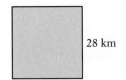

28 km

Find the perimeter of the square shown at the left.

Use the formula for the perimeter of a square.	$P = 4s$
Substitute 28 for s.	$P = 4(28)$
Multiply.	$P = 112$

The perimeter of the square is 112 km.

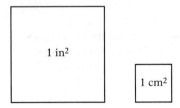

1 in²

1 cm²

Area is the amount of surface in a region. Area can be used to describe the size of a skating rink, the floor of a room, or a playground. Area is measured in square units.

A square that measures 1 inch on each side has an area of 1 square inch, which is written 1 in². A square that measures 1 centimeter on each side has an area of 1 square centimeter, which is written 1 cm².

Larger areas can be measured in square feet (ft²), square meters (m²), acres (43,560 ft²), square miles (mi²), or any other square unit.

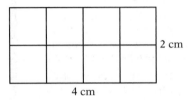

2 cm

4 cm

The area of the rectangle is 8 cm².

The area of a geometric figure is the number of squares that are necessary to cover the figure. In the figure at the left, a rectangle has been drawn and covered with squares. Eight squares, each of area 1 cm², were used to cover the rectangle. The area of the rectangle is 8 cm². Note from this figure that the area of a rectangle can be found by multiplying the length of the rectangle by its width.

Area of a Rectangle

The formula for the area of a rectangle is $A = LW$, where A is the area, L is the length, and W is the width of the rectangle.

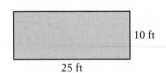

10 ft

25 ft

Find the area of the rectangle shown at the left.

Use the formula for the area of a rectangle.	$A = LW$
Substitute 25 for L and 10 for W.	$A = 25(10)$
Multiply.	$A = 250$

The area of the rectangle is 250 ft².

s

$A = s \cdot s = s^2$

A square is a rectangle in which all sides are the same length. Therefore, both the length and the width of a square can be represented by s, and $A = LW = s \cdot s = s^2$.

Area of a Square

The formula for the area of a square is $A = s^2$, where A is the area and s is the length of a side of a square.

Find the area of the square shown at the right.

Use the formula for the area of a square.	$A = s^2$
Substitute 8 for s.	$A = 8^2$
Multiply.	$A = 64$

8 mi

The area of the square is 64 mi².

In this section, some of the phrases used to indicate the operations of multiplication and division were presented. In solving application problems, you might also look for the following types of questions:

Multiplication	**Division**
per . . . How many altogether?	What is the hourly rate?
each . . . What is the total number of . . . ?	Find the amount per . . .
every . . . Find the total . . .	How many does each . . . ?

Calculator Note

Many scientific calculators have an $\boxed{x^2}$ key. This key is used to square the displayed number. For example, after pressing 8 $\boxed{x^2}$ $\boxed{=}$, the display reads 64.

> **Take Note**
>
> Each of the following problems indicates multiplication:
>
> "You purchased 6 boxes of doughnuts with 12 doughnuts *per* box. *How many* doughnuts did you purchase *altogether*?"
>
> "If *each* bottle of apple juice contains 32 oz, *what is the total number of* ounces in 8 bottles of the juice?"
>
> "You purchased 5 bags of oranges. *Every* bag contained 10 oranges. *Find the total number* of oranges purchased."

Figure 1.16 shows the cost of a first-class postage stamp from the 1950s to 2007. Use this graph for Example 20 and You Try It 20.

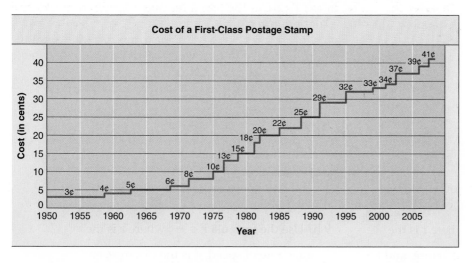

Figure 1.16   Cost of a First-Class Postage Stamp

EXAMPLE 20

How many times more expensive was a stamp in 1980 than in 1950? Use Figure 1.16.

Strategy

To find how many times more expensive a stamp was, divide the cost in 1980 (15) by the cost in 1950 (3).

Solution

$15 \div 3 = 5$

A stamp was 5 times more expensive in 1980.

YOU TRY IT 20

How many times more expensive was a stamp in 1997 than in 1960? Use Figure 1.16.

Your Strategy

32 ÷ 4 =

Your Solution

8 times more expensive

Solution on p. S3

EXAMPLE 21

Find the amount of sod needed to cover a football field. A football field measures 120 yd by 50 yd.

Strategy

Draw a diagram.

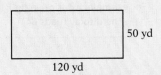

50 yd

120 yd

To find the amount of sod needed, use the formula for the area of a rectangle, $A = LW$. $L = 120$ and $W = 50$

Solution

$A = LW$

$A = 120(50)$

$A = 6,000$

6,000 ft² of sod are needed.

YOU TRY IT 21

A homeowner wants to carpet the family room. The floor is square and measures 6 m on each side. How much carpet should be purchased?

Your Strategy

Your Solution

EXAMPLE 22

At what rate of speed would you need to travel in order to drive a distance of 294 mi in 6 h? Use the formula $r = \dfrac{d}{t}$, where r is the average rate of speed, d is the distance, and t is the time.

Strategy

To find the rate of speed, replace d by 294 and t by 6 in the given formula and solve for r.

Solution

$r = \dfrac{d}{t}$

$r = \dfrac{294}{6} = 49$

You would need to travel at a speed of 49 mph.

YOU TRY IT 22

At what rate of speed would you need to travel in order to drive a distance of 486 mi in 9 h? Use the formula $r = \dfrac{d}{t}$, where r is the average rate of speed, d is the distance, and t is the time.

Your Strategy

Your Solution

Solutions on p. S3

1.3 Exercises

OBJECTIVE A **Multiplication of whole numbers**

1. ✏️ Explain how to rewrite the addition 6 + 6 + 6 + 6 + 6 as multiplication.

2. In the multiplication $7 \times 3 = 21$, the product is _____ and the factors are _____ and _____.

3. Find the product for the multiplication problem shown at the right.
    a. Multiply 3 times the number in the ones column: $3 \times$ _____ = _____.
        Write _____ in the ones column of the product and carry _____ to the tens column.
    b. Multiply 3 times the number in the tens column: $3 \times$ _____ = _____. To obtain the next digit in the product, add the carry digit you found in part **a** to the product in part **b**: _____ + _____ = _____.
    c. The product of 3 and 24 is _____.

$$\begin{array}{r} 24 \\ \times\ 3 \\ \hline \end{array}$$

Multiply.

4. (9)(127)

5. (4)(623)

6. (6,709)(7)

7. (3,608)(5)

8. $8 \cdot 58{,}769$

9. $7 \cdot 60{,}047$

10. $\begin{array}{r} 683 \\ \times\ 71 \\ \hline \end{array}$

11. $\begin{array}{r} 591 \\ \times\ 92 \\ \hline \end{array}$

12. $\begin{array}{r} 7{,}053 \\ \times\ \ \ 46 \\ \hline \end{array}$

13. $\begin{array}{r} 6{,}704 \\ \times\ \ \ 58 \\ \hline \end{array}$

14. $\begin{array}{r} 3{,}285 \\ \times\ \ 976 \\ \hline \end{array}$

15. $\begin{array}{r} 5{,}327 \\ \times\ \ 624 \\ \hline \end{array}$

16. Find the product of 500 and 3.

17. Find 30 multiplied by 80.

18. What is 40 times 50?

19. What is twice 700?

20. What is the product of 400, 3, 20, and 0?

21. Write the product of f and g.

22. *Health* The figure at the right shows the number of calories burned on three different exercise machines during 1 h of a light, moderate, or vigorous workout. How many calories would you burn by **a.** working out vigorously on a stair climber for a total of 6 h? **b.** working out moderately on a treadmill for a total of 12 h?

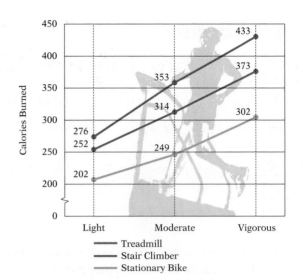

Calories Burned on Exercise Machines
Source: Journal of American Medical Association

Multiply. Then check by estimating the product.

23. $3{,}467 \cdot 359$

24. $8{,}745(63)$

25. $(39{,}246)(29)$

26. $64{,}409 \cdot 67$

27. $745(63)$

28. $432 \cdot 91$

29. $(8{,}941)(726)$

30. $2{,}837(216)$

Evaluate the expression for the given values of the variables.

31. ab, for $a = 465$ and $b = 32$

32. cd, for $c = 381$ and $d = 25$

33. $7a$, for $a = 465$

34. $6n$, for $n = 382$

35. xyz, for $x = 5$, $y = 12$, and $z = 30$

36. abc, for $a = 4$, $b = 20$, and $c = 50$

37. $2xy$, for $x = 67$ and $y = 23$

38. $4ab$, for $a = 95$ and $b = 33$

39. Find a one-digit number and a two-digit number whose product is a number that ends in two zeros.

40. Find a two-digit number that ends in a zero and a three-digit number that ends in two zeros whose product is a number that ends in four zeros.

Identify the property that justifies the statement.

41. $1 \cdot 29 = 29$

42. $(10 \cdot 5) \cdot 8 = 10 \cdot (5 \cdot 8)$

43. $43 \cdot 1 = 1 \cdot 43$

44. $0(76) = 0$

Use the given property of multiplication to complete the statement.

45. The Commutative Property of Multiplication
$19 \cdot ? = 30 \cdot 19$

46. The Associative Property of Multiplication
$(? \cdot 6)100 = 5(6 \cdot 100)$

47. The Multiplication Property of Zero
$45 \cdot 0 = ?$

48. The Multiplication Property of One
$? \cdot 77 = 77$

49. Is 6 a solution of the equation $4x = 24$?

50. Is 0 a solution of the equation $4 = 4n$?

51. Is 23 a solution of the equation $96 = 3z$?

52. Is 14 a solution of the equation $56 = 4c$?

53. Is 19 a solution of the equation $2y = 38$?

54. Is 11 a solution of the equation $44 = 3a$?

OBJECTIVE B Exponents

55. a. In the exponential expression 3^4, the base is _____ and the exponent is _____.
b. To evaluate 3^4, use 3 as a factor four times: _____ $\cdot$ _____ $\cdot$ _____ $\cdot$ _____ = _____.

56. State the base and the exponent of the exponential expression.
a. 5 squared
base = ____ , exponent = ____
b. 4 to the sixth power
base = ____ , exponent = ____
c. 7 cubed
base = ____ , exponent = ____

Write in exponential form.

57. $2 \cdot 2 \cdot 2 \cdot 7 \cdot 7 \cdot 7 \cdot 7 \cdot 7$

58. $3 \cdot 3 \cdot 3 \cdot 3 \cdot 3 \cdot 3 \cdot 5 \cdot 5 \cdot 5$

59. $2 \cdot 2 \cdot 3 \cdot 3 \cdot 3 \cdot 5 \cdot 5 \cdot 5 \cdot 5$

60. $7 \cdot 7 \cdot 11 \cdot 11 \cdot 11 \cdot 19 \cdot 19 \cdot 19 \cdot 19$

61. $c \cdot c$

62. $d \cdot d \cdot d$

63. $x \cdot x \cdot x \cdot y \cdot y \cdot y$

64. $a \cdot a \cdot b \cdot b \cdot b \cdot b$

Evaluate.

65. 2^5

66. 2^6

67. 10^6

68. 10^9

69. $2^3 \cdot 5^2$

70. $2^4 \cdot 3^2$

71. $3^2 \cdot 10^3$

72. $2^4 \cdot 10^2$

73. $0^2 \cdot 6^2$

74. $4^3 \cdot 0^3$

75. $2^2 \cdot 5 \cdot 3^3$

76. $5^2 \cdot 2 \cdot 3^4$

77. Find the square of 12.

78. What is the cube of 6?

79. Find the cube of 8.

80. What is the square of 11?

81. Write the fourth power of a.

82. Write the fifth power of t.

Evaluate the expression for the given values of the variables.

83. x^3y, for $x = 2$ and $y = 3$

84. x^2y, for $x = 3$ and $y = 4$

85. ab^6, for $a = 5$ and $b = 2$

86. ab^3, for $a = 7$ and $b = 4$

87. c^2d^2, for $c = 3$ and $d = 5$

88. m^3n^3, for $m = 5$ and $n = 10$

89.   Rewrite each expression using the numbers 2 and 6 exactly once. Then evaluate the expression.
   a. $2 + 2 + 2 + 2 + 2 + 2$ **b.** $2 \cdot 2 \cdot 2 \cdot 2 \cdot 2 \cdot 2$

OBJECTIVE C **Division of whole numbers**

90.   In what situation does a division problem have a remainder?

For Exercises 91 and 92, use the division problem $6\overline{)495}$.

91. Express the division problem using the symbol $\div$. _____ $\div$ _____

92. Express the division problem using a fraction. _____

Divide.

93. $9\overline{)2,763}$

94. $4\overline{)2,160}$

95. $5\overline{)1,549}$

96. $8\overline{)1,636}$

97. $15,300 \div 6$

98. $43,500 \div 5$

99. $681 \div 32$

100. $879 \div 41$

101. $9{,}152 \div 62$ **102.** $4{,}161 \div 23$ **103.** $7{,}408 \div 37$ **104.** $5{,}207 \div 26$

105. $31{,}546 \div 78$ **106.** $38{,}976 \div 64$ **107.** $7{,}713 \div 476$ **108.** $8{,}947 \div 223$

109. Find the quotient of 7,256 and 8.

110. What is the quotient of 8,172 and 9?

111. What is 6,168 divided by 7?

112. Find 4,153 divided by 9.

113. Write the quotient of c and d.

114. *Insurance* The table at the right shows the sources of laptop computer insurance claims in a recent year. Claims have been rounded to the nearest ten-thousand dollars. **a.** What was the average monthly claim for theft? **b.** For all sources combined, find the average claims per month.

Source	Claims (in dollars)
Accidents	560,000
Theft	300,000
Power Surge	80,000
Lightning	50,000
Transit	20,000
Water/flood	20,000
Other	110,000

Source: Safeware, The Insurance Company

Divide. Then check by estimating the quotient.

115. $36{,}472 \div 47$ **116.** $62{,}176 \div 58$ **117.** $389{,}804 \div 76$ **118.** $637{,}072 \div 29$

119. $79\overline{)38{,}984}$ **120.** $53\overline{)11{,}792}$ **121.** $219\overline{)332{,}004}$ **122.** $324\overline{)632{,}124}$

Evaluate the variable expression $\dfrac{x}{y}$ for the given values of x and y.

123. $x = 48; y = 1$ **124.** $x = 56; y = 56$ **125.** $x = 79; y = 0$

126. $x = 0; y = 23$ **127.** $x = 39{,}200; y = 4$ **128.** $x = 16{,}200; y = 3$

129. Is 9 a solution of the equation $\dfrac{36}{z} = 4$?

130. Is 60 a solution of the equation $\dfrac{n}{12} = 5$?

131. Is 49 a solution of the equation $56 = \dfrac{x}{7}$?

132. Is 16 a solution of the equation $6 = \dfrac{48}{y}$?

OBJECTIVE D Factors and prime factorization

133. Circle the numbers that divide evenly into 15. These numbers are called the _____ of 15.

1 2 ③ 4 ⑤ 6 7 8 9 10 11 12 13 14 15

134. The only factors of 11 are _____ and _____. The number 11 is called a _____ number.

Find all the factors of the number.

135. 10   **136.** 20   **137.** 12   **138.** 9   **139.** 8

140. 16   **141.** 13   **142.** 17   **143.** 18   **144.** 24

145. 25   **146.** 36   **147.** 56   **148.** 45   **149.** 28

150. 32   **151.** 48   **152.** 64   **153.** 54   **154.** 75

Find the prime factorization of the number.

155. 16   **156.** 24   **157.** 12   **158.** 27   **159.** 15

160. 36   **161.** 40   **162.** 50   **163.** 37   **164.** 83

165. 65   **166.** 80   **167.** 28   **168.** 49   **169.** 42

170. 81   **171.** 51   **172.** 89   **173.** 46   **174.** 120

OBJECTIVE E Applications and formulas

For Exercises 175 and 176, state whether you would use multiplication or division to find the specified amount.

175. Three friends want to share equally a restaurant bill of $37.95. To find how much each person should pay, use _____.

176. You drove at 60 mph for 4 h. To find the total distance you traveled, use _____.

177. *Nutrition* One ounce of cheddar cheese contains 115 calories. Find the number of calories in 4 oz of cheddar cheese.

Nutrition Facts	Amount/Serving	% DV*	Amount/Serving	% DV*
Serv. Size 1 oz.	Total Fat 9g	14%	Total Carb. 1g	0%
Servings Per Package 12	Sat Fat 5g	25%	Fiber 0g	0%
Calories 115	Cholest. 30mg	10%	Sugars 0g	
Fat Cal. 80	Sodium 170mg	7%	Protein 7g	
*Percent Daily Values (DV) are based on a 2,000 calorie diet	Vitamin A 6% · Vitamin C 0% · Calcium 20% · Iron 0%			

178. *Sports*   During his football career, John Riggins ran the ball 2,916 times. He averaged about 4 yd per carry. About how many total yards did he gain during his career?

179. *Aviation*   A plane flying from Los Angeles to Boston uses 865 gal of jet fuel each hour. How many gallons of jet fuel are used on a 5-hour flight?

180. *Geometry*   Find **a.** the perimeter and **b.** the area of a square that measures 16 mi on each side.

16 mi

181. *Geometry*   Find **a.** the perimeter and **b.** the area of a rectangle with a length of 24 m and a width of 15 m.

182. *Geometry*   Find the length of fencing needed to surround a square corral that measures 55 ft on each side.

183. *Geometry*   A fieldstone patio is in the shape of a square that measures 9 ft on each side. What is the area of the patio?

184. *Finances*   A computer analyst doing consulting work received $5,376 for working 168 h on a project. Find the hourly rate the consultant charged.

185. *Business*   A buyer for a department store purchased 215 suits at $83 each. Estimate the total cost of the order.

186. 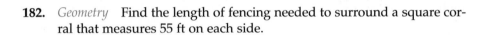 *Finances*   Financial advisors may predict how much money we should have saved for retirement by the ages of 35, 45, 55, and 65. One such prediction is included in the table below. **a.** A couple has earnings of $100,000 per year. According to the table, by how much should their savings grow per year from age 45 to 55? **b.** A couple has earnings of $50,000 per year. According to the table, by how much should their savings grow per year from age 55 to 65?

Minimum Levels of Savings Required for Married Couples to Be Prepared for Retirement				
	Savings Accumulation by Age			
Earnings	35	45	55	65
$50,000	8,000	23,000	90,000	170,000
$75,000	17,000	60,000	170,000	310,000
$100,000	34,000	110,000	280,000	480,000
$150,000	67,000	210,000	490,000	840,000

John Riggins

187. *Finances* Find the total amount paid on a loan when the monthly payment is $285 and the loan is paid off in 24 months. Use the formula $A = MN$, where A is the total amount paid, M is the monthly payment, and N is the number of payments.

188. *Finances* Find the total amount paid on a loan when the monthly payment is $187 and the loan is paid off in 36 months. Use the formula $A = MN$, where A is the total amount paid, M is the monthly payment, and N is the number of payments.

189. *Travel* Use the formula $t = \dfrac{d}{r}$, where t is the time, d is the distance, and r is the average rate of speed, to find the time it would take to drive 513 mi at an average speed of 57 mph.

190. *Travel* Use the formula $t = \dfrac{d}{r}$, where t is the time, d is the distance, and r is the average rate of speed, to find the time it would take to drive 432 mi at an average speed of 54 mph.

191. *Investments* The current value of the stocks in a mutual fund is $10,500,000. The number of shares outstanding is 500,000. Find the value per share of the fund. Use the formula $V = \dfrac{C}{S}$, where V is the value per share, C is the current value of the stocks in the fund, and S is the number of shares outstanding.

192. *Investments* The current value of the stocks in a mutual fund is $4,500,000. The number of shares outstanding is 250,000. Find the value per share of the fund. Use the formula $V = \dfrac{C}{S}$, where V is the value per share, C is the current value of the stocks in the fund, and S is the number of shares outstanding.

New York Stock Exchange

CRITICAL THINKING

193. *Time* There are 52 weeks in a year. Is this an exact figure or an approximation?

194. *Mathematics* 13,827 is not divisible by 4. By rearranging the digits, find the largest possible number that is divisible by 4.

195. *Mathematics* A **palindromic number** is a whole number that remains unchanged when its digits are written in reverse order. For example, 818 is a palindromic number. Find the smallest three-digit multiple of 6 that is a palindromic number.

196. Prepare a monthly budget for a family of four. Explain how you arrived at the cost of each item. Annualize the budget you prepared.

Monthly Budget	
Rent	$975
Electricity	
Telephone	
Gas	
Food	

1.4 Solving Equations with Whole Numbers

OBJECTIVE A   Solving equations

Recall that a **solution** of an equation is a number that, when substituted for the variable, produces a true equation.

The solution of the equation $x + 5 = 11$ is 6 because when 6 is substituted for x, the result is a true equation.

$$x + 5 = 11$$
$$6 + 5 = 11$$

If 2 is subtracted from each side of the equation $x + 5 = 11$, the resulting equation is $x + 3 = 9$. Note that the solution of this equation is also 6.

$$x + 5 = 11$$
$$x + 5 - 2 = 11 - 2$$
$$x + 3 = 9 \qquad 6 + 3 = 9$$

> **Take Note**
>
> An *equation* always has an equals sign (=). An *expression* does not have an equals sign.
> $x + 5 = 11$ is an equation.
> $x + 5$ is an expression.

This illustrates the **subtraction property of equations.**

> **The same number can be subtracted from each side of an equation without changing the solution of the equation.**

The subtraction property is used to *solve* an equation. To **solve an equation** means to find a solution of the equation. That is, to solve an equation you must find a number that, when substituted for the variable, produces a true equation.

An equation such as $x = 8$ is easy to solve. The solution is 8, the number that when substituted for the variable produces the true equation $8 = 8$. In solving an equation, the goal is to get the variable alone on one side of the equation; the number on the other side of the equation is the solution.

To solve an equation in which a number is added to a variable, use the subtraction property of equations: Subtract that number from each side of the equation.

Solve: $x + 5 = 11$

Note the effect of subtracting 5 from each side of the equation and then simplifying. The variable, x, is on one side of the equation; a number, 6, is on the other side.

$$x + 5 = 11$$
$$x + 5 - 5 = 11 - 5$$
$$x + 0 = 6$$
$$x = 6$$

The solution is 6.

Check:
$$\begin{array}{c|c} x + 5 & = 11 \\ \hline 6 + 5 & 11 \\ 11 & = 11 \end{array}$$

Note that we have checked the solution. You should always check the solution of an equation.

Solve: $19 = 11 + m$

11 is added to m. Subtract 11 from each side of the equation.

$$19 = 11 + m$$
$$19 - 11 = 11 - 11 + m$$
$$8 = 0 + m$$
$$8 = m$$

The solution is 8.

Check:
$$\begin{array}{c|c} 19 = 11 + m \\ \hline 19 & 11 + 8 \\ 19 = 19 \end{array}$$

> **Take Note**
>
> For this equation, the variable is on the right side. The goal is to get the variable alone on the right side.

The solution of the equation $4y = 12$ is 3 because when 3 is substituted for y, the result is a true equation.

$$4y = 12$$
$$4(3) = 12$$
$$12 = 12$$

If each side of the equation $4y = 12$ is divided by 2, the resulting equation is $2y = 6$. Note that the solution of this equation is also 3.

$$4y = 12$$
$$\frac{4y}{2} = \frac{12}{2}$$
$$2y = 6 \qquad 2(3) = 6$$

This illustrates the **division property of equations.**

Each side of an equation can be divided by the same number (except zero) without changing the solution of the equation.

Solve: $30 = 5a$

a is multiplied by 5. To get a alone on the right side, divide each side of the equation by 5.

$$30 = 5a$$
$$\frac{30}{5} = \frac{5a}{5}$$
$$6 = 1a$$
$$6 = a$$

Check:
$$30 = 5a$$
$$30 \mid 5(6)$$
$$30 = 30$$

The solution is 6.

EXAMPLE 1 Solve: $9 + n = 28$

Solution
$$9 + n = 18$$
$$9 - 9 + n = 28 - 9$$
$$0 + n = 19$$
$$n = 19$$

Check:
$$9 + n = 28$$
$$9 + 19 \mid 28$$
$$28 = 28$$

The solution is 19.

YOU TRY IT 1 Solve: $37 = a + 12$

Your Solution

EXAMPLE 2 Solve: $20 = 5c$

Solution
$$20 = 5c$$
$$\frac{20}{5} = \frac{5c}{5}$$
$$4 = 1c$$
$$4 = c$$

Check:
$$20 = 5c$$
$$20 \mid 5(4)$$
$$20 = 20$$

The solution is 4.

YOU TRY IT 2 Solve: $3z = 36$

Your Solution

Solutions on p. S3

OBJECTIVE B **Applications and formulas**

Recall that an equation states that two mathematical expressions are equal. To translate a sentence into an equation, you must recognize the words or phrases that mean "equals." Some of these phrases are

equals	is	was
is equal to	represents	is the same as

The number of scientific calculators sold by Evergreen Electronics last month is three times the number of graphing calculators the company sold this month. If it sold 225 scientific calculators last month, how many graphing calculators were sold this month?

Strategy To find the number of graphing calculators sold, write and solve an equation using x to represent the number of graphing calculators sold.

Solution

The number of scientific calculators sold last month	is	three times the number of graphing calculators sold this month
225	=	3x

$$\frac{225}{3} = \frac{3x}{3}$$

$$75 = x$$

Evergreen Electronics sold 75 graphing calculators this month.

> **Take Note**
>
> Sentences or phrases that begin "how many...," "how much...," "find...," and "what is..." are followed by a phrase that indicates what you are looking for. (In the problem at the left, the phrase is *graphing calculators:* "How many *graphing calculators*....") Look for these phrases to determine the unknown.

EXAMPLE 3

The product of seven and a number equals twenty-eight. Find the number.

Solution

The unknown number: n

The product of seven and a number	equals	twenty-eight

$$7n = 28$$

$$\frac{7n}{7} = \frac{28}{7}$$

$$n = 4$$

The number is 4.

YOU TRY IT 3

A number increased by four is seventeen. Find the number.

Your Solution

Solution on p. S3

EXAMPLE 4

A child born in 2000 was expected to live to the age of 77. This is 29 years longer than the life expectancy of a child born in 1900. (*Sources:* U.S. Department of Health and Human Services' Administration of Aging; Census Bureau; National Center for Health Statistics) Find the life expectancy of a child born in 1900.

Strategy

To find the life expectancy of a child born in 1900, write and solve an equation using x to represent the unknown life expectancy.

Solution

Life expectancy in 2000	is	29 years longer than the life expectancy in 1900

$$77 = x + 29$$
$$77 - 29 = x + 29 - 29$$
$$48 = x$$

The life expectancy of a child born in 1900 was 48 years.

EXAMPLE 5

Use the formula $A = P + I$, where A is the value of an investment, P is the original investment, and I is the interest earned, to find the interest earned on an original investment of $12,000 that now has a value of $14,280.

Strategy

To find the interest earned, replace A by 14,280 and P by 12,000 in the given formula and solve for I.

Solution

$$A = P + I$$
$$14,280 = 12,000 + I$$
$$14,280 - 12,000 = 12,000 - 12,000 + I$$
$$2,280 = I$$

The interest earned on the investment is $2,280.

YOU TRY IT 4

In a recent year, more than 7 million people had cosmetic plastic surgery. During that year, the number of liposuctions performed was 220,159 more than the number of face lifts performed. There were 354,015 liposuctions performed. (*Source:* American Society of Plastic Surgery) How many face lifts were performed that year?

Your Strategy

Your Solution

YOU TRY IT 5

Use the formula $A = P + I$, where A is the value of an investment, P is the original investment, and I is the interest earned, to find the interest earned on an original investment of $18,000 that now has a value of $21,060.

Your Strategy

Your Solution

Solutions on p. S3

1.4 Exercises

OBJECTIVE A Solving equations

1. To solve $5 + x = 20$, subtract _____ from each side of the equation. The solution is _____.

2. To solve $5x = 20$, _____ each side of the equation by 5. The solution is _____.

Solve.

3. $x + 9 = 23$ 4. $y + 17 = 42$ 5. $8 + b = 33$ 6. $15 + n = 54$

7. $3m = 15$ 8. $8z = 32$ 9. $52 = 4c$ 10. $60 = 5d$

11. $16 = w + 9$ 12. $72 = t + 44$ 13. $28 = 19 + p$ 14. $33 = 18 + x$

15. $10y = 80$ 16. $12n = 60$ 17. $41 = 41d$ 18. $93 = 93m$

19. $b + 7 = 7$ 20. $q + 23 = 23$ 21. $15 + t = 91$ 22. $79 + w = 88$

OBJECTIVE B Applications and formulas

For Exercises 23 and 24, translate each sentence into an equation. Use n to represent the unknown number.

23. A number   increased by   eight   is equal to   thirteen.

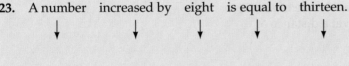

_____   _____   _____   _____   _____

24. Three times a number   is   forty-two.

_____   _____   _____

25. Sixteen added to a number is equal to forty. Find the number.

26. The sum of eleven and a number equals fifty-two. Find the number.

27. Five times a number is thirty. Find the number.

28. The product of ten and a number is equal to two hundred. Find the number.

29. Fifteen is three more than a number. Find the number.

30. A number multiplied by twenty equals four hundred. Find the number.

31. *Geometry* The length of a rectangle is 5 in. more than the width. The length is 17 in. Find the width of the rectangle.

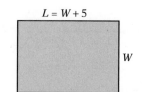

$L = W + 5$

W

32. *Temperature* The average daily low temperature in Duluth, Minnesota in June is eight times the average daily low temperature in Duluth in December. The average daily low temperature in Duluth in June is 48°. Find the average daily low temperature in Duluth in December.

33. *Geography* The table at the right lists the distances from four cities in Texas to Austin, Texas. The distance from Galveston to Austin is 22 miles more than the distance from Fort Worth, Texas, to Austin. Find the distance from Fort Worth to Austin.

City in Texas	Number of Miles to Austin, Texas
Corpus Christi	215
Dallas	195
Galveston	212
Houston	160

34. *Finances* Use the formula $A = MN$, where A is the total amount paid, M is the monthly payment, and N is the number of payments, to find the number of payments made on a loan for which the total amount paid is $13,968 and the monthly payment is $582.

35. *Finances* Use the formula $A = MN$, where A is the total amount paid, M is the monthly payment, and N is the number of payments, to find the number of payments made on a loan for which the total amount paid is $17,460 and the monthly payment is $485.

36. *Travel* Use the formula $d = rt$, where d is distance, r is rate of speed, and t is time, to find how long it would take to travel a distance of 1,120 mi at a speed of 140 mph.

37. *Travel* Use the formula $d = rt$, where d is distance, r is rate of speed, and t is time, to find how long it would take to travel a distance of 825 mi at a speed of 165 mph.

CRITICAL THINKING

38. Write two word problems for a classmate to solve, one that is a number problem (like Exercises 25 to 30 above) and another that involves using a formula (like Exercises 34 to 37 above).

1.5 The Order of Operations Agreement

OBJECTIVE A **The Order of Operations Agreement**

More than one operation may occur in a numerical expression. For example, the expression

$$4 + 3(5)$$

includes two arithmetic operations, addition and multiplication. The operations could be performed in different orders.

If we multiply first and then add, we have:

4 + 3(5)
4 + 15
19

If we add first and then multiply, we have:

4 + 3(5)
7(5)
35

To prevent more than one answer to the same problem, an Order of Operations Agreement is followed. By this agreement, 19 is the only correct answer.

The Order of Operations Agreement

Step 1   Do all operations inside parentheses.

Step 2   Simplify any numerical expressions containing exponents.

Step 3   Do multiplication and division as they occur from left to right.

Step 4   Do addition and subtraction as they occur from left to right.

Calculator Note

Many calculators use the Order of Operations Agreement shown at the left.

Enter 4 + 3 × 5 = into your calculator. If the answer is 19, your calculator uses the Order of Operations Agreement.

Simplify: $2(4 + 1) - 2^3 + 6 \div 2$

Perform operations in parentheses.	$2(4 + 1) - 2^3 + 6 \div 2$ $= 2(5) - 2^3 + 6 \div 2$
Simplify expressions with exponents.	$= 2(5) - 8 + 6 \div 2$
Do multiplication and division as they occur from left to right.	$= 10 - 8 + 6 \div 2$ $= 10 - 8 + 3$
Do addition and subtraction as they occur from left to right.	$= 2 + 3$ $= 5$

Calculator Note

Here is an example of using the parentheses keys on a calculator. To evaluate 28(103 − 78), enter:

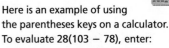

Note that × is required on most calculators.

One or more of the above steps may not be needed to simplify an expression. In that case, proceed to the next step in the Order of Operations Agreement.

Simplify: $8 + 9 \div 3$

There are no parentheses (Step 1).	$8 + 9 \div 3$
There are no exponents (Step 2).	
Do the division (Step 3).	$= 8 + 3$
Do the addition (Step 4).	$= 11$

Point of Interest

Try this: Use the same one-digit number three times to write an expression that is equal to 30.

Evaluate $5a - (b + c)^2$ for $a = 6$, $b = 1$, and $c = 3$.

$$5a - (b + c)^2$$

Replace a with 6, b with 1, and c with 3. $5(6) - (1 + 3)^2$

Use the Order of Operations Agreement to simplify the resulting numerical expression. Perform operations inside parentheses. $= 5(6) - (4)^2$

Simplify expressions with exponents. $= 5(6) - 16$

Do the multiplication. $= 30 - 16$

Do the subtraction. $= 14$

EXAMPLE 1

Simplify: $18 \div (6 + 3) \cdot 9 - 4^2$

Solution

$$\begin{aligned} 18 \div (6 + 3) \cdot 9 - 4^2 &= 18 \div 9 \cdot 9 - 4^2 \\ &= 18 \div 9 \cdot 9 - 16 \\ &= 2 \cdot 9 - 16 \\ &= 18 - 16 \\ &= 2 \end{aligned}$$

YOU TRY IT 1

Simplify: $4 \cdot (8 - 3) \div 5 - 2$

Your Solution

EXAMPLE 2

Simplify: $20 + 24(8 - 5) \div 2^2$

Solution

$$\begin{aligned} 20 + 24(8 - 5) \div 2^2 &= 20 + 24(3) \div 2^2 \\ &= 20 + 24(3) \div 4 \\ &= 20 + 72 \div 4 \\ &= 20 + 18 \\ &= 38 \end{aligned}$$

YOU TRY IT 2

Simplify: $16 + 3(6 - 1)^2 \div 5$

Your Solution

EXAMPLE 3

Evaluate $(a - b)^2 + 3c$ for $a = 6$, $b = 4$, and $c = 1$.

Solution

$$(a - b)^2 + 3c$$
$$\begin{aligned} (6 - 4)^2 + 3(1) &= (2)^2 + 3(1) \\ &= 4 + 3(1) \\ &= 4 + 3 \\ &= 7 \end{aligned}$$

YOU TRY IT 3

Evaluate $(a - b)^2 + 5c$ for $a = 7$, $b = 2$, and $c = 4$.

Your Solution

Solutions on pp. S3–S4

1.5 Exercises

OBJECTIVE A   **The Order of Operations Agreement**

1. The first step in simplifying the expression $18 - 7 \cdot 2$ is _____.

2. Simplify: $2^3 + 3 \cdot (1 + 4)$
   $\qquad 2^3 + 3 \cdot (1 + 4)$
   a. Perform operations in parentheses. $= 2^3 + 3 \cdot (\underline{\quad})$
   b. Simplify expressions with exponents. $= \underline{\quad} + 3 \cdot 5$
   c. Multiply. $= 8 + \underline{\quad}$
   d. Add. $= \underline{\quad}$

Simplify.

3. $8 \div 4 + 2$

4. $12 - 9 \div 3$

5. $6 \cdot 4 + 5$

6. $5 \cdot 7 + 3$

7. $4^2 - 3$

8. $6^2 - 14$

9. $5 \cdot (6 - 3) + 4$

10. $8 + (6 + 2) \div 4$

11. $9 + (7 + 5) \div 6$

12. $14 \cdot (3 + 2) \div 10$

13. $13 \cdot (1 + 5) \div 13$

14. $14 - 2^3 + 9$

15. $6 \cdot 3^2 + 7$

16. $18 + 5 \cdot 3^2$

17. $14 + 5 \cdot 2^3$

18. $20 + (9 - 4) \cdot 2$

19. $10 + (8 - 5) \cdot 3$

20. $3^2 + 5 \cdot (6 - 2)$

21. $2^3 + 4(10 - 6)$

22. $3^2 \cdot 2^2 + 3 \cdot 2$

23. $6(7) + 4^2 \cdot 3^2$

24. $14 - 2(6)$

25. $18 + 3(7)$

26. $2(9 - 2) + 5$

27. $6(8 - 3) - 12$

28. $15 - (7 - 1) \div 3$

29. $16 - (13 - 5) \div 4$

30. $11 + 2 - 3 \cdot 4 \div 3$

31. $17 + 1 - 8 \cdot 2 \div 4$

32. $3(5 + 3) \div 8$

Evaluate the expression for the given values of the variables.

33. $x - 2y$, for $x = 8$ and $y = 3$

34. $x + 6y$, for $x = 5$ and $y = 4$

35. $x^2 + 3y$, for $x = 6$ and $y = 7$

36. $3x^2 + y$, for $x = 2$ and $y = 9$

37. $x^2 + y \div x$, for $x = 2$ and $y = 8$

38. $x + y^2 \div x$, for $x = 4$ and $y = 8$

39. $4x + (x - y)^2$, for $x = 8$ and $y = 2$

40. $(x + y)^2 - 2y$, for $x = 3$ and $y = 6$

41. $x^2 + 3(x - y) + z^2$, for $x = 2$, $y = 1$, and $z = 3$

42. $x^2 + 4(x - y) \div z^2$, for $x = 8$, $y = 6$, and $z = 2$

43. Use the inequality symbol $>$ to compare the expressions $11 + (8 + 4) \div 6$ and $12 + (9 - 5) \cdot 3$.

44. Use the inequality symbol $<$ to compare the expressions $3^2 + 7(4 - 2)$ and $14 - 2^3 + 20$.

For Exercises 45 to 48, insert parentheses as needed in the expression $5 + 7 \cdot 3 - 1$ in order to make the equation true.

45. $5 + 7 \cdot 3 - 1 = 19$

46. $5 + 7 \cdot 3 - 1 = 24$

47. $5 + 7 \cdot 3 - 1 = 25$

48. $5 + 7 \cdot 3 - 1 = 35$

CRITICAL THINKING

49. What is the smallest prime number greater than $15 + (8 - 3)(2^4)$?

50. Simplify $(47 + 48 + 49 + 51 + 52 + 53) \div 100$. What do you notice that will allow you to calculate the answer mentally?

Focus on **Problem Solving**

Questions to Ask

Y ou encounter problem-solving situations every day. Some problems are easy to solve, and you may mentally solve these problems without considering the steps you are taking in order to draw a conclusion. Others may be more challenging and require more thought and consideration.

Suppose a friend suggests that you both take a trip over spring break. You'd like to go. What questions go through your mind? You might ask yourself some of the following questions:

> How much will the trip cost? What will be the cost for travel, lodging, meals, etc.?
> Are some costs going to be shared by both me and my friend?
> Can I afford it?
> How much money do I have in the bank?
> How much more money than I have now do I need?
> How much time is there to earn that much money?
> How much can I earn in that amount of time?
> How much money must I keep in the bank in order to pay the next tuition bill (or some other expense)?

These questions require different mathematical skills. Determining the cost of the trip requires *estimation;* for example, you must use your knowledge of air fares or the cost of gasoline to arrive at an estimate of these costs. If some of the costs are going to be shared, you need to *divide* those costs by 2 in order to determine your share of the expense. The question regarding how much more money you need requires *subtraction:* the amount needed minus the amount currently in the bank. To determine how much money you can earn in the given amount of time requires *multiplication*—for example, the amount you earn per week times the number of weeks to be worked. To determine whether the amount you can earn in the given amount of time is sufficient, you need to use your knowledge of *order relations* to compare the amount you can earn with the amount needed.

Facing the problem-solving situation described above may not seem difficult to you. The reason may be that you have faced similar situations before and, therefore, know how to work through this one. You may feel better prepared to deal with a circumstance such as this one because you know what questions to ask. An important aspect of learning to solve problems is learning what questions to ask. As you work through application problems in this text, try to become more conscious of the mental process you are going through. You might begin the process by asking yourself the following questions whenever you are solving an application problem:

1. Have I read the problem enough times to be able to understand the situation being described?
2. Will restating the problem in different words help me to understand the problem situation better?
3. What facts are given? (You might make a list of the information contained in the problem.)
4. What information is being asked for?
5. What relationship exists among the given facts? What relationship exists among the given facts and the solution?
6. What mathematical operations are needed in order to solve the problem?

Try to focus on the problem-solving situation, not on the computation or on getting the answer quickly. And remember, the more problems you solve, the better able you will be to solve other problems in the future, partly because you are learning what questions to ask.

Projects & Group Activities

Surveys

The circle graph on page 17 shows the results of a survey of 150 people who were asked, "What bothers you most about movie theaters?" Note that the responses included (1) people talking in the theater, (2) high ticket prices, (3) high prices for food purchased in the theater, (4) dirty floors, and (5) uncomfortable seats.

Conduct a similar survey in your class. Ask each classmate which of the five conditions stated above is most irritating. Record the number of students who choose each one of the five possible responses. Prepare a bar graph to display the results of the survey. A model is provided below to help you get started.

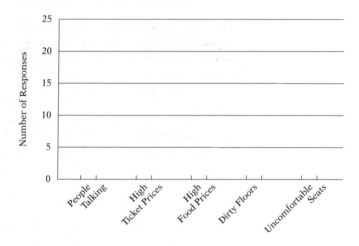

Responses to Theater-Goers Survey

Applications of Patterns in Mathematics

For the circle shown at the far left below, use a straight line to connect each dot on the circle with every other dot on the circle. How many different straight lines are needed?

Follow the same procedure for each of the other circles. How many different straight lines are needed for each?

Find a pattern to describe the number of dots on a circle and the corresponding number of different lines drawn. Use the pattern to determine the number of different lines that would need to be drawn in a circle with 7 dots and in a circle with 8 dots.

Now use the pattern to answer the following:

You are arranging a tennis tournament with nine players. How many singles matches will be played among the nine players if each player plays each of the other players once?

Salary Calculator

On the Internet, go to

 http://www.bankrate.com/brm/movecalc.asp

This website can be used to calculate the salary you would need in order to maintain your current standard of living if you were to move to another city. Select the city you live in now and the city you would like to move to.

1. Is the salary greater in the city you live in now or the city you would like to move to? What is the difference between the two salaries?
2. Select a few other cities you might like to move to. Perform the same calculations. Which of the cities you selected is the most expensive to live in? the least expensive?
3. How might you determine the salaries for people in your occupation in any of the cities you selected?

Subtraction Squares

Draw a square. Write the four numbers 7, 5, 9, and 2, one at each of the four corners. Draw a second square around the first so that it goes through each of the four corners. At each corner of the second square, write the difference of the numbers at the closest corners of the smaller square: $7 - 5 = 2$, $9 - 5 = 4$, $9 - 2 = 7$, and $7 - 2 = 5$.

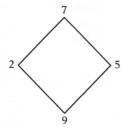 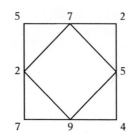

Repeat the process until you come to a pattern of four numbers that does not change. What is the pattern? Try the same procedure with any other four starting numbers. Do you end up with the same pattern? Provide an explanation for what happens.

Chapter 1 Summary

Key Words   **Examples**

The **natural numbers** or **counting numbers** are 1, 2, 3, 4, 5, 6, 7, 8, 9, 10,.... [1.1A, p. 3]

The **whole numbers** are 0, 1, 2, 3, 4, 5, 6, 7, 8, 9, 10,.... [1.1A, p. 3]

The symbol for "is less than" is <. The symbol for "is greater than" is >. A statement that uses the symbol < or > is an **inequality.** [1.1A, p. 3]	$3 < 7$ $9 > 2$
When a whole number is written using the digits 0, 1, 2, 3, 4, 5, 6, 7, 8, and 9, it is said to be in **standard form.** The position of each digit in the number determines that digit's **place value.** [1.1B, p. 4]	The number 598,317 is in standard form. The digit 8 is in the thousands place.
A **pictograph** represents data by using a symbol that is characteristic of the data. A **circle graph** represents data by the size of the sectors. A **bar graph** represents data by the height of the bars. A **broken-line graph** represents data by the position of the lines and shows trends or comparisons. [1.1D, pp. 8–10]	
Addition is the process of finding the total of two or more numbers. The numbers being added are called **addends.** The answer is the **sum.** [1.2A, pp. 19–20]	$\begin{array}{r} \overset{1\ \ 11}{8{,}762} \\ + 1{,}359 \\ \hline 10{,}121 \end{array}$
Subtraction is the process of finding the difference between two numbers. The **minuend** minus the **subtrahend** equals the **difference.** [1.2B, pp. 25–26]	$\begin{array}{r} \overset{4\ 11\ 11\ 6\ 13}{52{,}173} \\ -34{,}968 \\ \hline 17{,}205 \end{array}$
Multiplication is the repeated addition of the same number. The numbers that are multiplied are called **factors.** The answer is the **product.** [1.3A, p. 41]	$\begin{array}{r} \overset{4\ 5}{358} \\ \times\quad 7 \\ \hline 2{,}506 \end{array}$
Division is used to separate objects into equal groups. The **dividend** divided by the **divisor** equals the **quotient.** For any division problem, **(quotient · divisor) + remainder = dividend.** [1.3C, pp. 48–50]	$\begin{array}{r} 93\ \text{r}3 \\ 7)\overline{654} \\ -63 \\ \hline 24 \\ -21 \\ \hline 3 \end{array}$ Check: $(7 \cdot 93) + 3 = 651 + 3 = 654$
The expression 3^5 is in **exponential form.** The **exponent,** 5, indicates how many times the **base,** 3, occurs as a factor in the multiplication. [1.3B, p. 46]	$5^4 = 5 \cdot 5 \cdot 5 \cdot 5 = 625$
Natural number **factors** of a number divide that number evenly (there is no remainder). [1.3D, p. 53]	$18 \div 1 = 18$ $18 \div 2 = 9$ $18 \div 3 = 6$ $18 \div 4$ 4 does not divide 18 evenly. $18 \div 5$ 5 does not divide 18 evenly. $18 \div 6 = 3$ The factors are repeating. The factors of 18 are 1, 2, 3, 6, 9, and 18.
A number greater than 1 is a **prime number** if its only whole number factors are 1 and itself. If a number is not prime, it is a **composite number.** [1.3D, p. 53]	The prime numbers less than 20 are 2, 3, 5, 7, 11, 13, 17, and 19. The composite numbers less than 20 are 4, 6, 8, 9, 10, 12, 14, 15, 16, and 18.

Chapter 1 Test

1. Multiply: $3,297 \times 100$

2. Evaluate $2^4 \cdot 10^3$.

3. Find the difference between 4,902 and 873.

4. Write $x \cdot x \cdot x \cdot x \cdot y \cdot y \cdot y$ in exponential notation.

5. Is 7 a solution of the equation $23 = p + 16$?

6. Round 2,961 to the nearest hundred.

7. Place the correct symbol, $<$ or $>$, between the two numbers.

   7,177   7,717

8. Write eight thousand four hundred ninety in standard form.

9. Write 382,904 in words.

10. Estimate the sum of 392, 477, 519, and 648.

11. Find the product of 8 and 1,376.

12. Estimate the product of 36,479 and 58.

13. Find all the factors of 92.

14. Find the prime factorization of 240.

15. Evaluate $x - y$ for $x = 39,241$ and $y = 8,375$.

16. Identify the property that justifies the statement.

    $14 + y = y + 14$

17. Evaluate $\dfrac{x}{y}$ for $x = 3,588$ and $y = 4$.

18. Simplify: $27 - (12 - 3) \div 9$

19. *Education* The table at the right shows the average weekly earnings, based on level of education, for people aged 25 and older. What is the difference between average weekly earnings for an individual with some college, but no degree, and an individual with a bachelor's degree?

Educational Level	Average Weekly Earnings
No high school diploma	409
High school diploma	583
Some college, no degree	653
Associate degree	699
Bachelor's degree	937
Master's degree	1,129

Source: Bureau of Labor Statistics

20. Solve: $68 = 17 + d$

21. Solve: $176 = 4t$

22. Evaluate $5x + (x - y)^2$ for $x = 8$ and $y = 4$.

23. Complete the statement by using the Associative Property of Addition.

$$(3 + 7) + x = 3 + (? + x)$$

24. *Mathematics*   The sum of twelve and a number is equal to ninety. Find the number.

25. *Mathematics*   What is the product of all the natural numbers less than 7?

26. *Finances*   You purchase a computer system that includes an operating system priced at $850, a monitor that cost $270, an extended keyboard priced at $175, and a printer for $425. You pay for the purchase by check. You had $2,276 in your checking account before making the purchase. What was the balance in your account after making the purchase?

27. *Geometry*   The length of each side of a square is 24 cm. Find **a.** the perimeter and **b.** the area of the square.

28. *Finances*   A data processor receives a total salary of $5,690 per month. Deductions from the paycheck include $854 for taxes, $272 for retirement, and $108 for insurance. Find the data processor's monthly take-home pay.

29. *Automobiles*   The graph at the right shows hybrid car sales from 2001 to 2005. **a.** Between which two years did the number of hybrid cars sold increase the most? **b.** What was the amount of that increase?

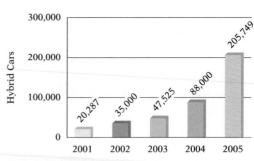

Hybrid Car Sales
Source: **hybridcars.com**

30. *Finances*   Use the formula $C = U \cdot R$, where C is the commission earned, U is the number of units sold, and R is the rate per unit, to find the commission earned from selling 480 boxes of greeting cards when the commission rate per box is $2.

31. *Investments*   The current value of the stocks in a mutual fund is $5,500,000. The number of shares outstanding is 500,000. Find the value per share of the fund. Use the formula $V = \dfrac{C}{S}$, where V is the value per share, C is the current value of the stocks in the fund, and S is the number of shares outstanding.

Integers

Stock market reports involve signed numbers. Positive numbers indicate an increase in the price of a share of stock, and negative numbers indicate a decrease. Positive and negative numbers are also used to indicate whether a company has experienced a profit or loss over a specified period of time. Examples are provided in **Exercise 41 on page 124** and **Exercises 96 and 97 on page 126.**

Student Website

*Need help? For online student resources,
visit college.hmco.com/pic/aufmannPA5e.*

DVD   SSM

1. Place the correct symbol, $<$ or $>$, between the two numbers.
   54   45

2. What is the distance from 4 to 8 on the number line?

For Exercises 3 to 6, add, subtract, multiply, or divide.

3. $7,654 + 8,193$

4. $6,097 - 2,318$

5. 472×56

6. $144 \div 24$

7. Solve: $22 = y + 9$

8. Solve: $12b = 60$

9. What is the price of a scooter that cost a business $129 and has a markup of $43? Use the formula $P = C + M$, where P is the price of a product to a consumer, C is the cost paid by the store for the product, and M is the markup.

10. Simplify: $(8 - 6)^2 + 12 \div 4 \cdot 3^2$

If you multiply the first 20 natural numbers $(1 \cdot 2 \cdot 3 \cdot 4 \cdot 5 \cdots \cdot 17 \cdot 18 \cdot 19 \cdot 20)$, how many zeros will be at the end of the product?

2.1 Introduction to Integers

OBJECTIVE A   **Integers and the number line**

In Chapter 1, only zero and numbers greater than zero were discussed. In this chapter, numbers less than zero are introduced. Phrases such as "7 degrees below zero," "$50 in debt," and "20 feet below sea level" refer to numbers less than zero.

Numbers greater than zero are called **positive numbers.** Numbers less than zero are called **negative numbers.**

> ## Positive and Negative Numbers
>
> A number n is positive if $n > 0$.
> A number n is negative if $n < 0$.

A positive number can be indicated by placing a plus sign (+) in front of the number. For example, we can write +4 instead of 4. Both +4 and 4 represent "positive 4." Usually, however, the plus sign is omitted and it is understood that the number is a positive number.

A negative number is indicated by placing a negative sign (−) in front of the number. The number −1 is read "negative one," −2 is read "negative two," and so on.

The number line can be extended to the left of zero to show negative numbers.

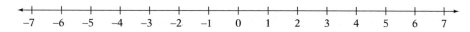

The **integers** are . . . $-4, -3, -2, -1, 0, 1, 2, 3, 4, \ldots$. The integers to the right of zero are the **positive integers.** The integers to the left of zero are the **negative integers.** Zero is an integer, but it is neither positive nor negative. The point corresponding to 0 on the number line is called the **origin.**

On a number line, the numbers get larger as we move from left to right. The numbers get smaller as we move from right to left. Therefore, a number line can be used to visualize the order relation between two integers.

A number that appears to the right of a given number is greater than (>) the given number. A number that appears to the left of a given number is less than (<) the given number.

2 is to the right of −3 on the number line.
2 is greater than −3.
$2 > -3$

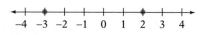

−4 is to the left of 1 on the number line.
−4 is less than 1.
$-4 < 1$

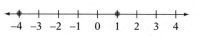

Point of Interest

Chinese manuscripts dating from about 250 B.C. contain the first recorded use of negative numbers. However, it was not until late in the fourteenth century that mathematicians generally accepted these numbers.

Order Relations

$a > b$ if a is to the right of b on the number line.
$a < b$ if a is to the left of b on the number line.

EXAMPLE 1 On the number line, what number is 5 units to the right of -2?

Solution

5 units

$$\begin{array}{ccccccccc} -4 & -3 & -2 & -1 & 0 & 1 & 2 & 3 & 4 \end{array}$$

3 is 5 units to the right of -2.

YOU TRY IT 1 On the number line, what number is 4 units to the left of 1?

Your Solution

EXAMPLE 2 If G is 2 and I is 4, what numbers are B and D?

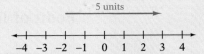
A B C D E F G H I

Solution

$$\begin{array}{ccccccccc} -4 & -3 & -2 & -1 & 0 & 1 & 2 & 3 & 4 \end{array}$$

B is -3 and D is -1.

YOU TRY IT 2 If G is 1 and H is 2, what numbers are A and C?

A B C D E F G H I

Your Solution

EXAMPLE 3 Place the correct symbol, $<$ or $>$, between the two numbers.

a. $-3 \quad -1$ **b.** $1 \quad -2$

Solution **a.** -3 is to the left of -1 on the number line.

$$-3 < -1$$

b. 1 is to the right of -2 on the number line.

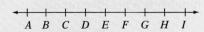

$$1 > -2$$

YOU TRY IT 3 Place the correct symbol, $<$ or $>$, between the two numbers.

a. $2 \quad -5$ **b.** $-4 \quad 3$

Your Solution

EXAMPLE 4 Write the given numbers in order from smallest to largest.

$5, -2, 3, 0, -6$

Solution $-6, -2, 0, 3, 5$

YOU TRY IT 4 Write the given numbers in order from smallest to largest.

$-7, 4, -1, 0, 8$

Your Solution

Solutions on p. S4

OBJECTIVE B   **Opposites**

The distance from 0 to 3 on the number line is 3 units. The distance from 0 to −3 on the number line is 3 units. 3 and −3 are the same distance from 0 on the number line, but 3 is to the right of 0 and −3 is to the left of 0.

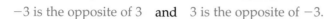

Two numbers that are the same distance from zero on the number line but on opposite sides of zero are called **opposites.**

$$-3 \text{ is the opposite of } 3 \quad \text{and} \quad 3 \text{ is the opposite of } -3.$$

For any number n, the opposite of n is $-n$ and the opposite of $-n$ is n.

We can now define the **integers** as the whole numbers and their opposites.

A negative sign can be read as "the opposite of."

    $-(3) = -3$ The opposite of positive 3 is negative 3.

    $-(-3) = 3$ The opposite of negative 3 is positive 3.

Therefore, $-(a) = -a$ and $-(-a) = a$.

Note that with the introduction of negative integers and opposites, the symbols + and − can be read in different ways.

$6 + 2$	"six plus two"	+ is read "plus"
$+2$	"positive two"	+ is read "positive"
$6 - 2$	"six minus two"	− is read "minus"
-2	"negative two"	− is read "negative"
$-(-6)$	"the opposite of negative six"	− is read first as "the opposite of" and then as "negative"

When the symbols + and − indicate the operations of addition and subtraction, spaces are inserted before and after the symbol. When the symbols + and − indicate the sign of a number (positive or negative), there is no space between the symbol and the number.

Calculator Note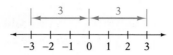

The +/− key on your calculator is used to find the opposite of a number. The − key is used to perform the operation of subtraction.

EXAMPLE 5   Find the opposite number.

    a. −8   **b.** 15   **c.** a

Solution **a.** 8   **b.** −15   **c.** $-a$

YOU TRY IT 5   Find the opposite number.

    a. 24   **b.** −13   **c.** $-b$

Your Solution

Solution on p. S4

EXAMPLE 6 Write the expression in words.

    a. $7 - (-9)$ **b.** $-4 + 10$

Solution **a.** seven minus negative nine

    b. negative four plus ten

YOU TRY IT 6 Write the expression in words.

    a. $-3 - 12$ **b.** $8 + (-5)$

Your Solution

EXAMPLE 7 Simplify.

    a. $-(-27)$ **b.** $-(-c)$

Solution **a.** $-(-27) = 27$

    b. $-(-c) = c$

YOU TRY IT 7 Simplify.

    a. $-(-59)$ **b.** $-(y)$

Your Solution

Solutions on p. S4

OBJECTIVE C **Absolute value**

The **absolute value** of a number is the distance from zero to the number on the number line. Distance is never a negative number. Therefore, the absolute value of a number is a positive number or zero. The symbol for absolute value is "| |."

The distance from 0 to 3 is 3 units. Thus $|3| = 3$ (the absolute value of 3 is 3).

The distance from 0 to -3 is 3 units. Thus $|-3| = 3$ (the absolute value of -3 is 3).

Because the distance from 0 to 3 and the distance from 0 to -3 are the same,

    $|3| = |-3| = 3$.

Absolute Value
....

The absolute value of a positive number is positive.   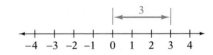

The absolute value of a negative number is positive.   

The absolute value of zero is zero. $|0| = 0$

Take Note

In the example at the right, it is important to be aware that the negative sign is *in front of the absolute value symbol.* This means $-|7| = -7$, but $|-7| = 7$.

Evaluate $-|7|$.

The negative sign is *in front of* the absolute value symbol.

Recall that a negative sign can be read as "the opposite of."

Therefore, $-|7|$ can be read "the opposite of the absolute value of 7."

$-|7| = -7$

EXAMPLE 8 Find the absolute value of **a.** 6 and **b.** −9.

Solution **a.** $|6| = 6$

b. $|-9| = 9$

YOU TRY IT 8 Find the absolute value of **a.** −8 and **b.** 12.

Your Solution

EXAMPLE 9 Evaluate **a.** $|-27|$ and **b.** $-|-14|$.

Solution **a.** $|-27| = 27$

b. $-|-14| = -14$

YOU TRY IT 9 Evaluate **a.** $|0|$ and **b.** $-|35|$.

Your Solution

EXAMPLE 10 Evaluate $|-x|$ for $x = -4$.

Solution $|-x| = |-(-4)| = |4| = 4$

YOU TRY IT 10 Evaluate $|-y|$ for $y = 2$.

Your Solution

EXAMPLE 11 Write the given numbers in order from smallest to largest.

$|-7|, -5, |0|, -(-4), -|-3|$

Solution $|-7| = 7, |0| = 0,$

$-(-4) = 4, -|-3| = -3$

$-5, -|-3|, |0|, -(-4), |-7|$

YOU TRY IT 11 Write the given numbers in order from smallest to largest.

$|6|, |-2|, -(-1), -4, -|-8|$

Your Solution

Solutions on p. S4

OBJECTIVE D **Applications**

Data that are represented by negative numbers on a bar graph are shown below the horizontal axis. For instance, Figure 2.1 shows the lowest recorded temperatures, in degrees Fahrenheit, for selected states in the United States. Hawaii's lowest recorded temperature is 12°F, which is a positive number, so the bar that represents that temperature is above the horizontal axis. The bars for the other states are below the horizontal axis and therefore represent negative numbers.

We can see from the graph that the state with the lowest recorded temperature is New York, with a temperature of −52°F.

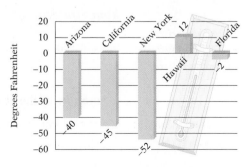

Figure 2.1 Lowest Recorded Temperatures

In a golf tournament, scores below par are recorded as negative numbers; scores above par are recorded as positive numbers. The winner of the tournament is the player who has the lowest score.

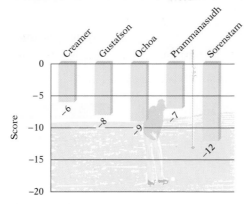

Figure 2.2   The Top Finishers in the 2006 Samsung World Championship

Figure 2.2 shows the number of strokes under par for the five best finishers in the 2006 Samsung World Championship in Palm Desert, California. Use this graph for Example 12 and You Try It 12.

EXAMPLE 12

Use Figure 2.2 to name the player who won the tournament.

Strategy

Use the bar graph and find the player with the lowest score.

Solution

$-12 < -9 < -8 < -7 < -6$

The lowest number among the scores is -12.

Sorenstam won the tournament.

YOU TRY IT 12

Use Figure 2.2 to name the player who came in third in the tournament.

Your Strategy

Your Solution

EXAMPLE 13

Which is the colder temperature, $-18°F$ or $-15°F$?

Strategy

To determine which is the colder temperature, compare the numbers -18 and -15. The lower number corresponds to the colder temperature.

Solution

$-18 < -15$

The colder temperature is $-18°F$.

YOU TRY IT 13

Which is closer to blastoff, -9 s and counting or -7 s and counting?

Your Strategy

Your Solution

Solutions on p. S4

2.1 Exercises

OBJECTIVE A **Integers and the number line**

1. Fill in the blank with *left* or *right*.

    a. On a number line, the number −8 is to the _____ of the number −3.

    b. On a number line, the number 0 is to the _____ of the number −4.

2. Fill in the blank with < or >.

    a. On a number line, −1 is to the right of −10, so −1 _____ −10.

    b. On a number line, −5 is to the left of 2, so −5 _____ 2.

Graph the number on the number line.

3. −5

    <+−−+−−+−−+−−+−−+−−+−−+−−+−−+−−+−−+−−+>
    −6 −5 −4 −3 −2 −1 0 1 2 3 4 5 6

4. −1

    <+−−+−−+−−+−−+−−+−−+−−+−−+−−+−−+−−+−−+>
    −6 −5 −4 −3 −2 −1 0 1 2 3 4 5 6

5. −6

    <+−−+−−+−−+−−+−−+−−+−−+−−+−−+−−+−−+−−+>
    −6 −5 −4 −3 −2 −1 0 1 2 3 4 5 6

6. −2

    <+−−+−−+−−+−−+−−+−−+−−+−−+−−+−−+−−+−−+>
    −6 −5 −4 −3 −2 −1 0 1 2 3 4 5 6

7. x, for $x = 5$

    <+−−+−−+−−+−−+−−+−−+−−+−−+−−+−−+−−+−−+>
    −6 −5 −4 −3 −2 −1 0 1 2 3 4 5 6

8. x, for $x = 0$

    <+−−+−−+−−+−−+−−+−−+−−+−−+−−+−−+−−+−−+>
    −6 −5 −4 −3 −2 −1 0 1 2 3 4 5 6

9. x, for $x = -4$

    <+−−+−−+−−+−−+−−+−−+−−+−−+−−+−−+−−+−−+>
    −6 −5 −4 −3 −2 −1 0 1 2 3 4 5 6

10. x, for $x = -3$

    <+−−+−−+−−+−−+−−+−−+−−+−−+−−+−−+−−+−−+>
    −6 −5 −4 −3 −2 −1 0 1 2 3 4 5 6

On the number line, which number is:

11. 3 units to the right of −2?

12. 5 units to the right of −3?

13. 4 units to the left of 3?

14. 2 units to the left of −1?

15. 6 units to the right of −3?

16. 4 units to the right of −4?

For Exercises 17 to 20, use the following number line.

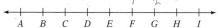

    <+−−+−−+−−+−−+−−+−−+−−+−−+−−+>
     A   B   C   D   E   F   G   H   I

17. If F is 1 and G is 2, what numbers are A and C?

    −4   −2

18. If G is 1 and H is 2, what numbers are B and D?

19. If H is 0 and I is 1, what numbers are A and D?

20. If G is 2 and I is 4, what numbers are B and E?

Place the correct symbol, < or >, between the two numbers.

21. −2 −5

22. −6 −1

23. 3 −7

24. −11 −8

25. −42 27

26. 21 −34

27. 53 −46

28. −27 −39

29. −51 −20

30. −136 0

31. −131 101

32. 127 −150

Write the given numbers in order from smallest to largest.

33. 3, −7, 0, −2

34. −4, 8, 6, −1

35. −3, 1, −5, 4

36. −6, 2, −8, 7

37. 9, −4, 5, 0

38. 6, −9, −12, 8

39. −10, 4, 12, −5, −7

40. 11, −8, −1, 7, −6

41. Determine whether each statement is always true, never true, or sometimes true.

a. A number that is to the right of the number 5 on the number line is a negative number.

b. A number that is to the left of the number 3 on the number line is a negative number.

c. A number that is to the right of the number −4 on the number line is a negative number.

d. A number that is to the left of the number −6 on the number line is a negative number.

OBJECTIVE B Opposites

42. The opposite of a positive number is a _____ number. The opposite of a negative number is a _____ number.

43. In the expression $8 - (-2)$, the first − sign is read as _____ and the second − sign is read as _____.

Find the opposite of the number.

44. 22

45. 45

46. −31

47. −88

48. c

49. n

50. $-w$

51. $-d$

Write the expression in words.

52. $-(-11)$

53. $-(-13)$

54. $-(-d)$

55. $-(-p)$

56. $-2 + (-5)$ **57.** $5 + (-10)$ **58.** $6 - (-7)$ **59.** $-14 - (-3)$

60. $9 - 12$ **61.** $-13 - 8$ **62.** $-a - b$ **63.** $m + (-n)$

Simplify.

64. $-(-5)$ **65.** $-(-7)$ **66.** $-(29)$ **67.** $-(46)$ **68.** $-(-52)$

69. $-(-73)$ **70.** $-(-m)$ **71.** $-(-z)$ **72.** $-(b)$ **73.** $-(p)$

74. Write the statement "the opposite of negative a is b" in symbols. Does a equal b, or are they opposites?

75. If $a < 0$, is $-(-a)$ positive or negative?

OBJECTIVE C **Absolute value**

76. The equation $|-5| = 5$ is read "the _____ of negative five is five."

77. Evaluate $|-y|$ for $y = -6$. $|-y|$
     a. Replace y with _____. $= |-(-6)|$
     b. The opposite of -6 is 6. $= |\_\_\_\_\_|$
     c. The absolute value of 6 is 6. $= \_\_\_\_\_$

Find the absolute value of the number.

78. 4   **79.** -4 **80.** -7 **81.** 9

82. -1 **83.** -11 **84.** 10   **85.** -12

Evaluate.

86. $|-15|$ **87.** $|-23|$ **88.** $-|33|$ **89.** $-|27|$

The opposite of absolute value is < 36.

90. $|32|$ **91.** $|25|$ **92.** $-|-36|$ **93.** $-|-41|$

= 36

-36

94. $-|-81|$ **95.** $-|-93|$ **96.** $|x|$, for $x = 7$ **97.** $|x|$, for $x = -10$

98. $|-x|$, for $x = 2$ **99.** $|-x|$, for $x = 8$ **100.** $|-y|$, for $y = -3$ **101.** $|-y|$, for $y = -6$

Place the correct symbol, $<$, $=$, or $>$, between the two numbers.

102. $|7|$ $|-9|$ **103.** $|-12|$ $|8|$ **104.** $|-5|$ $|-2|$ **105.** $|6|$ $|13|$

106. $|-8|$ $|3|$ **107.** $|-1|$ $|-17|$ **108.** $|-14|$ $|14|$ **109.** $|x|$ $|-x|$

Write the given numbers in order from smallest to largest.

110. $|-8|, -(-3), |2|, -|-5|$ **111.** $-|6|, -(4), |-7|, -(-9)$ **112.** $-(-1), |-6|, |0|, -|3|$

113. $-|-7|, -9, -(5), |4|$ **114.** $-|2|, -(-8), 6, |1|, -7$ **115.** $-(-3), -|-8|, |5|, -|10|, -(-2)$

116. Find the values of a for which $|a| = 7$.   **117.** Find the values of y for which $|y| = 11$.

118. Given that x is an integer, find all values of x for which $|x| < 5$.   **119.** Given that c is an integer, find all values of c for which $|c| < 7$.

120. Determine whether each statement is always true, never true, or sometimes true.

a. The absolute value of a negative number n is greater than n.

b. The absolute value of a number n is the opposite of n.

121. Find two numbers a and b such that $a < b$ and $|a| > |b|$.

OBJECTIVE D   **Applications**

The table below gives wind-chill temperatures for combinations of temperature and wind speed. For example, the combination of a temperature of 15°F and a wind blowing at 10 mph has a cooling power equal to 3°F. Use this table for Exercises 122 to 129.

Wind Speed (mph)	Wind Chill Factors														
	Thermometer Reading (degrees Fahrenheit)														
	25	20	15	10	5	0	−5	−10	−15	−20	−25	−30	−35	−40	−45
5	19	13	7	1	−5	−11	−16	−22	−28	−34	−40	−46	−52	−57	−63
10	15	9	3	−4	−10	−16	−22	−28	−35	−41	−47	−53	−59	−66	−72
15	13	6	0	−7	−13	−19	−26	−32	−39	−45	−51	−58	−64	−71	−77
20	11	4	−2	−9	−15	−22	−29	−35	−42	−48	−55	−61	−68	−74	−81
25	9	3	−4	−11	−17	−24	−31	−37	−44	−51	−58	−64	−71	−78	−84
30	8	1	−5	−12	−19	−26	−33	−39	−46	−53	−60	−67	−73	−80	−87
35	7	0	−7	−14	−21	−27	−34	−41	−48	−55	−62	−69	−76	−82	−89
40	6	−1	−8	−15	−22	−29	−36	−43	−50	−57	−64	−71	−78	−84	−91
45	5	−2	−9	−16	−23	−30	−37	−44	−51	−58	−65	−72	−79	−86	−93

122. To find the wind-chill factor when the temperature is 20°F and the wind speed is 10 mph, find the number that is both in the column under 20 and in the row to the right of 10. This number is _____, so the wind-chill factor is _____ °F.

123. To find the cooling power of a temperature of −35°F and a wind speed of 30 mph, find the number that is both in the column under −35 and in the row to the right of 30. This number is _____, so the cooling power is _____°F.

124. *Environmental Science*   Find the wind chill factor when the temperature is 5°F and the wind speed is 15 mph.

125. *Environmental Science*   Find the wind chill factor when the temperature is 10°F and the wind speed is 20 mph.

126. *Environmental Science*   Find the cooling power of a temperature of −10°F and a 5-mph wind.

127. *Environmental Science*   Find the cooling power of a temperature of −15°F and a 10-mph wind.

128. *Environmental Science*   Which feels colder, a temperature of 0°F with a 15-mph wind or a temperature of 10°F with a 25-mph wind?

129. *Environmental Science*   Which would feel colder, a temperature of −30°F with a 5-mph wind or a temperature of −20°F with a 10-mph wind?

130. *Rocketry*   Which is closer to blastoff, −12 min and counting or −17 min and counting?

One of the measures used by a financial analyst to evaluate the financial strength of a company is *earnings per share*. This number is found by taking the total profit of the company and dividing by the number of shares of stock that the company has sold to investors. If the company has a loss instead of a profit, the earnings per share is a negative number. In a bar graph, a profit is shown by a bar extending above the horizontal axis, and a loss is shown by a bar extending below the horizontal axis. The figure at the right shows the earnings per share for Mycopen for the years 2004 through 2009. Use this graph for Exercises 131 to 134.

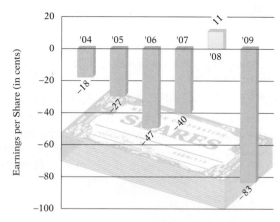

Mycopen Earnings per Share (in cents)

131. *Business* **a.** What were the earnings per share for Mycopen in 2005? **b.** What were the earnings per share for Mycopen in 2007?

132. *Business* For the years shown, in which year did Mycopen have the greatest loss?

133. *Business* For the years shown, did Mycopen ever have a profit? If so, in what year?

134. *Business* In which year was the Mycopen earnings per share lower, 2004 or 2006?

135. *Investments* In the stock market, the net change in the price of a share of stock is recorded as a positive or a negative number. If the price rises, the net change is positive. If the price falls, the net change is negative. If the net change for a share of Stock A is −2 and the net change for a share of Stock B is −1, which stock showed the least net change?

136. *Business* Some businesses show a profit as a positive number and a loss as a negative number. During the first quarter of this year, the loss experienced by a company was recorded as −12,575. During the second quarter of this year, the loss experienced by the company was −11,350. During which quarter was the loss greater?

137. *Business* Some businesses show a profit as a positive number and a loss as a negative number. During the third quarter of last year, the loss experienced by a company was recorded as −26,800. During the fourth quarter of last year, the loss experienced by the company was −24,900. During which quarter was the loss greater?

CRITICAL THINKING

138. *Mathematics* *A* is a point on the number line halfway between −9 and 3. *B* is a point halfway between *A* and the graph of 1 on the number line. *B* is the graph of what number?

139. **a.** Name two numbers that are 4 units from 2 on the number line.
b. Name two numbers that are 5 units from 3 on the number line.

2.2 Addition and Subtraction of Integers

OBJECTIVE A **Addition of integers**

Not only can an integer be graphed on a number line, an integer can be represented anywhere along a number line by an arrow. A positive number is represented by an arrow pointing to the right. A negative number is represented by an arrow pointing to the left. The absolute value of the number is represented by the length of the arrow. The integers 5 and -4 are shown on the number line in the figure below.

The sum of two integers can be shown on a number line. To add two integers, find the point on the number line corresponding to the first addend. At that point, draw an arrow representing the second addend. The sum is the number directly below the tip of the arrow.

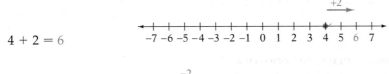

$4 + 2 = 6$

$-4 + (-2) = -6$

$-4 + 2 = -2$

$4 + (-2) = 2$

The sums shown above can be categorized by the signs of the addends.

The addends have the same sign.

$\quad 4 + 2 \qquad$ positive 4 plus positive 2
$-4 + (-2) \qquad$ negative 4 plus negative 2

The addends have different signs.

$-4 + 2 \qquad$ negative 4 plus positive 2
$\quad 4 + (-2) \qquad$ positive 4 plus negative 2

The rule for adding two integers depends on whether the signs of the addends are the same or different.

Rule for Adding Two Integers

To add two integers with the same sign, add the absolute values of the numbers. Then attach the sign of the addends.

To add two integers with different signs, find the absolute values of the numbers. Subtract the smaller absolute value from the larger absolute value. Then attach the sign of the addend with the larger absolute value.

Add: $(-4) + (-9)$

The signs of the addends are the same.
Add the absolute values of the numbers.
$|-4| = 4, |-9| = 9, 4 + 9 = 13$
Attach the sign of the addends.
(Both addends are negative.
The sum is negative.)

$(-4) + (-9) = -13$

Calculator Note

To add $-14 + (-47)$ with your calculator, enter the following:

$\underbrace{14 \boxed{+/-}}_{-14} + \underbrace{47 \boxed{+/-}}_{-47} \boxed{=}$

Add: $-14 + (-47)$

The signs are the same.
Add the absolute values of the numbers.
Attach the sign of the addends.

$-14 + (-47) = -61$

Add: $6 + (-13)$

The signs of the addends are different.
Find the absolute values of the numbers.
$|6| = 6, |-13| = 13$
Subtract the smaller absolute value from the larger absolute value.
$13 - 6 = 7$
Attach the sign of the number with the larger absolute value.
$|-13| > |6|$. Attach the negative sign.

$6 + (-13) = -7$

Add: $162 + (-247)$

The signs are different. Find the difference between the absolute values of the numbers.
$247 - 162 = 85$
Attach the sign of the number with the larger absolute value.

$162 + (-247) = -85$

Add: $-8 + 8$

The signs are different. Find the difference between the absolute values of the numbers.
$8 - 8 = 0$

$-8 + 8 = 0$

in this last example that we are adding a number and its opposite (-8 8), and the sum is 0. The opposite of a number is called its **additive verse.** The opposite or additive inverse of -8 is 8, and the opposite or additive inverse of 8 is -8. The sum of a number and its additive inverse is always zero. This is known as the Inverse Property of Addition.

The properties of addition presented in Chapter 1 hold true for integers as well as whole numbers. These properties are repeated below, along with the Inverse Property of Addition.

The Addition Property of Zero	$a + 0 = a$ or $0 + a = a$
The Commutative Property of Addition	$a + b = b + a$
The Associative Property of Addition	$(a + b) + c = a + (b + c)$
The Inverse Property of Addition	$a + (-a) = 0$ or $-a + a = 0$

> **Take Note**
>
> With the Commutative Properties, the order in which the numbers appear changes. With the Associative Properties, the order in which the numbers appear remains the same.

Add: $(-4) + (-6) + (-8) + 9$

Add the first two numbers.

Add the sum to the third number.

Continue until all the numbers have been added.

$$(-4) + (-6) + (-8) + 9$$
$$= (-10) + (-8) + 9$$
$$= (-18) + 9$$
$$= -9$$

> **Take Note**
>
> For the example at the left, check that the sum is the same if the numbers are added in a different order.

The price of Byplex Corporation's stock fell each trading day of the first week of June 2009. Use Figure 2.3 to find the change in the price of Byplex stock over the week's time.

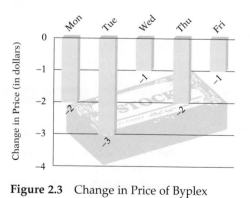

Figure 2.3   Change in Price of Byplex Corporation Stock

Add the five changes in price.

$$-2 + (-3) + (-1) + (-2) + (-1)$$
$$= (-5) + (-1) + (-2) + (-1)$$
$$= -6 + (-2) + (-1)$$
$$= -8 + (-1) = -9$$

The change in the price was -9.

This means that the price of the stock fell \$9 per share.

Evaluate $-x + y$ for $x = -15$ and $y = -5$.

Replace x with -15 and y with -5.

Simplify $-(-15)$.

Add.

$$-x + y$$
$$-(-15) + (-5)$$
$$= 15 + (-5)$$
$$= 10$$

Take Note

Recall that a solution of an equation is a number that, when substituted for the variable, produces a true equation.

Is -7 a solution of the equation $x + 4 = -3$?

Replace x by -7 and then simplify.

The results are equal.

$$\begin{array}{c|c} x + 4 & = -3 \\ \hline -7 + 4 & -3 \\ -3 & = -3 \end{array}$$

-7 is a solution of the equation.

EXAMPLE 1   Add: $-97 + (-45)$

Solution $-97 + (-45) = -142$

YOU TRY IT 1   Add: $-38 + (-62)$

Your Solution

EXAMPLE 2   Add: $81 + (-79)$

Solution $81 + (-79) = 2$

YOU TRY IT 2   Add: $47 + (-53)$

Your Solution

EXAMPLE 3   Add: $42 + (-12) + (-30)$

Solution $42 + (-12) + (-30)$
$$= 30 + (-30)$$
$$= 0$$

YOU TRY IT 3   Add: $-36 + 17 + (-21)$

Your Solution

EXAMPLE 4   What is -162 increased by 98?

Solution $-162 + 98 = -64$

YOU TRY IT 4   Find the sum of -154 and -37.

Your Solution

EXAMPLE 5   Evaluate $-x + y$ for $x = -11$ and $y = -2$.

Solution $-x + y$
$$-(-11) + (-2) = 11 + (-2)$$
$$= 9$$

YOU TRY IT 5   Evaluate $-x + y$ for $x = -3$ and $y = -10$.

Your Solution

Solutions on p. S4

EXAMPLE 6 Is −6 a solution of the equation 3 + y = −2?

YOU TRY IT 6 Is −9 a solution of the equation 2 = 11 + a?

Solution

$$3 + y = -2$$
$$\frac{3 + (-6) \mid -2}{-3 \ne -2}$$

No, −6 is not a solution of the equation.

Your Solution

Solution on p. S4

OBJECTIVE B **Subtraction of integers**

Before the rules for subtracting two integers are explained, look at the translation into words of expressions that represent the difference of two integers.

9 − 3	positive 9 minus positive 3
−9 − 3	negative 9 minus positive 3
9 − (−3)	positive 9 minus negative 3
−9 − (−3)	negative 9 minus negative 3

Note that the sign − is used in two different ways. One way is as a negative sign, as in −9 (negative 9). The second way is to indicate the operation of subtraction, as in 9 − 3 (9 minus 3).

Look at the next four expressions and decide whether the second number in each expression is a positive number or a negative number.

1. (−10) − 8 $(-10) + {}^-8$

2. (−10) − (−8) $(-10) + 8$

3. 10 − (−8) $10 + 8$

4. 10 − 8 $10 + {}^-8$

In expressions 1 and 4, the second number is positive 8. In expressions 2 and 3, the second number is negative 8.

Opposites are used to rewrite subtraction problems as related addition problems. Notice below that the subtraction of a whole number is the same as the addition of the opposite number.

Subtraction		*Addition of the Opposite*	
8 − 4	=	8 + (−4)	= 4
7 − 5	=	7 + (−5)	= 2
9 − 2	=	9 + (−2)	= 7

Subtraction of integers can be written as the addition of the opposite number. To subtract two integers, rewrite the subtraction expression as the first number plus the opposite of the second number. Some examples are shown below.

First number	−	second number	=	First number	+	opposite of the second number	
8	−	15	=	8	+	$(-15) = -7$	
8	−	(-15)	=	8	+	$15 = 23$	
-8	−	15	=	-8	+	$(-15) = -23$	
-8	−	(-15)	=	-8	+	$15 = 7$	

Rule for Subtracting Two Integers

To subtract two integers, add the opposite of the second integer to the first integer.

Subtract: $(-15) - 75$

Rewrite the subtraction operation as the sum of the first number and the opposite of the second number. The opposite of 75 is -75.

Add.

$$(-15) - 75$$
$$= (-15) + (-75)$$
$$= -90$$

Subtract: $6 - (-20)$

Rewrite the subtraction operation as the sum of the first number and the opposite of the second number. The opposite of -20 is 20.

$$6 - (-20)$$
$$= 6 + 20$$
$$= 26$$

Subtract: $11 - 42$

Rewrite the subtraction operation as the sum of the first number and the opposite of the second number. The opposite of 42 is -42.

$$11 - 42$$
$$= 11 + (-42)$$
$$= -31$$

Take Note

$42 - 11 = 31$

$11 - 42 = -31$

$42 - 11 \neq 11 - 42$

By the Commutative Property of Addition, the order in which two numbers are added does not affect the sum; $a + b = b + a$. However, note from this last example that the order in which two numbers are subtracted *does* affect the difference. The operation of subtraction is not commutative.

When subtraction occurs several times in an expression, rewrite each subtraction as addition of the opposite and then add.

Subtract: $-13 - 5 - (-8)$

Rewrite each subtraction as addition of the opposite.

$$-13 - 5 - (-8)$$
$$= -13 + (-5) + 8$$

Add.

$$= -18 + 8$$
$$= -10$$

Calculator Note

To subtract $-13 - 5 - (-8)$ with your calculator, enter the following:

13 [+/−] − 5 − 8 [+/−] =
    └─ −13 ─┘   └─ −8 ─┘

Simplify: $-14 + 6 - (-7)$

This problem involves both addition and subtraction. Rewrite the subtraction as addition of the opposite.

$$-14 + 6 - (-7)$$
$$= -14 + 6 + 7$$
$$= -8 + 7$$

Add.

$$= -1$$

Evaluate $a - b$ for $a = -2$ and $b = -9$.

Replace a with -2 and b with -9.

$$a - b$$
$$-2 - (-9)$$

Rewrite the subtraction as addition of the opposite.

$$= -2 + 9$$

Add.

$$= 7$$

Is -4 a solution of the equation $3 - a = 11 + a$?

$$3 - a = 11 + a$$

Replace a by -4 and then simplify.

$3 - (-4)$	$11 + (-4)$
$3 + 4$	7

The results are equal.

$$7 = 7$$

Yes, -4 is a solution of the equation.

EXAMPLE 7 Subtract: $-12 - (-17)$

Solution $-12 - (-17) = -12 + 17$
$$= 5$$

YOU TRY IT 7 Subtract: $-35 - (-34)$

Your Solution

EXAMPLE 8 Subtract: $66 - (-90)$

Solution $66 - (-90) = 66 + 90$
$$= 156$$

YOU TRY IT 8 Subtract: $83 - (-29)$

Your Solution

Solutions on p. S4

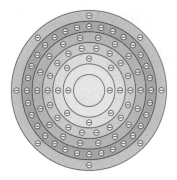

Radon

The table below shows the boiling point and the melting point in degrees Celsius of three chemical elements. Use this table for Example 9 and You Try It 9.

Chemical Element	Boiling Point	Melting Point
Mercury	357	−39
Radon	−62	−71
Xenon	−108	−112

EXAMPLE 9 Use the table above to find the difference between the boiling point and the melting point of mercury.

Solution The boiling point of mercury is 357.

The melting point of mercury is −39.

$$357 - (-39) = 357 + 39$$
$$= 396$$

The difference is 396°C.

YOU TRY IT 9 Use the table above to find the difference between the boiling point and the melting point of xenon.

Your Solution

EXAMPLE 10 What is −12 minus 8?

Solution $-12 - 8 = -12 + (-8)$
$$= -20$$

YOU TRY IT 10 What is 14 less than −8?

Your Solution

EXAMPLE 11 Subtract 91 from 43.

Solution $43 - 91 = 43 + (-91)$
$$= -48$$

YOU TRY IT 11 What is 25 decreased by 68?

Your Solution

EXAMPLE 12 Simplify:
$-8 - 30 - (-12) - 7 - (-14)$

Solution $-8 - 30 - (-12) - 7 - (-14)$
$$= -8 + (-30) + 12 + (-7) + 14$$
$$= -38 + 12 + (-7) + 14$$
$$= -26 + (-7) + 14$$
$$= -33 + 14$$
$$= -19$$

YOU TRY IT 12 Simplify:
$-4 - (-3) + 12 - (-7) - 20$

Your Solution

Solutions on p. S4

64. Is -3 a solution of the equation $x + 4 = 1$?

65. Is -8 a solution of the equation $6 = -3 + z$?

66. Is -6 a solution of the equation $6 = 12 + n$?

67. Is -8 a solution of the equation $-7 + m = -15$?

68. Is -2 a solution of the equation $3 + y = y + 3$?

69. Is -4 a solution of the equation $1 + z = z + 2$?

Determine whether each statement is always true, sometimes true, or never true. Assume a and b are integers.

70. If $a > 0$ and $b > 0$, then $a + b > 0$.

71. If $a > 0$ and $b < 0$, then $a + b > 0$.

72. If $a = -b$, then $a + b < 0$.

73. If $a < 0$ and $b < 0$, then $a + b < 0$.

OBJECTIVE B   **Subtraction of integers**

Rewrite each subtraction as addition of the opposite.

74. $-9 - 5 = -9 +$ _____

75. $6 - (-4) - 3 = 6 +$ _____ $+$ _____

Subtract.

76. $7 - 14$

77. $6 - 9$

78. $-7 - 2$

79. $-9 - 4$

80. $7 - (-2)$

81. $3 - (-4)$

82. $-6 - (-6)$

83. $-4 - (-4)$

84. $-12 - 16$

85. $-10 - 7$

86. $(-9) - (-3)$

87. $(-7) - (-4)$

88. $4 - (-14)$

89. $-4 - (-16)$

90. $(-14) - (-7)$

91. $3 - (-24)$

92. $9 - (-9)$

93. $(-41) - 65$

94. $57 - 86$

95. $-95 - (-28)$

96. How much larger is 5 than -11?

97. What is -10 decreased by -4?

98. Find -13 minus -8.

99. What is 6 less than -9?

The figure at the right shows the highest and lowest temperatures ever recorded for selected regions of the world. Use this graph for Exercises 100 to 102.

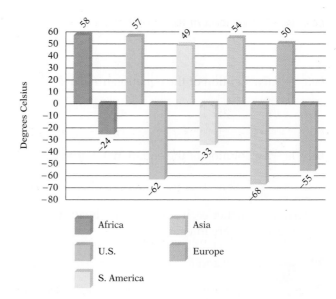

100. *Temperature* What is the difference between the highest and lowest temperatures ever recorded in Africa?

101. *Temperature* What is the difference between the highest and lowest temperatures ever recorded in South America?

102. *Temperature* What is the difference between the lowest temperature recorded in Europe and the lowest temperature recorded in Asia?

Highest and Lowest Temperatures Recorded
(in degrees Celsius)

Simplify.

103. $-4 - 3 - 2$

104. $4 - 5 - 12$

105. $12 - (-7) - 8$

106. $-12 - (-3) - (-15)$

107. $4 - 12 - (-8)$

108. $-30 - (-65) - 29 - 4$

109. $-16 - 47 - 63 - 12$

110. $42 - (-30) - 65 - (-11)$

111. $12 - (-6) + 8$

112. $-7 + 9 - (-3)$

113. $-8 - (-14) + 7$

114. $-4 + 6 - 8 - 2$

115. $9 - 12 + 0 - 5$

116. $11 - (-2) - 6 + 10$

117. $5 + 4 - (-3) - 7$

118. $-1 - 8 + 6 - (-2)$

119. $-13 + 9 - (-10) - 4$

120. $6 - (-13) - 14 + 7$

Evaluate the expression for the given values of the variables.

121. $-x - y$, for $x = -3$ and $y = 9$

122. $x - (-y)$, for $x = -3$ and $y = 9$

123. $-x - (-y)$, for $x = -3$ and $y = 9$

124. $a - (-b)$, for $a = -6$ and $b = 10$

125. $a - b - c$, for $a = 4$, $b = -2$, and $c = 9$

126. $a - b - c$, for $a = -1$, $b = 7$, and $c = -15$

127. $x - y - (-z)$, for $x = -9$, $y = 3$, and $z = 30$

128. $-x - (-y) - z$, for $x = 8$, $y = 1$, and $z = -14$

129. Is -3 a solution of the equation $x - 7 = -10$?

130. Is -4 a solution of the equation $1 = 3 - y$?

131. Is -2 a solution of the equation $-5 - w = 7$?

132. Is -8 a solution of the equation $-12 = m - 4$?

133. Is -6 a solution of the equation $-t - 5 = 7 + t$?

134. Is -7 a solution of the equation $5 + a = -9 - a$?

Determine whether each statement is always true, sometimes true, or never true. Assume a and b are integers.

135. 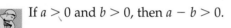 If $a > 0$ and $b > 0$, then $a - b > 0$.

136. If $a > 0$ and $b < 0$, then $a - b > 0$.

OBJECTIVE C   Applications and formulas

The elevation, or height, of places on Earth is measured in relation to sea level, or the average level of the ocean's surface. The table below shows height above sea level as a positive number and depth below sea level as a negative number. Use the table below for Exercises 137 to 140.

Mt. Everest

Continent	Highest Elevation (in meters)		Lowest Elevation (in meters)	
Africa	Mt. Kilimanjaro	5,895	Lake Assal	−156
Asia	Mt. Everest	8,850	Dead Sea	−411
Europe	Mt. Elbrus	5,642	Caspian Sea	−28
America	Mt. Aconcagua	6,960	Death Valley	−86

137. Circle the correct words and fill in the blanks to complete the sentence: To find the difference in elevation between Mt. Elbrus and the Caspian Sea, add/subtract the elevation −28 m to/from the elevation _____ m.

138. *Geography* What is the difference in elevation between **a.** Mt. Aconcagua and Death Valley and **b.** Mt. Kilimanjaro and Lake Assal?

139. *Geography* For which continent shown is the difference between the highest and lowest elevations greatest?

140. *Geography* For which continent shown is the difference between the highest and lowest elevations smallest?

141. Circle the correct words and fill in the blanks to complete the sentence: To find the temperature after a rise of 12°F from −5°F, add/subtract the temperature _____ °F to/from the temperature _____ °F.

142. *Temperature* Find the temperature after a rise of 9°C from −6°C.

The table at the right shows the average temperatures at different cruising altitudes for airplanes. Use the table for Exercises 143 to 145.

Cruising Altitude	Average Temperature
12,000 ft	16°
20,000 ft	−12°
30,000 ft	−48°
40,000 ft	−70°
50,000 ft	−70°

143. *Temperature* What is the difference between the average temperatures at 12,000 ft and at 40,000 ft?

144. *Temperature* What is the difference between the average temperatures at 40,000 ft and at 50,000 ft?

145. *Temperature* How much colder is the average temperature at 30,000 ft than at 20,000 ft?

146. *Sports* Use the equation $S = N - P$, where S is a golfer's score relative to par in a tournament, N is the number of strokes made by the golfer, and P is par, to find a golfer's score relative to par when the golfer made 196 strokes and par is 208.

147. *Sports* Use the equation $S = N - P$, where S is a golfer's score relative to par in a tournament, N is the number of strokes made by the golfer, and P is par, to find a golfer's score relative to par when the golfer made 49 strokes and par is 52.

148. *Mathematics* The distance, d, between point a and point b on the number line is given by the formula $d = |a - b|$. Find d when $a = 6$ and $b = -15$.

149. *Mathematics* The distance, d, between point a and point b on the number line is given by the formula $d = |a - b|$. Find d when $a = 7$ and $b = -12$.

CRITICAL THINKING

150. *Mathematics* Given the list of numbers at the right, find the largest difference that can be obtained by subtracting one number in the list from a different number in the list.   5, −2, −9, 11, 14

151. The sum of two negative integers is −7. Find the integers.

2.3 Multiplication and Division of Integers

OBJECTIVE A Multiplication of integers

When 5 is multiplied by a sequence of decreasing integers, each product decreases by 5.

$$5(3) = 15$$
$$5(2) = 10$$
$$5(1) = 5$$
$$5(0) = 0$$

The pattern developed can be continued so that 5 is multiplied by a sequence of negative numbers. To maintain the pattern of decreasing by 5, the resulting products must be negative.

$$5(-1) = -5$$
$$5(-2) = -10$$
$$5(-3) = -15$$
$$5(-4) = -20$$

This example illustrates that the product of a positive number and a negative number is negative.

When -5 is multiplied by a sequence of decreasing integers, each product increases by 5.

$$-5(3) = -15$$
$$-5(2) = -10$$
$$-5(1) = -5$$
$$-5(0) = 0$$

The pattern developed can be continued so that -5 is multiplied by a sequence of negative numbers. To maintain the pattern of increasing by 5, the resulting products must be positive.

$$-5(-1) = 5$$
$$-5(-2) = 10$$
$$-5(-3) = 15$$
$$-5(-4) = 20$$

This example illustrates that the product of two negative numbers is positive.

The pattern for multiplication shown above is summarized in the following rule for multiplying integers.

Point of Interest

Operations with negative numbers were not accepted until the late thirteenth century. One of the first attempts to prove that the product of two negative numbers is positive was made in the book *Ars Magna,* by Girolamo Cardan, in 1545.

Rule for Multiplying Two Integers

To multiply two integers with the same sign, multiply the absolute values of the factors. The product is **positive.**

To multiply two integers with different signs, multiply the absolute values of the factors. The product is **negative.**

Multiply: $-9(12)$

The signs are different. The product is negative. $-9(12) = -108$

Multiply: $(-6)(-15)$

The signs are the same. The product is positive. $(-6)(-15) = 90$

Calculator Note

To multiply $(-6)(-15)$ with your calculator, enter the following:

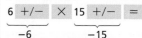

Figure 2.5 shows the melting points of bromine and mercury. The melting point of helium is 7 times the melting point of mercury. Find the melting point of helium.

Multiply the melting point of mercury (−39°C) by 7.

$$−39(7) = −273$$

The melting point of helium is −273°C.

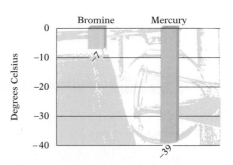

Figure 2.5 Melting Points of Chemical Elements (in degrees Celsius)

The properties of multiplication presented in Chapter 1 hold true for integers as well as whole numbers. These properties are repeated below.

The Multiplication Property of Zero	$a \cdot 0 = 0$ or $0 \cdot a = 0$
The Multiplication Property of One	$a \cdot 1 = a$ or $1 \cdot a = a$
The Commutative Property of Multiplication	$a \cdot b = b \cdot a$
The Associative Property of Multiplication	$(a \cdot b) \cdot c = a \cdot (b \cdot c)$

Multiply: $2(−3)(−5)(−7)$

Multiply the first two numbers.

Then multiply the product by the third number.

Continue until all the numbers have been multiplied.

$$2(−3)(−5)(−7)$$
$$= −6(−5)(−7)$$
$$= 30(−7)$$
$$= −210$$

Take Note

For the example at the right, the product is the same if the numbers are multiplied in a different order. For instance,

$2(−3)(−5)(−7)=$
$2(−3)(35)=$
$2(−105)=$
$−210$

By the Multiplication Property of One, $1 \cdot 6 = 6$ and $\mathbf{1} \cdot x = x$. Applying the rules for multiplication, we can extend this to $−1 \cdot 6 = −6$ and $\mathbf{−1} \cdot x = −x$.

Evaluate $−ab$ for $a = −2$ and $b = −9$.

Replace a with $−2$ and b with $−9$.

Simplify $−(−2)$.

Multiply.

$$−ab$$
$$−(−2)(−9)$$
$$= 2(−9)$$
$$= −18$$

Take Note

When variables are placed next to each other, it is understood that the operation is multiplication. $−ab$ means "the opposite of a times b."

Is $−4$ a solution of the equation $5x = −20$?

Replace x by $−4$ and then simplify.

The results are equal.

$$5x = −20$$
$$\begin{array}{c|c} 5(−4) & −20 \\ \hline −20 & = −20 \end{array}$$

Yes, $−4$ is a solution of the equation.

EXAMPLE 1   Find −42 times 62.

Solution   −42 · 62 = −2,604

YOU TRY IT 1   What is −38 multiplied by 51?

Your Solution

EXAMPLE 2   Multiply: −5(−4)(6)(−3)

Solution   −5(−4)(6)(−3) = 20(6)(−3)
                              = 120(−3)
                              = −360

YOU TRY IT 2   Multiply: −7(−8)(9)(−2)

Your Solution

EXAMPLE 3   Evaluate −5x for x = −11.

Solution   −5x
             −5(−11) = 55

YOU TRY IT 3   Evaluate −9y for y = 20.

Your Solution

EXAMPLE 4   Is 5 a solution of the equation
                30 = −6z?

Solution $\dfrac{30 = -6z}{30 \mid -6(5)}$
             30 ≠ −30
             No, 5 is not a solution of the
             equation.

YOU TRY IT 4   Is −3 a solution of the
                   equation 12 = −4a?

Your Solution

Solutions on p. S5

OBJECTIVE B   **Division of integers**

For every division problem, there is a related multiplication problem.

Division: $\dfrac{8}{2} = 4$ Related multiplication: 4(2) = 8

This fact can be used to illustrate a rule for dividing integers.

$\dfrac{12}{3} = 4$ because 4(3) = 12   and $\dfrac{-12}{-3} = 4$ because 4(−3) = −12.

These two division examples suggest that the quotient of two numbers with the same sign is positive. Now consider these two examples.

$\dfrac{12}{-3} = -4$ because −4(−3) = 12

$\dfrac{-12}{3} = -4$ because −4(3) = −12

These two division examples suggest that the quotient of two numbers with different signs is negative. This property is summarized next.

> **Take Note**
>
> Recall that the fraction bar can be read "divided by." Therefore,
>
> $\dfrac{8}{2}$ can be read "8 divided by 2."

Rule for Dividing Two Integers

To divide two numbers with the same sign, divide the absolute values of the numbers. The quotient is **positive.**

To divide two numbers with different signs, divide the absolute values of the numbers. The quotient is **negative.**

Note from this rule that $\dfrac{12}{-3}$, $\dfrac{-12}{3}$, and $-\dfrac{12}{3}$ are all equal to -4.

If a and b are integers ($b \neq 0$), then $\dfrac{a}{-b} = \dfrac{-a}{b} = -\dfrac{a}{b}$.

Calculator Note

To divide (-105) by (-5) with your calculator, enter the following:

105 [+/−] ÷ 5 [+/−] [=]

 $\underbrace{}_{-105}$ $\underbrace{}_{-5}$

Divide: $-36 \div 9$

The signs are different. The quotient is negative. $-36 \div 9 = -4$

Divide: $(-105) \div (-5)$

The signs are the same. The quotient is positive. $(-105) \div (-5) = 21$

Figure 2.6 shows the record high and low temperatures in the United States for the first four months of the year. We can read from the graph that the record low temperature for April is $-36°F$. This is four times the record low temperature for September. What is the record low temperature for September?

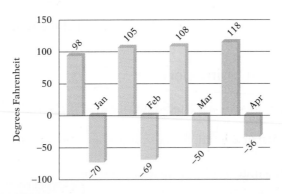

Figure 2.6   Record High and Low Temperatures, in Degrees Fahrenheit, in the United States for January, February, March, and April
Source: National Climatic Data Center, Asheville, NC, and Storm Phillips, STORMFAX, Inc.

To find the record low temperature for September, divide the record low for April (-36) by 4. $-36 \div 4 = -9$

The record low temperature in the United States for the month of September is $-9°F$.

The division properties of zero and one, which were presented in Chapter 1, hold true for integers as well as whole numbers. These properties are repeated here.

Division Properties of Zero and One

If $a \neq 0$, $\dfrac{0}{a} = 0$. | If $a \neq 0$, $\dfrac{a}{a} = 1$.

$\dfrac{a}{1} = a$ | $\dfrac{a}{0}$ is undefined.

Evaluate $a \div (-b)$ for $a = -28$ and $b = -4$.

	$a \div (-b)$
Replace a with -28 and b with -4.	$-28 \div (-(-4))$
Simplify $-(-4)$.	$= -28 \div (4)$
Divide.	$= -7$

Is -4 a solution of the equation $\dfrac{-20}{x} = 5$?

$$\frac{-20}{x} = 5$$

Replace x by -4 and then simplify.

$$\frac{-20}{-4} \;\bigg|\; 5$$

The results are equal.

$$5 = 5$$

Yes, -4 is a solution of the equation.

Point of Interest

Historical manuscripts indicate that mathematics is at least 4000 years old. Yet it was only 400 years ago that mathematicians started using variables to stand for numbers. Before that time, mathematics was written in words.

EXAMPLE 5   Find the quotient of -23 and -23.

Solution $-23 \div (-23) = 1$

YOU TRY IT 5   What is 0 divided by -17?

Your Solution

EXAMPLE 6   Divide: $\dfrac{95}{-5}$

Solution $\dfrac{95}{-5} = -19$

YOU TRY IT 6   Divide: $\dfrac{84}{-6}$

Your Solution

EXAMPLE 7   Divide: $x \div 0$

Solution   Division by zero is not defined.
$x \div 0$ is undefined.

YOU TRY IT 7   Divide: $x \div 1$

Your Solution

Solutions on p. S5

EXAMPLE 8 Evaluate $\dfrac{-a}{b}$ for $a = -6$ and $b = -3$.

Solution $\dfrac{-a}{b}$

$$\dfrac{-(-6)}{-3} = \dfrac{6}{-3} = -2$$

YOU TRY IT 8 Evaluate $\dfrac{a}{-b}$ for $a = -14$ and $b = -7$.

Your Solution

EXAMPLE 9 Is -9 a solution of the equation $-3 = \dfrac{x}{3}$?

Solution $-3 = \dfrac{x}{3}$

$$-3 \,\Big|\, \dfrac{-9}{3}$$

$$-3 = -3$$

Yes, -9 is a solution of the equation.

YOU TRY IT 9 Is -3 a solution of the equation $\dfrac{-6}{y} = -2$?

Your Solution

Solutions on p. S5

OBJECTIVE C Applications

EXAMPLE 10

The daily low temperatures during one week were: $-10°$, $2°$, $-1°$, $-9°$, $1°$, $0°$, and $3°$. Find the average daily low temperature for the week.

Strategy

To find the average daily low temperature:

▸ Add the seven temperature readings.
▸ Divide by 7.

Solution

$-10 + 2 + (-1) + (-9) + 1 + 0 + 3 = -14$

$-14 \div 7 = -2$

The average daily low temperature was $-2°$.

YOU TRY IT 10

The daily high temperatures during one week were: $-7°$, $-8°$, $0°$, $-1°$, $-6°$, $-11°$, and $-2°$. Find the average daily high temperature for the week.

Your Strategy

Your Solution

Solution on p. S5

2.3 EXERCISES

OBJECTIVE A **Multiplication of integers**

1. Name the operation in each expression and explain how you determined that it was that operation.
    a. $8(-7)$ **b.** $8 - 7$ **c.** $8 - (-7)$ **d.** $-xy$ **e.** $x(-y)$ **f.** $-x - y$

In Exercise 2, circle the correct words to complete each sentence.

2. **a.** In the multiplication problem $15(-3)$, the signs of the factors are the same/different, so the sign of the product will be positive/negative.

    b. In the multiplication problem $-7(-12)$, the signs of the factors are the same/different, so the sign of the product will be positive/negative.

Multiply.

3. $-4 \cdot 6$	**4.** $-7 \cdot 3$	**5.** $-2(-3)$	**6.** $-5(-1)$
7. $(9)(2)$	**8.** $(3)(8)$	**9.** $5(-4)$	**10.** $4(-7)$
11. $-8(2)$	**12.** $-9(3)$	**13.** $(-5)(-5)$	**14.** $(-3)(-6)$
15. $(-7)(0)$	**16.** $-11(1)$	**17.** $14(3)$	**18.** $62(9)$
19. $-32(4)$	**20.** $-24(3)$	**21.** $(-8)(-26)$	**22.** $(-4)(-35)$
23. $9(-27)$	**24.** $8(-40)$	**25.** $-5 \cdot (23)$	**26.** $-6 \cdot (38)$
27. $-7(-34)$	**28.** $-4(-51)$	**29.** $4 \cdot (-8) \cdot 3$	**30.** $5 \cdot 7 \cdot (-2)$
31. $(-6)(5)(7)$	**32.** $(-9)(-9)(2)$	**33.** $-8(-7)(-4)$	**34.** $-1(4)(-9)$

35. What is twice -20?

36. Find the product of 100 and -7.

37. What is -30 multiplied by -6?

38. What is -9 times -40?

39. Write the product of $-q$ and r.

40. Write the product of $-f$, g, and h.

41. *Business* The table at the right shows the net income for the first quarter of 2006 for three companies in the recreational vehicles sector. (*Note:* Negative net income indicates a loss.) If net income continued at the same level throughout 2006, what would be the 2006 annual net income for **a.** Arctic Cat, Inc., **b.** Coach Industries Group, and **c.** National RV Holdings?

Company	Net Income 1st Quarter of 2006
Arctic Cat	–592,000
Coach Industries Group	–385,000
National RV Holdings	–2,054,000

Source: **finance.yahoo.com**

Identify the property that justifies the statement.

42. $0(-7) = 0$

43. $1p = p$

44. $-8(-5) = -5(-8)$

45. $-3(9 \cdot 4) = (-3 \cdot 9)4$

Use the given property of multiplication to complete the statement.

46. The Commutative Property of Multiplication
$-3(-9) = -9(?)$

47. The Associative Property of Multiplication
$?(5 \cdot 10) = (-6 \cdot 5)10$

48. The Multiplication Property of Zero
$-81 \cdot ? = 0$

49. The Multiplication Property of One
$?(-14) = -14$

Evaluate the expression for the given values of the variables.

50. xy, for $x = -3$ and $y = -8$

51. $-xy$, for $x = -3$ and $y = -8$

52. $x(-y)$, for $x = -3$ and $y = -8$

53. $-xyz$, for $x = -6$, $y = 2$, and $z = -5$

54. $-8a$, for $a = -24$

55. $-7n$, for $n = -51$

56. $5xy$, for $x = -9$ and $y = -2$

57. $8ab$, for $a = 7$ and $b = -1$

58. $-4cd$, for $c = 25$ and $d = -8$

59. $-5st$, for $s = -40$ and $t = -8$

60. Is -4 a solution of the equation $6m = -24$?

61. Is -3 a solution of the equation $-5x = -15$?

62. Is -6 a solution of the equation $48 = -8y$?

63. Is 0 a solution of the equation $-8 = -8a$?

64. Is 7 a solution of the equation $-3c = 21$?

65. Is 9 a solution of the equation $-27 = -3c$?

66. Will the product of three negative numbers be positive or negative?

67. Will the product of three positive numbers and two negative numbers be positive or negative?

OBJECTIVE B **Division of integers**

68. The fraction that represents the quotient -63 and 9 is _____.

69. Circle the correct words to complete the sentence: The signs of the numbers in the division problem $28 \div (-4)$ are the same/different, so the sign of the quotient will be positive/negative.

Divide.

70. $12 \div (-6)$

71. $18 \div (-3)$

72. $(-72) \div (-9)$

73. $(-64) \div (-8)$

74. $0 \div (-6)$

75. $-49 \div 1$

76. $81 \div (-9)$

77. $-40 \div (-5)$

78. $\dfrac{72}{-3}$

79. $\dfrac{44}{-4}$

80. $\dfrac{-93}{-3}$

81. $\dfrac{-98}{-7}$

82. $-114 \div (-6)$

83. $-91 \div (-7)$

84. $-53 \div 0$

85. $(-162) \div (-162)$

86. $-128 \div 4$ **87.** $-130 \div (-5)$ **88.** $(-200) \div 8$ **89.** $(-92) \div (-4)$

90. Find the quotient of -700 and 70.

91. Find 550 divided by -5.

92. What is -670 divided by -10?

93. What is the quotient of -333 and -3?

94. Write the quotient of $-a$ and b.

95. Write -9 divided by x.

Business The figure at the right shows the net income for the first quarter of 2006 for three airlines. (*Note:* Negative income indicates a loss. One quarter of the year is three months.) Use this figure for Exercises 96 and 97.

96. For the quarter shown, what was the average monthly net income for Continental Airlines?

97. For the quarter shown, what was the average monthly net income for Northwest Airlines?

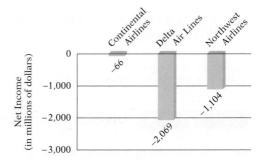

Net Income for First Quarter of 2006
Source: **finance.yahoo.com**

Evaluate the expression for the given values of the variables.

98. $a \div b$, for $a = -36$ and $b = -4$

99. $-a \div b$, for $a = -36$ and $b = -4$

100. $a \div (-b)$, for $a = -36$ and $b = -4$

101. $(-a) \div (-b)$, for $a = -36$ and $b = -4$

102. $\dfrac{x}{y}$, for $x = -42$ and $y = -7$

103. $\dfrac{-x}{y}$, for $x = -42$ and $y = -7$

104. $\dfrac{x}{-y}$, for $x = -42$ and $y = -7$

105. $\dfrac{-x}{-y}$, for $x = -42$ and $y = -7$

106. Is 20 a solution of the equation $\dfrac{m}{-2} = -10$?

107. Is 18 a solution of the equation $6 = \dfrac{-c}{-3}$?

108. Is 0 a solution of the equation $0 = \dfrac{a}{-4}$?

109. Is -3 a solution of the equation $\dfrac{21}{n} = 7$?

110. Is -6 a solution of the equation $\dfrac{x}{2} = \dfrac{-18}{x}$?

111. Is 8 a solution of the equation $\dfrac{m}{-4} = \dfrac{-16}{m}$?

 For Exercises 112 to 115, state whether the expression is equivalent to $\dfrac{a}{b}$ or $-\dfrac{a}{b}$. Assume a and b are nonzero integers.

112. $a \div (-b)$

113. $-\dfrac{-a}{b}$

114. $(-a) \div (-b)$

115. $-\dfrac{-a}{-b}$

OBJECTIVE C Applications

116. To find the average of eight numbers, find the _____ of the numbers and divide the result by _____.

117. To find the average of the numbers 8, -5, 22, 13, and -42, find the _____ of the numbers and divide the result by _____.

118. *Sports* The combined scores of the top five golfers in a tournament equaled -10 (10 under par). What was the average score of the five golfers?

119. *Sports* The combined scores of the top four golfers in a tournament equaled -12 (12 under par). What was the average score of the four golfers?

120. *Temperature* The daily high temperatures during one week were $-6°$, $-11°$, $1°$, $5°$, $-3°$, $-9°$, and $-5°$. Find the average daily high temperature for the week.

121. *Temperature* The daily low temperatures during one week were $4°$, $-5°$, $8°$, $-1°$, $-12°$, $-14°$, and $-8°$. Find the average daily low temperature for the week.

The following figure shows the record low temperatures, in degrees Fahrenheit, in the United States for each month. Use this figure for Exercises 122 to 124.

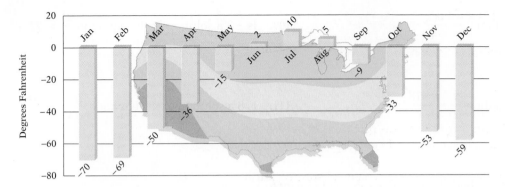

Record Low Temperatures, in Degrees Fahrenheit, in the United States
Source: National Climatic Data Center, Asheville, NC, and Storm Phillips, STORMFAX, Inc.

122.   *Temperature*   What is the average record low temperature for July, August, and September?

123.   *Temperature*   What is the average record low temperature for the first three months of the year?

124.   *Temperature*   What is the average record low temperature for the three months with the lowest record low temperatures?

Mathematics   A geometric sequence is a list of numbers in which each number after the first is found by multiplying the preceding number in the list by the same number. For example, in the sequence 1, 3, 9, 27, 81, ... , each number after the first is found by multiplying the preceding number in the list by 3. To find the multiplier in a geometric sequence, divide the second number in the sequence by the first number; for the example above, 3 ÷ 1 = 3.

125.   Find the next three numbers in the geometric sequence −5, 15, −45, ...

126.   Find the next three numbers in the geometric sequence 2, −4, 8, ...

127.   Find the next three numbers in the geometric sequence −3, −12, −48, ...

128.   Find the next three numbers in the geometric sequence −1, −5, −25, ...

CRITICAL THINKING

129.   *Mathematics*   **a.** Find the largest possible product of two negative integers whose sum is −18. **b.** Find the smallest possible sum of two negative integers whose product is 16.

130.   Use repeated addition to show that the product of two integers with different signs is a negative number.

2.4 Exercises

OBJECTIVE A **Solving equations**

1. To solve $-13 = x - 3$, add _____ to each side of the equation. The solution is _____.

2. To solve $-14 = 7y$, divide each side of the equation by _____. The solution is _____.

Solve.

3. $x - 6 = 9$ **4.** $m - 4 = 6$ **5.** $8 = y - 3$ **6.** $12 = t - 4$

7. $x - 5 = -12$ **8.** $n - 7 = -21$ **9.** $-10 = z + 6$ **10.** $-21 = c + 4$

11. $x + 12 = 4$ **12.** $y + 7 = 2$ **13.** $-12 = c - 12$ **14.** $n - 9 = -9$

15. $6 + x = 4$ **16.** $12 + y = 7$ **17.** $12 = n - 8$ **18.** $19 = b - 23$

19. $3m = -15$ **20.** $6p = -54$ **21.** $-10 = 5v$ **22.** $-20 = 2z$

23. $-8x = -40$ **24.** $-4y = -28$ **25.** $-60 = -6v$ **26.** $3x = -39$

27. $5x = -100$ **28.** $-4n = 0$ **29.** $4x = 0$ **30.** $-15 = -15z$

OBJECTIVE B **Applications and formulas**

For Exercises 1 and 2, translate each sentence into an equation. Use n to represent the unknown number.

31. Negative six is the sum of a number and twelve.

32. Negative two times some number equals ten.
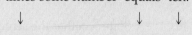

33. Ten less than a number is fifteen. Find the number.

34. The difference between a number and five is twenty-two. Find the number.

35. Zero is equal to fifteen more than some number. Find the number.

36. Twenty equals the sum of a number and thirty-one. Find the number.

37. Sixteen equals negative two times a number. Find the number.

38. The product of negative six and a number is negative forty-two. Find the number.

Economics Use the table at the right for Exercises 39 and 40.

39. The U.S. balance of trade in 2000 was $43,508 million more than the U.S. balance of trade in 2002. What was the U.S. balance of trade in 2002?

40. The U.S. balance of trade in 1995 was $266,411 million more than the U.S. balance of trade in 2001. What was the U.S. balance of trade in 2001?

Year	U.S. Balance of Trade (in millions of dollars)
1985	−121,880
1990	−80,864
1995	−96,384
2000	−377,559
2005	−716,730

Source: U.S. Census Bureau, Foreign Trade Division

41. *Temperature* The temperature now is 5° higher than it was this morning. The temperature now is 8°C. What was the temperature this morning?

42. *Business* A car dealer wants to make a profit of $925 on the sale of a car that cost the dealer $12,600. Use the equation $P = S - C$, where P is the profit on an item, S is the selling price, and C is the cost, to find the selling price of the car.

43. *Business* An office supplier wants to make a profit of $95 on the sale of a software package that cost the supplier $385. Use the equation $P = S - C$, where P is the profit on an item, S is the selling price, and C is the cost, to find the selling price of the software.

44. *Business* The net worth of a business is given by the formula $N = A - L$, where N is the net worth, A is the assets of the business (or the amount owned), and L is the liabilities of the business (or the amount owed). Use this formula to find the assets of a business that has a net worth of $11 million and liabilities of $4 million.

45. *Business* The net worth of ABL Electronics is $43 million and it has liabilities of $14 million. Use the net worth formula $N = A - L$, where N is the net worth, A is the assets of the business (or the amount owned), and L is the liabilities of the business (or the amount owed), to find the assets of ABL Electronics.

CRITICAL THINKING

46. State whether the sentence is true or false. Explain your answer.
   a. Zero cannot be the solution of an equation.
   b. If an equation contains a negative number, then the solution of the equation is a negative number.

47. Find the value of $3y - 8$ given that $-3y = -36$.

2.5 The Order of Operations Agreement

OBJECTIVE A The Order of Operations Agreement

The Order of Operations Agreement, used in Chapter 1, is repeated here for your reference.

> ### The Order of Operations Agreement
>
> **Step 1** Do all operations inside parentheses.
>
> **Step 2** Simplify any numerical expressions containing exponents.
>
> **Step 3** Do multiplication and division as they occur from left to right.
>
> **Step 4** Do addition and subtraction as they occur from left to right.

Note how the following expressions containing exponents are simplified.

$(-3)^2 = (-3)(-3) = 9$ The (-3) is squared. Multiply -3 by -3.

$-(3)^2 = -(3 \cdot 3) = -9$ Read $-(3^2)$ as "the opposite of three squared." 3^2 is 9. The opposite of 9 is -9.

$-3^2 = -(3^2) = -9$ The expression -3^2 is the same as $-(3^2)$.

> ### Take Note
>
> The -3 is squared only when the negative sign is *inside* the parentheses. In $(-3)^2$, we are squaring -3; in -3^2, we are finding the opposite of 3^2.

Simplify: $8 - 4 \div (-2)$

There are no operations inside parentheses (Step 1).

There are no exponents (Step 2).

Do the division (Step 3). $8 - 4 \div (-2) = 8 - (-2)$

Do the subtraction (Step 4). $= 8 + 2 = 10$

> ### Calculator Note
>
> As shown above and at the left, the value of -3^2 is different from the value of $(-3)^2$. The keystrokes to evaluate each of these on your calculator are different.
>
> To evaluate -3^2, enter
>
> 3 $\boxed{x^2}$ $\boxed{+/-}$
>
> To evaluate $(-3)^2$, enter
>
> 3 $\boxed{+/-}$ $\boxed{x^2}$

Simplify: $(-3)^2 - 2(8 - 3) + (-5)$

Perform operations inside parentheses.

$(-3)^2 - 2(8 - 3) + (-5)$
$= (-3)^2 - 2(5) + (-5)$

Simplify expressions with exponents.

$= 9 - 2(5) + (-5)$

Do multiplication and division as they occur from left to right.

$= 9 - 10 + (-5)$

Do addition and subtraction as they occur from left to right.

$= 9 + (-10) + (-5)$
$= -1 + (-5)$
$= -6$

Evaluate $ab - b^2$ for $a = 2$ and $b = -6$.

$ab - b^2$

Replace a with 2 and each b with -6.

$2(-6) - (-6)^2$

Use the Order of Operations Agreement to simplify the resulting numerical expression. Simplify the exponential expression.

$= 2(-6) - 36$

Do the multiplication.

$= -12 - 36$

Do the subtraction.

$= -12 + (-36)$

$= -48$

EXAMPLE 1 Simplify $(-4)^2$ and -4^2.

Solution $(-4)^2 = (-4)(-4) = 16$
$-4^2 = -(4 \cdot 4) = -16$

YOU TRY IT 1 Simplify $(-5)^2$ and -5^2.

Your Solution

EXAMPLE 2 Simplify: $12 \div (-2)^2 - 5$

Solution $12 \div (-2)^2 - 5 = 12 \div 4 - 5$
$= 3 - 5$
$= 3 + (-5)$
$= -2$

YOU TRY IT 2 Simplify: $8 \div 4 \cdot 4 - (-2)^2$

Your Solution

EXAMPLE 3 Simplify:
$(-3)^2(5 - 7)^2 - (-9) \div 3$

Solution $(-3)^2(5 - 7)^2 - (-9) \div 3$
$= (-3)^2(-2)^2 - (-9) \div 3$
$= (9)(4) - (-9) \div 3$
$= 36 - (-9) \div 3$
$= 36 - (-3)$
$= 36 + 3$
$= 39$

YOU TRY IT 3 Simplify:
$(-2)^2(3 - 7)^2 - (-16) \div (-4)$

Your Solution

EXAMPLE 4 Evaluate $6a \div (-b)$ for $a = -2$ and $b = -3$.

Solution $6a \div (-b)$
$6(-2) \div (-(-3))$
$= 6(-2) \div (3)$
$= -12 \div 3$
$= -4$

YOU TRY IT 4 Evaluate $3a - 4b$ for $a = -2$ and $b = 5$.

Your Solution

Solutions on p. S6

2.5 Exercises

OBJECTIVE A   **The Order of Operations Agreement**

1. To simplify the expression $6 - 4 \div (-2)$, the first operation to perform is _____ .

2. Simplify $(-7)^2 - 5(2 - 3)$

   a. Perform operations in parentheses.

   b. Simplify expressions with exponents.

   c. Multiply.

   d. Rewrite subtraction as addition of the opposite.

   e. Add.

   $(-7)^2 - 5(2 - 3)$

   $= (-7)^2 - 5(\_\_\_\_\_)$

   $= \_\_\_\_\_ - 5(-1)$

   $= 49 - \_\_\_\_\_$

   $= 49 + \_\_\_\_\_$

   $= \_\_\_\_\_$

Simplify.

3. $3 - 12 \div 2$

4. $-16 \div 2 + 8$

5. $2(3 - 5) - 2$

6. $2 - (8 - 10) \div 2$

7. $4 - (-3)^2$

8. $-2^2 - 6$

9. $4 \cdot (2 - 4) - 4$

10. $6 - 2 \cdot (1 - 3)$

11. $4 - (-2)^2 + (-3)$

12. $-3 + (-6)^2 - 1$

13. $3^3 - 4(2)$

14. $9 \div 3 - (-3)^2$

15. $3 \cdot (6 - 2) \div 6$

16. $4 \cdot (2 - 7) \div 5$

17. $2^3 - (-3)^2 + 2$

18. $6(8 - 2) \div 4$

19. $6 - 2(1 - 5)$

20. $(-2)^2 - (-3)^2 + 1$

21. $6 - (-4)(-3)^2$

22. $4 - (-5)(-2)^2$

23. $4 \cdot 2 - 3 \cdot 7$

24. $16 \div 2 - 9 \div 3$

25. $-2^2 - 5(3) - 1$

26. $4 - 2 \cdot 7 - 3^2$

27. $3 \cdot 2^3 + 5 \cdot (3 + 2) - 17$

28. $3 \cdot 4^2 - 16 - 4 + 3 - (1 - 2)^2$

29. $-12(6 - 8) + 1^3 \cdot 3^2 \cdot 2 - 6(2)$

30. $-3 \cdot (-2)^2 \cdot 4 \div 8 - (-12)$

31. $-27 - (-3)^2 - 2 - 7 + 6 \cdot 3$

32. $(-1) \cdot (4 - 7)^2 \div 9 + 6 - 3 - 4(2)$

33. $16 - 4 \cdot 8 + 4^2 - (-18) - (-9)$

34. $(-3)^2 \cdot (5 - 7)^2 - (-9) \div 3$

Evaluate the variable expression for $a = -2$, $b = 4$, $c = -1$, and $d = 3$.

35. $3a + 2b$

36. $a - 2c$

37. $16 \div (ac)$

38. $6b \div (-a)$

39. $bc \div (2a)$

40. $a^2 - b^2$

41. $b^2 - c^2$

42. $2a - (c + a)^2$

43. $(b - a)^2 + 4c$

44. $\dfrac{b + c}{d}$

45. $\dfrac{d - b}{c}$

46. $\dfrac{2d + b}{-a}$

47. $\dfrac{b - d}{c - a}$

48. $(d - a)^2 \div 5$

49. $(d - a)^2 - 3c$

50. $(b + d)^2 - 4a$

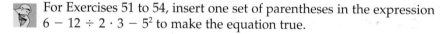 For Exercises 51 to 54, insert one set of parentheses in the expression $6 - 12 \div 2 \cdot 3 - 5^2$ to make the equation true.

51. $6 - 12 \div 2 \cdot 3 - 5^2 = -34$

52. $6 - 12 \div 2 \cdot 3 - 5^2 = -18$

53. $6 - 12 \div 2 \cdot 3 - 5^2 = -21$

54. $6 - 12 \div 2 \cdot 3 - 5^2 = -37$

CRITICAL THINKING

55. What is the smallest integer greater than $-2^2 - (-3)^2 + 5(4) \div 10 - (-6)$?

56. **a.** Is -4 a solution of the equation $x^2 - 2x - 8 = 0$?
  b. Is -3 a solution of the equation $x^3 + 3x^2 - 5x - 15 = 0$?

57. Evaluate $a \div bc$ and $a \div (bc)$ for $a = 16$, $b = 2$, and $c = -4$. Explain why the answers are not the same.

Focus on **Problem Solving**

Drawing Diagrams

How do you best remember something? Do you remember best what you hear? The word *aural* means "pertaining to the ear"; people with a strong aural memory remember best those things that they hear. The word *visual* means "pertaining to the sense of sight"; people with a strong visual memory remember best that which they see written down. Some people claim that their memory is in their writing hand—they remember something only if they write it down! The method by which you best remember something is probably also the method by which you can best learn something new.

In problem-solving situations, try to capitalize on your strengths. If you tend to understand the material better when you hear it spoken, read application problems aloud or have someone else read them to you. If writing helps you to organize ideas, rewrite application problems in your own words.

No matter what your main strength, visualizing a problem can be a valuable aid in problem solving. A drawing, sketch, diagram, or chart can be a useful tool in problem solving, just as calculators and computers are tools. A diagram can be helpful in gaining an understanding of the relationships inherent in a problem-solving situation. A sketch will help you to organize the given information and can lead to your being able to focus on the method by which the solution can be determined.

A tour bus drives 5 mi south, then 4 mi west, then 3 mi north, then 4 mi east. How far is the tour bus from the starting point?

Draw a diagram of the given information.

From the diagram, we can see that the solution can be determined by subtracting 3 from 5: $5 - 3 = 2$.

The bus is 2 mi from the starting point.

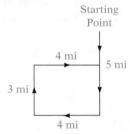

If you roll two ordinary six-sided dice and multiply the two numbers that appear on top, how many different possible products are there?

Make a chart of the possible products. In the chart below, repeated products are marked with an asterisk.

$1 \cdot 1 = 1$	$2 \cdot 1 = 2$ (*)	$3 \cdot 1 = 3$ (*)	$4 \cdot 1 = 4$ (*)	$5 \cdot 1 = 5$ (*)	$6 \cdot 1 = 6$ (*)
$1 \cdot 2 = 2$	$2 \cdot 2 = 4$ (*)	$3 \cdot 2 = 6$ (*)	$4 \cdot 2 = 8$ (*)	$5 \cdot 2 = 10$ (*)	$6 \cdot 2 = 12$ (*)
$1 \cdot 3 = 3$	$2 \cdot 3 = 6$ (*)	$3 \cdot 3 = 9$	$4 \cdot 3 = 12$ (*)	$5 \cdot 3 = 15$ (*)	$6 \cdot 3 = 18$ (*)
$1 \cdot 4 = 4$	$2 \cdot 4 = 8$	$3 \cdot 4 = 12$ (*)	$4 \cdot 4 = 16$	$5 \cdot 4 = 20$ (*)	$6 \cdot 4 = 24$ (*)
$1 \cdot 5 = 5$	$2 \cdot 5 = 10$	$3 \cdot 5 = 15$	$4 \cdot 5 = 20$	$5 \cdot 5 = 25$	$6 \cdot 5 = 30$ (*)
$1 \cdot 6 = 6$	$2 \cdot 6 = 12$	$3 \cdot 6 = 18$	$4 \cdot 6 = 24$	$5 \cdot 6 = 30$	$6 \cdot 6 = 36$

By counting the products that are not repeats, we can see that there are 18 different possible products.

Look at Sections 1 and 2 in this chapter. You will notice that number lines are used to help you visualize the integers, as an aid in ordering integers, to help you understand the concepts of opposite and absolute value, and to illustrate addition of integers. As you begin your work with integers, you may find that sketching a number line proves helpful in coming to understand a problem or in working through a calculation that involves integers.

Projects & Group Activities

Multiplication of Integers

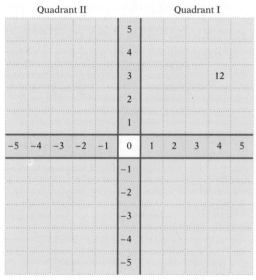

Quadrant II   Quadrant I

Quadrant III   Quadrant IV

The grid at the left has four regions, or quadrants, numbered counterclockwise, starting at the upper right, with the Roman numerals I, II, III, IV.

1. Complete Quadrant I by multiplying each of the horizontal numbers 1 through 5 by each of the vertical numbers 1 through 5. The product 4(3) has been filled in for you. Complete Quadrants II, III, and IV by again multiplying each horizontal number by each vertical number.

2. What is the sign of all the products in Quadrant I? Quadrant II? Quadrant III? Quadrant IV?

3. Describe at least three patterns that you observe in the completed grid.

4. How does the grid show that multiplication of integers is commutative?

5. How can you use the grid to find the quotient of two integers? Provide at least two examples of division of integers.

Closure

The whole numbers are said to be *closed* with respect to addition because when two whole numbers are added, the result is a whole number. The whole numbers are not closed with respect to subtraction because, for example, 4 and 7 are whole numbers, but $4 - 7 = -3$ and -3 is not a whole number. Complete the table below by entering a Y if the operation is closed with respect to those numbers and an N if it is not closed. When we discuss whether multiplication and division are closed, zero is not included because division by zero is not defined.

	Addition	*Subtraction*	*Multiplication*	*Division*
Whole numbers	Y	N		
Integers				

Chapter 2 Summary

Key Words	Examples
A number n is a **positive number** if $n > 0$. A number n is a **negative number** if $n < 0$. [2.1A, p. 89]	Positive numbers are numbers greater than zero. 9, 87, and 603 are positive numbers. Negative numbers are numbers less than zero. -5, -41, and -729 are negative numbers.
The **integers** are . . . $-4, -3, -2, -1, 0, 1, 2, 3, 4, \ldots$. The integers can be defined as the whole numbers and their opposites. **Positive integers** are to the right of zero on the number line. **Negative integers** are to the left of zero on the number line. [2.1A, p. 89]	$-729, -41, -5, 9, 87$, and 603 are integers. 0 is an integer, but it is neither a positive nor a negative integer.
Opposite numbers are two numbers that are the same distance from zero on the number line but on opposite sides of zero. The opposite of a number is called its **additive inverse**. [2.1B, p. 91; 2.2A, p. 103]	8 is the opposite, or additive inverse, of -8. -2 is the opposite, or additive inverse, of 2.
The **absolute value** of a number is the distance from zero to the number on the number line. The absolute value of a number is a positive number or zero. The symbol for absolute value is "$\| \|$". [2.1C, p. 92]	$\|9\| = 9$ $\|-9\| = 9$ $-\|9\| = -9$

Essential Rules and Procedures

To add integers with the same sign, add the absolute values of the numbers. Then attach the sign of the addends. [2.2A, p. 102]	$6 + 4 = 10$ $-6 + (-4) = -10$
To add integers with different signs, find the absolute values of the numbers. Subtract the lesser absolute value from the greater absolute value. Then attach the sign of the addend with the greater absolute value. [2.2A, p. 102]	$-6 + 4 = -2$ $6 + (-4) = 2$
To subtract two integers, add the opposite of the second integer to the first integer. [2.2B, p. 106]	$6 - 4 = 6 + (-4) = 2$ $6 - (-4) = 6 + 4 = 10$ $-6 - 4 = -6 + (-4) = -10$ $-6 - (-4) = -6 + 4 = -2$
To multiply integers with the same sign, multiply the absolute values of the factors. The product is positive. [2.3A, p. 117]	$3 \cdot 5 = 15$ $-3(-5) = 15$

To multiply integers with different signs, multiply the absolute values of the factors. The product is negative. [2.3A, p. 117]	$-3(5) = -15$ $3(-5) = -15$
To divide two numbers with the same sign, divide the absolute values of the numbers. The quotient is positive. [2.3B, p. 120]	$15 \div 3 = 5$ $(-15) \div (-3) = 5$
To divide two numbers with different signs, divide the absolute values of the numbers. The quotient is negative. [2.3B, p. 120]	$-15 \div 3 = -5$ $15 \div (-3) = -5$

Order Relations $a > b$ if a is to the right of b on the number line.
  $a < b$ if a is to the left of b on the number line.
  [2.1A, p. 90]
	$-6 > -12$
	$-8 < 4$

Properties of Addition [2.2A, p. 103]

Addition Property of Zero $a + 0 = a$ or $0 + a = a$	$-6 + 0 = -6$
Commutative Property of Addition $a + b = b + a$	$-8 + 4 = 4 + (-8)$
Associative Property of Addition $(a + b) + c = a + (b + c)$	$(-5 + 4) + 6 = -5 + (4 + 6)$
Inverse Property of Addition $a + (-a) = 0$ or $-a + a = 0$	$7 + (-7) = 0$

Properties of Multiplication [2.3A, p. 118]

Multiplication Property of Zero $a \cdot 0 = 0$ or $0 \cdot a = 0$	$-9(0) = 0$
Multiplication Property of One $a \cdot 1 = a$ or $1 \cdot a = a$	$-3(1) = -3$
Commutative Property of Multiplication $a \cdot b = b \cdot a$	$-2(6) = 6(-2)$
Associative Property of Multiplication $(a \cdot b) \cdot c = a \cdot (b \cdot c)$	$(-2 \cdot 4) \cdot 5 = -2 \cdot (4 \cdot 5)$

Division Properties of Zero and One [2.3B, p. 121]

If $a \neq 0, 0 \div a = 0$.	$0 \div (-5) = 0$
If $a \neq 0, a \div a = 1$.	$-5 \div (-5) = 1$
$a \div 1 = a$	$-5 \div 1 = -5$
$a \div 0$ is undefined.	$-5 \div 0$ is undefined.

Addition Property of Equations [2.4A, p. 129]

The same number can be added to each side of an equation without changing the solution of the equation.	$x - 4 = 12$ $x - 4 + 4 = 12 + 4$ $x = 16$

The Order of Operations Agreement [2.5A, p. 135]

Step 1 Do all operations inside parentheses.	$(-4)^2 - 3(1 - 5) = (-4)^2 - 3(-4)$
Step 2 Simplify any numerical expressions containing exponents.	$= 16 - 3(-4)$
Step 3 Do multiplication and division as they occur from left to right.	$= 16 - (-12)$
Step 4 Do addition and subtraction as they occur from left to right.	$= 16 + 12$
	$= 28$

Chapter 2 Review Exercises

1. Write the expression $8 - (-1)$ in words.

2. Evaluate $-|-36|$.

3. Find the product of -40 and -5.

4. Evaluate $-a \div b$ for $a = -27$ and $b = -3$.

5. Add: $-28 + 14$

6. Simplify: $-(-13)$

7. Graph -2 on the number line.

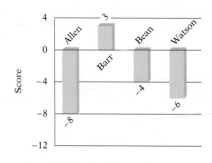

8. Solve: $-24 = -6y$

9. Divide: $-51 \div (-3)$

10. Find the quotient of 840 and -4.

11. Subtract: $-6 - (-7) - 15 - (-12)$

12. Evaluate $-ab$ for $a = -2$ and $b = -9$.

13. Find the sum of 18, -13, and -6.

14. Multiply: $-18(4)$

15. Simplify: $(-2)^2 - (-3)^2 \div (1 - 4)^2 \cdot 2 - 6$

16. Evaluate $-x - y$ for $x = -1$ and $y = 3$.

17. *Sports* The scores of four golfers after the final round of the 2006 U.S. Senior Open are shown in the figure at the right. What is the difference between Barr's score and Allen's score?

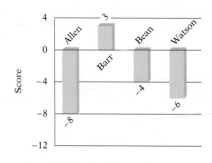

Golfers' Scores in 2006 U.S. Senior Open

18. Find the difference between -15 and -28.

19. Identify the property that justifies the statement. $-11(-50) = -50(-11)$

20. Is -9 a solution of $-6 - t = 3$?

21. Simplify: $-9 + 16 - (-7)$

22. Divide: $\dfrac{0}{-17}$

23. Multiply: $-5(2)(-6)(-1)$

24. Add: $3 + (-9) + 4 + (-10)$

25. Evaluate $(a - b)^2 - 2a$ for $a = -2$ and $b = -3$.

26. Place the correct symbol, $<$ or $>$, between the two numbers.

$-8 \quad -10$

27. Complete the statement by using the Inverse Property of Addition.

$-21 + ? = 0$

28. Find the absolute value of -27.

29. Forty-eight is the product of negative six and some number. Find the number.

30. *Temperature* Which is colder, a temperature of $-4°C$ or a temperature of $-12°C$?

31. *Chemistry* The figure at the right shows the boiling points in degrees Celsius of three chemical elements. The boiling point of neon is 7 times the highest boiling point shown in the table. What is the boiling point of neon?

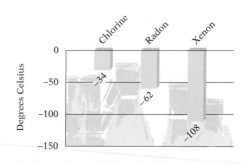

Boiling Points of Chemical Elements

32. *Temperature* Find the temperature after an increase of $5°C$ from $-8°C$.

33. *Mathematics* The distance, d, between point a and point b on the number line is given by the formula $d = |a - b|$. Find d for $a = 7$ and $b = -5$.

Chapter 2 Test

1. Write the expression $-3 + (-5)$ in words.

2. Evaluate $-|-34|$.

3. What is 3 minus -15?

4. Evaluate $a + b$ for $a = -11$ and $b = -9$.

5. Evaluate $(-x)(-y)$ for $x = -4$ and $y = -6$.

6. Identify the property that justifies the statement.
   $-23 + 4 = 4 + (-23)$

7. What is -360 divided by -30?

8. Find the sum of -3, -6, and 11.

9. Place the correct symbol between the two numbers.
   16 $>$ -19

10. Subtract: $7 + (-3) - 12$

11. Evaluate $a - b - c$ for $a = 6$, $b = -2$, and $c = 11$.

12. Simplify: $-(-49)$

13. Find the product of 50 and -5.

14. Write the given numbers in order from smallest to largest.
    $-|5|, -(-11), |-9|, -(3)$

15. Is -9 a solution of the equation $17 - x = 8$?

16. On the number line, which number is 2 units to the right of -5?

17. *Sports* The scores of four golfers after the final round of the 2006 Masters are shown in the figure at the right. What is the difference between Mickelson's score and Allenby's score?

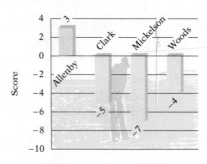

Golfers' Scores in 2006 Masters

18. Divide: $\dfrac{0}{-16}$

19. Evaluate $2bc - (c + a)^3$ for $a = -2$, $b = 4$, and $c = -1$.

20. Find the opposite of 25.

21. Solve: $c - 11 = 5$

22. Subtract: $0 - 11$

23. Divide: $-96 \div (-4)$

24. Simplify: $16 \div 4 - 12 \div (-2)$

25. Evaluate $\dfrac{-x}{y}$ for $x = -56$ and $y = -8$.

26. Evaluate $3xy$ for $x = -2$ and $y = -10$.

27. Solve: $-11w = 121$

28. What is 14 less than 4?

29. *Temperature*   Find the temperature after an increase of 11°C from −6°C.

30.   *Environmental Science*   The wind chill factor when the temperature is −25°F and the wind is blowing at 40 mph is four times the wind chill factor when the temperature is −5°F and the wind is blowing at 5 mph. If the wind chill factor at −5°F with a 5-mph wind is −16°F, what is the wind chill factor at −25°F with a 40-mph wind?

31. *Temperature*   The high temperature today is 8° lower than the high temperature yesterday. The high temperature today is −13°C. What was the high temperature yesterday?

32. *Mathematics*   The distance, d, between point a and b on the number line is given by the formula $d = |a - b|$. Find d for $a = 4$ and $b = -12$.

33. *Business*   The net worth of a business is given by the formula $N = A - L$, where N is the net worth, A is the assets of the business (or the amount owned), and L is the liabilities of the business (or the amount owed). Use this formula to find the assets of a business that has a net worth of $18 million and liabilities of $6 million.

Cumulative Review Exercises

1. Find the difference between -27 and -32.

2. Estimate the product of 439 and 28.

3. Divide: $19{,}254 \div 6$

4. Simplify: $16 \div (3 + 5) \cdot 9 - 2^4$

5. Evaluate $-|-82|$.

6. Write three hundred nine thousand four hundred eighty in standard form.

7. Evaluate $5xy$ for $x = 80$ and $y = 6$.

8. What is -294 divided by -14?

9. Subtract: $-28 - (-17)$

10. Find the sum of -24, 16, and -32.

11. Find all the factors of 44.

12. Evaluate $x^4 y^2$ for $x = 2$ and $y = 11$.

13. Round 629,874 to the nearest thousand.

14. Estimate the sum of 356, 481, 294, and 117.

15. Evaluate $-a - b$ for $a = -4$ and $b = -5$.

16. Find the product of -100 and 25.

17. Find the prime factorization of 69.

18. Solve: $3x = -48$

19. Simplify: $(1 - 5)^2 \div (-6 + 4) + 8(-3)$

20. Evaluate $-c \div d$ for $c = -32$ and $d = -8$.

21. Evaluate $\dfrac{a}{b}$ for $a = 39$ and $b = -13$.

22. Place the correct symbol, $<$ or $>$, between the two numbers.

    $-62 \quad 26$

23. What is −18 multiplied by −7?

24. Solve: 12 + p = 3

25. Write 2 · 2 · 2 · 2 · 2 · 7 · 7 in exponential notation.

26. Evaluate $4a + (a - b)^3$ for a = 5 and b = 2.

27. Add: 5,971 + 482 + 3,609

28. What is 5 less than −21?

29. Estimate the difference between 7,352 and 1,986.

30. Evaluate $3^4 \cdot 5^2$.

31. *History* The land area of the United States prior to the Louisiana Purchase was 891,364 mi². The land area of the Louisiana Purchase, which was purchased from France in 1803, was 831,321 mi². What was the land area of the United States immediately after the Louisiana Purchase?

32. *History* Albert Einstein was born on March 14, 1879. He died on April 18, 1955. How old was Albert Einstein when he died?

33. *Finances* A customer makes a down payment of $3,550 on a car costing $17,750. Find the amount that remains to be paid.

Albert Einstein

34. *Real Estate* A construction company is considering purchasing a 25-acre tract of land on which to build single-family homes. If the price is $3,690 per acre, what is the total cost of the land?

35. *Temperature* Find the temperature after an increase of 7°C from −12°C.

36. *Temperature* Record temperatures, in degrees Fahrenheit, for four states in the United States are shown at the right. **a.** What is the difference between the record high and record low temperatures in Arizona? **b.** For which state is the difference between the record high and record low temperatures greatest?

Record Temperatures (in degrees Fahrenheit)		
State	Lowest	Highest
Alabama	−27	112
Alaska	−80	100
Arizona	−40	128
Arkansas	−29	120

37. *Business* As a sales representative, your goal is to sell $120,000 in merchandise during the year. You sold $28,550 in merchandise during the first quarter of the year, $34,850 during the second quarter, and $31,700 during the third quarter. What must your sales for the fourth quarter be if you are to meet your goal for the year?

38. *Sports* Use the equation S = N − P, where S is a golfer's score relative to par in a tournament, N is the number of strokes made by the golfer, and P is par, to find a golfer's score relative to par when the golfer made 198 strokes and par is 206.

CHAPTER **3**

Fractions

DVD

SSM

Student Website

Need help? For online student resources, visit college.hmco.com/pic/aufmannPA5e.

The flute is a woodwind instrument. It has a cylindrical shape and is approximately 66 centimeters in length. It has a range of about three octaves. Sound is produced by blowing into it, causing air in the tube to vibrate.

A flutist, or any musician, must learn to read music. This involves learning about notes, rests, clefs, measures, and time signatures. The time signature appears as a fraction at the beginning of a piece of music and tells the musician how many beats to play per measure. The **Project on page 225** demonstrates how to interpret the time signature.

For Exercises 1 to 6, add, subtract, multiply, or divide.

1. 4×5

2. $2 \cdot 2 \cdot 2 \cdot 3 \cdot 5$

3. 9×1

4. $-6 + 4$

5. $-10 - 3$

6. $63 \div 30$

7. What is the smallest number into which both 8 and 12 divide evenly?

8. What is the greatest number that divides evenly into both 16 and 20?

9. Simplify: $8 \times 7 + 3$

10. Complete: $8 = ? + 1$

11. Place the correct symbol, $<$ or $>$, between the two numbers.

44   48

GO Figure

Maria and Pedro are siblings. Pedro has as many brothers as sisters. Maria has twice as many brothers as sisters. How many children are in the family?

3.1 Least Common Multiple and Greatest Common Factor

OBJECTIVE A   Least common multiple (LCM)

The **multiples** of a number are the products of that number and the numbers 1, 2, 3, 4, 5, ...

$$4 \cdot 1 = 4$$
$$4 \cdot 2 = 8$$
$$4 \cdot 3 = 12$$
$$4 \cdot 4 = 16$$
$$4 \cdot 5 = 20$$ The multiples of 4 are 4, 8, 12, 16, 20, ...

.
.
.

A number that is a multiple of two or more numbers is a **common multiple** of those numbers.

The multiples of 6 are 6, 12, 18, 24, 30, 36, 42, 48, 54, 60, 66, 72, ...
The multiples of 8 are 8, 16, 24, 32, 40, 48, 56, 64, 72, 80, 88, 96, ...
Some common multiples of 6 and 8 are 24, 48, and 72.

The **least common multiple (LCM)** is the smallest common multiple of two or more numbers.

The least common multiple of 6 and 8 is 24.

Listing the multiples of each number is one way to find the LCM. Another way to find the LCM uses the prime factorization of each number.

To find the LCM of 6 and 8 using prime factorization:

Write the prime factorization of each number and circle the highest power of each prime factor.

$$6 = 2 \cdot ③$$
$$8 = ②^3$$

The LCM is the product of the circled factors.

$$2^3 \cdot 3 = 8 \cdot 3 = 24$$

The LCM of 6 and 8 is 24.

Find the LCM of 32 and 36.

Write the prime factorization of each number and circle the highest power of each prime factor.

$$32 = ②^5$$
$$36 = 2^2 \cdot ③^2$$

The LCM is the product of the circled factors.

$$2^5 \cdot 3^2 = 32 \cdot 9 = 288$$

The LCM of 32 and 36 is 288.

EXAMPLE 1 Find the LCM of 12, 18, and 40.

Solution $12 = 2^2 \cdot 3$
$18 = 2 \cdot \boxed{3^2}$
$40 = \boxed{2^3} \cdot \boxed{5}$

$\text{LCM} = 2^3 \cdot 3^2 \cdot 5$
$= 8 \cdot 9 \cdot 5 = 360$

YOU TRY IT 1 Find the LCM of 16, 24, and 28.

Your Solution

Solution on p. S6

OBJECTIVE B Greatest common factor (GCF)

Recall that a number that divides another number evenly is a **factor** of the number.

18 can be evenly divided by 1, 2, 3, 6, 9, and 18.
1, 2, 3, 6, 9, and 18 are factors of 18.

A number that is a factor of two or more numbers is a **common factor** of those numbers.

The factors of 24 are **1**, **2**, **3**, **4**, **6**, 8, **12**, and 24.
The factors of 36 are **1**, **2**, **3**, **4**, **6**, 9, **12**, 18, and 36.
The common factors of 24 and 36 are **1**, **2**, **3**, **4**, **6**, and **12**.

The **greatest common factor (GCF)** is the largest common factor of two or more numbers.

The greatest common factor of 24 and 36 is **12**.

> **Take Note**
>
> 12 is the GCF of 24 and 36 because 12 is the largest natural number that divides evenly into both 24 and 36.

Listing the factors of each number is one way to find the GCF. Another way to find the GCF uses the prime factorization of each number.

To find the GCF of 24 and 36 using prime factorization:

Write the prime factorization of each number and circle the lowest power of each prime factor that occurs in *both* factorizations.

$24 = 2^3 \cdot \boxed{3}$
$36 = \boxed{2^2} \cdot 3^2$

The GCF is the product of the circled factors.

$2^2 \cdot 3 = 4 \cdot 3 = 12$

Find the GCF of 12 and 30.

Write the prime factorization of each number and circle the lowest power of each prime factor that occurs in both factorizations. The prime factor 5 occurs in the prime factorization of 30 but not in the prime factorization of 12. Since 5 is not a factor in both factorizations, do not circle 5.

$12 = 2^2 \cdot \boxed{3}$
$30 = \boxed{2} \cdot 3 \cdot 5$

The GCF is the product of the circled factors.

$2 \cdot 3 = 6$

The GCF of 12 and 30 is 6.

EXAMPLE 2   Find the GCF of 14 and 27.

Solution $14 = 2 \cdot 7$

$27 = 3^3$

No common prime factor occurs in the factorizations.

GCF = 1

YOU TRY IT 2   Find the GCF of 25 and 52.

Your Solution

EXAMPLE 3   Find the GCF of 16, 20, and 28.

Solution $16 = 2^4$

$20 = \boxed{2^2} \cdot 5$

$28 = 2^2 \cdot 7$

GCF = 2^2 = 4

YOU TRY IT 3   Find the GCF of 32, 40, and 56.

Your Solution

Solutions on p. S6

OBJECTIVE C   **Applications**

EXAMPLE 4

Each month, copies of a national magazine are delivered to three different stores that have ordered 50, 75, and 125 copies, respectively. How many copies should be packaged together so that no package needs to be opened during delivery?

Strategy

To find the numbers of copies to be packaged together, find the GCF of 50, 75, and 125.

Solution

$50 = 2 \cdot \boxed{5^2}$

$75 = 3 \cdot 5^2$

$125 = 5^3$

GCF = 5^2 = 25

Each package should contain 25 copies of the magazine.

YOU TRY IT 4

A discount catalog offers blank CDs at reduced prices. The customer must order 20, 50, or 100 CDs. How many CDs should be packaged together so that no package needs to be opened when a clerk is filling an order?

Your Strategy

Your Solution

Solution on p. S6

EXAMPLE 5

To accommodate several activity periods and science labs after the lunch period and before the closing homeroom period, a high school wants to have both 25-minute class periods and 40-minute class periods running simultaneously in the afternoon class schedule. There is a 5-minute passing time between each class. How long a period of time must be scheduled if all students are to be in the closing homeroom period at the same time? How many 25-minute classes and 40-minute classes will be scheduled in that amount of time?

YOU TRY IT 5

You and a friend are running laps at the track. You run one lap every 3 min. Your friend runs one lap every 4 min. If you start at the same time from the same place on the track, in how many minutes will both of you be at the starting point again? Will you have passed each other at some other point on the track prior to that time?

Strategy

To find the amount of time to be scheduled:

► Add the passing time (5 min) to the 25-minute class period and to the 40-minute class period to find the length of each period including the passing time.
► Find the LCM of the two time periods found in Step 1.

To find the number of 25-minute and 40-minute classes:

► Divide the LCM by each time period found in Step 1.

Your Strategy

Solution

$25 + 5 = 30$

$40 + 5 = 45$

$30 = ②\cdot 3 \cdot ⑤$

$45 = ③^2 \cdot 5$

$\text{LCM} = 2 \cdot 3^2 \cdot 5 = 90$

A 90-minute time period must be scheduled.

$90 \div 30 = 3$

$90 \div 45 = 2$

There will be three 25-minute class periods and two 40-minute class periods in the 90-minute period.

Your Solution

Solution on p. S6

3.1 Exercises

OBJECTIVE A Least common multiple (LCM)

1. **a.** Circle each number in the list that is a multiple of 6.
   3 6 9 12 18 24 27 30 36 45

   b. Underline each number in the list that is a multiple of 9.

   c. Use the list to identify the least common multiple of 6 and 9: _____.

2. Find the LCM of 18 and 30.

   a. Write the prime factorization of each number.

   $18 = $ _____ $30 = $ _____

   b. Use the highest power of each factor in part (a) to write the prime factorization of the LCM.

   LCM = _____

   c. Find each product of the factors.

   LCM = _____

Find the LCM of the numbers.

3. 4 and 8

4. 3 and 9

5. 2 and 7

6. 5 and 11

7. 6 and 10

8. 8 and 12

9. 9 and 15

10. 14 and 21

11. 12 and 16

12. 8 and 14

13. 4 and 10

14. 9 and 30

15. 14 and 42

16. 16 and 48

17. 24 and 36

18. 16 and 28

19. 30 and 40

20. 45 and 60

21. 3, 5, and 10

22. 5, 10, and 20

23. 4, 8, and 12

24. 3, 12, and 18

25. 9, 36, and 45

26. 9, 36, and 72

27. 6, 9, and 15

28. 30, 40, and 60

29. 13, 26, and 39

30. 12, 48, and 72

31. True or false? If two numbers have no common factors, then the LCM of the two numbers is their product.

32. True or false? If one number is a factor of a second number, then the LCM of the two numbers is the second number.

OBJECTIVE B **Greatest common factor (GCF)**

33.   **a.** Circle each number in the list that is a factor of 8.   2   4   7   8   14   16   28   56

      b. Underline each number in the list that is a factor of 28.

      c. Use the list to identify the greatest common factor of 8 and 28: _____.

34.   Find the GCF of 30 and 45.

      a. Write the prime factorization of each number.   30 = _____   45 = _____

      b. Use the lowest power of each common factor in part (a)
         to write the prime factorization of the GCF.   GCF = _____

      c. Find the product of the factors.   GCF = _____

Find the GCF of the numbers.

35. 9 and 12	**36.** 6 and 15	**37.** 18 and 30	**38.** 15 and 35
39. 14 and 42	**40.** 25 and 50	**41.** 16 and 80	**42.** 17 and 51
43. 21 and 55	**44.** 32 and 35	**45.** 8 and 36	**46.** 12 and 80
47. 12 and 76	**48.** 16 and 60	**49.** 24 and 30	**50.** 16 and 28
51. 24 and 36	**52.** 30 and 40	**53.** 45 and 75	**54.** 12 and 54
55. 6, 10, and 12	**56.** 8, 12, and 20	**57.** 6, 15, and 36	**58.** 15, 20, and 30
59. 21, 63, and 84	**60.** 12, 28, and 48	**61.** 24, 36, and 60	**62.** 32, 56, and 72

63.   **a.** Think of two different numbers. Find the GCF of your numbers and the
          LCM of your numbers.

      b. Which statement is true about your numbers?
         (i) The GCF is a factor of the LCM.
         (ii) The LCM is a factor of the GCF.

64.  Repeat Exercise 63 for two more pairs of numbers. Are your answers to part **b.** the same as in Exercise 63?

OBJECTIVE C   Applications

For Exercises 65 to 68, read the given exercise and state whether you will use an LCM or a GCF to solve the problem.

65. Exercise 69   **66.** Exercise 70   **67.** Exercise 71   **68.** Exercise 72

69. *Business*   Two machines are filling cereal boxes. One machine, which is filling 12-ounce boxes, fills one box every 2 min. The second machine, which is filling 18-ounce boxes, fills one box every 3 min. How often are the two machines starting to fill a box at the same time?

70. *Business*   A discount catalog offers stockings at reduced prices. The customer must order 3 pairs, 6 pairs, or 12 pairs of stockings. How many pairs should be packaged together so that no package needs to be opened when a clerk is filling an order?

71. *Business*   Each week, copies of a national magazine are delivered to three different stores that have ordered 75 copies, 100 copies, and 150 copies, respectively. How many copies should be packaged together so that no package needs to be opened during delivery?

72. *Sports*   You and a friend are swimming laps at a pool. You swim one lap every 4 min. Your friend swims one lap every 5 min. If you start at the same time from the same end of the pool, in how many minutes will both of you be at the starting point again? How many times will you have passed each other in the pool prior to that time?

73. *Scheduling*   A mathematics conference is scheduling 30-minute sessions and 40-minute sessions. There will be a 10-minute break after each session. The sessions at the conference start at 9 A.M. At what time will all sessions begin at the same time once again? At what time should lunch be scheduled if all participants are to eat at the same time?

CRITICAL THINKING

74. Find the LCM of x and $2x$. Find the GCF of x and $2x$.

75. In your own words, define the least common multiple of two numbers and the greatest common factor of two numbers.

3.2 Introduction to Fractions

OBJECTIVE A Proper fractions, improper fractions, and mixed numbers

A recipe calls for $\frac{1}{2}$ cup of butter; a carpenter uses a $\frac{3}{8}$-inch screw; and a stock broker might say that Sears closed down $\frac{3}{4}$. The numbers $\frac{1}{2}$, $\frac{3}{8}$, and $\frac{3}{4}$ are fractions.

A **fraction** can represent the number of equal parts of a whole. The circle at the right is divided into 8 equal parts. 3 of the 8 parts are shaded. The shaded portion of the circle is represented by the fraction $\frac{3}{8}$.

Each part of a fraction has a name.

$$\text{Fraction bar} \longrightarrow \frac{3}{8} \begin{array}{l} \longleftarrow \textbf{Numerator} \\ \longleftarrow \textbf{Denominator} \end{array}$$

In a **proper fraction,** the numerator is smaller than the denominator. A proper fraction is less than 1.

$$\frac{1}{2} \quad \frac{3}{8} \quad \frac{3}{4}$$
Proper fractions

In an **improper fraction,** the numerator is greater than or equal to the denominator. An improper fraction is a number greater than or equal to 1.

$$\frac{7}{3} \quad \frac{4}{4}$$
Improper fractions

The shaded portion of the circles at the right is represented by the improper fraction $\frac{7}{3}$.

The shaded portion of the square at the right is represented by the improper fraction $\frac{4}{4}$.

A fraction bar can be read "divided by." Therefore, the fraction $\frac{4}{4}$ can be read "$4 \div 4$." Because a number divided by itself is equal to 1, $4 \div 4 = 1$ and $\frac{4}{4} = 1$.

The shaded portion of the square above can be represented as $\frac{4}{4}$ or 1.

Since the fraction bar can be read as "divided by" and any number divided by 1 is the number, any whole number can be represented as an improper fraction. For example, $5 = \frac{5}{1}$ and $7 = \frac{7}{1}$.

Because zero divided by any number other than zero is zero, the numerator of a fraction can be zero.

For example, $\frac{0}{6} = 0$ because $0 \div 6 = 0$.

Recall that division by zero is not defined. Therefore, the denominator of a fraction cannot be zero.

For example, $\frac{9}{0}$ is not defined because $\frac{9}{0} = 9 \div 0$, and division by zero is not defined.

A **mixed number** is a number greater than 1 with a whole number part and a fractional part.

The shaded portion of the circles at the right is represented by the mixed number $2\frac{1}{2}$.

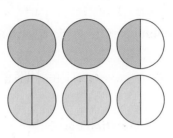

Note from the diagram at the right that the improper fraction $\frac{5}{2}$ is equal to the mixed number $2\frac{1}{2}$.

$$\frac{5}{2} = 2\frac{1}{2}$$

An improper fraction can be written as a mixed number.

To write $\frac{5}{2}$ as a mixed number, read the fraction bar as "divided by."

$\frac{5}{2}$ means $5 \div 2$.

Divide the numerator by the denominator.	To write the fractional part of the mixed number, write the remainder over the divisor.	Write the answer.
$\begin{array}{r} 2 \\ 2\overline{)5} \\ -4 \\ \hline 1 \end{array}$	$\begin{array}{r} 2\frac{1}{2} \\ 2\overline{)5} \\ -4 \\ \hline 1 \end{array}$	$\dfrac{5}{2} = 2\dfrac{1}{2}$

To write a mixed number as an improper fraction, multiply the denominator of the fractional part of the mixed number by the whole number part. The sum of this product and the numerator of the fractional part is the numerator of the improper fraction. The denominator remains the same.

Write $4\frac{5}{6}$ as an improper fraction.

$$\begin{array}{c} + \\ \times \end{array} 4\frac{5}{6} = \frac{(6 \cdot 4) + 5}{6} = \frac{24 + 5}{6} = \frac{29}{6}$$

EXAMPLE 1 Express the shaded portion of the circles as an improper fraction and as a mixed number.

Solution $\frac{19}{4}; 4\frac{3}{4}$

YOU TRY IT 1 Express the shaded portion of the circles as an improper fraction and as a mixed number.

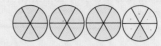

Your Solution

EXAMPLE 2 Write $\frac{14}{5}$ as a mixed number.

Solution
$$\begin{array}{r} 2 \\ 5\overline{)14} \\ -10 \\ \hline 4 \end{array} \qquad \frac{14}{5} = 2\frac{4}{5}$$

YOU TRY IT 2 Write $\frac{26}{3}$ as a mixed number.

Your Solution

EXAMPLE 3 Write $\frac{35}{7}$ as a whole number.

Solution
$$\begin{array}{r} 5 \\ 7\overline{)35} \\ -35 \\ \hline 0 \end{array} \qquad \frac{35}{7} = 5$$
▶ *Note:* The remainder is zero.

YOU TRY IT 3 Write $\frac{36}{4}$ as a whole number.

Your Solution

EXAMPLE 4 Write $12\frac{5}{8}$ as an improper fraction.

Solution $12\frac{5}{8} = \frac{(8 \cdot 12) + 5}{8} = \frac{96 + 5}{8}$

$= \frac{101}{8}$

YOU TRY IT 4 Write $9\frac{4}{7}$ as an improper fraction.

Your Solution

EXAMPLE 5 Write 9 as an improper fraction.

Solution $9 = \frac{9}{1}$

YOU TRY IT 5 Write 3 as an improper fraction.

Your Solution

Solutions on p. S6

OBJECTIVE B Equivalent fractions

Fractions can be graphed as points on a number line. The number lines at the right show thirds, sixths, and ninths graphed from 0 to 1.

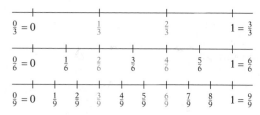

A particular point on the number line may be represented by different fractions, all of which are equal.

For example, $\dfrac{0}{3} = \dfrac{0}{6} = \dfrac{0}{9}$, $\dfrac{1}{3} = \dfrac{2}{6} = \dfrac{3}{9}$, $\dfrac{2}{3} = \dfrac{4}{6} = \dfrac{6}{9}$, and $\dfrac{3}{3} = \dfrac{6}{6} = \dfrac{9}{9}$.

Equal fractions with different denominators are called **equivalent fractions.**

$\dfrac{1}{3}, \dfrac{2}{6}$, and $\dfrac{3}{9}$ are equivalent fractions. $\dfrac{2}{3}, \dfrac{4}{6}$, and $\dfrac{6}{9}$ are equivalent fractions.

Note that we can rewrite $\dfrac{2}{3}$ as $\dfrac{4}{6}$ by multiplying both the numerator and denominator of $\dfrac{2}{3}$ by 2.

$$\dfrac{2}{3} = \dfrac{2 \cdot 2}{3 \cdot 2} = \dfrac{4}{6}$$

Also, we can rewrite $\dfrac{4}{6}$ as $\dfrac{2}{3}$ by dividing both the numerator and denominator of $\dfrac{4}{6}$ by 2.

$$\dfrac{4}{6} = \dfrac{4 \div 2}{6 \div 2} = \dfrac{2}{3}$$

This suggests the following property of fractions.

Point of Interest

Leonardo of Pisa, who was also called Fibonacci (c. 1175–1250), is credited with bringing the Hindu–Arabic number system to the Western world and promoting its use instead of the cumbersome Roman numeral system. He was also influential in promoting the idea of the fraction bar. His notation, however, was very different from what we use today. For instance, he wrote

$\dfrac{3}{4}\,\dfrac{5}{7}$ to mean $\dfrac{5}{7} + \dfrac{3}{7 \cdot 4}$.

Equivalent Fractions

The numerator and denominator of a fraction can be multiplied by or divided by the same nonzero number. The resulting fraction is equivalent to the original fraction.

$$\dfrac{a}{b} = \dfrac{a \cdot c}{b \cdot c}, \quad \dfrac{a}{b} = \dfrac{a \div c}{b \div c}, \qquad \text{where} \quad b \neq 0 \quad \text{and} \quad c \neq 0$$

Write an equivalent fraction with the given denominator.	$\dfrac{3}{8} = \dfrac{}{40}$
Divide the larger denominator by the smaller one.	$40 \div 8 = 5$
Multiply the numerator and denominator of the given fraction by the quotient (5).	$\dfrac{3}{8} = \dfrac{3 \cdot 5}{8 \cdot 5} = \dfrac{15}{40}$

A fraction is in **simplest form** when the numerator and denominator have no common factors other than 1. The fraction $\dfrac{3}{8}$ is in simplest form because 3 and 8 have no common factors other than 1. The fraction $\dfrac{15}{50}$ is not in simplest form because the numerator and denominator have a common factor of 5.

To write a fraction in simplest form, divide the numerator and denominator of the fraction by their common factors.

Write $\frac{12}{15}$ in simplest form.

12 and 15 have a common factor of 3. Divide the numerator and denominator by 3.

$$\frac{12}{15} = \frac{12 \div 3}{15 \div 3} = \frac{4}{5}$$

Simplifying a fraction requires that you recognize the common factors of the numerator and denominator. One way to do this is to write the prime factorization of the numerator and denominator and then divide by the common prime factors.

Write $\frac{30}{42}$ in simplest form.

Write the prime factorization of the numerator and denominator. Divide by the common factors.

$$\frac{30}{42} = \frac{2 \cdot \overset{1}{3} \cdot \overset{1}{5}}{\underset{1}{2} \cdot \underset{1}{3} \cdot 7} = \frac{5}{7}$$

Write $\frac{2x}{6}$ in simplest form.

Factor the numerator and denominator. Then divide by the common factors.

$$\frac{2x}{6} = \frac{\overset{1}{2} \cdot x}{\underset{1}{2} \cdot 3} = \frac{x}{3}$$

EXAMPLE 6 Write an equivalent fraction with the given denominator:

$\frac{2}{5} = \frac{}{30}$.

Solution $30 \div 5 = 6$

$$\frac{2}{5} = \frac{2 \cdot 6}{5 \cdot 6} = \frac{12}{30}$$

$\frac{12}{30}$ is equivalent to $\frac{2}{5}$.

YOU TRY IT 6 Write an equivalent fraction with the given denominator:

$\frac{5}{8} = \frac{}{48}$.

Your Solution

EXAMPLE 7 Write an equivalent fraction with the given denominator:

$3 = \frac{}{15}$.

Solution $3 = \frac{3}{1}$ $15 \div 1 = 15$

$$3 = \frac{3}{1} = \frac{3 \cdot 15}{1 \cdot 15} = \frac{45}{15}$$

$\frac{45}{15}$ is equivalent to 3.

YOU TRY IT 7 Write an equivalent fraction with the given denominator:

$8 = \frac{}{12}$.

Your Solution

Solutions on p. S6

<u>EXAMPLE 8</u>   Write $\dfrac{18}{54}$ in simplest form.

Solution $\dfrac{18}{54} = \dfrac{2 \cdot \overset{1}{\cancel{3}} \cdot \overset{1}{\cancel{3}}}{\underset{1}{2} \cdot \underset{1}{\cancel{3}} \cdot \underset{1}{\cancel{3}} \cdot 3} = \dfrac{1}{3}$

<u>YOU TRY IT 8</u>   Write $\dfrac{21}{84}$ in simplest form.

Your Solution

<u>EXAMPLE 9</u>   Write $\dfrac{36}{20}$ in simplest form.

Solution $\dfrac{36}{20} = \dfrac{\overset{1}{\cancel{2}} \cdot \overset{1}{\cancel{2}} \cdot 3 \cdot 3}{\underset{1}{\cancel{2}} \cdot \underset{1}{\cancel{2}} \cdot 5} = \dfrac{9}{5}$

<u>YOU TRY IT 9</u>   Write $\dfrac{32}{12}$ in simplest form.

Your Solution

<u>EXAMPLE 10</u>   Write $\dfrac{10m}{12}$ in simplest form.

Solution $\dfrac{10m}{12} = \dfrac{\overset{1}{\cancel{2}} \cdot 5 \cdot m}{\underset{1}{\cancel{2}} \cdot 2 \cdot 3} = \dfrac{5m}{6}$

<u>YOU TRY IT 10</u>   Write $\dfrac{11t}{11}$ in simplest form.

Your Solution

Solutions on pp. S6–S7

OBJECTIVE C   Order relations between two fractions

The number line can be used to determine the order relation between two fractions.

A fraction that appears to the left of a given fraction is less than the given fraction.

$\dfrac{3}{8}$ is to the left of $\dfrac{5}{8}$.

$\dfrac{3}{8} < \dfrac{5}{8}$

A fraction that appears to the right of a given fraction is greater than the given fraction.

$\dfrac{7}{8}$ is to the right of $\dfrac{3}{8}$.

$\dfrac{7}{8} > \dfrac{3}{8}$

To find the order relation between two fractions with the *same* denominator, compare the numerators. The fraction with the smaller numerator is the smaller fraction. The larger fraction is the fraction with the larger numerator.

$\dfrac{3}{8}$ and $\dfrac{5}{8}$ have the same denominator. $\dfrac{3}{8} < \dfrac{5}{8}$ because $3 < 5$.

$\dfrac{7}{8}$ and $\dfrac{3}{8}$ have the same denominator. $\dfrac{7}{8} > \dfrac{3}{8}$ because $7 > 3$.

Point of Interest

Archimedes (c. 287–212 B.C.) is the person who calculated that $\pi \approx 3\frac{1}{7}$. He actually showed that $3\frac{10}{71} < \pi < 3\frac{1}{7}$. The approximation $3\frac{10}{71}$ is more accurate than $3\frac{1}{7}$ but more difficult to use.

Before comparing two fractions with *different* denominators, rewrite the fractions with a common denominator. The common denominator is the least common multiple (LCM) of the denominators of the fractions. The LCM of the denominators is sometimes called the least common denominator or LCD.

Find the order relation between $\frac{5}{12}$ and $\frac{7}{18}$.

Find the LCM of the denominators.

The LCM of 12 and 18 is 36.

Write each fraction as an equivalent fraction with the LCM as the denominator.

$$\frac{5}{12} = \frac{5 \cdot 3}{12 \cdot 3} = \frac{15}{36} \longleftarrow \text{Larger numerator}$$

$$\frac{7}{18} = \frac{7 \cdot 2}{18 \cdot 2} = \frac{14}{36} \longleftarrow \text{Smaller numerator}$$

Compare the fractions.

$$\frac{15}{36} > \frac{14}{36}$$

$$\frac{5}{12} > \frac{7}{18}$$

EXAMPLE 11 Place the correct symbol, < or >, between the two numbers.

$$\frac{2}{3} \qquad \frac{4}{7}$$

Solution The LCM of 3 and 7 is 21.

$$\frac{2}{3} = \frac{14}{21} \qquad \frac{4}{7} = \frac{12}{21}$$

$$\frac{14}{21} > \frac{12}{21}$$

$$\frac{2}{3} > \frac{4}{7}$$

YOU TRY IT 11 Place the correct symbol, < or >, between the two numbers.

$$\frac{4}{9} \qquad \frac{8}{21}$$

Your Solution

EXAMPLE 12 Place the correct symbol, < or >, between the two numbers.

$$\frac{7}{12} \qquad \frac{11}{18}$$

Solution The LCM of 12 and 18 is 36.

$$\frac{7}{12} = \frac{21}{36} \qquad \frac{11}{18} = \frac{22}{36}$$

$$\frac{21}{36} < \frac{22}{36}$$

$$\frac{7}{12} < \frac{11}{18}$$

YOU TRY IT 12 Place the correct symbol, < or >, between the two numbers.

$$\frac{17}{24} \qquad \frac{7}{9}$$

Your Solution

Solutions on p. S7

OBJECTIVE D **Applications**

The graph at the right shows the U.S. population distribution by age. Use this graph for Example 13 and You Try It 13.

U.S. Population Distribution by Age (in millions)
Source: U.S. Bureau of the Census

EXAMPLE 13

What fraction of the total U.S. population is age 65 or older?

Strategy

To find the fraction:

▶ Add the populations of all the segments to find the total U.S. population.
▶ Write a fraction with the population age 65 or older in the numerator and the total population in the denominator. Write the fraction in simplest form.

Solution

$20 + 61 + 105 + 73 + 37 = 296$

$\dfrac{37}{296} = \dfrac{1}{8}$

$\frac{1}{8}$ of the U.S. population is age 65 or older.

YOU TRY IT 13

What fraction of the total U.S. population is under 5 years of age?

Your Strategy

Your Solution

EXAMPLE 14

Of every dollar spent for gasoline, 10 cents goes to gas stations. (*Source:* Oil Price Information Services) What fraction of every dollar spent for gasoline goes to gas stations?

Strategy

To find the fraction, write a fraction with the amount that goes to gas stations in the numerator and the number of cents in one dollar (100) in the denominator. Simplify the fraction.

Solution

$\dfrac{10}{100} = \dfrac{1}{10}$

$\frac{1}{10}$ of every dollar spent for gasoline goes to gas stations.

YOU TRY IT 14

Of every dollar spent for gasoline, 6 cents goes to refineries. (*Source:* Oil Price Information Services) What fraction of every dollar spent for gasoline goes to refineries?

Your Strategy

Your Solution

Solutions on p. S7

3.2 Exercises

OBJECTIVE A Proper fractions, improper fractions, and mixed numbers

1. Use the fraction $\frac{9}{4}$.

   a. The numerator of the fraction is _____. The denominator of the fraction is _____.

   b. Because the numerator is greater than the denominator, this fraction is called a(n) _____ fraction.

   c. The fraction bar can be read "_____," so the fraction also represents the division problem _____ ÷ _____.

2. Fill in each blank with 0, 1, 6, or *undefined*.

   a. $\frac{6}{0} =$ _____

   b. $\frac{6}{1} =$ _____

   c. $\frac{6}{6} =$ _____

   d. $\frac{0}{6} =$ _____

Express the shaded portion of the circle as a fraction.

3. 4. 5.  6.

Express the shaded portion of the circles as an improper fraction and as a mixed number.

7. 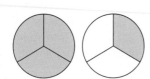 8. 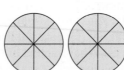 10.

9. 10.

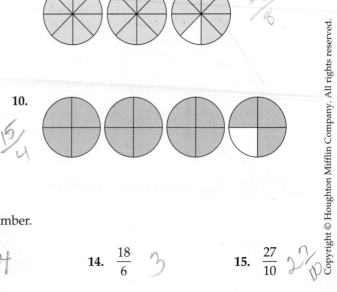

Write the improper fraction as a mixed number or a whole number.

11. $\frac{13}{4}$ 12. $\frac{14}{3}$ 13. $\frac{20}{5}$ 14. $\frac{18}{6}$ 15. $\frac{27}{10}$

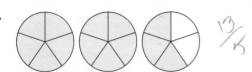

16. $\frac{31}{3}$ $10\frac{1}{3}$

17. $\frac{56}{8}$ 7

18. $\frac{27}{9}$ 3

19. $\frac{17}{9}$ $1\frac{8}{9}$

20. $\frac{8}{3}$ $2\frac{2}{3}$

21. $\frac{12}{5}$ $2\frac{2}{5}$

22. $\frac{19}{8}$ $2\frac{3}{8}$

23. $\frac{18}{1}$ 18

24. $\frac{21}{1}$ 21

25. $\frac{32}{15}$ $2\frac{2}{15}$

26. $\frac{39}{14}$ $2\frac{11}{14}$

27. $\frac{8}{8}$ 1

28. $\frac{12}{12}$ 1

29. $\frac{28}{3}$ $9\frac{1}{3}$

30. $\frac{43}{5}$ $8\frac{3}{5}$

Write the mixed number or whole number as an improper fraction.

31. $2\frac{1}{4}$ $\frac{9}{4}$

32. $4\frac{2}{5}$ $\frac{22}{5}$

33. $5\frac{1}{2}$ $\frac{11}{2}$

34. $3\frac{2}{3}$ $\frac{11}{3}$

35. $2\frac{4}{5}$ $\frac{14}{5}$

36. $6\frac{3}{8}$ $\frac{51}{8}$

37. $7\frac{5}{6}$ $\frac{47}{6}$

38. $9\frac{1}{5}$ $\frac{46}{5}$

39. 7 $\frac{7}{1}$

40. 4 $\frac{4}{1}$

41. $8\frac{1}{4}$ $\frac{33}{4}$

42. $1\frac{7}{9}$ $\frac{16}{9}$

43. $10\frac{1}{3}$ $\frac{31}{3}$

44. $6\frac{3}{7}$ $\frac{45}{7}$

45. $4\frac{7}{12}$ $\frac{55}{12}$

46. $5\frac{4}{9}$ $\frac{49}{9}$

47. 8 $\frac{8}{1}$

48. 6 $\frac{6}{1}$

49. $12\frac{4}{5}$ $\frac{64}{5}$

50. $11\frac{5}{8}$ $\frac{93}{8}$

51. When a mixed number is written as an improper fraction $\frac{a}{b}$, is $a < b$ or is $a > b$? $a > b$

52. If an improper fraction can be written as a whole number, is the numerator a multiple of the denominator, or is the denominator a multiple of the numerator?  numerator

OBJECTIVE B   Equivalent fractions

53. To write $\frac{5}{6}$ as an equivalent fraction with a denominator of 24, multiply the numerator and the denominator by $24 \div 6 = $ _____. Thus, $\frac{5}{6} = \frac{5 \cdot \square}{6 \cdot \square} = \frac{\square}{24}$.

54. The fraction $\frac{5}{6}$ is in simplest form because the only common factor of the numerator and the denominator is _____. The fraction $\frac{10}{12}$ is not in simplest form because the numerator and the denominator have a common factor of _____.

Write an equivalent fraction with the given denominator.

55. $\frac{1}{2} = \frac{}{12}$

56. $\frac{1}{4} = \frac{}{20}$

57. $\frac{3}{8} = \frac{}{24}$

58. $\frac{9}{11} = \frac{}{44}$

59. $\frac{2}{17} = \frac{}{51}$

60. $\frac{9}{10} = \frac{}{80}$

61. $\frac{3}{4} = \frac{}{32}$

62. $\frac{5}{8} = \frac{}{32}$

63. $6 = \frac{108}{18}$

64. $5 = \frac{}{35}$

65. $\frac{1}{3} = \frac{}{90}$

66. $\frac{3}{16} = \frac{}{48}$

67. $\frac{2}{3} = \frac{}{21}$

68. $\frac{4}{9} = \frac{}{36}$

69. $\frac{6}{7} = \frac{}{49}$

70. $\frac{7}{8} = \frac{}{40}$

71. $\frac{4}{9} = \frac{}{18}$

72. $\frac{11}{12} = \frac{}{48}$

73. $7 = \frac{28}{4}$

74. $9 = \frac{}{6}$

Write the fraction in simplest form.

75. $\frac{3}{12}$

76. $\frac{10}{22}$

77. $\frac{33}{44}$

78. $\frac{6}{14}$

79. $\frac{4}{24}$

80. $\frac{25}{75}$

81. $\frac{8}{33}$

82. $\frac{9}{25}$

83. $\frac{0}{8}$

84. $\frac{0}{11}$

85. $\frac{42}{36}$

86. $\frac{30}{18}$

87. $\frac{16}{16}$

88. $\frac{24}{24}$

89. $\frac{21}{35}$

90. $\frac{11}{55}$

91. $\frac{16}{60}$

92. $\frac{8}{84}$

93. $\frac{12}{20}$

94. $\frac{24}{36}$

95. $\dfrac{12m}{18}$ **96.** $\dfrac{20x}{25}$ **97.** $\dfrac{4y}{8}$ **98.** $\dfrac{14z}{28}$ **99.** $\dfrac{24a}{36}$

100. $\dfrac{28z}{21}$ **101.** $\dfrac{8c}{8}$ **102.** $\dfrac{9w}{9}$ **103.** $\dfrac{18k}{3}$ **104.** $\dfrac{24t}{4}$

For Exercises 105 to 107, for the given condition, state whether the fraction (i) must be in simplest form, (ii) cannot be in simplest form, or (iii) might be in simplest form. If (iii) is true, then name two fractions that meet the given condition, one that is in simplest form and one that is not in simplest form.

105. The numerator and denominator are both even numbers.

106. The numerator and denominator are both odd numbers.

107. The numerator is an odd number and the denominator is an even number.

OBJECTIVE C   Order relations between two fractions

108. a. To decide the order relation between two fractions, first write the fractions with a common _____. The lowest common denominator (LCD) of $\frac{3}{10}$ and $\frac{1}{6}$ is the LCM of _____ and _____, which is _____.

b. $\dfrac{3}{10} = \dfrac{}{30}$ and $\dfrac{1}{6} = \dfrac{}{30}$. Because $9 > 5$, $\dfrac{3}{10}$ _____ $\dfrac{1}{6}$.

Place the correct symbol, $<$ or $>$, between the two numbers.

109. $\dfrac{3}{8}$ $\dfrac{2}{5}$ **110.** $\dfrac{5}{7}$ $\dfrac{2}{3}$ **111.** $\dfrac{3}{4}$ $\dfrac{7}{9}$ **112.** $\dfrac{7}{12}$ $\dfrac{5}{8}$

113. $\dfrac{2}{3}$ $\dfrac{7}{11}$ **114.** $\dfrac{11}{14}$ $\dfrac{3}{4}$ **115.** $\dfrac{17}{24}$ $\dfrac{11}{16}$ **116.** $\dfrac{11}{12}$ $\dfrac{7}{9}$

117. $\dfrac{7}{15}$ $\dfrac{5}{12}$ **118.** $\dfrac{5}{8}$ $\dfrac{4}{7}$ **119.** $\dfrac{5}{9}$ $\dfrac{11}{21}$ **120.** $\dfrac{11}{30}$ $\dfrac{7}{24}$

121. $\dfrac{7}{12}$ $\dfrac{13}{18}$

122. $\dfrac{9}{11}$ $\dfrac{7}{8}$

123. $\dfrac{4}{5}$ $\dfrac{7}{9}$

124. $\dfrac{3}{4}$ $\dfrac{11}{13}$

125. $\dfrac{9}{16}$ $\dfrac{5}{9}$

126. $\dfrac{2}{3}$ $\dfrac{7}{10}$

127. $\dfrac{5}{8}$ $\dfrac{13}{20}$

128. $\dfrac{3}{10}$ $\dfrac{7}{25}$

 For Exercises 129 and 130, find an example of two fractions in simplest form, $\dfrac{a}{b}$ and $\dfrac{c}{d}$, that fit the given conditions.

129. $a > c$, $b \neq d$, and $\dfrac{a}{b} > \dfrac{c}{d}$.

130. $a > c$, $b \neq d$, and $\dfrac{a}{b} < \dfrac{c}{d}$.

OBJECTIVE D **Applications**

131. The test grades in a class consisted of five A's, three B's, and six C's.

 a. The total number of test grades was _____ + _____ + _____ = _____.

 b. The fraction of the class that received A's was $\dfrac{\text{number of A's}}{\text{total number of grades}}$ = ☐.

132. *Measurement* A ton is equal to 2,000 lb. What fractional part of a ton is 250 lb?

133. *Measurement* A pound is equal to 16 oz. What fractional part of a pound is 6 oz?

134. *Measurement* If a history class lasts 50 min, what fractional part of an hour is the history class?

135. *Measurement* If you sleep for 8 h one night, what fractional part of one day did you spend sleeping?

136. *Jewelry* Gold is designated by karats. Pure gold is 24 karats. What fractional part of an 18-karat gold bracelet is pure gold?

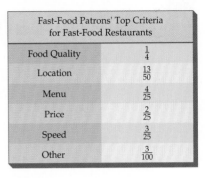

The Food Industry   The table at the right shows the results of a survey that asked fast-food patrons their criteria for choosing where to go for fast food. Three out of every 25 people surveyed said that the speed of the service was most important. Use this table for Exercises 137 and 138.

137. According to the survey, do more people choose a fast-food restaurant on the basis of its location or on the basis of the quality of its food?

Source: Maritz Marketing Research, Inc.

138. Which criterion was cited by most people?

139. *Card Games*   A standard deck of playing cards consists of 52 cards. **a.** What fractional part of a standard deck of cards is spades? **b.** What fractional part of a standard deck of cards is aces?

140. *Education*   You answer 42 questions correctly on an exam of 50 questions. Did you answer more or less than $\frac{8}{10}$ of the questions correctly?

141. *Education*   To pass a real estate examination, you must answer at least $\frac{7}{10}$ of the questions correctly. If the exam has 200 questions and you answer 150 correctly, do you pass the exam?

142. *Sports*   Wilt Chamberlain held the record for the most field goals in a basketball game. He had 36 field goals in 63 attempts. What fraction of the number of attempts did he not have a field goal?

Wilt Chamberlain

CRITICAL THINKING

143. Is the expression $x < \frac{4}{9}$ true when $x = \frac{3}{8}$? Is it true when $x = \frac{5}{12}$? Is it true for any negative number?

144. *Geography*   What fraction of the states in the United States begin with the letter A?

145. a.   On the number line, what fraction is halfway between $\frac{2}{a}$ and $\frac{4}{a}$?

   b.   Find two fractions evenly spaced between $\frac{5}{b}$ and $\frac{8}{b}$.

3.3 Multiplication and Division of Fractions

OBJECTIVE A Multiplication of fractions

To multiply two fractions, multiply the numerators and multiply the denominators.

> ## Multiplication of Fractions
>
> The product of two fractions is the product of the numerators over the product of the denominators.
>
> $$\frac{a}{b} \cdot \frac{c}{d} = \frac{ac}{bd}, \quad \text{where} \quad b \neq 0 \quad \text{and} \quad d \neq 0$$

Note that fractions do not need to have the same denominator in order to be multiplied.

After multiplying two fractions, write the product in simplest form.

Multiply: $\frac{2}{5} \cdot \frac{1}{3}$

Multiply the numerators.
Multiply the denominators.

$$\frac{2}{5} \cdot \frac{1}{3} = \frac{2 \cdot 1}{5 \cdot 3} = \frac{2}{15}$$

The product $\frac{2}{5} \cdot \frac{1}{3}$ can be read "$\frac{2}{5}$ times $\frac{1}{3}$" or "$\frac{2}{5}$ of $\frac{1}{3}$."

Reading the times sign as "of" is useful in diagraming the product of two fractions.

$\frac{1}{3}$ of the bar at the right is shaded.

We want to shade $\frac{2}{5}$ of the $\frac{1}{3}$ already shaded.

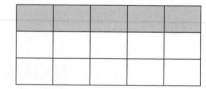

$\frac{2}{15}$ of the bar is now shaded.

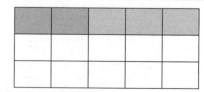

$$\frac{2}{5} \text{ of } \frac{1}{3} = \frac{2}{5} \cdot \frac{1}{3} = \frac{2}{15}$$

If a is a natural number, then $\frac{1}{a}$ is called the **reciprocal** or **multiplicative inverse** of a. Note that $a \cdot \frac{1}{a} = \frac{a}{1} \cdot \frac{1}{a} = \frac{a}{a} = 1$.

The product of a number and its multiplicative inverse is 1.

$$\frac{1}{8} \cdot 8 = 8 \cdot \frac{1}{8} = 1$$

Multiply: $\dfrac{3}{8} \cdot \dfrac{4}{9}$

Multiply the numerators.
Multiply the denominators.

$$\dfrac{3}{8} \cdot \dfrac{4}{9} = \dfrac{3 \cdot 4}{8 \cdot 9}$$

Express the fraction in simplest form by first writing the prime factorization of each number.

$$= \dfrac{3 \cdot 2 \cdot 2}{2 \cdot 2 \cdot 2 \cdot 3 \cdot 3}$$

Divide by the common factors and write the product in simplest form.

$$= \dfrac{1}{6}$$

The sign rules for multiplying positive and negative fractions are the same rules used to multiply integers.

The product of two numbers with the same sign is positive.
The product of two numbers with different signs is negative.

Multiply: $-\dfrac{3}{4} \cdot \dfrac{8}{15}$

The signs are different.
The product is negative.

$$-\dfrac{3}{4} \cdot \dfrac{8}{15} = -\left(\dfrac{3}{4} \cdot \dfrac{8}{15}\right)$$

Multiply the numerators.
Multiply the denominators.

$$= -\dfrac{3 \cdot 8}{4 \cdot 15}$$

Write the product in simplest form.

$$= -\dfrac{3 \cdot 2 \cdot 2 \cdot 2}{2 \cdot 2 \cdot 3 \cdot 5}$$

$$= -\dfrac{2}{5}$$

Multiply: $-\dfrac{3}{8}\left(-\dfrac{2}{5}\right)\left(-\dfrac{10}{21}\right)$

$$-\dfrac{3}{8}\left(-\dfrac{2}{5}\right)\left(-\dfrac{10}{21}\right)$$

Multiply the first two fractions. The product is positive.

$$= \left(\dfrac{3}{8} \cdot \dfrac{2}{5}\right)\left(-\dfrac{10}{21}\right)$$

The product of the first two fractions and the third fraction is negative.

$$= -\left(\dfrac{3}{8} \cdot \dfrac{2}{5} \cdot \dfrac{10}{21}\right)$$

Multiply the numerators.
Multiply the denominators.

$$= -\dfrac{3 \cdot 2 \cdot 10}{8 \cdot 5 \cdot 21}$$

Write the product in simplest form.

$$= -\dfrac{3 \cdot 2 \cdot 2 \cdot 5}{2 \cdot 2 \cdot 2 \cdot 5 \cdot 3 \cdot 7}$$

$$= -\dfrac{1}{14}$$

Copyright © Houghton Mifflin Company. All rights reserved.

Point of Interest

Try this: What is the result if you take one-third of a half-dozen and add to it one-fourth of the product of the result and 8?

Thus, the product of three negative fractions is negative. We can modify the rule for multiplying positive and negative fractions to say that the product of an odd number of negative fractions is negative and the product of an even number of negative fractions is positive.

To multiply a whole number by a fraction or a mixed number, first write the whole number as a fraction with a denominator of 1.

Multiply: $3 \cdot \frac{5}{8}$

Write the whole number 3 as the fraction $\frac{3}{1}$.

$$3 \cdot \frac{5}{8} = \frac{3}{1} \cdot \frac{5}{8}$$

Multiply the fractions. There are no common factors in the numerator and denominator.

$$= \frac{3 \cdot 5}{1 \cdot 8}$$

Write the improper fraction as a mixed number.

$$= \frac{15}{8} = 1\frac{7}{8}$$

Multiply: $\frac{x}{7} \cdot \frac{y}{5}$

Multiply the numerators. Multiply the denominators.

$$\frac{x}{7} \cdot \frac{y}{5} = \frac{x \cdot y}{7 \cdot 5}$$

Write the product in simplest form.

$$= \frac{xy}{35}$$

When a factor is a mixed number, first write the mixed number as an improper fraction. Then multiply.

Find the product of $-4\frac{1}{6}$ and $2\frac{7}{10}$.

The signs are different. The product is negative.

$$-4\frac{1}{6} \cdot 2\frac{7}{10} = -\left(4\frac{1}{6} \cdot 2\frac{7}{10}\right)$$

Write each mixed number as an improper fraction.

$$= -\left(\frac{25}{6} \cdot \frac{27}{10}\right)$$

Multiply the fractions.

$$= -\frac{25 \cdot 27}{6 \cdot 10}$$

$$= -\frac{5 \cdot 5 \cdot 3 \cdot 3 \cdot 3}{2 \cdot 3 \cdot 2 \cdot 5}$$

Write the product in simplest form.

$$= -\frac{45}{4} = -11\frac{1}{4}$$

Is $-\frac{2}{3}$ a solution of the equation $\frac{3}{4}x = -\frac{1}{2}$?

$$\frac{3}{4}x = -\frac{1}{2}$$

Replace x by $-\frac{2}{3}$ and then simplify.

$$\frac{3}{4}\left(-\frac{2}{3}\right) \;\Big|\; -\frac{1}{2}$$

$$-\frac{3 \cdot 2}{4 \cdot 3} \;\Big|\; -\frac{1}{2}$$

$$-\frac{3 \cdot 2}{2 \cdot 2 \cdot 3} \;\Big|\; -\frac{1}{2}$$

The results are equal.

$$-\frac{1}{2} = -\frac{1}{2}$$

Yes, $-\frac{2}{3}$ is a solution of the equation.

EXAMPLE 1   Multiply: $\frac{7}{9} \cdot \frac{3}{14} \cdot \frac{2}{5}$

Solution

$$\frac{7}{9} \cdot \frac{3}{14} \cdot \frac{2}{5} = \frac{7 \cdot 3 \cdot 2}{9 \cdot 14 \cdot 5}$$

$$= \frac{7 \cdot 3 \cdot 2}{3 \cdot 3 \cdot 2 \cdot 7 \cdot 5} = \frac{1}{15}$$

YOU TRY IT 1   Multiply: $\frac{5}{12} \cdot \frac{9}{35} \cdot \frac{7}{8}$

Your Solution

EXAMPLE 2   Multiply: $\frac{6}{x} \cdot \frac{8}{y}$

Solution

$$\frac{6}{x} \cdot \frac{8}{y} = \frac{6 \cdot 8}{x \cdot y}$$

$$= \frac{48}{xy}$$

YOU TRY IT 2   Multiply: $\frac{y}{10} \cdot \frac{z}{7}$

Your Solution

EXAMPLE 3   Multiply: $-\frac{3}{4}\left(\frac{1}{2}\right)\left(-\frac{8}{9}\right)$

Solution

$$-\frac{3}{4}\left(\frac{1}{2}\right)\left(-\frac{8}{9}\right)$$

$$= \frac{3}{4} \cdot \frac{1}{2} \cdot \frac{8}{9} \quad \blacktriangleright \text{ The product of}$$
two negative
fractions is
positive.

$$= \frac{3 \cdot 1 \cdot 8}{4 \cdot 2 \cdot 9}$$

$$= \frac{3 \cdot 1 \cdot 2 \cdot 2 \cdot 2}{2 \cdot 2 \cdot 2 \cdot 3 \cdot 3} = \frac{1}{3}$$

YOU TRY IT 3   Multiply: $-\frac{1}{3}\left(-\frac{5}{12}\right)\left(\frac{8}{15}\right)$

Your Solution

Solutions on p. S7

EXAMPLE 4 What is the product of $\frac{7}{12}$ and 4?

Solution

$$\frac{7}{12} \cdot 4 = \frac{7}{12} \cdot \frac{4}{1}$$

$$= \frac{7 \cdot 4}{12 \cdot 1}$$

$$= \frac{7 \cdot 2 \cdot 2}{2 \cdot 2 \cdot 3 \cdot 1}$$

$$= \frac{7}{3}$$

$$= 2\frac{1}{3}$$

YOU TRY IT 4 Find the product of $\frac{8}{9}$ and 6.

Your Solution

EXAMPLE 5 Multiply: $-7\frac{1}{2} \cdot 4\frac{2}{5}$

Solution

$$-7\frac{1}{2} \cdot 4\frac{2}{5} = -\left(\frac{15}{2} \cdot \frac{22}{5}\right)$$

$$= -\frac{15 \cdot 22}{2 \cdot 5}$$

$$= -\frac{3 \cdot 5 \cdot 2 \cdot 11}{2 \cdot 5}$$

$$= -\frac{33}{1} = -33$$

YOU TRY IT 5 Multiply: $3\frac{6}{7} \cdot 2\frac{4}{9}$

Your Solution

EXAMPLE 6 Evaluate the variable expression xy for $x = 1\frac{4}{5}$ and $y = -\frac{5}{6}$.

Solution xy

$$1\frac{4}{5}\left(-\frac{5}{6}\right) = -\left(\frac{9}{5} \cdot \frac{5}{6}\right)$$

$$= -\frac{9 \cdot 5}{5 \cdot 6}$$

$$= -\frac{3 \cdot 3 \cdot 5}{5 \cdot 2 \cdot 3}$$

$$= -\frac{3}{2} = -1\frac{1}{2}$$

YOU TRY IT 6 Evaluate the variable expression xy for $x = 5\frac{1}{8}$ and $y = \frac{2}{3}$.

Your Solution

Solutions on p. S7

OBJECTIVE B **Division of fractions**

The **reciprocal** of a fraction is that fraction with the numerator and denominator interchanged.

The reciprocal of $\frac{3}{4}$ is $\frac{4}{3}$.

The reciprocal of $\frac{a}{b}$ is $\frac{b}{a}$.

The process of interchanging the numerator and denominator of a fraction is called **inverting** the fraction.

To find the reciprocal of a whole number, first rewrite the whole number as a fraction with a denominator of 1. Then invert the fraction.

$6 = \frac{6}{1}$

The reciprocal of 6 is $\frac{1}{6}$.

Reciprocals are used to rewrite division problems as related multiplication problems. Look at the following two problems:

$6 \div 2 = 3$ $6 \cdot \frac{1}{2} = 3$

6 divided by 2 equals 3.   6 times the reciprocal of 2 equals 3.

Division is defined as multiplication by the reciprocal. Therefore, "divided by 2" is the same as "times $\frac{1}{2}$." Fractions are divided by making this substitution.

Division of Fractions

To divide two fractions, multiply by the reciprocal of the divisor.

$$\frac{a}{b} \div \frac{c}{d} = \frac{a}{b} \cdot \frac{d}{c}, \quad \text{where} \quad b \neq 0, \quad c \neq 0, \quad \text{and} \quad d \neq 0$$

Divide: $\frac{2}{5} \div \frac{3}{4}$

Rewrite the division as multiplication by the reciprocal.

$\frac{2}{5} \div \frac{3}{4} = \frac{2}{5} \cdot \frac{4}{3}$

Multiply the fractions.

$= \frac{2 \cdot 4}{5 \cdot 3}$

$= \frac{2 \cdot 2 \cdot 2}{5 \cdot 3} = \frac{8}{15}$

Point of Interest

Try this: What number when multiplied by its reciprocal is equal to 1?

The sign rules for dividing positive and negative fractions are the same rules used to divide integers.

The quotient of two numbers with the same sign is positive.
The quotient of two numbers with different signs is negative.

Simplify: $-\frac{7}{10} \div \left(-\frac{14}{15}\right)$

The signs are the same.
The quotient is positive.
$$-\frac{7}{10} \div \left(-\frac{14}{15}\right) = \frac{7}{10} \div \frac{14}{15}$$

Rewrite the division as multiplication by the reciprocal.
$$= \frac{7}{10} \cdot \frac{15}{14}$$

Multiply the fractions.
$$= \frac{7 \cdot 15}{10 \cdot 14}$$

$$= \frac{7 \cdot 3 \cdot 5}{2 \cdot 5 \cdot 2 \cdot 7}$$

$$= \frac{3}{4}$$

To divide a fraction and a whole number, first write the whole number as a fraction with a denominator of 1.

Take Note

$\frac{3}{4} \div 6 = \frac{1}{8}$ means that if $\frac{3}{4}$ is divided into 6 equal parts, each equal part is $\frac{1}{8}$. For example, if 6 people share $\frac{3}{4}$ of a pizza, each person eats $\frac{1}{8}$ of the pizza.

Find the quotient of $\frac{3}{4}$ and 6.

Write the whole number 6 as the fraction $\frac{6}{1}$.
$$\frac{3}{4} \div 6 = \frac{3}{4} \div \frac{6}{1}$$

Rewrite the division as multiplication by the reciprocal.
$$= \frac{3}{4} \cdot \frac{1}{6}$$

Multiply the fractions.
$$= \frac{3 \cdot 1}{4 \cdot 6}$$

$$= \frac{3 \cdot 1}{2 \cdot 2 \cdot 2 \cdot 3}$$

$$= \frac{1}{8}$$

When a number in a quotient is a mixed number, first write the mixed number as an improper fraction. Then divide the fractions.

Divide: $\frac{2}{3} \div 1\frac{1}{4}$

Write the mixed number $1\frac{1}{4}$ as an improper fraction.
$$\frac{2}{3} \div 1\frac{1}{4} = \frac{2}{3} \div \frac{5}{4}$$

Rewrite the division as multiplication by the reciprocal.
$$= \frac{2}{3} \cdot \frac{4}{5}$$

Multiply the fractions.
$$= \frac{2 \cdot 4}{3 \cdot 5} = \frac{8}{15}$$

EXAMPLE 7 Divide: $\frac{4}{5} \div \frac{8}{15}$

Solution $\quad \frac{4}{5} \div \frac{8}{15} = \frac{4}{5} \cdot \frac{15}{8}$

$$= \frac{4 \cdot 15}{5 \cdot 8}$$

$$= \frac{2 \cdot 2 \cdot 3 \cdot 5}{5 \cdot 2 \cdot 2 \cdot 2}$$

$$= \frac{3}{2} = 1\frac{1}{2}$$

YOU TRY IT 7 Divide: $\frac{5}{6} \div \frac{10}{27}$

Your Solution

EXAMPLE 8 Divide: $\frac{x}{2} \div \frac{y}{4}$

Solution $\quad \frac{x}{2} \div \frac{y}{4} = \frac{x}{2} \cdot \frac{4}{y}$

$$= \frac{x \cdot 4}{2 \cdot y}$$

$$= \frac{x \cdot 2 \cdot 2}{2 \cdot y} = \frac{2x}{y}$$

YOU TRY IT 8 Divide: $\frac{x}{8} \div \frac{y}{6}$

Your Solution

EXAMPLE 9 What is the quotient of 6 and $-\frac{3}{5}$?

Solution $\quad 6 \div \left(-\frac{3}{5}\right) = -\left(\frac{6}{1} \div \frac{3}{5}\right)$

$$= -\left(\frac{6}{1} \cdot \frac{5}{3}\right)$$

$$= -\frac{6 \cdot 5}{1 \cdot 3}$$

$$= -\frac{2 \cdot 3 \cdot 5}{1 \cdot 3}$$

$$= -\frac{10}{1} = -10$$

YOU TRY IT 9 Find the quotient of 4 and $-\frac{6}{7}$.

Your Solution

Solutions on p. S7

EXAMPLE 10 Divide: $3\frac{4}{15} \div 2\frac{1}{10}$

Solution

$$3\frac{4}{15} \div 2\frac{1}{10} = \frac{49}{15} \div \frac{21}{10}$$

$$= \frac{49}{15} \cdot \frac{10}{21}$$

$$= \frac{49 \cdot 10}{15 \cdot 21}$$

$$= \frac{7 \cdot 7 \cdot 2 \cdot 5}{3 \cdot 5 \cdot 3 \cdot 7}$$

$$= \frac{14}{9} = 1\frac{5}{9}$$

YOU TRY IT 10 Divide: $4\frac{3}{8} \div 3\frac{1}{2}$

Your Solution

EXAMPLE 11 Evaluate $x \div y$ for $x = 3\frac{1}{8}$ and $y = 5$.

Solution $x \div y$

$$3\frac{1}{8} \div 5 = \frac{25}{8} \div \frac{5}{1}$$

$$= \frac{25}{8} \cdot \frac{1}{5}$$

$$= \frac{25 \cdot 1}{8 \cdot 5}$$

$$= \frac{5 \cdot 5 \cdot 1}{2 \cdot 2 \cdot 2 \cdot 5} = \frac{5}{8}$$

YOU TRY IT 11 Evaluate $x \div y$ for $x = 2\frac{1}{4}$ and $y = 9$.

Your Solution

Solutions on p. S8

OBJECTIVE C **Applications and formulas**

Figure *ABC* is a triangle. *AB* is the **base,** *b,* of the triangle. The line segment from *C* that forms a right angle with the base is the **height,** *h,* of the triangle. The formula for the area of a triangle is given below. Use this formula for Example 12 and You Try It 12.

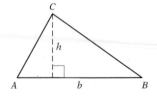

Area of Triangle
....

The formula for the area of a triangle is $A = \frac{1}{2}bh$, where *A* is the area of the triangle, *b* is the base, and *h* is the height.

EXAMPLE 12

A riveter uses metal plates that are in the shape of a triangle and have a base of 12 cm and a height of 6 cm. Find the area of one metal plate.

Strategy

To find the area, use the formula for the area of a triangle, $A = \frac{1}{2}bh$. $b = 12$ and $h = 6$.

Solution

$$A = \frac{1}{2}bh$$

$$A = \frac{1}{2}(12)(6)$$

$$A = 36$$

The area is 36 cm².

YOU TRY IT 12

Find the amount of felt needed to make a banner that is in the shape of a triangle with a base of 18 in. and a height of 9 in.

Your Strategy

Your Solution

EXAMPLE 13

A 12-foot board is cut into pieces $2\frac{1}{2}$ ft long for use as bookshelves. What is the length of the remaining piece after as many shelves as possible are cut?

Strategy

To find the length of the remaining piece:

▶ Divide the total length (12) by the length of each shelf $\left(2\frac{1}{2}\right)$. The quotient is the number of shelves cut, with a certain fraction of a shelf left over.

▶ Multiply the fraction left over by the length of a shelf.

Solution

$$12 \div 2\frac{1}{2} = \frac{12}{1} \div \frac{5}{2} = \frac{12}{1} \cdot \frac{2}{5} = \frac{12 \cdot 2}{1 \cdot 5} = \frac{24}{5} = 4\frac{4}{5}$$

4 shelves, each $2\frac{1}{2}$ ft long, can be cut from the board. The piece remaining is $\frac{4}{5}$ of $2\frac{1}{2}$ ft long.

$$\frac{4}{5} \cdot 2\frac{1}{2} = \frac{4}{5} \cdot \frac{5}{2} = \frac{4 \cdot 5}{5 \cdot 2} = 2$$

The length of the remaining piece is 2 ft.

YOU TRY IT 13

The Booster Club is making 22 sashes for the high school band members. Each sash requires $1\frac{3}{8}$ yd of material at a cost of $12 per yard. Find the total cost of the material.

Your Strategy

Your Solution

Solutions on p. S8

3.3 Exercises

OBJECTIVE A **Multiplication of fractions**

1. Circle the correct word to complete the sentence.

   a. Fractions <u>can/cannot</u> be multiplied when their denominators are not the same.

   b. To multiply two fractions, write the <u>sum/product</u> of the numerators over the <u>sum/product</u> of the denominators.

2. The product of 1 and a number is $\frac{3}{8}$. Find the number. Explain how you arrived at the answer.

Multiply.

3. $\frac{2}{3} \cdot \frac{9}{10}$

4. $\frac{3}{8} \cdot \frac{4}{5}$

5. $-\frac{6}{7} \cdot \frac{11}{12}$

6. $\frac{5}{6} \cdot \left(-\frac{2}{5}\right)$

7. $\frac{14}{15} \cdot \frac{6}{7}$

8. $\frac{15}{16} \cdot \frac{4}{9}$

9. $-\frac{6}{7} \cdot \frac{0}{10}$

10. $\frac{5}{12} \cdot \frac{3}{0}$

11. $\left(-\frac{4}{15}\right) \cdot \left(-\frac{3}{8}\right)$

12. $\left(-\frac{3}{4}\right) \cdot \left(-\frac{2}{9}\right)$

13. $-\frac{3}{4} \cdot \frac{1}{2}$

14. $-\frac{8}{15} \cdot \frac{5}{12}$

15. $\frac{9}{x} \cdot \frac{7}{y}$

16. $\frac{4}{c} \cdot \frac{8}{d}$

17. $-\frac{y}{5} \cdot \frac{z}{6}$

18. $-\frac{a}{10} \cdot \left(-\frac{b}{6}\right)$

19. $\frac{2}{3} \cdot \frac{3}{8} \cdot \frac{4}{9}$

20. $\frac{5}{7} \cdot \frac{1}{6} \cdot \frac{14}{15}$

21. $-\frac{7}{12} \cdot \frac{5}{8} \cdot \frac{16}{25}$

22. $\frac{5}{12} \cdot \left(-\frac{1}{3}\right) \cdot \left(-\frac{8}{15}\right)$

23. $\left(-\frac{3}{5}\right) \cdot \frac{1}{2} \cdot \left(-\frac{5}{8}\right)$

24. $\frac{5}{6} \cdot \left(-\frac{2}{3}\right) \cdot \frac{3}{25}$

25. $6 \cdot \frac{1}{6}$

26. $\frac{1}{10} \cdot 10$

27. $\frac{3}{4} \cdot 8$

28. $\frac{5}{7} \cdot 14$

29. $12 \cdot \left(-\frac{5}{8}\right)$

30. $24 \cdot \left(-\frac{3}{8}\right)$

31. $-16 \cdot \dfrac{7}{30}$ **32.** $-9 \cdot \dfrac{7}{15}$ **33.** $\dfrac{6}{7} \cdot 0$ **34.** $0 \cdot \dfrac{9}{11}$

35. $\dfrac{5}{22} \cdot 2\dfrac{1}{5}$ **36.** $\dfrac{4}{15} \cdot 1\dfrac{7}{8}$ **37.** $3\dfrac{1}{2} \cdot 5\dfrac{3}{7}$ **38.** $2\dfrac{1}{4} \cdot 1\dfrac{1}{3}$

39. $3\dfrac{1}{3} \cdot \left(-\dfrac{7}{10}\right)$ **40.** $2\dfrac{1}{4} \cdot \left(-\dfrac{7}{9}\right)$ **41.** $-1\dfrac{2}{3} \cdot \left(-\dfrac{3}{5}\right)$ **42.** $-2\dfrac{1}{8} \cdot \left(-\dfrac{4}{17}\right)$

43. $3\dfrac{1}{3} \cdot 2\dfrac{1}{3}$ **44.** $3\dfrac{1}{4} \cdot 2\dfrac{2}{3}$ **45.** $3\dfrac{1}{3} \cdot (-9)$ **46.** $-2\dfrac{1}{2} \cdot 4$

47. $8 \cdot 5\dfrac{1}{4}$ **48.** $3 \cdot 2\dfrac{1}{9}$ **49.** $3\dfrac{1}{2} \cdot 1\dfrac{5}{7} \cdot \dfrac{11}{12}$ **50.** $2\dfrac{2}{3} \cdot \dfrac{8}{9} \cdot 1\dfrac{5}{16}$

51. Find the product of $\dfrac{3}{4}$ and $\dfrac{14}{15}$.

52. Find the product of $\dfrac{12}{25}$ and $\dfrac{5}{16}$.

53. Find $-\dfrac{9}{16}$ multiplied by $\dfrac{4}{27}$.

54. Find $\dfrac{3}{7}$ multiplied by $-\dfrac{14}{15}$.

55. What is the product of $-\dfrac{7}{24}, \dfrac{8}{21}$, and $\dfrac{3}{7}$?

56. What is the product of $-\dfrac{5}{13}, -\dfrac{26}{75}$, and $\dfrac{5}{8}$?

57. What is $4\dfrac{4}{5}$ times $\dfrac{3}{8}$?

58. What is $5\dfrac{1}{3}$ times $\dfrac{3}{16}$?

59. Find the product of $-2\dfrac{2}{3}$ and $-1\dfrac{11}{16}$.

60. Find the product of $1\dfrac{3}{11}$ and $5\dfrac{1}{2}$.

Cost of Living A typical household in the United States has an average after-tax income of $45,000. The graph at the right represents how this annual income is spent. Use this graph for Exercises 61 and 62.

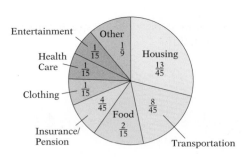

61. Find the amount of money a typical household in the United States spends on housing per year.

62. How much money does a typical household in the United States spend annually on food?

How a Typical U.S. Household Spends Its Annual Income

Source: Based on data from American Demographics

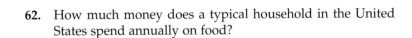

Evaluate the variable expression xy for the given values of x and y.

63. $x = -\dfrac{5}{16}, y = \dfrac{7}{15}$

64. $x = -\dfrac{2}{5}, y = -\dfrac{5}{6}$

65. $x = \dfrac{4}{7}, y = 6\dfrac{1}{8}$

66. $x = 6\dfrac{3}{5}, y = 3\dfrac{1}{3}$

67. $x = -49, y = \dfrac{5}{14}$

68. $x = -\dfrac{3}{10}, y = -35$

69. $x = 1\dfrac{3}{13}, y = -6\dfrac{1}{2}$

70. $x = -3\dfrac{1}{2}, y = -2\dfrac{2}{7}$

Evaluate the variable expression xyz for the given values of x, y, and z.

71. $x = \dfrac{3}{8}, y = \dfrac{2}{3}, z = \dfrac{4}{5}$

72. $x = 4, y = \dfrac{0}{8}, z = 1\dfrac{5}{9}$

73. $x = 2\dfrac{3}{8}, y = -\dfrac{3}{19}, z = -\dfrac{4}{9}$

74. $x = \dfrac{4}{5}, y = -15, z = \dfrac{7}{8}$

75. $x = \dfrac{5}{6}, y = -3, z = 1\dfrac{7}{15}$

76. $x = 4\dfrac{1}{2}, y = 3\dfrac{5}{9}, z = 1\dfrac{7}{8}$

77. Is $-\dfrac{1}{3}$ a solution of the equation $\dfrac{3}{4}y = -\dfrac{1}{4}$?

78. Is $\dfrac{2}{5}$ a solution of the equation $-\dfrac{5}{6}z = \dfrac{1}{3}$?

79. Is $\dfrac{3}{4}$ a solution of the equation $\dfrac{4}{5}x = \dfrac{5}{3}$?

80. Is $\dfrac{1}{2}$ a solution of the equation $\dfrac{3}{4}p = \dfrac{3}{2}$?

81. Is $-\dfrac{1}{6}$ a solution of the equation $6x = 1$?

82. Is $-\dfrac{4}{5}$ a solution of the equation $\dfrac{5}{4}n = -1$?

83. Is the product $\frac{2}{3} \cdot n$ greater than n or less than n when n is a proper fraction?

84. Give an example of a proper fraction and an improper fraction whose product is -1.

OBJECTIVE B **Division of fractions**

85. The reciprocal of $\frac{7}{3}$ is _____.

86. The reciprocal of $-\frac{5}{6}$ is _____.

Divide.

87. $\dfrac{5}{7} \div \dfrac{2}{5}$

88. $\dfrac{3}{8} \div \dfrac{2}{3}$

89. $\dfrac{4}{7} \div \left(-\dfrac{4}{7}\right)$

90. $-\dfrac{5}{7} \div \left(-\dfrac{5}{6}\right)$

91. $0 \div \dfrac{7}{9}$

92. $0 \div \dfrac{4}{5}$

93. $\left(-\dfrac{1}{3}\right) \div \dfrac{1}{2}$

94. $\left(-\dfrac{3}{8}\right) \div \dfrac{7}{8}$

95. $-\dfrac{5}{16} \div \left(-\dfrac{3}{8}\right)$

96. $\left(-\dfrac{3}{4}\right) \div \left(-\dfrac{5}{6}\right)$

97. $\dfrac{0}{1} \div \dfrac{1}{9}$

98. $\dfrac{1}{2} \div \left(-\dfrac{8}{0}\right)$

99. $6 \div \dfrac{3}{4}$

100. $8 \div \dfrac{2}{3}$

101. $\dfrac{3}{4} \div (-6)$

102. $-\dfrac{2}{3} \div 8$

103. $\dfrac{9}{10} \div 0$

104. $\dfrac{2}{11} \div 0$

105. $\dfrac{5}{12} \div \left(-\dfrac{15}{32}\right)$

106. $\dfrac{3}{8} \div \left(-\dfrac{5}{12}\right)$

107. $\left(-\dfrac{2}{3}\right) \div (-4)$

108. $\left(-\dfrac{4}{9}\right) \div (-6)$

109. $\dfrac{8}{x} \div \left(-\dfrac{y}{4}\right)$

110. $-\dfrac{9}{m} \div \dfrac{n}{7}$

111. $\dfrac{b}{6} \div \dfrac{5}{d}$ **112.** $\dfrac{y}{10} \div \dfrac{4}{z}$ **113.** $3\dfrac{1}{3} \div \dfrac{5}{8}$ **114.** $5\dfrac{1}{2} \div \dfrac{1}{4}$

115. $5\dfrac{3}{5} \div \left(-\dfrac{7}{10}\right)$ **116.** $6\dfrac{8}{9} \div \left(-\dfrac{31}{36}\right)$ **117.** $-1\dfrac{1}{2} \div 1\dfrac{3}{4}$ **118.** $-1\dfrac{3}{5} \div 3\dfrac{1}{10}$

119. $5\dfrac{1}{2} \div 11$ **120.** $4\dfrac{2}{3} \div 7$ **121.** $5\dfrac{2}{7} \div 1$ **122.** $9\dfrac{5}{6} \div 1$

123. $-16 \div 1\dfrac{1}{3}$ **124.** $-9 \div \left(-3\dfrac{3}{5}\right)$ **125.** $2\dfrac{4}{13} \div 1\dfrac{5}{26}$ **126.** $3\dfrac{3}{8} \div 2\dfrac{7}{16}$

127. Find the quotient of $\dfrac{9}{10}$ and $\dfrac{3}{4}$.   **128.** Find the quotient of $\dfrac{3}{5}$ and $\dfrac{12}{25}$.

129. What is $-\dfrac{15}{24}$ divided by $\dfrac{3}{5}$?   **130.** What is $\dfrac{5}{6}$ divided by $-\dfrac{10}{21}$?

131. Find $\dfrac{7}{8}$ divided by $3\dfrac{1}{4}$.   **132.** Find $-\dfrac{3}{8}$ divided by $2\dfrac{1}{4}$.

133. What is the quotient of $-3\dfrac{5}{11}$ and $3\dfrac{4}{5}$?   **134.** What is the quotient of $-10\dfrac{1}{5}$ and $-1\dfrac{7}{10}$?

Evaluate the variable expression $x \div y$ for the given values of x and y.

135. $x = -\dfrac{5}{8}, y = -\dfrac{15}{2}$ **136.** $x = -\dfrac{14}{3}, y = -\dfrac{7}{9}$ **137.** $x = \dfrac{1}{7}, y = 0$ **138.** $x = \dfrac{4}{0}, y = 12$

139. $x = -18, y = \dfrac{3}{8}$ **140.** $x = 20, y = -\dfrac{5}{6}$ **141.** $x = -\dfrac{1}{2}, y = -3\dfrac{5}{8}$ **142.** $x = 4\dfrac{3}{8}, y = 7$

143. $x = 6\dfrac{2}{5}, y = -4$ **144.** $x = -2\dfrac{5}{8}, y = 1\dfrac{3}{4}$ **145.** $x = -3\dfrac{2}{5}, y = -1\dfrac{7}{10}$ **146.** $x = -5\dfrac{2}{5}, y = -9$

147. Is the quotient $n \div \dfrac{1}{2}$ greater than n or less than n when n is a proper fraction?

148. Give an example of a proper fraction and an improper fraction whose quotient is -1.

OBJECTIVE C   **Applications and formulas**

149. Fill in the blank with the correct operation: A gardener wants to space his rows of vegetables $1\dfrac{1}{4}$ ft apart. His garden is 12 ft long. To find how many rows he can fit in the garden, use _____.

150. A car used $10\dfrac{1}{4}$ gal of gas to travel 246 mi. To find the number of miles the car travels on 1 gal of gas, divide _____ by _____.

Solve.

 The Food Industry   The table at the right shows the net weights of four different boxes of cereal. Use this table for Exercises 151 and 152.

Cereal	Net Weight
Kellogg Honey Crunch Corn Flakes	24 oz
Nabisco Instant Cream of Wheat	28 oz
Post Shredded Wheat	18 oz
Quaker Oats	41 oz

151. Find the number of $\dfrac{3}{4}$-ounce servings in a box of Kellogg Honey Crunch Corn Flakes.

152. Find the number of $1\dfrac{1}{4}$-ounce servings in a box of Shredded Wheat.

153. *Sports*   A chukker is one period of play in a polo match. A chukker lasts $7\frac{1}{2}$ min. Find the length of time in four chukkers.

154. *History*   The Assyrian calendar was based on the phases of the moon. One lunation was $29\frac{1}{2}$ days long. There were 12 lunations in one year. Find the number of days in one year in the Assyrian calendar.

155. *Measurement*   One rod is equal to $5\frac{1}{2}$ yd. How many feet are in one rod? How many inches are in one rod?

156. *Travel*   A car used $12\frac{1}{2}$ gal of gasoline on a 275-mile trip. How many miles can this car travel on 1 gal of gasoline?

157. *Housework*   According to a national survey, the average couple spends $4\frac{1}{2}$ h cleaning house each week. How many hours does the average couple spend cleaning house each year?

158. *Business*   A factory worker can assemble a product in $7\frac{1}{2}$ min. How many products can the worker assemble in one hour?

159. *Real Estate*   A developer purchases $25\frac{1}{2}$ acres of land and plans to set aside 3 acres for an entranceway to a housing development to be built on the property. Each house will be built on a $\frac{3}{4}$-acre plot of land. How many houses does the developer plan to build on the property?

160. *Consumerism*   You are planning a barbecue for 25 people. You want to serve $\frac{1}{4}$-pound hamburger patties to your guests and you estimate each person will eat two hamburgers. How much hamburger meat should you buy for the barbecue?

161. *Board Games*   A wooden travel game board has hinges that allow the board to be folded in half. If the dimensions of the open board are 14 in. by 14 in. by $\frac{7}{8}$ in., what are the dimensions of the board when it is closed?

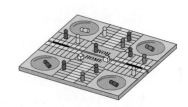

162. *Wages*   Find the total wages of an employee who worked $26\frac{1}{2}$ h this week and who earns an hourly wage of $12.

To add two mixed numbers, first write the fractional parts as equivalent fractions with a common denominator. Then add the fractional parts and add the whole numbers.

Add: $3\dfrac{5}{8} + 4\dfrac{7}{12}$

Write the fractions as equivalent fractions with a common denominator. The common denominator is the LCM of 8 and 12 (24).

$$3\dfrac{5}{8} + 4\dfrac{7}{12} = 3\dfrac{15}{24} + 4\dfrac{14}{24}$$

Add the fractional parts and add the whole numbers.

$$= 7\dfrac{29}{24}$$

Write the sum in simplest form.

$$= 7 + \dfrac{29}{24}$$

$$= 7 + 1\dfrac{5}{24}$$

$$= 8\dfrac{5}{24}$$

Evaluate $x + y$ for $x = 2\dfrac{3}{4}$ and $y = 7\dfrac{5}{6}$.

$$x + y$$

Replace x with $2\dfrac{3}{4}$ and y with $7\dfrac{5}{6}$.

$$2\dfrac{3}{4} + 7\dfrac{5}{6}$$

Write the fractions as equivalent fractions with a common denominator.

$$= 2\dfrac{9}{12} + 7\dfrac{10}{12}$$

Add the fractional parts and add the whole numbers.

$$= 9\dfrac{19}{12}$$

Write the sum in simplest form.

$$= 10\dfrac{7}{12}$$

EXAMPLE 1   Add: $\dfrac{9}{16} + \dfrac{5}{12}$

Solution $\dfrac{9}{16} + \dfrac{5}{12} = \dfrac{27}{48} + \dfrac{20}{48}$

$$= \dfrac{27 + 20}{48} = \dfrac{47}{48}$$

YOU TRY IT 1   Add: $\dfrac{7}{12} + \dfrac{3}{8}$

Your Solution

Solution on p. S8

EXAMPLE 2

Add: $\frac{4}{5} + \frac{3}{4} + \frac{5}{8}$

Solution

$\frac{4}{5} + \frac{3}{4} + \frac{5}{8} = \frac{32}{40} + \frac{30}{40} + \frac{25}{40} = \frac{87}{40} = 2\frac{7}{40}$

YOU TRY IT 2

Add: $\frac{3}{5} + \frac{2}{3} + \frac{5}{6}$

Your Solution

EXAMPLE 3

Find the sum of $12\frac{4}{7}$ and 19.

Solution

$12\frac{4}{7} + 19 = 31\frac{4}{7}$

YOU TRY IT 3

What is the sum of 16 and $8\frac{5}{9}$?

Your Solution

EXAMPLE 4

Add: $-\frac{3}{8} + \frac{3}{4} + \left(-\frac{5}{6}\right)$

Solution

$-\frac{3}{8} + \frac{3}{4} + \left(-\frac{5}{6}\right) = \frac{-3}{8} + \frac{3}{4} + \frac{-5}{6}$

$= \frac{-9}{24} + \frac{18}{24} + \frac{-20}{24}$

$= \frac{-9 + 18 + (-20)}{24}$

$= \frac{-11}{24} = -\frac{11}{24}$

YOU TRY IT 4

Add: $-\frac{5}{12} + \frac{5}{8} + \left(-\frac{1}{6}\right)$

Your Solution

EXAMPLE 5

Evaluate $x + y + z$ for $x = 2\frac{1}{6}$, $y = 4\frac{3}{8}$, and $z = 7\frac{5}{9}$.

Solution

$x + y + z$

$2\frac{1}{6} + 4\frac{3}{8} + 7\frac{5}{9} = 2\frac{12}{72} + 4\frac{27}{72} + 7\frac{40}{72}$

$= 13\frac{79}{72}$

$= 14\frac{7}{72}$

YOU TRY IT 5

Evaluate $x + y + z$ for $x = 3\frac{5}{6}$, $y = 2\frac{1}{9}$, and $z = 5\frac{5}{12}$.

Your Solution

Solutions on p. S8

OBJECTIVE B   Subtraction of fractions

In the last objective, it was stated that in order for fractions to be added, the fractions must have the same denominator. The same is true for subtracting fractions: The two fractions must have the same denominator.

Point of Interest

The first woman mathematician for whom documented evidence exists is Hypatia (370–415). She lived in Alexandria, Egypt, and lectured at the Museum, the forerunner of our modern university. She made important contributions in mathematics, astronomy, and philosophy.

Subtraction of Fractions

To subtract fractions with the same denominator, subtract the numerators and place the difference over the common denominator.

$$\frac{a}{b} - \frac{c}{b} = \frac{a - c}{b}, \qquad \text{where} \quad b \neq 0$$

Subtract: $\dfrac{5}{8} - \dfrac{3}{8}$

The denominators are the same. Subtract the numerators and place the difference over the common denominator.

$$\frac{5}{8} - \frac{3}{8} = \frac{5 - 3}{8}$$

Write the answer in simplest form.

$$= \frac{2}{8} = \frac{1}{4}$$

To subtract fractions with different denominators, first rewrite the fractions as equivalent fractions with a common denominator. The common denominator is the least common multiple (LCM) of the denominators of the fractions.

Subtract: $\dfrac{5}{12} - \dfrac{3}{8}$

The common denominator is the LCM of 12 and 8.

The LCM of 12 and 8 is 24.

Write the fractions as equivalent fractions with the common denominator.

$$\frac{5}{12} - \frac{3}{8} = \frac{10}{24} - \frac{9}{24}$$

Subtract the fractions.

$$= \frac{10 - 9}{24} = \frac{1}{24}$$

To subtract fractions with negative signs, first rewrite the fractions with the negative signs in the numerators.

Simplify: $-\dfrac{2}{9} - \dfrac{5}{12}$

Rewrite the negative fraction with the negative sign in the numerator.

$$-\frac{2}{9} - \frac{5}{12} = \frac{-2}{9} - \frac{5}{12}$$

Write the fractions as equivalent fractions with a common denominator.

$$= \frac{-8}{36} - \frac{15}{36}$$

Subtract the numerators and place the difference over the common denominator.

$$= \frac{-8 - 15}{36} = \frac{-23}{36}$$

Write the negative sign in front of the fraction.

$$= -\frac{23}{36}$$

Subtract: $\dfrac{2}{3} - \left(-\dfrac{4}{5}\right)$

Rewrite subtraction as addition of the opposite.

$$\dfrac{2}{3} - \left(-\dfrac{4}{5}\right) = \dfrac{2}{3} + \dfrac{4}{5}$$

Write the fractions as equivalent fractions with a common denominator.

$$= \dfrac{10}{15} + \dfrac{12}{15}$$

Add the fractions.

$$= \dfrac{10 + 12}{15}$$

$$= \dfrac{22}{15} = 1\dfrac{7}{15}$$

To subtract mixed numbers when borrowing is not necessary, subtract the fractional parts and then subtract the whole numbers.

Find the difference between $5\dfrac{8}{9}$ and $2\dfrac{5}{6}$.

The LCM of 9 and 6 is 18.

Write the fractions as equivalent fractions with the LCM as the common denominator.

$$5\dfrac{8}{9} - 2\dfrac{5}{6} = 5\dfrac{16}{18} - 2\dfrac{15}{18}$$

Subtract the fractional parts and subtract the whole numbers.

$$= 3\dfrac{1}{18}$$

As in subtraction with whole numbers, subtraction of mixed numbers may involve borrowing.

Subtract: $7 - 4\dfrac{2}{3}$

Borrow 1 from 7. Write the 1 as a fraction with the same denominator as the fractional part of the mixed number (3).

$$7 - 4\dfrac{2}{3} = 6\dfrac{3}{3} - 4\dfrac{2}{3}$$

Note: $7 = 6 + 1 = 6 + \dfrac{3}{3} = 6\dfrac{3}{3}$

Subtract the fractional parts and subtract the whole numbers.

$$= 2\dfrac{1}{3}$$

Subtract: $9\dfrac{1}{8} - 2\dfrac{5}{6}$

Write the fractions as equivalent fractions with a common denominator.

$$9\dfrac{1}{8} - 2\dfrac{5}{6} = 9\dfrac{3}{24} - 2\dfrac{20}{24}$$

$3 < 20$. Borrow 1 from 9. Add the 1 to $\dfrac{3}{24}$.

Note: $9\dfrac{3}{24} = 9 + \dfrac{3}{24} = 8 + 1 + \dfrac{3}{24}$

$$= 8 + \dfrac{24}{24} + \dfrac{3}{24} = 8 + \dfrac{27}{24} = 8\dfrac{27}{24}$$

$$= 8\dfrac{27}{24} - 2\dfrac{20}{24}$$

Subtract.

$$= 6\dfrac{7}{24}$$

Evaluate $x - y$ for $x = 7\frac{2}{9}$ and $y = 3\frac{5}{12}$.

$x - y$

Replace x with $7\frac{2}{9}$ and y with $3\frac{5}{12}$.

$7\frac{2}{9} - 3\frac{5}{12}$

Write the fractions as equivalent fractions with a common denominator.

$= 7\frac{8}{36} - 3\frac{15}{36}$

$8 < 15$. Borrow 1 from 7. Add the 1 to $\frac{8}{36}$.

Note: $7\frac{8}{36} = 6 + \frac{36}{36} + \frac{8}{36} = 6\frac{44}{36}$

$= 6\frac{44}{36} - 3\frac{15}{36}$

Subtract.

$= 3\frac{29}{36}$

EXAMPLE 6

Subtract: $-\frac{5}{6} - \left(-\frac{3}{8}\right)$

Solution

$-\frac{5}{6} - \left(-\frac{3}{8}\right) = -\frac{5}{6} + \frac{3}{8} = \frac{-20}{24} + \frac{9}{24}$

$= \frac{-20 + 9}{24}$

$= \frac{-11}{24} = -\frac{11}{24}$

YOU TRY IT 6

Subtract: $-\frac{5}{6} - \frac{7}{9}$

Your Solution

EXAMPLE 7

Find the difference between $8\frac{5}{6}$ and $2\frac{3}{4}$.

Solution

$8\frac{5}{6} - 2\frac{3}{4} = 8\frac{10}{12} - 2\frac{9}{12} = 6\frac{1}{12}$

YOU TRY IT 7

Find the difference between $9\frac{7}{8}$ and $5\frac{2}{3}$.

Your Solution

EXAMPLE 8

Subtract: $7 - 3\frac{5}{13}$

Solution

$7 - 3\frac{5}{13} = 6\frac{13}{13} - 3\frac{5}{13} = 3\frac{8}{13}$

YOU TRY IT 8

Subtract: $6 - 4\frac{2}{11}$

Your Solution

Solutions on p. S8

EXAMPLE 9

Is $\frac{3}{8}$ a solution of the equation $\frac{2}{3} = w - \frac{5}{6}$?

Solution

$$\frac{2}{3} = w - \frac{5}{6}$$

$\dfrac{2}{3}$	$\dfrac{3}{8} - \dfrac{5}{6}$
$\dfrac{2}{3}$	$\dfrac{9}{24} - \dfrac{20}{24}$
$\dfrac{2}{3}$	$\dfrac{-11}{24}$

$$\frac{2}{3} \neq -\frac{11}{24}$$

No, $\frac{3}{8}$ is not a solution of the equation.

YOU TRY IT 9

Is $-\frac{1}{4}$ a solution of the equation $\frac{2}{3} - v = \frac{11}{12}$?

Your Solution

Solution on p. S8

OBJECTIVE C **Applications and formulas**

EXAMPLE 10

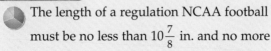

The length of a regulation NCAA football must be no less than $10\frac{7}{8}$ in. and no more than $11\frac{7}{16}$ in. What is the difference between the minimum and maximum lengths of an NCAA regulation football?

Strategy

To find the difference, subtract the minimum length $\left(10\frac{7}{8}\right)$ from the maximum length $\left(11\frac{7}{16}\right)$.

Solution

$$11\frac{7}{16} - 10\frac{7}{8} = 11\frac{7}{16} - 10\frac{14}{16} = 10\frac{23}{16} - 10\frac{14}{16} = \frac{9}{16}$$

The difference is $\frac{9}{16}$ in.

YOU TRY IT 10

The Heller Research Group conducted a survey to determine favorite doughnut flavors. $\frac{2}{5}$ of the respondents named glazed doughnuts, $\frac{8}{25}$ named filled doughnuts, and $\frac{3}{20}$ named frosted doughnuts. What fraction of the respondents did not name glazed, filled, or frosted as their favorite type of doughnut?

Your Strategy

Your Solution

Solution on pp. S8–S9

3.4 Exercises

OBJECTIVE A **Addition of fractions**

Circle the correct phrase to complete the sentence.

1. Fractions cannot be added unless their <u>numerators/denominators</u> are the same.

2. To add two fractions with the same denominator, place the sum of the numerators over the <u>sum of the denominators/common denominator.</u>

Add.

3. $\dfrac{4}{11} + \dfrac{5}{11}$

4. $\dfrac{3}{7} + \dfrac{2}{7}$

5. $\dfrac{2}{3} + \dfrac{1}{3}$

6. $\dfrac{1}{2} + \dfrac{1}{2}$

7. $\dfrac{5}{6} + \dfrac{5}{6}$

8. $\dfrac{3}{8} + \dfrac{7}{8}$

9. $\dfrac{7}{18} + \dfrac{13}{18} + \dfrac{1}{18}$

10. $\dfrac{8}{15} + \dfrac{2}{15} + \dfrac{11}{15}$

11. $\dfrac{7}{b} + \dfrac{9}{b}$

12. $\dfrac{3}{y} + \dfrac{6}{y}$

13. $\dfrac{5}{c} + \dfrac{4}{c}$

14. $\dfrac{2}{a} + \dfrac{8}{a}$

15. $\dfrac{1}{x} + \dfrac{4}{x} + \dfrac{6}{x}$

16. $\dfrac{8}{n} + \dfrac{5}{n} + \dfrac{3}{n}$

17. $\dfrac{1}{4} + \dfrac{2}{3}$

18. $\dfrac{2}{3} + \dfrac{1}{2}$

19. $\dfrac{7}{15} + \dfrac{9}{20}$

20. $\dfrac{4}{9} + \dfrac{1}{6}$

21. $\dfrac{2}{3} + \dfrac{1}{12} + \dfrac{5}{6}$

22. $\dfrac{3}{8} + \dfrac{1}{2} + \dfrac{5}{12}$

23. $\dfrac{7}{12} + \dfrac{3}{4} + \dfrac{4}{5}$

24. $\dfrac{7}{11} + \dfrac{1}{2} + \dfrac{5}{6}$

25. $-\dfrac{3}{4} + \dfrac{2}{3}$

26. $-\dfrac{7}{12} + \dfrac{5}{8}$

27. $\dfrac{2}{5} + \left(-\dfrac{11}{15}\right)$

28. $\dfrac{1}{4} + \left(-\dfrac{1}{7}\right)$

29. $\dfrac{3}{8} + \left(-\dfrac{1}{2}\right) + \dfrac{7}{12}$

30. $-\dfrac{7}{12} + \dfrac{2}{3} + \left(-\dfrac{4}{5}\right)$

31. $\frac{2}{3} + \left(-\frac{5}{6}\right) + \frac{1}{4}$ **32.** $-\frac{5}{8} + \frac{3}{4} + \frac{1}{2}$ **33.** $8 + 7\frac{2}{3}$ **34.** $6 + 9\frac{3}{5}$

35. $2\frac{1}{6} + 3\frac{1}{2}$ **36.** $1\frac{3}{10} + 4\frac{3}{5}$ **37.** $8\frac{3}{5} + 6\frac{9}{20}$ **38.** $7\frac{5}{12} + 3\frac{7}{9}$

39. $5\frac{5}{12} + 4\frac{7}{9}$ **40.** $2\frac{11}{12} + 3\frac{7}{15}$ **41.** $2\frac{1}{4} + 3\frac{1}{2} + 1\frac{2}{3}$ **42.** $1\frac{2}{3} + 2\frac{5}{6} + 4\frac{7}{9}$

Solve.

43. What is $-\frac{5}{6}$ added to $\frac{4}{9}$?   **44.** What is $\frac{7}{12}$ added to $-\frac{11}{16}$?

45. Find the total of $\frac{2}{7}$, $\frac{3}{14}$, and $\frac{1}{4}$.   **46.** Find the total of $\frac{1}{3}$, $\frac{5}{18}$, and $\frac{2}{9}$.

47. What is $-\frac{2}{3}$ more than $-\frac{5}{6}$?   **48.** What is $-\frac{7}{12}$ more than $-\frac{5}{9}$?

49. Find $3\frac{7}{12}$ plus $2\frac{5}{8}$.   **50.** Find the sum of $7\frac{11}{15}$, $2\frac{7}{10}$, and $5\frac{2}{5}$.

Evaluate the variable expression $x + y$ for the given values of x and y.

51. $x = \frac{3}{5}, y = \frac{4}{5}$ **52.** $x = \frac{5}{8}, y = \frac{3}{8}$ **53.** $x = \frac{2}{3}, y = -\frac{3}{4}$ **54.** $x = -\frac{3}{8}, y = \frac{2}{9}$

55. $x = \frac{5}{6}, y = \frac{8}{9}$ **56.** $x = \frac{3}{10}, y = -\frac{7}{15}$ **57.** $x = -\frac{5}{8}, y = -\frac{1}{6}$ **58.** $x = -\frac{3}{8}, y = -\frac{5}{6}$

Evaluate the variable expression $x + y + z$ for the given values of x, y, and z.

59. $x = \dfrac{3}{8}, y = \dfrac{1}{4}, z = \dfrac{7}{12}$

60. $x = \dfrac{5}{6}, y = \dfrac{2}{3}, z = \dfrac{7}{24}$

61. $x = 1\dfrac{1}{2}, y = 3\dfrac{3}{4}, z = 6\dfrac{5}{12}$

62. $x = 7\dfrac{2}{3}, y = 2\dfrac{5}{6}, z = 5\dfrac{4}{9}$

63. $x = 4\dfrac{3}{5}, y = 8\dfrac{7}{10}, z = 1\dfrac{9}{20}$

64. $x = 2\dfrac{3}{14}, y = 5\dfrac{5}{7}, z = 3\dfrac{1}{2}$

65. Is $-\dfrac{3}{5}$ a solution of the equation $z + \dfrac{1}{4} = -\dfrac{7}{20}$?

66. Is $\dfrac{3}{8}$ a solution of the equation $\dfrac{3}{4} = t + \dfrac{3}{8}$?

67. Is $-\dfrac{5}{6}$ a solution of the equation $\dfrac{1}{4} + x = -\dfrac{7}{12}$?

68. Is $-\dfrac{4}{5}$ a solution of the equation $0 = q + \dfrac{4}{5}$?

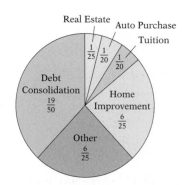

Loans The figure at the right shows how the money borrowed on home equity loans is spent. Use this graph for Exercises 69 and 70.

69. What fractional part of the money borrowed on home equity loans is spent on debt consolidation and home improvement?

70. What fractional part of the money borrowed on home equity loans is spent on home improvement, cars, and tuition?

How Money Borrowed on Home Equity Loans Is Spent
Source: Consumer Bankers Association

71. Which expression is equivalent to $\dfrac{3}{4} + \dfrac{1}{5}$?

(i) $\dfrac{3 + 1}{4 + 5}$ (ii) $\dfrac{(3 + 5) + (1 + 4)}{4 + 5}$ (iii) $\dfrac{(3 \cdot 5) + (1 \cdot 4)}{4 \cdot 5}$ (iv) $\dfrac{3 + 1}{4 \cdot 5}$

72. Estimate the sum to the nearest integer.

a. $\dfrac{7}{8} + \dfrac{4}{5}$ b. $\dfrac{1}{3} + \left(-\dfrac{1}{2}\right)$ c. $-\dfrac{1}{8} + 1\dfrac{1}{4}$ d. $-1\dfrac{1}{3} + \dfrac{1}{5}$

OBJECTIVE B **Subtraction of fractions**

73. **a.** Write "the difference of $\frac{1}{2}$ and $-\frac{3}{7}$" as a subtraction problem:

_____ − _____.

b. Rewrite the subtraction problem in part (a) as an addition problem:

_____ + _____

74. Complete the subtraction: $8 - 3\frac{4}{5} = 7\dfrac{\boxed{}}{5} - 3\frac{4}{5} = 4\dfrac{\boxed{}}{5}$

Subtract.

75. $\dfrac{7}{12} - \dfrac{5}{12}$

76. $\dfrac{17}{20} - \dfrac{9}{20}$

77. $\dfrac{11}{24} - \dfrac{7}{24}$

78. $\dfrac{39}{48} - \dfrac{23}{48}$

79. $\dfrac{8}{d} - \dfrac{3}{d}$

80. $\dfrac{12}{y} - \dfrac{7}{y}$

81. $\dfrac{5}{n} - \dfrac{10}{n}$

82. $\dfrac{6}{c} - \dfrac{13}{c}$

83. $\dfrac{3}{7} - \dfrac{5}{14}$

84. $\dfrac{7}{8} - \dfrac{5}{16}$

85. $\dfrac{2}{3} - \dfrac{1}{6}$

86. $\dfrac{5}{21} - \dfrac{1}{6}$

87. $\dfrac{11}{12} - \dfrac{2}{3}$

88. $\dfrac{9}{20} - \dfrac{1}{30}$

89. $-\dfrac{1}{2} - \dfrac{3}{8}$

90. $-\dfrac{5}{6} - \dfrac{1}{9}$

91. $-\dfrac{3}{10} - \dfrac{4}{5}$

92. $-\dfrac{7}{15} - \dfrac{3}{10}$

93. $-\dfrac{5}{12} - \left(-\dfrac{2}{3}\right)$

94. $-\dfrac{3}{10} - \left(-\dfrac{5}{6}\right)$

95. $-\dfrac{5}{9} - \left(-\dfrac{11}{12}\right)$

96. $-\dfrac{5}{8} - \left(-\dfrac{7}{12}\right)$

97. $4\dfrac{11}{18} - 2\dfrac{5}{18}$

98. $3\dfrac{7}{12} - 1\dfrac{1}{12}$

99. $8\dfrac{3}{4} - 2$

100. $6\dfrac{5}{9} - 4$

101. $8\dfrac{5}{6} - 7\dfrac{3}{4}$

102. $5\dfrac{7}{8} - 3\dfrac{2}{3}$

103. $7 - 3\dfrac{5}{8}$

104. $6 - 2\dfrac{4}{5}$

105. $10 - 4\dfrac{8}{9}$

106. $5 - 2\dfrac{7}{18}$

107. $7\dfrac{3}{8} - 4\dfrac{5}{8}$

108. $11\dfrac{1}{6} - 8\dfrac{5}{6}$

109. $12\dfrac{5}{12} - 10\dfrac{17}{24}$

110. $16\dfrac{1}{3} - 11\dfrac{5}{12}$

111. $6\dfrac{2}{3} - 1\dfrac{7}{8}$

112. $7\dfrac{7}{12} - 2\dfrac{5}{6}$

113. $10\dfrac{2}{5} - 8\dfrac{7}{10}$

114. $5\dfrac{5}{6} - 4\dfrac{7}{8}$

Solve.

115. What is $-\dfrac{7}{12}$ minus $\dfrac{7}{9}$?

116. What is $\dfrac{3}{5}$ decreased by $-\dfrac{7}{10}$?

117. What is $-\dfrac{2}{3}$ less than $-\dfrac{7}{8}$?

118. Find the difference between $-\dfrac{1}{6}$ and $-\dfrac{8}{9}$.

119. Find 8 less $1\dfrac{7}{12}$.

120. Find 9 minus $5\dfrac{3}{20}$.

Evaluate the variable expression $x - y$ for the given values of x and y.

121. $x = \dfrac{8}{9}, y = \dfrac{5}{9}$

122. $x = \dfrac{5}{6}, y = \dfrac{1}{6}$

123. $x = -\dfrac{11}{12}, y = \dfrac{5}{12}$

124. $x = -\dfrac{15}{16}, y = \dfrac{5}{16}$

125. $x = -\dfrac{2}{3}, y = -\dfrac{3}{4}$

126. $x = -\dfrac{5}{12}, y = -\dfrac{5}{9}$

127. $x = -\dfrac{3}{10}, y = -\dfrac{7}{15}$

128. $x = -\dfrac{5}{6}, y = -\dfrac{2}{15}$

129. $x = 5\dfrac{7}{9}, y = 4\dfrac{2}{3}$

130. $x = 9\dfrac{5}{8}, y = 2\dfrac{3}{16}$

131. $x = 7\dfrac{9}{10}, y = 3\dfrac{1}{2}$

132. $x = 6\dfrac{4}{9}, y = 1\dfrac{1}{6}$

133. $x = 5, y = 2\dfrac{7}{9}$ **134.** $x = 8, y = 4\dfrac{5}{6}$ **135.** $x = 10\dfrac{1}{2}, y = 5\dfrac{7}{12}$ **136.** $x = 9\dfrac{2}{15}, y = 6\dfrac{11}{15}$

137. Is $-\dfrac{3}{4}$ a solution of the equation $\dfrac{4}{5} = \dfrac{31}{20} - y$?   **138.** Is $\dfrac{5}{8}$ a solution of the equation $-\dfrac{1}{4} = x - \dfrac{7}{8}$?

139. Is $-\dfrac{3}{5}$ a solution of the equation $x - \dfrac{1}{4} = -\dfrac{17}{20}$?   **140.** Is $-\dfrac{2}{3}$ a solution of the equation $\dfrac{2}{3} - x = 0$?

For Exercises 141 to 144, give an example of a subtraction problem that meets the described condition. The fractions in your examples must be proper fractions with different denominators. If it is not possible to write a subtraction problem that meets the given condition, write "not possible."

141. A positive fraction is subtracted from a positive fraction and the result is a negative fraction.

142. A negative fraction is subtracted from a positive fraction and the result is a negative fraction.

143. A positive fraction is subtracted from a negative fraction and the result is a positive fraction.

144. A negative fraction is subtracted from a negative fraction and the result is a positive fraction.

OBJECTIVE C **Applications and formulas**

For Exercises 145 and 146, state whether you would use addition or subtraction to find the specified amount.

145. You have $3\dfrac{1}{2}$ h available to do an English assignment and to study for a math test. You spend $1\dfrac{2}{3}$ h on the English assignment. To find the amount of time you have left to study for the math test, use _____.

146. This morning, you studied for $1\dfrac{1}{4}$ h and this afternoon you studied for $1\dfrac{1}{2}$ h. To find the total amount of time you spent studying today use _____.

147. *Real Estate* You purchased $3\dfrac{1}{4}$ acres of land and then sold $1\dfrac{1}{2}$ acres of the property. How many acres of the property do you own now?

EXAMPLE 1 Solve: $y + \dfrac{2}{3} = \dfrac{3}{4}$

Solution

$$y + \dfrac{2}{3} = \dfrac{3}{4}$$

$$y + \dfrac{2}{3} - \dfrac{2}{3} = \dfrac{3}{4} - \dfrac{2}{3}$$ ▶ $\dfrac{2}{3}$ is added to

$$y = \dfrac{9}{12} - \dfrac{8}{12}$$ y. Subtract $\dfrac{2}{3}$.

$$y = \dfrac{1}{12}$$

The solution is $\dfrac{1}{12}$.

YOU TRY IT 1 Solve: $-\dfrac{1}{5} = z - \dfrac{5}{6}$

Your Solution

EXAMPLE 2 Solve: $-\dfrac{3}{5} = \dfrac{6}{7}c$

Solution

$$-\dfrac{3}{5} = \dfrac{6}{7}c$$

$$\dfrac{7}{6}\left(-\dfrac{3}{5}\right) = \dfrac{7}{6} \cdot \dfrac{6}{7}c$$

$$-\dfrac{7}{10} = c$$

The solution is $-\dfrac{7}{10}$.

YOU TRY IT 2 Solve: $26 = 4x$

Your Solution

Solutions on p. S9

OBJECTIVE B **Applications**

EXAMPLE 3

Three-eighths times a number is equal to negative one-fourth. Find the number.

Solution

The unknown number: y

| Three-eighths times a number | is equal to | negative one-fourth |

$$\dfrac{3}{8}y = -\dfrac{1}{4}$$

$$\dfrac{8}{3} \cdot \dfrac{3}{8}y = \dfrac{8}{3}\left(-\dfrac{1}{4}\right)$$

$$y = -\dfrac{2}{3}$$

The number is $-\dfrac{2}{3}$.

YOU TRY IT 3

Negative five-sixths is equal to ten-thirds of a number. Find the number.

Your Solution

Solution on p. S9

EXAMPLE 4

One-third of all of the sugar produced by Sucor, Inc. is brown sugar. This year Sucor produced 250,000 lb of brown sugar. How many pounds of sugar were produced by Sucor?

Strategy

To find the number of pounds of sugar produced, write and solve an equation using x to represent the number of pounds of sugar produced.

Solution

One-third of the sugar produced	is	brown sugar

$$\frac{1}{3}x = 250{,}000$$

$$3 \cdot \frac{1}{3}x = 3 \cdot 250{,}000$$

$$x = 750{,}000$$

Sucor produced 750,000 lb of sugar.

YOU TRY IT 4

The number of computer software games sold by BAL Software in January was three-fifths of all the software products sold by the company. BAL Software sold 450 computer software games in January. Find the total number of software products sold in January.

Your Strategy

Your Solution

EXAMPLE 5

The average score on exams taken during a semester is given by $A = \frac{T}{N}$, where A is the average score, T is the total number of points scored on all tests, and N is the number of tests. Find the total number of points scored by a student whose average score for 6 tests was 84.

Strategy

To find the total number of points scored, replace A with 84 and N with 6 in the given formula and solve for T.

Solution

$$A = \frac{T}{N}$$

$$84 = \frac{T}{6}$$

$$6 \cdot 84 = 6 \cdot \frac{T}{6}$$

$$504 = T$$

The total number of points scored was 504.

YOU TRY IT 5

The average score on exams taken during a semester is given by $A = \frac{T}{N}$, where A is the average score, T is the total number of points scored on all tests, and N is the number of tests. Find the total number of points scored by a student whose average score for 5 tests was 73.

Your Strategy

Your Solution

Solutions on p. S9

3.5 Exercises

OBJECTIVE A Solving equations

1. To solve $5 = \dfrac{x}{8}$, multiply each side of the equation by _____.

2. To solve $-\dfrac{6}{7}a = 14$, multiply each side of the equation by _____.

3. To solve $\dfrac{1}{5} + n = \dfrac{3}{8}$, subtract _____ from each side of the equation.

4. To solve $\dfrac{2}{3} = y - \dfrac{3}{7}$, add _____ to each side of the equation.

Solve.

5. $\dfrac{x}{4} = 9$

6. $8 = \dfrac{y}{2}$

7. $-3 = \dfrac{m}{4}$

8. $\dfrac{n}{5} = -2$

9. $\dfrac{2}{5}x = 10$

10. $\dfrac{3}{4}z = 12$

11. $-\dfrac{5}{6}w = 10$

12. $-\dfrac{1}{2}x = 3$

13. $\dfrac{1}{4} + y = \dfrac{3}{4}$

14. $\dfrac{5}{9} = t - \dfrac{1}{9}$

15. $x + \dfrac{1}{4} = \dfrac{5}{6}$

16. $\dfrac{7}{8} = y - \dfrac{1}{6}$

17. $-\dfrac{2x}{3} = -\dfrac{1}{2}$

18. $-\dfrac{4a}{5} = \dfrac{2}{3}$

19. $\dfrac{5n}{6} = -\dfrac{2}{3}$

20. $\dfrac{7z}{8} = -\dfrac{5}{16}$

21. $-\dfrac{3}{8}t = -\dfrac{1}{4}$

22. $-\dfrac{3}{4}t = -\dfrac{7}{8}$

23. $4a = 6$

24. $6z = 10$

25. $-9c = 12$

26. $-10z = 28$

27. $-2x = \dfrac{8}{9}$

28. $-5y = -\dfrac{15}{16}$

For Exercises 29 and 30, use the following expressions.

(i) $\dfrac{2}{7} \cdot \dfrac{5}{3}$ (ii) $\dfrac{3}{5} \cdot \dfrac{7}{2}$ (iii) $\left(-\dfrac{7}{2}\right) \cdot \dfrac{3}{5}$ (iv) $\dfrac{3}{5} - \dfrac{2}{7}$ (v) $\dfrac{3}{5} + \dfrac{2}{7}$

29. Which expression represents the solution of the equation $\dfrac{3}{5} = -\dfrac{2}{7}x$?

30. Which expression represents the solution of the equation $\dfrac{2}{7}x = \dfrac{3}{5}$?

OBJECTIVE B   Applications

For Exercises 31 and 32, translate the sentence into an equation. Use n to represent the unknown number.

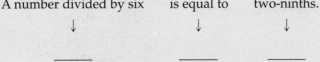

31.   A number divided by six   is equal to   two-ninths.
       ↓   ↓   ↓

       ____   ____   ____

32.   Two-thirds less than a number   is   one-half.
       ↓   ↓   ↓

       ____   ____   ____

33. A number minus one-third equals one-half. Find the number.

34. The sum of a number and one-fourth is one-sixth. Find the number.

35. Three-fifths times a number is nine-tenths. Find the number.

36. The product of negative two-thirds and a number is five-sixths. Find the number.

37. The quotient of a number and negative four is three-fourths. Find the number.

38. A number divided by negative two equals two-fifths. Find the number.

39. Negative three-fourths of a number is equal to one-sixth. Find the number.

40. Negative three-eighths equals the product of two-thirds and some number. Find the number.

41.   *Populations*   The population of the Chippewa tribe is one-half the poulation of the Navajo tribe. The population of the Chippewa tribe is 149,000 people. (*Source:* Census Bureau) What is the population of the Navajo tribe?

42. 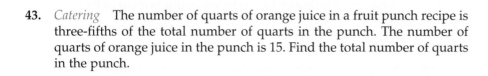 *Education* During the 2006–2007 academic year, the average cost for tuition and fees at a four-year public college was $\frac{3}{11}$ the average cost for tuition and fees at a four-year private college. The average cost for tuition and fees at a four-year public college was $6,000. (*Source:* College Board) What was the average cost for tuition and fees at a four-year private college?

43. *Catering* The number of quarts of orange juice in a fruit punch recipe is three-fifths of the total number of quarts in the punch. The number of quarts of orange juice in the punch is 15. Find the total number of quarts in the punch.

44. *The Electorate* The number of people who voted in an election for mayor of a city was two-thirds of the total number of eligible voters. There were 24,416 people who voted in the election. Find the number of eligible voters.

45. *Cost of Living* The amount of rent paid by a mechanic is $\frac{2}{5}$ of the mechanic's monthly income. Using the figure at the right, determine the mechanic's monthly income.

46. *Travel* The average number of miles per gallon for a car is calculated using the formula $a = \frac{m}{g}$, where a is the average number of miles per gallon and m is the number of miles traveled on g gallons of gas. Use this formula to find the number of miles a car can travel on 16 gal of gas if the car averages 26 mi per gallon.

47. *Travel* The average number of miles per gallon for a truck is calculated using the formula $a = \frac{m}{g}$, where a is the average number of miles per gallon and m is the number of miles traveled on g gallons of gas. Use this formula to find the number of miles a truck can travel on 38 gal of diesel fuel if the truck averages 14 mi per gallon.

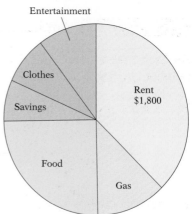

CRITICAL THINKING

48. If $\frac{3}{8}x = -\frac{1}{4}$, is $6x$ greater than -1 or less than -1?

49. Given $-\frac{x}{2} = \frac{2}{3}$, select the best answer from the choices below.
   a. $-9x > 10$ **b.** $-6x < 8$ **c.** $-9x > 10$ and $-6x < 8$

50. Explain why dividing each side of $3x = 6$ by 3 is the same as multiplying each side of the equation by $\frac{1}{3}$.

3.6 Exponents, Complex Fractions, and the Order of Operations Agreement

OBJECTIVE A Exponents

Recall that an exponent indicates the repeated multiplication of the same factor. For example,

$$3^5 = 3 \cdot 3 \cdot 3 \cdot 3 \cdot 3$$

The exponent, 5, indicates how many times the base, 3, occurs as a factor in the multiplication.

The base of an exponential expression can be a fraction; for example, $\left(\dfrac{2}{3}\right)^4$. To evaluate this expression, write the factor as many times as indicated by the exponent and then multiply.

$$\left(\frac{2}{3}\right)^4 = \frac{2}{3} \cdot \frac{2}{3} \cdot \frac{2}{3} \cdot \frac{2}{3} = \frac{2 \cdot 2 \cdot 2 \cdot 2}{3 \cdot 3 \cdot 3 \cdot 3} = \frac{16}{81}$$

Evaluate $\left(-\dfrac{3}{5}\right)^2 \cdot \left(\dfrac{5}{6}\right)^3$.

$$\left(-\frac{3}{5}\right)^2 \cdot \left(\frac{5}{6}\right)^3$$

Write each factor as many times as indicated by the exponent.

$$= \left(-\frac{3}{5}\right) \cdot \left(-\frac{3}{5}\right) \cdot \frac{5}{6} \cdot \frac{5}{6} \cdot \frac{5}{6}$$

Multiply. The product of two negative numbers is positive.

$$= \frac{3}{5} \cdot \frac{3}{5} \cdot \frac{5}{6} \cdot \frac{5}{6} \cdot \frac{5}{6}$$

$$= \frac{3 \cdot 3 \cdot 5 \cdot 5 \cdot 5}{5 \cdot 5 \cdot 6 \cdot 6 \cdot 6}$$

Write the product in simplest form.

$$= \frac{5}{24}$$

Evaluate x^3 for $x = 2\dfrac{1}{2}$.

$$x^3$$

Replace x with $2\dfrac{1}{2}$.

$$\left(2\frac{1}{2}\right)^3$$

Write the mixed number as an improper fraction.

$$= \left(\frac{5}{2}\right)^3$$

Write the base as many times as indicated by the exponent.

$$= \frac{5}{2} \cdot \frac{5}{2} \cdot \frac{5}{2}$$

Multiply.

$$= \frac{125}{8}$$

Write the improper fraction as a mixed number.

$$= 15\frac{5}{8}$$

EXAMPLE 1 Evaluate $\left(-\dfrac{3}{4}\right)^3 \cdot 8^2$.

Solution $\left(-\dfrac{3}{4}\right)^3 \cdot 8^2$

$$= \left(-\dfrac{3}{4}\right)\left(-\dfrac{3}{4}\right)\left(-\dfrac{3}{4}\right) \cdot 8 \cdot 8$$

$$= -\left(\dfrac{3}{4} \cdot \dfrac{3}{4} \cdot \dfrac{3}{4} \cdot \dfrac{8}{1} \cdot \dfrac{8}{1}\right)$$

$$= -\dfrac{3 \cdot 3 \cdot 3 \cdot 8 \cdot 8}{4 \cdot 4 \cdot 4 \cdot 1 \cdot 1} = \boxed{-27}$$

YOU TRY IT 1 Evaluate $\left(\dfrac{2}{9}\right)^2 \cdot (-3)^4$.

Your Solution

EXAMPLE 2 Evaluate x^2y^2 for
$x = 1\dfrac{1}{2}$ and $y = \dfrac{2}{3}$.

Solution x^2y^2

$$\left(1\dfrac{1}{2}\right)^2 \cdot \left(\dfrac{2}{3}\right)^2 = \left(\dfrac{3}{2}\right)^2 \cdot \left(\dfrac{2}{3}\right)^2$$

$$= \dfrac{3}{2} \cdot \dfrac{3}{2} \cdot \dfrac{2}{3} \cdot \dfrac{2}{3}$$

$$= \dfrac{3 \cdot 3 \cdot 2 \cdot 2}{2 \cdot 2 \cdot 3 \cdot 3} = 1$$

YOU TRY IT 2 Evaluate x^4y^3 for
$x = 2\dfrac{1}{3}$ and $y = \dfrac{3}{7}$.

Your Solution

Solutions on p. S9

OBJECTIVE B   Complex fractions

A **complex fraction** is a fraction whose numerator or denominator contains one or more fractions. Examples of complex fractions are shown below.

Main fraction bar ⟶ $\dfrac{\dfrac{3}{4}}{\dfrac{7}{8}}$ $\qquad \dfrac{4}{3 - \dfrac{1}{2}}$ $\qquad \dfrac{\dfrac{9}{10} + \dfrac{3}{5}}{\dfrac{5}{6}}$ $\qquad \dfrac{3\dfrac{1}{2} \cdot 2\dfrac{5}{8}}{\left(4\dfrac{2}{3}\right) \div \left(3\dfrac{1}{5}\right)}$

Look at the first example given above and recall that the fraction bar can be read "divided by."

Therefore, $\dfrac{\dfrac{3}{4}}{\dfrac{7}{8}}$ can be read "$\dfrac{3}{4}$ divided by $\dfrac{7}{8}$" and can be written $\dfrac{3}{4} \div \dfrac{7}{8}$. This

is the division of two fractions and can be simplified by multiplying by the reciprocal, as shown at the top of the next page.

$$\frac{\frac{3}{4}}{\frac{7}{8}} = \frac{3}{4} \div \frac{7}{8} = \frac{3}{4} \cdot \frac{8}{7} = \frac{3 \cdot 8}{4 \cdot 7} = \frac{6}{7}$$

To simplify a complex fraction, first simplify the expression above the main fraction bar and the expression below the main fraction bar; the result is one number in the numerator and one number in the denominator. Then rewrite the complex fraction as a division problem by reading the main fraction bar as "divided by."

Simplify: $\dfrac{4}{3 - \frac{1}{2}}$

The numerator (4) is already simplified. Simplify the expression in the denominator.

Note: $3 - \frac{1}{2} = \frac{6}{2} - \frac{1}{2} = \frac{5}{2}$

Rewrite the complex fraction as division.

Divide.

Write the answer in simplest form.

$$\frac{4}{3 - \frac{1}{2}} = \frac{4}{\frac{5}{2}}$$
$$= 4 \div \frac{5}{2}$$
$$= \frac{4}{1} \div \frac{5}{2}$$
$$= \frac{4}{1} \cdot \frac{2}{5}$$
$$= \frac{8}{5} = 1\frac{3}{5}$$

Simplify: $\dfrac{-\frac{9}{10} + \frac{3}{5}}{1\frac{1}{4}}$

Simplify the expression in the numerator.

Note: $-\frac{9}{10} + \frac{3}{5} = \frac{-9}{10} + \frac{6}{10} = \frac{-3}{10} = -\frac{3}{10}$

Write the mixed number in the denominator as an improper fraction.

Rewrite the complex fraction as division. The quotient will be negative.

Divide by multiplying by the reciprocal.

$$\frac{-\frac{9}{10} + \frac{3}{5}}{1\frac{1}{4}} = \frac{-\frac{3}{10}}{\frac{5}{4}}$$
$$= -\left(\frac{3}{10} \div \frac{5}{4}\right)$$
$$= -\left(\frac{3}{10} \cdot \frac{4}{5}\right)$$
$$= -\frac{6}{25}$$

Evaluate $\dfrac{wx}{yz}$ for $w = 1\frac{1}{3}$, $x = 2\frac{5}{8}$, $y = 4\frac{1}{2}$, and $z = 3\frac{1}{3}$.

$$\dfrac{wx}{yz}$$

Replace each variable with its given value.

$$\dfrac{1\frac{1}{3} \cdot 2\frac{5}{8}}{4\frac{1}{2} \cdot 3\frac{1}{3}}$$

Simplify the numerator.

Note: $1\frac{1}{3} \cdot 2\frac{5}{8} = \frac{4}{3} \cdot \frac{21}{8} = \frac{7}{2}$

Simplify the denominator.

$$= \dfrac{\frac{7}{2}}{15}$$

Note: $4\frac{1}{2} \cdot 3\frac{1}{3} = \frac{9}{2} \cdot \frac{10}{3} = 15$

Rewrite the complex fraction as division.

$$= \frac{7}{2} \div 15$$

Divide by multiplying by the reciprocal.

$$= \frac{7}{2} \cdot \frac{1}{15} = \frac{7}{30}$$

Note: $15 = \frac{15}{1}$; the reciprocal of $\frac{15}{1}$ is $\frac{1}{15}$.

EXAMPLE 3 Is $\frac{2}{3}$ a solution of $\dfrac{x + \frac{1}{2}}{x} = \frac{7}{4}$?

YOU TRY IT 3 Is $-\frac{1}{2}$ a solution of $\frac{2y - 3}{y} \stackrel{?}{=} -2$?

Solution

$$\dfrac{x + \frac{1}{2}}{x} = \frac{7}{4}$$

$$\dfrac{\frac{2}{3} + \frac{1}{2}}{\frac{2}{3}} \ \Big| \ \frac{7}{4}$$

▶ Replace x with $\frac{2}{3}$.

$$\dfrac{\frac{7}{6}}{\frac{2}{3}} \ \Big| \ \frac{7}{4}$$

▶ Simplify the complex fraction.

$$\frac{7}{6} \div \frac{2}{3} \ \Big| \ \frac{7}{4}$$

$$\frac{7}{6} \cdot \frac{3}{2} \ \Big| \ \frac{7}{4}$$

$$\frac{7}{4} = \frac{7}{4}$$

Yes, $\frac{2}{3}$ is a solution of the equation.

Your Solution

Solution on p. S10

EXAMPLE 4

Evaluate the variable expression $\frac{x-y}{z}$ for $x = 4\frac{1}{8}$, $y = 2\frac{5}{8}$, and $z = \frac{3}{4}$.

Solution

$$\frac{x-y}{z}$$

$$\frac{4\frac{1}{8} - 2\frac{5}{8}}{\frac{3}{4}} = \frac{\frac{3}{2}}{\frac{3}{4}} = \frac{3}{2} \div \frac{3}{4} = \frac{3}{2} \cdot \frac{4}{3} = 2$$

YOU TRY IT 4

Evaluate the variable expression $\frac{x}{y-z}$ for $x = 2\frac{4}{9}$, $y = 3$, and $z = 1\frac{1}{3}$.

Your Solution

Solution on p. S10

OBJECTIVE C **The Order of Operations Agreement**

The Order of Operations Agreement applies in simplifying expressions containing fractions.

The Order of Operations Agreement

Step 1 Do all operations inside parentheses.

Step 2 Simplify any numerical expressions containing exponents.

Step 3 Do multiplication and division as they occur from left to right.

Step 4 Do addition and subtraction as they occur from left to right.

Simplify: $\left(\frac{1}{2}\right)^2 + \left(\frac{2}{3} \div \frac{5}{9}\right) \cdot \frac{5}{6}$

$$\left(\frac{1}{2}\right)^2 + \left(\frac{2}{3} \div \frac{5}{9}\right) \cdot \frac{5}{6}$$

Do the operation inside the parentheses (Step 1).

$$= \left(\frac{1}{2}\right)^2 + \left(\frac{6}{5}\right) \cdot \frac{5}{6}$$

Simplify the exponential expression (Step 2).

$$= \frac{1}{4} + \left(\frac{6}{5}\right) \cdot \frac{5}{6}$$

Do the multiplication (Step 3).

$$= \frac{1}{4} + 1$$

Do the addition (Step 4).

$$= 1\frac{1}{4}$$

A fraction bar acts like parentheses. Therefore, simplify the numerator and denominator of a fraction as part of Step 1 in the Order of Operations Agreement.

Simplify: $6 - \dfrac{2 + 1}{15 - 8} \div \dfrac{3}{14}$

$$6 - \frac{2 + 1}{15 - 8} \div \frac{3}{14}$$

Perform operations above and below the fraction bar.

$$= 6 - \frac{3}{7} \div \frac{3}{14}$$

Do the division.

$$= 6 - \left(\frac{3}{7} \cdot \frac{14}{3}\right)$$

$$= 6 - 2$$

Do the subtraction.

$$= 4$$

Evaluate $\dfrac{w + x}{y} - z$ for $w = \dfrac{3}{4}$, $x = \dfrac{1}{4}$, $y = 2$, and $z = \dfrac{1}{3}$.

$$\frac{w + x}{y} - z$$

Replace each variable with its given value.

$$\frac{\dfrac{3}{4} + \dfrac{1}{4}}{2} - \frac{1}{3}$$

Simplify the numerator of the complex fraction.

$$= \frac{1}{2} - \frac{1}{3}$$

Do the subtraction.

$$= \frac{1}{6}$$

EXAMPLE 5 Simplify: $\left(-\dfrac{2}{3}\right)^2 \div \dfrac{7 - 2}{13 - 4} - \dfrac{1}{3}$

Solution

$$\left(-\frac{2}{3}\right)^2 \div \frac{7 - 2}{13 - 4} - \frac{1}{3}$$

$$= \left(-\frac{2}{3}\right)^2 \div \frac{5}{9} - \frac{1}{3}$$ ▶ Simplify $\dfrac{7 - 2}{13 - 4}$.

$$= \frac{4}{9} \div \frac{5}{9} - \frac{1}{3}$$ ▶ Simplify $\left(-\dfrac{2}{3}\right)^2$.

$$= \frac{4}{9} \cdot \frac{9}{5} - \frac{1}{3}$$ ▶ Rewrite division as multiplication by the reciprocal.

$$= \frac{4}{5} - \frac{1}{3} = \frac{7}{15}$$

YOU TRY IT 5 Simplify: $\left(-\dfrac{1}{2}\right)^3 \cdot \dfrac{7 - 3}{4 - 9} + \dfrac{4}{5}$

Your Solution

Solution on p. S10

3.6 Exercises

OBJECTIVE A Exponents

1. To evaluate $\left(-\dfrac{1}{5}\right)^3$, write $-\dfrac{1}{5}$ as a factor _____ times and then multiply.

2. To evaluate $\left(-\dfrac{4}{5}\right)^2$, write $-\dfrac{4}{5}$ as a factor _____ times and then multiply.

Evaluate.

3. $\left(\dfrac{3}{4}\right)^2$

4. $\left(\dfrac{5}{8}\right)^2$

5. $\left(-\dfrac{1}{6}\right)^3$

6. $\left(-\dfrac{2}{7}\right)^3$

7. $\left(2\dfrac{1}{4}\right)^2$

8. $\left(3\dfrac{1}{2}\right)^2$

9. $\left(\dfrac{5}{8}\right)^3 \cdot \left(\dfrac{2}{5}\right)^2$

10. $\left(\dfrac{3}{5}\right)^3 \cdot \left(\dfrac{1}{3}\right)^2$

11. $\left(\dfrac{18}{25}\right)^2 \cdot \left(\dfrac{5}{9}\right)^3$

12. $\left(\dfrac{2}{3}\right)^3 \cdot \left(\dfrac{5}{6}\right)^2$

13. $\left(\dfrac{4}{5}\right)^4 \cdot \left(-\dfrac{5}{8}\right)^3$

14. $\left(-\dfrac{9}{11}\right)^2 \cdot \left(\dfrac{1}{3}\right)^4$

15. $7^2 \cdot \left(\dfrac{2}{7}\right)^3$

16. $4^3 \cdot \left(\dfrac{5}{12}\right)^2$

17. $4 \cdot \left(\dfrac{4}{7}\right)^2 \cdot \left(-\dfrac{3}{4}\right)^3$

18. $3 \cdot \left(\dfrac{2}{5}\right)^2 \cdot \left(-\dfrac{1}{6}\right)^2$

Evaluate the variable expression for the given values of x and y.

19. x^4, for $x = \dfrac{2}{3}$

20. y^3, for $y = -\dfrac{3}{4}$

21. $x^4 y^2$, for $x = \dfrac{5}{6}$ and $y = -\dfrac{3}{5}$

22. $x^5 y^3$, for $x = -\dfrac{5}{8}$ and $y = \dfrac{4}{5}$

23. $x^3 y^2$, for $x = \dfrac{2}{3}$ and $y = 1\dfrac{1}{2}$

24. $x^2 y^4$, for $x = 2\dfrac{1}{3}$ and $y = \dfrac{3}{7}$

25. True or false? If a is positive and b is negative, then $\left(-\dfrac{a}{b}\right)^5$ is a positve number.

OBJECTIVE B **Complex fractions**

26. To simplify the complex fraction $\dfrac{\frac{1}{3}}{\frac{5}{6}}$, first write it as the division problem

_____ ÷ _____.

Simplify.

27. $\dfrac{\frac{9}{16}}{\frac{3}{4}}$

28. $\dfrac{\frac{7}{24}}{\frac{3}{8}}$

29. $\dfrac{-\frac{5}{6}}{\frac{15}{16}}$

30. $\dfrac{\frac{7}{12}}{-\frac{5}{18}}$

31. $\dfrac{\frac{2}{3}+\frac{1}{2}}{7}$

32. $\dfrac{-5}{\frac{3}{8}-\frac{1}{4}}$

33. $\dfrac{2+\frac{1}{4}}{\frac{3}{8}}$

34. $\dfrac{1-\frac{3}{4}}{\frac{5}{12}}$

35. $\dfrac{\frac{9}{25}}{\frac{4}{5}-\frac{1}{10}}$

36. $\dfrac{-\frac{5}{7}}{\frac{4}{7}-\frac{3}{14}}$

37. $\dfrac{\frac{1}{3}-\frac{3}{4}}{\frac{1}{6}+\frac{2}{3}}$

38. $\dfrac{\frac{9}{14}-\frac{1}{7}}{\frac{9}{14}+\frac{1}{7}}$

39. $\dfrac{3+2\frac{1}{3}}{5\frac{1}{6}-1}$

40. $\dfrac{4-3\frac{5}{8}}{2\frac{1}{2}-\frac{3}{4}}$

41. $\dfrac{5\frac{2}{3}-1\frac{1}{6}}{3\frac{5}{8}-2\frac{1}{4}}$

42. $\dfrac{3\frac{1}{4}-2\frac{1}{2}}{4\frac{3}{4}+1\frac{1}{2}}$

Evaluate the expression for the given values of the variables.

43. $\dfrac{x+y}{z}$, for $x=\dfrac{2}{3}$, $y=\dfrac{3}{4}$, and $z=\dfrac{1}{12}$

44. $\dfrac{x}{y+z}$, for $x=\dfrac{8}{15}$, $y=\dfrac{3}{5}$, and $z=\dfrac{2}{3}$

45. $\dfrac{xy}{z}$, for $x=\dfrac{3}{4}$, $y=-\dfrac{2}{3}$, and $z=\dfrac{5}{8}$

46. $\dfrac{x}{yz}$, for $x=-\dfrac{5}{12}$, $y=\dfrac{8}{9}$, and $z=-\dfrac{3}{4}$

47. $\dfrac{x-y}{z}$, for $x = 2\dfrac{5}{8}$, $y = 1\dfrac{1}{4}$, and $z = 1\dfrac{3}{8}$

48. $\dfrac{x}{y-z}$, for $x = 2\dfrac{3}{10}$, $y = 3\dfrac{2}{5}$, and $z = 1\dfrac{4}{5}$

49. Is $-\dfrac{3}{4}$ a solution of the equation $\dfrac{4x}{x+5} = -\dfrac{4}{3}$?

50. Is $-\dfrac{4}{5}$ a solution of the equation

$$\dfrac{15y}{\dfrac{3}{10} + y} = -24?$$

State whether the given expression is equivalent to 0, equivalent to 1, equivalent to $\left(\dfrac{a}{b}\right)^2$, or undefined.

51. $\dfrac{\dfrac{a}{b}}{\dfrac{a}{b}}$

52. $\dfrac{\dfrac{a}{b}}{\dfrac{0}{b}}$

53. $\dfrac{\dfrac{0}{b}}{\dfrac{a}{b}}$

54. $\dfrac{\dfrac{a}{b}}{\dfrac{b}{a}}$

OBJECTIVE C The Order of Operations Agreement

55. Simplifying the expression $\dfrac{2}{3} - \dfrac{4}{3 + \dfrac{3}{8}}$ involves performing three operations: subtraction, division, and addition. List these three operations in the order in which they must be performed.

56. Simplifying the expression $\dfrac{2}{9} \cdot \left(\dfrac{3}{4}\right)^2 + \dfrac{5}{6}$ involves performing three operations: multiplication, squaring, and addition. List these three operations in the order in which they must be performed.

Simplify.

57. $-\dfrac{3}{7} \cdot \dfrac{14}{15} + \dfrac{4}{5}$

58. $\dfrac{3}{5} \div \dfrac{6}{7} + \dfrac{4}{5}$

59. $\left(\dfrac{5}{6}\right)^2 - \dfrac{5}{9}$

60. $\left(\dfrac{3}{5}\right)^2 - \dfrac{3}{10}$

61. $\dfrac{3}{4} \cdot \left(\dfrac{11}{12} - \dfrac{7}{8}\right) + \dfrac{5}{16}$

62. $-\dfrac{7}{18} + \dfrac{5}{6} \cdot \left(\dfrac{2}{3} - \dfrac{1}{6}\right)$

63. $\dfrac{11}{16} - \left(\dfrac{3}{4}\right)^2 + \dfrac{7}{8}$

64. $\left(-\dfrac{2}{3}\right)^2 - \dfrac{7}{18} + \dfrac{5}{6}$

65. $\left(1\dfrac{1}{3} - \dfrac{5}{6}\right) + \dfrac{7}{8} \div \left(-\dfrac{1}{2}\right)^2$

66. $\left(\dfrac{1}{4}\right)^2 \div \left(2\dfrac{1}{2} - \dfrac{3}{4}\right) + \dfrac{5}{7}$

67. $\left(\dfrac{2}{3}\right)^2 + \dfrac{8-7}{3-9} \div \dfrac{3}{8}$

68. $\left(\dfrac{1}{3}\right)^2 \cdot \dfrac{14-5}{6-10} + \dfrac{3}{4}$

69. $-\dfrac{1}{2} + \dfrac{\dfrac{13}{25}}{4 - \dfrac{3}{4}} \div \dfrac{1}{5}$

70. $\dfrac{4}{5} - \dfrac{3 - \dfrac{7}{9}}{\dfrac{5}{6}} \cdot \dfrac{3}{8}$

71. $\left(\dfrac{2}{3}\right)^2 + \dfrac{\dfrac{5}{8} - \dfrac{1}{4}}{\dfrac{2}{3} - \dfrac{1}{6}} \cdot \dfrac{8}{9}$

Evaluate the expression for the given values of the variables.

72. $x^2 + \dfrac{y}{z}$, for $x = -\dfrac{2}{3}$, $y = \dfrac{5}{8}$, and $z = \dfrac{3}{4}$

73. $\dfrac{x}{y} - z^2$, for $x = \dfrac{5}{6}$, $y = \dfrac{1}{3}$, and $z = -\dfrac{3}{4}$

74. $x - y^3 z$, for $x = \dfrac{5}{6}$, $y = \dfrac{1}{2}$, and $z = \dfrac{8}{9}$

75. $xy^3 + z$, for $x = \dfrac{9}{10}$, $y = \dfrac{1}{3}$, and $z = \dfrac{7}{15}$

76. $\dfrac{wx}{y} + z$, for $w = \dfrac{4}{5}$, $x = \dfrac{5}{8}$, $y = \dfrac{3}{4}$, and $z = \dfrac{2}{3}$

77. $\dfrac{w}{xy} - z$, for $w = 2\dfrac{1}{2}$, $x = 4$, $y = \dfrac{3}{8}$, and $z = \dfrac{2}{3}$

78. Is $-\dfrac{1}{2}$ a solution of the equation $\dfrac{-8z}{z + \dfrac{5}{6}} - 4z = -14$?

79. Is $-\dfrac{1}{3}$ a solution of the equation $\dfrac{12w}{\dfrac{1}{6} - w} = -7$?

CRITICAL THINKING

80. *Computers*   A computer can perform 600,000 operations in one second. To the nearest minute, how many minutes will it take for the computer to perform 10^8 operations?

81. Given that x is a whole number, for what value of x will the expression $\left(\dfrac{3}{4}\right)^2 + x^5 \div \dfrac{7}{8}$ have a minimum value? What is the minimum value?

Focus on **Problem Solving**

Common Knowledge

An application problem may not provide all the information that is needed to solve the problem. Sometimes, however, the necessary information is common knowledge.

> You are traveling by bus from Boston to New York. The trip is 4 hours long. If the bus leaves Boston at 10 A.M., what time should you arrive in New York?
>
> What other information do you need to solve this problem?
>
> You need to know that, using a 12-hour clock, the hours run

10 A.M.
11 A.M.
12 P.M.
1 P.M.
2 P.M.

Four hours after 10 A.M. is 2 P.M.

You should arrive in New York at 2 P.M.

> You purchase a 41¢ stamp at the post office and hand the clerk a one-dollar bill. How much change do you receive?
>
> What information do you need to solve this problem?
>
> You need to know that there are 100¢ in one dollar.
>
> Your change is 100¢ − 41¢.

$100 - 41 = 59$

You receive 59¢ in change.

What information do you need to know to solve each of the following problems?

1. You sell a dozen tickets to a fundraiser. Each ticket costs $10. How much money do you collect?

2. The weekly lab period for your science course is one hour and twenty minutes long. Find the length of the science lab period in minutes.

3. An employee's monthly salary is $3750. Find the employee's annual salary.

4. A survey revealed that eighth graders spend an average of 3 hours each day watching television. Find the total time an eighth grader spends watching TV each week.

5. You want to buy a carpet for a room that is 15 ft wide and 18 ft long. Find the amount of carpet that you need.

Decimals and Real Numbers

Visitors to China may exchange their money for the local currency, which is the yuan. Generally, currency rates are listed in the newspaper by country and currency type. You can also find currency rates on the Internet. The values go beyond the usual hundredth decimal place to increase the accuracy of the exchange. **Exercises 28 and 29 on page 270** illustrate calculating money equivalences.

DVD   SSM

Student Website
Need help? For online student resources,
visit college.hmco.com/pic/aufmannPA5e.

1. Express the shaded portion of the rectangle as a fraction.

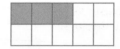

2. Round 36,852 to the nearest hundred.

3. Write 4,791 in words.

4. Write six thousand eight hundred forty-two in standard form.

5. Graph −3 on the number line.

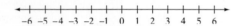

For Exercises 6 to 9, add, subtract, multiply, or divide.

6. −37 + 8,892 + 465

7. 2,403 − (−765)

8. −844(−91)

9. 23)6,412

10. Evaluate 8^2.

GO Figure

Super Yeast causes bread to double in volume each minute. If it takes one loaf of bread made with Super Yeast 30 min to fill the oven, how long does it take two loaves of bread made with Super Yeast to fill one-half the oven?

4.1 Introduction to Decimals

OBJECTIVE A **Place value**

The price tag on a sweater reads $61.88. The number 61.88 is in **decimal notation.** A number written in decimal notation is often called simply a **decimal.**

A number written in decimal notation has three parts.

61	.	88
Whole number part	**Decimal point**	**Decimal part**

The decimal part of the number represents a number less than 1. For example, $.88 is less than one dollar. The decimal point (.) separates the whole number part from the decimal part.

The position of a digit in a decimal determines the digit's place value. The place-value chart is extended to the right to show the place values of digits to the right of a decimal point.

In the decimal 458.302719, the position of the digit 7 determines that its place value is ten-thousandths.

Note the relationship between fractions and numbers written in decimal notation.

seven tenths	seven hundredths	seven thousandths
$\frac{7}{10} = 0.7$	$\frac{7}{100} = 0.07$	$\frac{7}{1,000} = 0.007$
1 zero in 10	2 zeros in 100	3 zeros in 1,000
1 decimal place in 0.7	2 decimal places in 0.07	3 decimal places in 0.007

To write a decimal in words, write the decimal part of the number as though it were a whole number, and then name the place value of the last digit.

0.9684   nine thousand six hundred eighty-four ten-thousandths

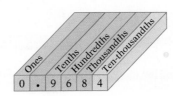

The decimal point in a decimal is read as "and."

372.516   three hundred seventy-two and five hundred sixteen thousandths

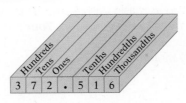

Point of Interest

The idea that all fractions should be represented in tenths, hundredths, and thousandths was presented in 1585 in Simon Stevin's publication *De Thiende.* Its French translation, *La Disme,* was well read and accepted by the French. This may help to explain why the French accepted the metric system so easily two hundred years later.

In *De Thiende,* Stevin argued in favor of his notation by including examples for astronomers, tapestry makers, surveyors, tailors, and the like. He stated that using decimals would enable calculations to be "performed . . . with as much ease as counter-reckoning."

To write a decimal in standard form when it is written in words, write the whole number part, replace the word *and* with a decimal point, and write the decimal part so that the last digit is in the given place-value position.

four and twenty-three <u>hundredths</u>

3 is in the hundredths place.   4.23

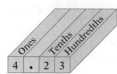

When writing a decimal in standard form, you may need to insert zeros after the decimal point so that the last digit is in the given place-value position.

ninety-one and eight <u>thousandths</u>

8 is in the thousandths place.   91.008
Insert two zeros so that the 8 is in
the thousandths place.

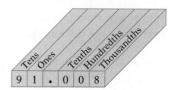

sixty-five <u>ten-thousandths</u>

5 is in the ten-thousandths place.   0.0065
Insert two zeros so that the 5 is in
the ten-thousandths place.

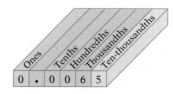

EXAMPLE 1 Name the place value of the digit 8 in the number 45.687.

Solution The digit 8 is in the hundredths place.

YOU TRY IT 1 Name the place value of the digit 4 in the number 907.1342.

Your Solution

EXAMPLE 2 Write $\frac{43}{100}$ as a decimal.

Solution $\frac{43}{100} = 0.43$ ▸ forty-three hundredths

YOU TRY IT 2 Write $\frac{501}{1,000}$ as a decimal.

Your Solution

EXAMPLE 3 Write 0.289 as a fraction.

Solution $0.289 = \frac{289}{1,000}$ ▸ 289 thousandths

YOU TRY IT 3 Write 0.67 as a fraction.

Your Solution

EXAMPLE 4 Write 293.50816 in words.

Solution two hundred ninety-three and fifty thousand eight hundred sixteen hundred-thousandths

YOU TRY IT 4 Write 55.6083 in words.

Your Solution

Solutions on p. S10

EXAMPLE 5 Write twenty-three and two hundred forty-seven millionths in standard form.

YOU TRY IT 5 Write eight hundred six and four hundred ninety-one hundred-thousandths in standard form.

Solution   23.000247
↑
└── millionths place

Your Solution

Solution on p. S10

OBJECTIVE B   **Order relations between decimals**

A whole number can be written as a decimal by writing a decimal point to the right of the last digit. For example,

$62 = 62.$ $497 = 497.$

You know that \$62 and \$62.00 both represent sixty-two dollars. Any number of zeros may be written to the right of the decimal point in a whole number without changing the value of the number.

$62 = 62.00 = 62.0000$ $497 = 497.0 = 497.000$

Also, any number of zeros may be written to the right of the last digit in a decimal without changing the value of the number.

$0.8 = 0.80 = 0.800$ $1.35 = 1.350 = 1.3500 = 1.35000 = 1.350000$

This fact is used to find the order relation between two decimals.

To compare two decimals, write the decimal part of each number so that each has the same number of decimal places. Then compare the two numbers.

Place the correct symbol, $<$ or $>$, between the two numbers 0.693 and 0.71.

0.693 has 3 decimal places.
0.71 has 2 decimal places.
Write 0.71 with 3 decimal places. $0.71 = 0.710$

Compare 0.693 and 0.710.
693 thousandths $<$ 710 thousandths $0.693 < 0.710$

Remove the zero written in 0.710. $0.693 < 0.71$

Place the correct symbol, $<$ or $>$, between the two numbers 5.8 and 5.493.

Write 5.8 with 3 decimal places. $5.8 = 5.800$

Compare 5.800 and 5.493.
The whole number part (5) is the same.
800 thousandths $>$ 493 thousandths $5.800 > 5.493$

Remove the extra zeros written in 5.800. $5.8 > 5.493$

Point of Interest

The decimal point did not make its appearance until the early 1600s. Stevin's notation used subscripts with circles around them after each digit: 0 for ones, 1 for tenths (which he called "primes"), 2 for hundredths (called "seconds"), 3 for thousandths ("thirds"), and so on. For example, 1.375 would have been written

1   3   7   5
⓪   ①   ②   ③

EXAMPLE 6 Place the correct symbol, < or >, between the two numbers.

0.039   0.1001

Solution   0.039 = 0.0390

0.0390 < 0.1001
0.039 < 0.1001

YOU TRY IT 6 Place the correct symbol, < or >, between the two numbers.

0.065   0.0802

Your Solution

EXAMPLE 7 Write the given numbers in order from smallest to largest.

1.01, 1.2, 1.002, 1.1, 1.12

Solution   1.010, 1.200, 1.002, 1.100, 1.120
1.002, 1.010, 1.100, 1.120, 1.200

1.002, 1.01, 1.1, 1.12, 1.2

YOU TRY IT 7 Write the given numbers in order from smallest to largest.

3.03, 0.33, 0.3, 3.3, 0.03

Your Solution

Solutions on p. S10

OBJECTIVE C   Rounding

In general, rounding decimals is similar to rounding whole numbers except that the digits to the right of the given place value are dropped instead of being replaced by zeros.

If the digit to the right of the given place value is less than 5, that digit and all digits to the right are dropped.

Round 6.9237 to the nearest hundredth.

Given place value (hundredths)

6.9237

3 < 5   Drop the digits 3 and 7.

6.9237 rounded to the nearest hundredth is 6.92.

If the digit to the right of the given place value is greater than or equal to 5, increase the digit in the given place value by 1, and drop all digits to its right.

Round 12.385 to the nearest tenth.

Given place value (tenths)

12.385

8 > 5   Increase 3 by 1 and drop all digits to the right of 3.

12.385 rounded to the nearest tenth is 12.4.

Round 0.46972 to the nearest thousandth.

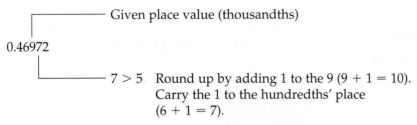

——————— Given place value (thousandths)

0.46972

——————— 7 > 5   Round up by adding 1 to the 9 (9 + 1 = 10).
Carry the 1 to the hundredths' place
(6 + 1 = 7).

0.46972 rounded to the nearest thousandth is 0.470.

Note that in this example, the zero in the given place value is not dropped. This indicates that the number is rounded to the nearest thousandth. If we dropped the zero and wrote 0.47, it would indicate that the number was rounded to the nearest hundredth.

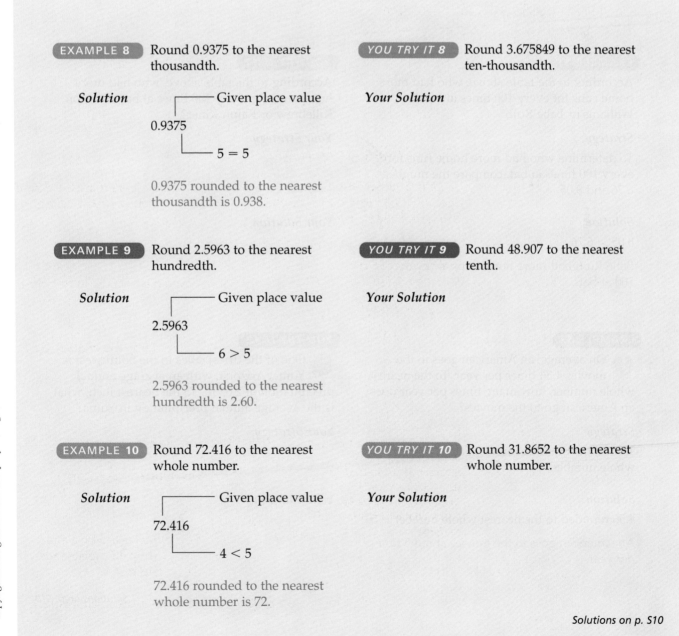

EXAMPLE 8   Round 0.9375 to the nearest thousandth.

Solution

——— Given place value

0.9375

——— 5 = 5

0.9375 rounded to the nearest thousandth is 0.938.

YOU TRY IT 8   Round 3.675849 to the nearest ten-thousandth.

Your Solution

EXAMPLE 9   Round 2.5963 to the nearest hundredth.

Solution

——— Given place value

2.5963

——— 6 > 5

2.5963 rounded to the nearest hundredth is 2.60.

YOU TRY IT 9   Round 48.907 to the nearest tenth.

Your Solution

EXAMPLE 10   Round 72.416 to the nearest whole number.

Solution

——— Given place value

72.416

——— 4 < 5

72.416 rounded to the nearest whole number is 72.

YOU TRY IT 10   Round 31.8652 to the nearest whole number.

Your Solution

Solutions on p. S10

Babe Ruth

OBJECTIVE D Applications

The table below shows the number of home runs hit, for every 100 times at bat, by four Major League baseball players. Use this table for Example 11 and You Try It 11.

Home Runs Hit for Every 100 At-Bats	
Harmon Killebrew	7.03
Ralph Kiner	7.09
Babe Ruth	8.05
Ted Williams	6.76

Source: Major League Baseball

EXAMPLE 11

According to the table above, who had more home runs for every 100 times at bat, Ted Williams or Babe Ruth?

Strategy

To determine who had more home runs for every 100 times at bat, compare the numbers 6.76 and 8.05.

Solution

8.05 > 6.76

Babe Ruth had more home runs for every 100 at-bats.

YOU TRY IT 11

According to the table above, who had more home runs for every 100 times at bat, Harmon Killebrew or Ralph Kiner?

Your Strategy

Your Solution

EXAMPLE 12

On average, an American goes to the movies 4.56 times per year. To the nearest whole number, how many times per year does an American go to the movies?

Strategy

To find the number, round 4.56 to the nearest whole number.

Solution

4.56 rounded to the nearest whole number is 5.

An American goes to the movies about 5 times per year.

YOU TRY IT 12

One of the driest cities in the Southwest is Yuma, Arizona, with an average annual precipitation of 2.65 in. To the nearest inch, what is the average annual precipitation in Yuma?

Your Strategy

Your Solution

Solutions on p. S10

4.1 Exercises

OBJECTIVE A Place value

1. In a decimal, the place values of the first six digits to the right of the decimal point are tenths, _____, _____, ten-thousandths, _____, and _____.

2. The place value of the digit 3 in 0.53 is _____, so when 0.53 is written as a fraction, the denominator is _____. The numerator is _____.

3. To write 85.102 in words, first write *eighty-five*. Replace the decimal point with the word _____ and then write *one hundred two* _____.

4. To write seventy-three millionths in standard form, insert _____ zeros between the decimal point and 73 so that the digit 3 is in the millionths place.

Name the place value of the digit 5.

5. 76.31587

6. 291.508

7. 432.09157

8. 0.0006512

9. 38.2591

10. 0.0000853

Write the fraction as a decimal.

11. $\dfrac{3}{10}$

12. $\dfrac{9}{10}$

13. $\dfrac{21}{100}$

14. $\dfrac{87}{100}$

15. $\dfrac{461}{1,000}$

16. $\dfrac{853}{1,000}$

17. $\dfrac{93}{1,000}$

18. $\dfrac{61}{1,000}$

Write the decimal as a fraction.

19. 0.1

20. 0.3

21. 0.47

22. 0.59

23. 0.289

24. 0.601

25. 0.09

26. 0.013

Write the number in words.

27. 0.37

28. 25.6

29. 9.4

30. 1.004　　　　　　　　　　**31.** 0.0053　　　　　　　　　　**32.** 41.108

33. 0.045　　　　　　　　　　**34.** 3.157　　　　　　　　　　**35.** 26.04

Write the number in standard form.

36. six hundred seventy-two thousandths

37. three and eight hundred six ten-thousandths

38. nine and four hundred seven ten-thousandths

39. four hundred seven and three hundredths

40. six hundred twelve and seven hundred four thousandths

41. two hundred forty-six and twenty-four thousandths

42. two thousand sixty-seven and nine thousand two ten-thousandths

43. seventy-three and two thousand six hundred eighty-four hundred-thousandths

OBJECTIVE B **Order relations between decimals**

For Exercises 44 and 45, fill in each blank with < or >.

44. To decide on the order relation between 0.017 and 0.107, compare 17 thousandths and 107 thousandths. Because 17 thousandths _____ 107 thousandths, 0.017 _____ 0.107.

45. To decide on the order relation between 3.4 and 3.05, write 3.4 as 3.40. The numbers have the same whole number parts, so compare 40 hundredths and 5 hundredths. Because 40 hundredths _____ 5 hundredths, 3.4 _____ 3.05.

Place the correct symbol, < or >, between the two numbers.

46. 0.16　0.6　　　　**47.** 0.7　0.56　　　　**48.** 5.54　5.45　　　　**49.** 3.605　3.065

50. 0.047　0.407　　　**51.** 9.004　9.04　　　**52.** 1.0008　1.008　　　**53.** 9.31　9.031

54. 7.6005　7.605　　　**55.** 4.6　40.6　　　**56.** 0.31502　0.3152　　　**57.** 0.07046　0.07036

Write the given numbers in order from smallest to largest.

58. 0.39, 0.309, 0.399　　　　**59.** 0.66, 0.699, 0.696, 0.609　　　　**60.** 0.24, 0.024, 0.204, 0.0024

61. 1.327, 1.237, 1.732, 1.372   **62.**   0.06, 0.059, 0.061, 0.0061   **63.** 21.87, 21.875, 21.805, 21.78

64. Use the inequality symbol < to rewrite the order relation expressed by the inequality 9.4 > 0.94.

65. Use the inequality symbol > to rewrite the order relation expressed by the inequality 0.062 < 0.62.

OBJECTIVE C   Rounding

66. A decimal rounded to the nearest thousandth will have _____ digits to the right of the decimal point.

67. Suppose you are rounding 5.13274 to the nearest hundredth. The digit in the hundredths place is _____. The digit that you use to decide whether this digit remains the same or is increased by 1 is _____. The digits that you drop are _____.

Round the number to the given place value.

68. 6.249; tenths   **69.** 5.398; tenths   **70.** 21.007; tenths

71. 30.0092; tenths   **72.** 18.40937; hundredths   **73.** 413.5972; hundredths

74. 72.4983; hundredths   **75.** 6.061745; thousandths   **76.** 936.2905; thousandths

77. 96.8027; whole number   **78.** 47.3192; whole number   **79.** 5,439.83; whole number

80. 7,014.96; whole number   **81.** 0.023591; ten-thousandths   **82.** 2.975268; hundred-thousandths

OBJECTIVE D   Applications

83. *Measurement*   A nickel weighs about 0.1763668 oz. Find the weight of a nickel to the nearest hundredth of an ounce.

84. *Sports*   Runners in the Boston Marathon run a distance of 26.21875 mi. To the nearest tenth of a mile, find the distance an entrant who completes the Boston Marathon runs.

Sports   The table at the right lists National Football League leading lifetime rushers. Use the table for Exercises 85 and 86.

85. Who had the greater average number of yards per carry, Walter Payton or Emmitt Smith?

Football Player	Average Number of Yards per Carry
Jerome Bettis	3.93
Curtis Martin	4.01
Walter Payton	4.36
Barry Sanders	4.99
Emmitt Smith	4.16

86. Of all the players listed in the table, who has the greatest average number of yards per carry?

Source: Pro Football Hall of Fame

87. *Consumerism*   Charge accounts generally require a minimum payment on the balance in the account each month. Use the minimum payment schedule shown below to determine the minimum payment due on the given account balances.

	Account Balance	Minimum Payment
a.	$187.93	
b.	$342.55	
c.	$261.48	
d.	$16.99	
e.	$310.00	
f.	$158.32	
g.	$200.10	

If the New Balance Is:	The Minimum Required Payment Is:
Up to $20.00	The new balance
$20.01 to $200.00	$20.00
$200.01 to $250.00	$25.00
$250.01 to $300.00	$30.00
$300.01 to $350.00	$35.00
$350.01 to $400.00	$40.00

88. *Consumerism*   Shipping and handling charges when ordering online generally are based on the dollar amount of the order. Use the table shown below to determine the cost of shipping each order.

	Amount of Order	Shipping Cost
a.	$12.42	
b.	$23.56	
c.	$47.80	
d.	$66.91	
e.	$35.75	
f.	$20.00	
g.	$18.25	

If the Amount Ordered Is:	The Shipping and Handling Charge Is:
$10.00 and under	$1.60
$10.01 to $20.00	$2.40
$20.01 to $30.00	$3.60
$30.01 to $40.00	$4.70
$40.01 to $50.00	$6.00
$50.01 and up	$7.00

CRITICAL THINKING

89. Indicate which digits of the number, if any, need not be entered on a calculator.
   a. 1.500   **b.** 0.908   **c.** 60.07   **d.** 0.0032

90. Find a number between **a.** 0.1 and 0.2, **b.** 1 and 1.1, and **c.** 0 and 0.005.

4.2 Addition and Subtraction of Decimals

OBJECTIVE A   Addition and subtraction of decimals

To add decimals, write the numbers so that the decimal points are on a vertical line. Add as you would with whole numbers. Then write the decimal point in the sum directly below the decimal points in the addends.

Add: $0.326 + 4.8 + 57.23$

Note that placing the decimal points on a vertical line ensures that digits of the same place value are added.

Point of Interest

Try this: Six different numbers are added together and their sum is 11. Four of the six numbers are 4, 3, 2, and 1. Find the other two numbers.

Find the sum of 0.64, 8.731, 12, and 5.9.

Arrange the numbers vertically, placing the decimal points on a vertical line.

Add the numbers in each column.

Write the decimal point in the sum directly below the decimal points in the addends.

```
  1 2
  0.64
  8.731
 12.
+ 5.9
──────
 27.271
```

To subtract decimals, write the numbers so that the decimal points are on a vertical line. Subtract as you would with whole numbers. Then write the decimal point in the difference directly below the decimal point in the subtrahend.

Subtract and check: $31.642 - 8.759$

Note that placing the decimal points on a vertical line ensures that digits of the same place value are subtracted.

Check: Subtrahend   8.759
         + Difference   + 22.883
         ──────────────────────────
         = Minuend   31.642

Subtract and check: $5.4 - 1.6832$

Insert zeros in the minuend so that it has the same number of decimal places as the subtrahend.

$$5.4000$$
$$- 1.6832$$

Subtract and then check.

$$\overset{4\ \ 13\ 9\ \ 9\,10}{\cancel{5}.\cancel{4}\cancel{0}\cancel{0}\cancel{0}}$$
$$- 1.6832$$
$$\overline{3.7168}$$

Check: 1.6832
$$+ 3.7168$$
$$\overline{5.4000}$$

Figure 4.1 shows the prices of an adult one-day passport to Walt Disney World for various years. Find the increase in price from 1995 to 2005.

Figure 4.1 Price of Adult One-Day Passport to Walt Disney World
Source: The Walt Disney Company

To find the increase in price, subtract the price in 1995 from the price in 2005.

$$63.64$$
$$- 39.22$$
$$\overline{24.42}$$

From 1995 to 2005, the price of an adult one-day passport to Walt Disney World increased by $24.42.

The sign rules for adding and subtracting decimals are the same rules used to add and subtract integers.

Simplify: $-36.087 + 54.29$

The signs of the addends are different. Subtract the smaller absolute value from the larger absolute value.

$54.29 - 36.087 = 18.203$

Attach the sign of the number with the larger absolute value.

$|54.29| > |-36.087|$

The sum is positive.

$-36.087 + 54.29 = 18.203$

Recall that the opposite or additive inverse of n is $-n$ and the opposite of $-n$ is n. To find the opposite of a number, change the sign of the number.

Simplify: $-2.86 - 10.3$

Rewrite subtraction as addition of the opposite. The opposite of 10.3 is -10.3.

$-2.86 - 10.3$

$= -2.86 + (-10.3)$

The signs of the addends are the same. Add the absolute values of the numbers. Attach the sign of the addends.

$= -13.16$

Evaluate $c - d$ when $c = 6.731$ and $d = -2.48$.

Replace c with 6.731 and d with -2.48.

Rewrite subtraction as addition of the opposite.

Add.

$$c - d$$
$$6.731 - (-2.48)$$
$$= 6.731 + 2.48$$
$$= 9.211$$

Point of Interest

Try this brain teaser. You have two U.S. coins that add up to $.55. One is not a nickel. What are these two coins?

Recall that to estimate the answer to a calculation, round each number to the highest place value of the number; the first digit of each number will be nonzero and all other digits will be zero. Perform the calculation using the rounded numbers.

Estimate the sum of 23.037 and 16.7892.

Round each number to the nearest ten.

Add the rounded numbers.

40 is an estimate of the sum of 23.037 and 16.7892. Note that 40 is very close to the actual sum of 39.8262.

$$23.037 \longrightarrow 20$$
$$16.7892 \longrightarrow +20$$
$$\overline{40}$$

$$23.037$$
$$+ 16.7892$$
$$\overline{39.8262}$$

When a number in an estimation is a decimal less than 1, round the decimal so that there is one nonzero digit.

Estimate the difference between 4.895 and 0.6193.

Round 4.895 to the nearest one.
Round 0.6193 to the nearest tenth.
Subtract the rounded numbers.

$$4.895 \longrightarrow 5.0$$
$$0.6193 \longrightarrow -\ 0.6$$
$$\overline{4.4}$$

4.4 is an estimate of the difference between 4.895 and 0.6193.
It is close to the actual difference of 4.2757.

$$4.8950$$
$$- 0.6193$$
$$\overline{4.2757}$$

EXAMPLE 1   Add: $35.8 + 182.406 + 71.0934$

Solution
$$\begin{array}{r} {\scriptstyle 1\ \ 1} \\ 35.8 \\ 182.406 \\ + \ 71.0934 \\ \hline 289.2994 \end{array}$$

YOU TRY IT 1   Add: $8.64 + 52.7 + 0.39105$

Your Solution

EXAMPLE 2   What is -251.49 more than -638.7?

Solution $-638.7 + (-251.49) = -890.19$

YOU TRY IT 2   What is 4.002 minus 9.378?

Your Solution

Solutions on p. S10

EXAMPLE 3 Subtract and check: $73 - 8.16$

Solution

$$
\begin{array}{r}
\overset{6\ \ 12\ \ 9\ \ 10}{7\cancel{3}.\cancel{0}\cancel{0}} \\
-\ \ \ 8.16 \\
\hline
64.84
\end{array}
$$

Check:
$$
\begin{array}{r}
8.16 \\
+\ 64.84 \\
\hline
73.00
\end{array}
$$

YOU TRY IT 3 Subtract and check: $25 - 4.91$

Your Solution

EXAMPLE 4 Estimate the sum of 0.3927, 0.4856, and 0.2104.

Solution

$$
\begin{array}{l}
0.3927 \longrightarrow\ \ \ 0.4 \\
0.4856 \longrightarrow\ \ \ 0.5 \\
0.2104 \longrightarrow\ +0.2 \\
\hline
1.1
\end{array}
$$

YOU TRY IT 4 Estimate the sum of 6.514, 8.903, and 2.275.

Your Solution

EXAMPLE 5 Evaluate $x + y + z$ for $x = -1.6$, $y = 7.9$, and $z = -4.8$.

Solution

$x + y + z$
$-1.6 + 7.9 + (-4.8) = 6.3 + (-4.8)$
$= 1.5$

YOU TRY IT 5 Evaluate $x + y + z$ for $x = -7.84$, $y = -3.05$, and $z = 2.19$.

Your Solution

EXAMPLE 6 Is -4.3 a solution of the equation $9.7 - b = 5.4$?

Solution

$$
\begin{array}{c|c}
9.7 - b = 5.4 & \\
\hline
9.7 - (-4.3) & 5.4 \\
9.7 + 4.3 & 5.4 \\
14.0 \ne 5.4 &
\end{array}
$$

▶ Replace b with -4.3.

No, -4.3 is not a solution of the equation.

YOU TRY IT 6 Is -23.8 a solution of the equation $-m + 16.9 = 40.7$?

Your Solution

Solutions on pp. S10–S11

OBJECTIVE B **Applications and formulas**

Figure 4.2 shows the breakdown by age group of Americans who are hearing impaired. Use this graph for Example 7 and You Try It 7.

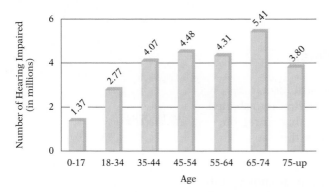

Figure 4.2   Breakdown by Age Group of Hearing-Impaired Americans
Source: American Speech-Language-Hearing Association

EXAMPLE 7

Use Figure 4.2 to determine whether the number of hearing-impaired individuals under the age of 45 is more or less than the number of hearing impaired who are over the age of 64.

Strategy

To make the comparison:

▶ Find the number of hearing-impaired individuals under the age of 45 by adding the numbers who are aged 0–17 (1.37 million), aged 18–34 (2.77 million), and aged 35–44 (4.07 million).
▶ Find the number of hearing-impaired individuals over the age of 64 by adding the numbers who are aged 65–74 (5.41 million) and aged 75 or older (3.80 million).
▶ Compare the two sums.

Solution

1.37 + 2.77 + 4.07 = 8.21

5.41 + 3.80 = 9.21

8.21 < 9.21

The number of hearing-impaired individuals under the age of 45 is less than the number of hearing impaired who are over the age of 64.

YOU TRY IT 7

Use Figure 4.2 to determine whether the number of hearing-impaired individuals under the age of 55 is more or less than the number of hearing impaired who are 55 or older.

Your Strategy

Your Solution

Solution on p. S11

4.2 Exercises

OBJECTIVE A **Addition and subtraction of decimals**

1. Set up the addition problem 2.391 + 45 + 13.0784 in a vertical format, as
   shown at the right. One addend is already placed. Fill in the first two shaded
   areas with the other addends lined up correctly. In the third shaded region,
   show the placement of the decimal point in the sum. Then add.

   + 13.0784

2. Set up the subtraction problem 34 − 18.21 in a vertical format, as shown at the
   right. Fill in the first two shaded regions with the minuend and subtrahend lined
   up correctly and zeros inserted as needed. In the third shaded region, show the
   placement of the decimal point in the difference. Then subtract.

Add or subtract.

3. 1.864 + 39 + 25.0781

4. 2.04 + 35.6 + 4.918

5. 35.9 + 8.217 + 146.74

6. 12 + 73.59 + 6.482

7. 36.47 − 15.21

8. 85.69 − 2.13

9. 28 − 6.74

10. 5 − 1.386

11. 6.02 − 3.252

12. 0.92 − 0.0037

13. −42.1 − 8.6

14. −6.57 − 8.933

15. 5.73 − 9.042

16. −31.894 + 7.5

17. −9.37 + 3.465

18. 1.09 − (−8.3)

19. −19 − (−2.65)

20. 3.18 − 5.72 − 6.4

21. −12.3 − 4.07 + 6.82

22. −8.9 + 7.36 − 14.2

23. −5.6 − (−3.82) − 17.409

24. Find the sum of 2.536, 14.97, 8.014, and 21.67.

25. Find the total of 6.24, 8.573, 19.06, and 22.488.

26. What is 6.9217 decreased by 3.4501?

27. What is 8.9 less than 62.57?

28. How much greater is 5 than 1.63?

29. What is the sum of −65.47 and −32.91?

30. Find 382.9 more than −430.6.

31. Find −138.72 minus 510.64.

32. What is 4.793 less than −6.82?

33. How much greater is −31 than −62.09?

Add or subtract. Then check by estimating the sum or difference.

34. 45.06 + 80.71

35. 6.408 + 5.917

36. 0.24 + 0.38 + 0.96

37. 56.87 − 23.24

38. 6.272 − 1.848

39. 0.931 − 0.628

40. 5.37 + 26.49

41. 87.65 − 49.032

42. 387.6 − 54.92

43. Education The graph at the right shows where U.S. children in grades K–12 are being educated. Figures are in millions of children.
  a. Find the total number of children in grades K–12.
  b. How many more children are being educated in public school than in private school?

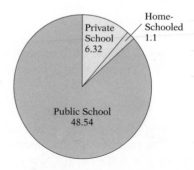

Where Children in Grades K–12 are Being Educated in the United States
Source: National Center for Education Statistics

Evaluate the variable expression $x + y$ for the given values of x and y.

44. $x = 62.97; y = -43.85$

45. $x = 5.904; y = -7.063$

46. $x = -125.41; y = 361.55$

47. $x = -6.175; y = -19.49$

Evaluate the variable expression $x + y + z$ for the given values of x, y, and z.

48. $x = 41.33; y = -26.095; z = 70.08$

49. $x = -6.059; y = 3.884; z = 15.71$

50. $x = 81.72; y = 36.067; z = -48.93$

51. $x = -16.219; y = 47; z = -2.3885$

Evaluate the variable expression $x - y$ for the given values of x and y.

52. $x = 43.29; y = 18.76$

53. $x = 6.029; y = -4.708$

54. $x = -16.329; y = 4.54$

55. $x = -21.073; y = 6.48$

56. $x = -3.69; y = -1.527$

57. $x = -8.21; y = -6.798$

58. Is -1.2 a solution of the equation $6.4 = 5.2 + a$?

59. Is -2.8 a solution of the equation $0.8 - p = 3.6$?

60. Is -0.5 a solution of the equation $x - 0.5 = 1$?

61. Is 36.8 a solution of the equation $27.4 = y - 9.4$?

62. Suppose n is a decimal number for which the difference $4.83 - n$ is a negative number. Which statement *must* be true about n?
(i) $n < 4.83$ **(ii)** $n > 4.83$ **(iii)** $n < -4.83$ **(iv)** $n > -4.83$

63. Suppose n is a decimal number for which the sum $6.875 + n$ is a negative number. Which statement *must* be true about n?
(i) $n < 6.875$ **(ii)** $n > 6.875$ **(iii)** $n < -6.875$ **(iv)** $n > -6.875$

OBJECTIVE B **Applications and formulas**

For Exercises 64 and 65, use Figure 4.2 on page 251. State whether you would use addition or subtraction to find the specified amount.

64. To find how many more Americans aged 65 to 74 are hearing impaired than aged 55 to 64, use _____.

65. To find the number of Americans aged 18 to 44 who are hearing impaired, use _____.

66. You have $20 to spend, and you make purchases for the following amounts: $4.24, $8.66, and $.54. Which of the following expressions correctly represent the amount of money you have left?
(i) $20 - 4.24 + 8.66 + 0.54$ (ii) $(4.24 + 8.66 + 0.54) - 20$
(iii) $20 - (4.24 + 8.66 + 0.54)$ (iv) $20 - 4.24 - 8.66 - 0.54$

67. You had $859.12 in your bank account at the beginning of the month. During the month you made deposits of $25 and $180.50 and withdrawals of $20, $75, and $10.78. Write a verbal description of what each expression below represents.
a. $25 + 180.50$
b. $20 + 75 + 10.78$
c. $859.12 + (25 + 180.50) - (20 + 75 + 10.78)$

68. *Temperature*   On January 22, 1943, in Spearfish, South Dakota, the temperature fell from 12.22°C at 9:00 A.M. to −20°C at 9:27 A.M. How many degrees did the temperature fall during the 27-minute period?

69. *Temperature*   On January 10, 1911, in Rapid City, South Dakota, the temperature fell from 12.78°C at 7:00 A.M. to −13.33°C at 7:15 A.M. How many degrees did the temperature fall during the 15-minute period?

Net Income   The graph at the right shows the net income, in billions, for Ford Motor Company for the years 2001 through 2005. Use this graph for Exercises 70 and 71.

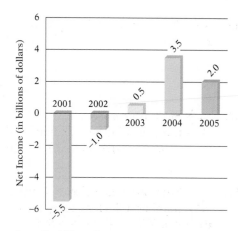

Source: Ford Motor Company

70. a. What was the increase in the net income from 2002 to 2004?
b. Between which two years shown in the graph was the increase in net income greatest?

71. a. What was the difference between the net income in 2003 and the net income in 2001? b. How much greater was the increase in net income from 2001 to 2002 than from 2002 to 2003?

72. *Consumerism* Using the menu shown below, estimate the bill for the following order: 1 soup, 1 cheese sticks, 1 blackened swordfish, 1 chicken divan, and 1 carrot cake.

Appetizers
Soup of the Day $5.75
Cheese Sticks $8.25
Potato Skins $8.50

Entrees
Roast Prime Rib $28.95
Blackened Swordfish $26.95
Chicken Divan $24.95

Desserts
Carrot Cake $7.25
Ice Cream Pie $8.50
Cheese Cake $9.75

73. *Consumerism* Using the menu shown above, estimate the bill for the following order: 1 potato skins, 1 cheese sticks, 1 roast prime rib, 1 chicken divan, 1 ice cream pie, and 1 cheese cake.

74. *Life Expectancy* The graph below shows the life expectancy at birth for males and females in the United States.
   a. Has life expectancy increased for both males and females with every 10-year period shown in the graph?
   b. Did males or females have a longer life expectancy in 2000? How much longer?
   c. During which year shown in the graph was the difference between male life expectancy and female life expectancy greatest?

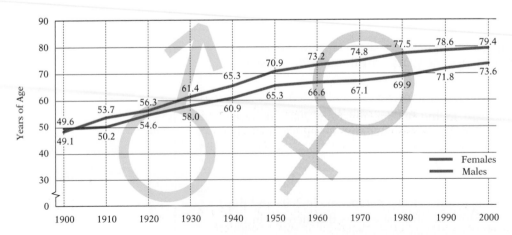

Life Expectancies of Males and Females in the United States

EXAMPLE 11 Evaluate $\frac{x}{y}$ for $x = -76.8$ and $y = 0.8$.

Solution $\frac{x}{y}$

$\frac{-76.8}{0.8} = -96$

YOU TRY IT 11 Evaluate $\frac{x}{y}$ for $x = -40.6$ and $y = -0.7$.

Your Solution

EXAMPLE 12 Is -0.4 a solution of the equation $\frac{8}{x} = -20$?

Solution $\frac{8}{x} = -20$

$\begin{array}{c|c} \dfrac{8}{-0.4} & -20 \end{array}$ ▶ Replace x by -0.4.

$-20 = -20$

Yes, -0.4 is a solution of the equation.

YOU TRY IT 12 Is -1.2 a solution of the equation $-2 = \frac{d}{-0.6}$?

Your Solution

Solutions on p. S11

OBJECTIVE C   **Fractions and decimals**

Because the fraction bar can be read "divided by," any fraction can be written as a decimal. To write a fraction as a decimal, divide the numerator of the fraction by the denominator.

Convert $\frac{3}{4}$ to a decimal.

$$\begin{array}{r} 0.75 \\ 4\overline{)3.00} \\ -2\,8 \\ \hline 20 \\ -20 \\ \hline 0 \end{array}$$

0.75 ◀—— This is a **terminating decimal.**

The remainder is zero.

$\frac{3}{4} = 0.75$

Take Note

The fraction bar can be read "divided by."

$\frac{3}{4} = 3 \div 4$

Dividing the numerator by the denominator results in a remainder of 0. The decimal 0.75 is a terminating decimal.

Convert $\frac{5}{11}$ to a decimal.

$$
\begin{array}{r}
0.4545 \\
11\overline{)5.0000} \\
-4\,4 \\
\hline
60 \\
-55 \\
\hline
50 \\
-44 \\
\hline
60 \\
-55 \\
\hline
5
\end{array}
$$

← This is a **repeating decimal.**

← The remainder is never zero.

$\dfrac{5}{11} = 0.\overline{45}$ The bar over the digits 45 is used to show that these digits repeat.

Convert $2\frac{4}{9}$ to a decimal.

Write the fractional part of the mixed number as a decimal. Divide the numerator by the denominator.

$$
\begin{array}{r}
0.444 = 0.\overline{4} \\
9\overline{)4.000}
\end{array}
$$

The whole number part of the mixed number is the whole number part of the decimal.

$2\dfrac{4}{9} = 2.\overline{4}$

To convert a decimal to a fraction, remove the decimal point and place the decimal part over a denominator equal to the place value of the last digit in the decimal.

→ hundredths → hundredths → tenths

$0.57 = \dfrac{57}{100}$ $\qquad$ $7.65 = 7\dfrac{65}{100} = 7\dfrac{13}{20}$ $\qquad$ $8.6 = 8\dfrac{6}{10} = 8\dfrac{3}{5}$

Convert 4.375 to a fraction.

The 5 in 4.375 is in the thousandths place. Write 0.375 as a fraction with a denominator of 1,000.

$4.375 = 4\dfrac{375}{1,000}$

Simplify the fraction.

$= 4\dfrac{3}{8}$

To find the order relation between a fraction and a decimal, first rewrite the fraction as a decimal. Then compare the two decimals.

Find the order relation between $\frac{6}{7}$ and 0.855.

Write the fraction as a decimal. Round to one more place value than the given decimal. (0.855 has 3 decimal places; round to 4 decimal places.)

$\dfrac{6}{7} \approx 0.8571$

Compare the two decimals.

$0.8571 > 0.8550$

Replace the decimal approximation of $\frac{6}{7}$ with $\frac{6}{7}$.

$\dfrac{6}{7} > 0.855$

EXAMPLE 13   Convert $\frac{5}{8}$ to a decimal.

Solution $\begin{array}{r} 0.625 \\ 8\overline{)5.000} \end{array}$

$\frac{5}{8} = 0.625$

YOU TRY IT 13   Convert $\frac{4}{5}$ to a decimal.

Your Solution

EXAMPLE 14   Convert $3\frac{1}{3}$ to a decimal.

Solution   Write $\frac{1}{3}$ as a decimal.

$\begin{array}{r} 0.333 = 0.\overline{3} \\ 3\overline{)1.000} \end{array}$

$3\frac{1}{3} = 3.\overline{3}$

YOU TRY IT 14   Convert $1\frac{5}{6}$ to a decimal.

Your Solution

EXAMPLE 15   Convert 7.25 to a fraction.

Solution $7.25 = 7\frac{25}{100} = 7\frac{1}{4}$

YOU TRY IT 15   Convert 6.2 to a fraction.

Your Solution

EXAMPLE 16   Place the correct symbol, < or >, between the two numbers.

$\quad 0.845 \qquad \frac{5}{6}$

Solution $\frac{5}{6} \approx 0.8333$

$0.8450 > 0.8333$

$0.845 > \frac{5}{6}$

YOU TRY IT 16   Place the correct symbol, < or >, between the two numbers.

$\quad 0.588 \qquad \frac{7}{12}$

Your Solution

Solutions on p. S11

OBJECTIVE D **Applications and formulas**

EXAMPLE 17

A one-year subscription to a monthly magazine costs $93. The price of each issue at the newsstand is $9.80. How much would you save per issue by buying a year's subscription rather than buying each issue at the newsstand?

Strategy

To find the amount saved:

▶ Find the subscription price per issue by dividing the cost of the subscription (93) by the number of issues (12).

▶ Subtract the subscription price per issue from the newsstand price (9.80).

Solution

$$
\begin{array}{r}
7.75 \\
12\overline{)93.00} \\
-84 \\
\hline
9\,0 \\
-8\,4 \\
\hline
60 \\
-60 \\
\hline
0
\end{array}
\qquad
\begin{array}{r}
9.80 \\
-7.75 \\
\hline
2.05
\end{array}
$$

The savings would be $2.05 per issue.

YOU TRY IT 17

You hand a postal clerk a ten-dollar bill to pay for the purchase of twelve 41¢ stamps. How much change do you receive?

Your Strategy

Your Solution

EXAMPLE 18

Use the formula $P = BF$, where P is the insurance premium, B is the base rate, and F is the rating factor, to find the insurance premium due on an insurance policy with a base rate of $342.50 and a rating factor of 2.2.

Strategy

To find the insurance premium due, replace B by 342.50 and F by 2.2 in the given formula and solve for P.

Solution

$P = BF$
$P = 342.50(2.2)$
$P = 753.50$

The insurance premium due is $753.50.

YOU TRY IT 18

Use the formula $P = BF$, where P is the insurance premium, B is the base rate, and F is the rating factor, to find the insurance premium due on an insurance policy with a base rate of $276.25 and a rating factor of 1.8.

Your Strategy

Your Solution

Solutions on p. S12

4.4 Exercises

OBJECTIVE A **Solving equations**

For Exercises 1 to 4, fill in the blank with *add 3.4 to each side, subtract 3.4 from each side, multiply each side by 3.4,* or *divide each side by 3.4.*

1. To solve $3.4 + x = -7.1$, _____.

2. To solve $3.4x = -7.1$, _____.

3. To solve $\dfrac{x}{3.4} = -7.1$, _____.

4. To solve $x - 3.4 = -7.1$, _____.

Solve. Write the answer as a decimal.

5. $y + 3.96 = 8.45$ **6.** $x - 2.8 = 1.34$ **7.** $-9.3 = c - 15$ **8.** $-28 = x - 3.27$

9. $7.3 = -\dfrac{n}{1.1}$ **10.** $-5.1 = \dfrac{y}{3.2}$ **11.** $-7x = 8.4$ **12.** $1.44 = -0.12t$

13. $y - 0.234 = -0.09$ **14.** $9 = z + 0.98$ **15.** $6.21r = -1.863$ **16.** $-78.1a = 85.91$

17. $-0.001 = x + 0.009$ **18.** $5 = 43.5 + c$ **19.** $\dfrac{x}{2} = -0.93$ **20.** $-1.03 = -\dfrac{z}{3}$

21. $-6v = 15$ **22.** $-55 = -40x$ **23.** $0.908 = 2.913 + x$ **24.** $\dfrac{t}{-2.1} = -7.8$

25. If the equation $x + 8.754 = n$ has a negative solution for x, is $n > 8.754$ or is $n < 8.754$?

26. If the equation $-5.73x = n$ has a positive solution for x, is $n > 0$ or is $n < 0$?

OBJECTIVE B **Applications**

27. JD Office Supplies wants to make a profit of $8.50 on the sale of a calendar that costs the store $15.23. To find the selling price, use the equation $P = S - C$, where P is the profit on an item, S is the selling price, and C is the cost.

    a. Replace P by _____ and replace C by _____ in the given formula:
    _____ $= S -$ _____.

    b. Solve the equation you wrote in part (a) by adding _____ to each side.

    c. The selling price of the calendar is _____.

28. *Cost of Living* The average cost per mile to operate a car is given by the equation $M = \frac{C}{N}$, where M is the average cost per mile, C is the total cost of operating the car, and N is the number of miles the car is driven. Use this formula to find the total cost of operating a car for 25,000 mi when the average cost per mile is $.42.

29. *Physics* The average acceleration of an object is given by $a = \frac{v}{t}$, where a is the average acceleration, v is the velocity, and t is the time. Find the velocity after 6.3 s of an object whose acceleration is 16 ft/s² (feet per second squared).

30. *Accounting* The fundamental accounting equation is $A = L + S$, where A is the assets of a company, L is the liabilities of the company, and S is the stockholders' equity. Find the stockholders' equity in a company whose assets are $34.8 million and whose liabilities are $29.9 million.

31. *Cost of Living* The cost of operating an electrical appliance is given by the formula $c = 0.001wtk$, where c is the cost of operating the appliance, w is the number of watts, t is the number of hours, and k is the cost per kilowatt-hour. Find the cost per kilowatt-hour if it costs $.04 to operate a 1,000-watt microwave for 4 h.

32. *Business* The markup on an item in a store equals the difference between the selling price of the item and the cost of the item. Find the selling price of a package of golf balls for which the cost is $9.81 and the markup is $5.19.

33. *Consumerism* The total of the monthly payments for a car lease is the product of the number of months of the lease and the monthly lease payment. The total of the monthly payments for a 60-month car lease is $21,387. Find the monthly lease payment.

34. *Geometry* The area of a rectangle is 210 in². If the width of the rectangle is 10.5 in., what is the length? Use the formula $A = LW$.

$A = 210 \text{ in}^2$

35. *Geometry* The length of a rectangle is 18 ft. If the area is 225 ft², what is the width of the rectangle? Use the formula $A = LW$.

CRITICAL THINKING

36. Solve: $0.\overline{33}x = 7$

37. **a.** Make up an equation of the form $x - b = c$ for which $x = -0.96$.
   b. Make up an equation of the form $ax = b$ for which $x = 2.1$.

38. For the equation $0.375x = 0.6$, a student offered the solution shown at the right. Is this a correct method of solving the equation? Explain your answer.

39. Consider the equation $12 = \frac{x}{a}$, where a is any positive number. Explain how increasing values of a affect the solution, x, of the equation.

$$0.375x = 0.6$$
$$\frac{375}{1,000}x = \frac{6}{10}$$
$$\frac{3}{8}x = \frac{3}{5}$$
$$\frac{8}{3} \cdot \frac{3}{8}x = \frac{8}{3} \cdot \frac{3}{5}$$
$$x = \frac{8}{5} = 1.6$$

4.5 Radical Expressions

OBJECTIVE A   Square roots of perfect squares

Recall that the square of a number is equal to the number multiplied times itself.

$$3^2 = 3 \cdot 3 = 9$$

The square of an integer is called a **perfect square.**

9 is a perfect square because 9 is the square of 3: $3^2 = 9$.

The numbers $1, 4, 9, 16, 25, 36, 49, 64, 81,$ and 100 are perfect squares.

$$1^2 = 1$$
$$2^2 = 4$$
$$3^2 = 9$$
$$4^2 = 16$$
$$5^2 = 25$$
$$6^2 = 36$$
$$7^2 = 49$$
$$8^2 = 64$$
$$9^2 = 81$$
$$10^2 = 100$$

Larger perfect squares can be found by squaring 11, squaring 12, squaring 13, and so on.

Note that squaring the negative integers results in the same list of numbers.

$$(-1)^2 = 1$$
$$(-2)^2 = 4$$
$$(-3)^2 = 9$$
$$(-4)^2 = 16, \text{ and so on.}$$

Perfect squares are used in simplifying square roots. The symbol for square root is $\sqrt{}$.

Square Root

A square root of a positive number x is a number whose square is x.

If $a^2 = x$, then $\sqrt{x} = a$.

The expression $\sqrt{9}$, read "the square root of 9," is equal to the number that when squared is equal to 9.

Since $3^2 = 9$, $\sqrt{9} = 3$.

Every positive number has two square roots, one a positive number and one a negative number. The symbol $\sqrt{}$ is used to indicate the positive square root of a number. When the negative square root of a number is to be found, a negative sign is placed in front of the square root symbol. For example,

$$\sqrt{9} = 3 \quad \text{and} \quad -\sqrt{9} = -3$$

Point of Interest

The square root symbol, $\sqrt{}$, is also called a **radical.** The number under the radical is called the **radicand.** In the radical expression $\sqrt{9}$, 9 is the radicand.

Simplify: $\sqrt{49}$

$\sqrt{49}$ is equal to the number that when squared equals 49. $7^2 = 49$.

$$\sqrt{49} = 7$$

Simplify: $-\sqrt{49}$

The negative sign in front of the square root symbol indicates the negative square root of 49. $(-7)^2 = 49$.

$$-\sqrt{49} = -7$$

Simplify: $\sqrt{25} + \sqrt{81}$

Simplify each radical expression.

Since $5^2 = 25$, $\sqrt{25} = 5$.

Since $9^2 = 81$, $\sqrt{81} = 9$.

Add.

$$\sqrt{25} + \sqrt{81} = 5 + 9$$
$$= 14$$

Simplify: $5\sqrt{64}$

The expression $5\sqrt{64}$ means 5 times $\sqrt{64}$.

Simplify $\sqrt{64}$.

Multiply.

$$5\sqrt{64} = 5 \cdot 8$$
$$= 40$$

Simplify: $6 + 4\sqrt{9}$

Simplify $\sqrt{9}$.

Use the Order of Operations Agreement.

$$6 + 4\sqrt{9} = 6 + 4 \cdot 3$$
$$= 6 + 12$$
$$= 18$$

Simplify: $\sqrt{\dfrac{1}{9}}$

$\sqrt{\dfrac{1}{9}}$ is equal to the number that when squared equals $\dfrac{1}{9}$. $\left(\dfrac{1}{3}\right)^2 = \dfrac{1}{9}$.

$$\sqrt{\dfrac{1}{9}} = \dfrac{1}{3}$$

Note that the square root of $\dfrac{1}{9}$ is equal to the square root of the numerator $(\sqrt{1} = 1)$ over the square root of the denominator $(\sqrt{9} = 3)$.

Evaluate $\sqrt{xy}$ when $x = 5$ and $y = 20$.

$$\sqrt{xy}$$

Replace x with 5 and y with 20. $\sqrt{5 \cdot 20}$

Simplify under the radical. $= \sqrt{100}$

Take the square root of 100. $10^2 = 100$. $= 10$

> **Take Note**
>
> The radical is a grouping symbol. Therefore, when simplifying numerical expressions, simplify the radicand as part of Step 1 of the Order of Operations Agreement.

EXAMPLE 1   Simplify: $\sqrt{121}$

Solution   Since $11^2 = 121$, $\sqrt{121} = 11$.

YOU TRY IT 1   Simplify: $-\sqrt{144}$

Your Solution

EXAMPLE 2   Simplify: $\sqrt{\dfrac{4}{25}}$

Solution   Since $\left(\dfrac{2}{5}\right)^2 = \dfrac{4}{25}$, $\sqrt{\dfrac{4}{25}} = \dfrac{2}{5}$.

YOU TRY IT 2   Simplify: $\sqrt{\dfrac{81}{100}}$

Your Solution

EXAMPLE 3   Simplify: $\sqrt{36} - 9\sqrt{4}$

Solution $\sqrt{36} - 9\sqrt{4} = 6 - 9 \cdot 2$
$\qquad\qquad\qquad\quad = 6 - 18$
$\qquad\qquad\qquad\quad = 6 + (-18)$
$\qquad\qquad\qquad\quad = -12$

YOU TRY IT 3   Simplify: $4\sqrt{16} - \sqrt{9}$

Your Solution

EXAMPLE 4   Evaluate $6\sqrt{ab}$ for $a = 2$ and $b = 8$.

Solution $6\sqrt{ab}$
$6\sqrt{2 \cdot 8} = 6\sqrt{16}$
$\qquad\qquad\; = 6(4)$
$\qquad\qquad\; = 24$

YOU TRY IT 4   Evaluate $5\sqrt{a + b}$ for $a = 17$ and $b = 19$.

Your Solution

Solutions on p. S12

OBJECTIVE B **Square roots of whole numbers**

In the last objective, the radicand in each radical expression was a perfect square. Since the square root of a perfect square is an integer, the exact value of each radical expression could be found.

If the radicand is not a perfect square, the square root can only be approximated. For example, the radicand in the radical expression $\sqrt{2}$ is 2, and 2 is not a perfect square. The square root of 2 can be approximated to any desired place value.

To the nearest tenth:	$\sqrt{2} \approx 1.4$	$(1.4)^2 = 1.96$
To the nearest hundredth:	$\sqrt{2} \approx 1.41$	$(1.41)^2 = 1.9881$
To the nearest thousandth:	$\sqrt{2} \approx 1.414$	$(1.414)^2 = 1.999396$
To the nearest ten-thousandth:	$\sqrt{2} \approx 1.4142$	$(1.4142)^2 = 1.99996164$

The square of each approximation gets closer and closer to 2 as the number of place values in the decimal approximation increases. But no matter how many place values are used to approximate $\sqrt{2}$, the digits never terminate or repeat. In general, the square root of any number that is not a perfect square can only be approximated.

Calculator Note

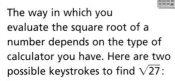

The way in which you evaluate the square root of a number depends on the type of calculator you have. Here are two possible keystrokes to find $\sqrt{27}$:

27 $\sqrt{\ }$ =

or

$\sqrt{\ }$ 27 **ENTER**

The first method is used on many scientific calculators. The second method is used on many graphing calculators.

Calculator Note

To evaluate $3\sqrt{5}$ on a calculator, enter either

3 × 5 $\sqrt{\ }$ =

or

3 $\sqrt{\ }$ 5 **ENTER**

Round the number in the display to the desired place value.

Approximate $\sqrt{11}$ to the nearest ten-thousandth.

11 is not a perfect square.
Use a calculator to approximate $\sqrt{11}$. $\sqrt{11} \approx 3.3166$

Approximate $3\sqrt{5}$ to the nearest ten-thousandth.

$3\sqrt{5}$ means 3 times $\sqrt{5}$. $3\sqrt{5} \approx 6.7082$

Between what two whole numbers is the value of $\sqrt{41}$?

Since the number 41 is between the perfect squares 36 and 49, the value of $\sqrt{41}$ is between $\sqrt{36}$ and $\sqrt{49}$.

$\sqrt{36} = 6$ and $\sqrt{49} = 7$,

so the value of $\sqrt{41}$ is between the whole numbers 6 and 7.

This can be written using inequality symbols as $6 < \sqrt{41} < 7$, which is read

"the square root of 41 is greater than 6 and less than 7."

Use a calculator to verify that $\sqrt{41} \approx 6.4$, which is between 6 and 7.

Sometimes we are not interested in an approximation of the square root of a number, but rather the exact value in simplest form.

A radical expression is in simplest form when the radicand contains no factor, other than 1, that is a perfect square. The Product Property of Square Roots is used to simplify radical expressions.

> **Product Property of Square Roots**
> ···
> If a and b are positive numbers, then $\sqrt{a \cdot b} = \sqrt{a} \cdot \sqrt{b}$.

The Product Property of Square Roots states that the square root of a product is equal to the product of the square roots. For example,

$$\sqrt{4 \cdot 9} = \sqrt{4} \cdot \sqrt{9}$$

Note that $\sqrt{4 \cdot 9} = \sqrt{36} = 6$ and $\sqrt{4} \cdot \sqrt{9} = 2 \cdot 3 = 6$.

Simplify: $\sqrt{50}$

Think: What perfect square is a factor of 50?

Begin with a perfect square that is larger than 50.

Then test each successively smaller perfect square.

$8^2 = 64$; 64 is too big.
$7^2 = 49$; 49 is not a factor of 50.
$6^2 = 36$; 36 is not a factor of 50.
$5^2 = 25$; 25 is a factor of 50. $(50 = 25 \cdot 2)$

Write $\sqrt{50}$ as $\sqrt{25 \cdot 2}$. $\qquad\qquad \sqrt{50} = \sqrt{25 \cdot 2}$

Use the Product Property of Square Roots. $\qquad = \sqrt{25} \cdot \sqrt{2}$

Simplify $\sqrt{25}$. $\qquad\qquad\qquad\qquad\qquad = 5 \cdot \sqrt{2}$

The radicand 2 contains no factor other than 1 that $\qquad = 5\sqrt{2}$
is a perfect square. The radical expression $5\sqrt{2}$ is in
simplest form.

Remember that $5\sqrt{2}$ means 5 times $\sqrt{2}$. Using a calculator, $5\sqrt{2} \approx 5(1.4142) = 7.071$, and $\sqrt{50} \approx 7.071$.

Calculator Note

The keystrokes to evaluate $5\sqrt{2}$ on a calculator are either

5 ⊠ 2 √ =

or

5 √ 2 ENTER

Round the number in the display to the desired place value.

EXAMPLE 5 Approximate $4\sqrt{17}$ to the nearest ten-thousandth.

Solution $4\sqrt{17} \approx 16.4924$ ▸ Use a calculator.

YOU TRY IT 5 Approximate $5\sqrt{23}$ to the nearest ten-thousandth.

Your Solution

Solution on p. S12

EXAMPLE 6 Between what two whole numbers is the value of $\sqrt{79}$?

Solution 79 is between the perfect squares 64 and 81.

$\sqrt{64} = 8$ and $\sqrt{81} = 9$.

$8 < \sqrt{79} < 9$

YOU TRY IT 6 Between what two whole numbers is the value of $\sqrt{57}$?

Your Solution

EXAMPLE 7 Simplify: $\sqrt{32}$

Solution $6^2 = 36$; 36 is too big.
$5^2 = 25$; 25 is not a factor of 32.
$4^2 = 16$; 16 is a factor of 32.

$\sqrt{32} = \sqrt{16 \cdot 2}$
$= \sqrt{16} \cdot \sqrt{2}$
$= 4 \cdot \sqrt{2}$
$= 4\sqrt{2}$

YOU TRY IT 7 Simplify: $\sqrt{80}$

Your Solution

Solutions on p. S12

OBJECTIVE C Applications and formulas

EXAMPLE 8

Find the range of a submarine periscope that is 8 ft above the surface of the water. Use the formula $R = 1.4\sqrt{h}$, where R is the range in miles and h is the height in feet of the periscope above the surface of the water. Round to the nearest hundredth.

Strategy

To find the range, replace h by 8 in the given formula and solve for R.

Solution

$R = 1.4\sqrt{h}$
$R = 1.4\sqrt{8}$
$R \approx 3.96$ ▶ Use a calculator.

The range of the periscope is 3.96 mi.

YOU TRY IT 8

Find the range of a submarine periscope that is 6 ft above the surface of the water. Use the formula $R = 1.4\sqrt{h}$, where R is the range in miles and h is the height in feet of the periscope above the surface of the water. Round to the nearest hundredth.

Your Strategy

Your Solution

Solution on p. S12

4.5 Exercises

OBJECTIVE A **Square roots of perfect squares**

1. A perfect square is the square of an _____. Circle each number in the list below that is a perfect square.

   1   2   3   4   8   9   20   48   49   50   75   81   90   100

2. The expression $\sqrt{64}$ is read "the _____ of sixty-four." The symbol $\sqrt{}$ is called the _____, and 64 is called the _____.

3. **a.** The expression $\sqrt{64}$ is used to mean the positive number whose square root is 64. The positive number whose square is 64 is _____, so we write $\sqrt{64} =$ _____.

   b. There is also a negative integer whose square is 64. This integer is _____. We write $-\sqrt{64} =$ _____.

4. Simplify: $\sqrt{81} - 3\sqrt{25}$ $\qquad\qquad$ $\sqrt{81} - 3\sqrt{25}$

   a. Simplify each radical expression. Remember that $81 = ($ _____ $)^2$ and $25 = ($ _____ $)^2$. $\qquad$ $=$ _____ $- 3 \cdot$ _____

   b. Use the Order of Operations Agreement. The next step is to _____. $\qquad$ $=$ _____ $-$ _____

   c. Subtract. $\qquad\qquad\qquad\qquad\qquad\qquad$ $=$ _____

5. $\sqrt{36}$ $\qquad\qquad$ 6. $\sqrt{1}$ $\qquad\qquad$ 7. $-\sqrt{9}$ $\qquad\qquad$ 8. $-\sqrt{1}$

9. $\sqrt{169}$ $\qquad\qquad$ 10. $\sqrt{196}$ $\qquad\qquad$ 11. $\sqrt{225}$ $\qquad\qquad$ 12. $\sqrt{81}$

13. $-\sqrt{25}$ $\qquad\qquad$ 14. $-\sqrt{64}$ $\qquad\qquad$ 15. $-\sqrt{100}$ $\qquad\qquad$ 16. $-\sqrt{4}$

17. $\sqrt{8 + 17}$ $\qquad\qquad$ 18. $\sqrt{40 + 24}$ $\qquad\qquad$ 19. $\sqrt{49} + \sqrt{9}$ $\qquad\qquad$ 20. $\sqrt{100} + \sqrt{16}$

21. $\sqrt{121} - \sqrt{4}$ $\qquad\qquad$ 22. $\sqrt{144} - \sqrt{25}$ $\qquad\qquad$ 23. $3\sqrt{81}$ $\qquad\qquad$ 24. $8\sqrt{36}$

25. $-2\sqrt{49}$

26. $-6\sqrt{121}$

27. $5\sqrt{16} - 4$

28. $7\sqrt{64} + 9$

29. $3 + 10\sqrt{1}$

30. $14 - 3\sqrt{144}$

31. $\sqrt{4} - 2\sqrt{16}$

32. $\sqrt{144} + 3\sqrt{9}$

33. $5\sqrt{25} + \sqrt{49}$

34. $20\sqrt{1} - \sqrt{36}$

35. $\sqrt{\dfrac{1}{100}}$

36. $\sqrt{\dfrac{1}{81}}$

37. $\sqrt{\dfrac{9}{16}}$

38. $\sqrt{\dfrac{25}{49}}$

39. $\sqrt{\dfrac{1}{4}} + \sqrt{\dfrac{1}{64}}$

40. $\sqrt{\dfrac{1}{36}} - \sqrt{\dfrac{1}{144}}$

Evaluate the expression for the given values of the variables.

41. $-4\sqrt{xy}$, for $x = 3$ and $y = 12$

42. $-3\sqrt{xy}$, for $x = 20$ and $y = 5$

43. $8\sqrt{x + y}$, for $x = 19$ and $y = 6$

44. $7\sqrt{x + y}$, for $x = 34$ and $y = 15$

45. $5 + 2\sqrt{ab}$, for $a = 27$ and $b = 3$

46. $6\sqrt{ab} - 9$, for $a = 2$ and $b = 32$

47. $\sqrt{a^2 + b^2}$, for $a = 3$ and $b = 4$

48. $\sqrt{c^2 - a^2}$, for $a = 6$ and $c = 10$

49. $\sqrt{c^2 - b^2}$, for $b = 12$ and $c = 13$

50. $\sqrt{b^2 - 4ac}$, for $a = 1$, $b = -4$, and $c = -5$

51. What is the sum of five and the square root of nine?

52. Find eight more than the square root of four.

53. Find the difference between six and the square root of twenty-five.

54. What is seven decreased by the square root of sixteen?

55. What is negative four times the square root of eighty-one?

56. Find the product of negative three and the square root of forty-nine.

57. Simplify. **a.** $\sqrt{\sqrt{16}}$ **b.** $-\sqrt{\sqrt{81}}$

58. Given that x is a positive number, state whether the expression represents a positive or a negative number.

a. $-3 - \sqrt{x}$ **b.** $-3(-\sqrt{x})$

OBJECTIVE B Square roots of whole numbers

59. Describe **a.** how to find the square root of a perfect square and **b.** how to simplify the square root of a number that is not a perfect square.

60. Explain why $2\sqrt{2}$ is in simplest form and $\sqrt{8}$ is not in simplest form.

61. **a.** 33 is between the perfect squares _____ and _____, so $\sqrt{33}$ is between $\sqrt{\underline{\hspace{0.5cm}}} = \underline{\hspace{0.5cm}}$ and $\sqrt{\underline{\hspace{0.5cm}}} = \underline{\hspace{0.5cm}}$.

b. Express the fact that $\sqrt{33}$ is between 5 and 6 as an inequality:

_____ < _____ < _____

62. Simplify: $\sqrt{128}$

    $\sqrt{128}$

a. Write 128 as the product of a perfect-square factor and a factor that does not contain a perfect square.

    $= \sqrt{\underline{\hspace{0.8cm}} \cdot 2}$

b. Use the Product Property of Square Roots to write the expression as the product of two square roots.

    $= \sqrt{\underline{\hspace{0.8cm}}} \cdot \sqrt{\underline{\hspace{0.8cm}}}$

c. Simplify $\sqrt{64}$.

    $= \underline{\hspace{0.8cm}} \cdot \sqrt{2} = \underline{\hspace{0.8cm}}$

Approximate to the nearest ten-thousandth.

63. $\sqrt{3}$ **64.** $\sqrt{7}$ **65.** $\sqrt{10}$ **66.** $\sqrt{19}$

67. $2\sqrt{6}$ **68.** $10\sqrt{21}$ **69.** $3\sqrt{14}$ **70.** $6\sqrt{15}$

71. $-4\sqrt{2}$ **72.** $-5\sqrt{13}$ **73.** $-8\sqrt{30}$ **74.** $-12\sqrt{53}$

Between what two whole numbers is the value of the radical expression?

75. $\sqrt{23}$ **76.** $\sqrt{47}$ **77.** $\sqrt{29}$ **78.** $\sqrt{71}$

79. $\sqrt{62}$ **80.** $\sqrt{103}$ **81.** $\sqrt{130}$ **82.** $\sqrt{95}$

Simplify.

83. $\sqrt{8}$ **84.** $\sqrt{12}$ **85.** $\sqrt{45}$ **86.** $\sqrt{18}$ **87.** $\sqrt{20}$

88. $\sqrt{44}$ **89.** $\sqrt{27}$ **90.** $\sqrt{56}$ **91.** $\sqrt{48}$ **92.** $\sqrt{28}$

93. $\sqrt{75}$ **94.** $\sqrt{96}$ **95.** $\sqrt{63}$ **96.** $\sqrt{72}$ **97.** $\sqrt{98}$

98. $\sqrt{108}$ **99.** $\sqrt{112}$ **100.** $\sqrt{200}$ **101.** $\sqrt{175}$ **102.** $\sqrt{180}$

103. True or false?

If $0 < \sqrt{a} < 1$, then $0 < a < 1$.

104. True or false?

For a positive number a, $\sqrt{a^3} = a\sqrt{a}$.

OBJECTIVE C **Applications and formulas**

Earth Science A tsunami is a great wave produced by underwater earthquakes or volcanic eruption. For Exercises 105 to 108, use the formula $v = 3\sqrt{d}$, where v is the velocity in feet per second of a tsunami as it approaches land and d is the depth in feet of the water.

105. To find the velocity of a tsunami when the depth of the water is 81 ft, replace _____ in the given formula with 81 and solve for _____:

$$v = 3\sqrt{\underline{\hspace{1cm}}} = 3(\underline{\hspace{1cm}}) = \underline{\hspace{1cm}} \text{ ft/s}$$

106. To find the velocity of a tsunami when the depth of the water is 121 ft, replace d in the given formula with _____ and solve for _____:

$$v = 3\sqrt{\underline{\hspace{1cm}}} = 3(\underline{\hspace{1cm}}) = \underline{\hspace{1cm}} \text{ ft/s}$$

107. Find the velocity of a tsunami when the depth of the water is 100 ft.

108. Find the velocity of a tsunami when the depth of the water is 144 ft.

EXAMPLE 3   Graph the real numbers
between −3 and 0.

Solution   Draw a left parenthesis at −3
and a right parenthesis at 0.
Draw a heavy line between −3
and 0.

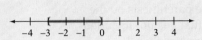

YOU TRY IT 3   Graph the real numbers
between −1 and 4.

Your Solution

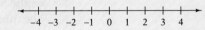

Solution on p. S12

OBJECTIVE B   **Inequalities in one variable**

Recall that the symbol for "is greater than" is >, and the symbol for "is less than" is <. The symbol ≥ means "is greater than or equal to." The symbol ≤ means "is less than or equal to."

The statement 5 < 5 is a false statement because 5 is not less than 5.

$5 < 5$ False

The statement 5 ≤ 5 is a true statement because 5 is "less than <u>or</u> equal to" 5; 5 is equal to 5.

$5 \leq 5$ True

An **inequality** contains the symbol >, <, ≥, or ≤, and expresses the relative order of two mathematical expressions.

$$4 > -3$$
$$-9.7 < 0$$
$$6 + 2 \geq 1$$
$$x \leq 5$$

Inequalities

The inequality $x \leq 5$ is read "x is less than or equal to 5."

For the inequality $x > -3$, which values of the variable listed below make the inequality true?

a. −6   **b.** −3.9   **c.** 0   **d.** $\sqrt{7}$

Replace x in $x > -3$ with each number, and determine whether each inequality is true.

a. $x > -3$
$-6 > -3$
False

b. $x > -3$
$-3.9 > -3$
False

c. $x > -3$
$0 > -3$
True

d. $x > -3$
$\sqrt{7} > -3$
True

The numbers 0 and $\sqrt{7}$ make the inequality true.

There are many values of the variable x that will make the inequality $x > -3$ true; any number greater than −3 makes the inequality true. Replacing x with any number less than −3 will result in a false statement.

What values of the variable x make the inequality $x \leq 4$ true?

All real numbers less than or equal to 4 make the inequality true.

The numbers that make an inequality true can be graphed on the real number line.

Graph $x > 1$.

The numbers that, when substituted for x, make this inequality true are all the real numbers greater than 1. The numbers greater than 1 are all the numbers to the right of 1 on the number line. The parenthesis on the graph indicates that 1 is not included in the numbers greater than 1.

Graph $x \geq 1$.

The numbers that make this inequality true are all the real numbers greater than or equal to 1. The bracket at 1 indicates that 1 is included in the numbers greater than or equal to 1.

Note: For $<$ or $>$, draw a parenthesis on the graph. For $\leq$ or $\geq$, draw a bracket.

EXAMPLE 4   For the inequality $x \leq -6$, which values of the variable listed below make the inequality true?

     a. -12 **b.** -6 **c.** 0 **d.** $\sqrt{5}$

Solution   **a.** $x \leq -6$
           $-12 \leq -6$ True

        b. $x \leq -6$
           $-6 \leq -6$ True

        c. $x \leq -6$
           $0 \leq -6$ False

        d. $x \leq -6$
           $\sqrt{5} \leq -6$ False

The numbers -12 and -6 make the inequality true.

YOU TRY IT 4   For the inequality $x \geq 4$, which values of the variable listed below make the inequality true?

     a. -1 **b.** 0 **c.** 4 **d.** $\sqrt{26}$

Your Solution

EXAMPLE 5   What values of the variable x make the inequality $x < 8$ true?

Solution   All real numbers less than 8 make the inequality true.

YOU TRY IT 5   What values of the variable x make the inequality $x > -7$ true?

Your Solution

Solutions on p. S12

EXAMPLE 6   Graph $x \leq 3$.

Solution   Draw a right bracket at 3.
Draw an arrow to the left of 3.

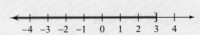

YOU TRY IT 6   Graph $x \geq -4$.

Your Solution

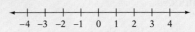

Solution on p. S13

OBJECTIVE C   **Applications**

Solving application problems requires recognition of the verbal phrases that translate into mathematical symbols. Below is a partial list of the phrases used to indicate each of the four inequality symbols.

$<$	is less than	$>$	is greater than
			is more than
			exceeds
$\leq$	is less than or equal to	$\geq$	is greater than or equal to
	maximum		minimum
	at most		at least
	or less		or more

EXAMPLE 7

The minimum wage at the company you work for is $9.25 an hour. Write an inequality for the wages at the company. Is it possible for an employee to earn $9.15 an hour?

Strategy

▶ To write the inequality, let w represent the wages. Since $9.25 is a minimum wage, all wages are greater than or equal to $9.25.

▶ To determine whether a wage of $9.15 is possible, replace w in the inequality by 9.15. If the inequality is true, it is possible. If the inequality is false, it is not possible.

Solution

$w \geq 9.25$

$9.15 \geq 9.25$ ▶ False

It is not possible for an employee to earn $9.15 an hour.

YOU TRY IT 7

On the highway near your home, motorists who exceed a speed of 55 mph are ticketed. Write an inequality for the speeds at which a motorist is ticketed. Will a motorist traveling at 58 mph be ticketed?

Your Strategy

Your Solution

Solution on p. S13

4.6 Exercises

OBJECTIVE A **Real numbers and the real number line**

For Exercises 1 to 4, fill in the blanks and circle the correct words to complete the sentences.

1. On the real number line, the graph of −1.5 is a solid dot halfway between _____ and _____. It is to the <u>left/right</u> of the number −1.

2. The graph at the right is the graph of the real numbers <u>less than/greater than</u> −2. The parenthesis at the number −2 is used to show that the −2 <u>is/is not</u> included in the graph.

3. The graph of the real numbers less than 5 is a heavy arrow that begins with a parenthesis at the number _____ and points to the <u>left/right</u>.

4. The graph at the right is the graph of the real numbers <u>between/greater than</u> −3 and 2.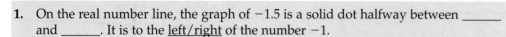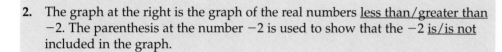

Graph the number on the real number line.

5. $2\dfrac{1}{2}$

6. $-2\dfrac{1}{2}$

7. −3.5

8. −0.5

9. $-4\dfrac{1}{2}$

10. $\dfrac{1}{2}$

11. 1.5

12. 5.5

Graph.

13. the real numbers greater than 6
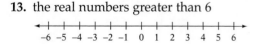

14. the real numbers greater than 1
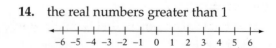

15. the real numbers less than 0
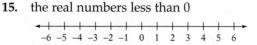

16. the real numbers less than 2
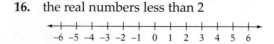

Focus on **Problem Solving**

From Concrete to Abstract

As you progress in your study of algebra, you will find that the problems become less concrete and more abstract. Problems that are concrete provide information pertaining to a specific instance. Abstract problems are theoretical; they are stated without reference to a specific instance. Let's look at an example of an abstract problem.

How many cents are in d dollars?

How can you solve this problem? Are you able to solve the same problem if the information given is concrete?

How many cents are in 5 dollars?

You know that there are 100 cents in 1 dollar. To find the number of cents in 5 dollars, multiply 5 by 100.

$$100 \cdot 5 = 500 \qquad \text{There are 500 cents in 5 dollars.}$$

Use the same procedure to find the number of cents in d dollars: multiply d by 100.

$$100 \cdot d = 100d \qquad \text{There are } 100d \text{ cents in } d \text{ dollars.}$$

This problem might be taken a step further:

If one pen costs c cents, how many pens can be purchased with d dollars?

Consider the same problem using numbers in place of the variables.

If one pen costs 25 cents, how many pens can be purchased with 2 dollars?

To solve this problem, you need to calculate the number of cents in 2 dollars (multiply 2 by 100) and divide the result by the cost per pen (25 cents).

$$\frac{100 \cdot 2}{25} = \frac{200}{25} = 8 \qquad \begin{array}{l} \text{If one pen costs 25 cents,} \\ \text{8 pens can be purchased with 2 dollars.} \end{array}$$

Use the same procedure to solve the related abstract problem. Calculate the number of cents in d dollars (multiply d by 100), and divide the result by the cost per pen (c cents).

$$\frac{100 \cdot d}{c} = \frac{100d}{c} \qquad \begin{array}{l} \text{If one pen costs } c \text{ cents, } \frac{100d}{c} \text{ pens} \\ \text{can be purchased with } d \text{ dollars.} \end{array}$$

At the heart of the study of algebra is the use of variables. It is the variables in the problems above that make them abstract. But it is variables that allow us to generalize situations and state rules about mathematics.

Try each of the following.

1. How many nickels are in d dollars?

2. How many gumballs can you buy if you have only d dollars and each gumball costs c cents?

3. If you travel m miles on one gallon of gasoline, how far can you travel on g gallons of gasoline?

4. If you walk one mile in *x* minutes, how far can you walk in *h* hours?

5. If one photocopy costs *n* nickels, how many photocopies can you make for *q* quarters?

Projects & Group Activities

Customer Billing   Chris works at B & W Garage as an auto mechanic and has just completed an engine overhaul for a customer. To determine the cost of the repair job, Chris keeps a list of times worked and parts used. A price list and a list of parts used and times worked are shown below. Use these tables, and the fact that the charge for labor is $46.75 per hour, to determine the total cost for parts and labor.

Parts Used		Time Spent	
Item	Quantity	Day	Hours
Gasket set	1	Monday	7.0
Ring set	1	Tuesday	7.5
Valves	8	Wednesday	6.5
Wrist pins	8	Thursday	8.5
Valve springs	16	Friday	9.0
Rod bearings	8		
Main bearings	5		
Valve seals	16		
Timing chain	1		

Price List		
Item Number	Description	Unit Price
27345	Valve spring	$9.25
41257	Main bearing	$17.49
54678	Valve	$16.99
29753	Ring set	$169.99
45837	Gasket set	$174.90
23751	Timing chain	$50.49
23765	Fuel pump	$229.99
28632	Wrist pin	$23.55
34922	Rod bearing	$13.69
2871	Valve seal	$1.69

Chapter 4 Summary

Key Words	**Examples**
A number written in **decimal notation** has three parts: a whole number part, a decimal point, and a decimal part. The **decimal part** of a number represents a number less than 1. A number written in decimal notation is often simply called a **decimal.** [4.1A, p. 237]	For the decimal 31.25, 31 is the whole number part and 25 is the decimal part.
The square of an integer is called a **perfect square.** [4.5A, p. 283]	$1^2 = 1, 2^2 = 4, 3^2 = 9, 4^2 = 16, 5^2 = 25, \ldots$, so 1, 4, 9, 16, 25, . . . are perfect squares.
A **square root** of a positive number *x* is a number whose square is *x*. The symbol for square root is $\sqrt{}$, which is called a **radical sign.** The number under the radical is called the **radicand.** [4.5A, pp. 283–284]	$\sqrt{25} = 5$ because $5^2 = 25$. In the expression $\sqrt{25}$, 25 is the radicand.

A radical expression is in **simplest form** when the radicand contains no factor, other than 1, that is a perfect square. [4.5B, p. 287]

$\sqrt{18}$ is not in simplest form because the radicand, 18, contains the factor 9, and 9 is a perfect square.

A **rational number** is a number that can be written in the form $\frac{a}{b}$, where a and b are integers and $b \neq 0$. Every rational number can be written as either a terminating decimal or a repeating decimal. All terminating and repeating decimals are rational numbers. [4.6A, p. 294]

$\frac{7}{16} = 0.4375$, a terminating decimal.

$\frac{4}{15} = 0.2\overline{6}$, a repeating decimal.

An **irrational number** is a number whose decimal representation never terminates or repeats. [4.6A, p. 295]

π, $\sqrt{3}$, and $0.23233233323333\ldots$ are irrational numbers.

The **real numbers** are all the rational numbers together with all the irrational numbers. [4.6A, p. 295]

An **inequality** contains the symbol $>$, $<$, $\geq$, or $\leq$ and expresses the relative order of two mathematical expressions. [4.6B, p. 297]

$3.9 \leq 5$ $8.3 \geq 8$ $x < 7$

$5 \leq 5$ $8 \geq 8$

Essential Rules and Procedures

To write a decimal in words, write the decimal part as though it were a whole number. Then name the place value of the last digit. The decimal point is read as "and." [4.1A, p. 237]

The decimal 12.875 is written in words as twelve and eight hundred seventy-five thousandths.

To write a decimal in standard form when it is written in words, write the whole number part, replace the word *and* with a decimal point, and write the decimal part so that the last digit is in the given place-value position. [4.1A, p. 238]

The decimal forty-nine and sixty-three thousandths is written in standard form as 49.063.

To compare two decimals, write the decimal part of each number so that each has the same number of decimal places. Then compare the two numbers. [4.1B, p. 239]

$1.790 > 1.789$

$0.8130 < 0.8315$

To round a decimal, use the same rules used with whole numbers, except drop the digits to the right of the given place value instead of replacing them with zeros. [4.1C, p. 240]

2.7134 rounded to the nearest tenth is 2.7. 0.4687 rounded to the nearest hundredth is 0.47.

To add or subtract decimals, write the decimals so that the decimal points are on a vertical line. Add or subtract as you would with whole numbers. Then write the decimal point in the answer directly below the decimal points in the given numbers. [4.2A, p. 247]

$$
\begin{array}{r}
{\scriptstyle 1\ 1} \\
1.35 \\
20.8 \\
+\ 0.76 \\
\hline
22.91
\end{array}
\qquad
\begin{array}{r}
{\scriptstyle 2\ 15\ \ \ 6\ 10} \\
3\!\!\!/\,3.8\!\!\!/7\!\!\!/0\!\!\!/ \\
-\ 9.641 \\
\hline
26.229
\end{array}
$$

To estimate the answer to a calculation, round each number to the highest place value of the number; the first digit of each number will be nonzero, and all other digits will be zero. If a number is a decimal less than 1, round the decimal so that there is one nonzero digit. Perform the calculation using the rounded numbers. [4.2A, p. 249]

$$35.87 \longrightarrow 40$$
$$61.09 \longrightarrow \underline{+\ 60}$$
$$100$$

$$0.3876 \longrightarrow 0.4$$
$$0.5472 \longrightarrow \underline{+\ 0.5}$$
$$0.9$$

To multiply decimals, multiply the numbers as you would whole numbers. Then write the decimal point in the product so that the number of decimal places in the product is the sum of the decimal places in the factors. [4.3A, p. 258]

$$
\begin{array}{lr}
26.83 & \text{2 decimal places} \\
\times\ 0.45 & \text{2 decimal places} \\
\hline
13415 & \\
10732 & \\
\hline
12.0735 & \text{4 decimal places}
\end{array}
$$

To multiply a decimal by a power of 10, move the decimal point to the right the same number of places as there are zeros in the power of 10. If the power of 10 is written in exponential notation, the exponent indicates how many places to move the decimal point. [4.3A, p. 259]

$$3.97 \cdot 10{,}000 = 39{,}700$$
$$0.641 \cdot 10^5 = 64{,}100$$

To divide decimals, move the decimal point in the divisor to the right so that the divisor is a whole number. Move the decimal point in the dividend the same number of places to the right. Place the decimal point in the quotient directly above the decimal point in the dividend. Then divide as you would with whole numbers. [4.3B, p. 261]

$$
\begin{array}{r}
6.2 \\
0.39\overline{)2.41.8} \\
\underline{-2\ 34} \\
78 \\
\underline{-78} \\
0
\end{array}
$$

To divide a decimal by a power of 10, move the decimal point to the left the same number of places as there are zeros in the power of 10. If the power of 10 is written in exponential notation, the exponent indicates how many places to move the decimal point. [4.3B, pp. 262–263]

$$972.8 \div 1{,}000 = 0.9728$$
$$61.305 \div 10^4 = 0.0061305$$

To write a fraction as a decimal, divide the numerator of the fraction by the denominator. [4.3C, p. 265]

$$\frac{7}{8} = 7 \div 8 = 0.875$$

To convert a decimal to a fraction, remove the decimal point and place the decimal part over a denominator equal to the place value of the last digit in the decimal. [4.3C, p. 266]

0.85 is eighty-five <u>hundredths</u>.

$$0.85 = \frac{85}{100} = \frac{17}{20}$$

To find the order relation between a decimal and a fraction, first rewrite the fraction as a decimal. Then compare the two decimals. [4.3C, p. 266]

Because $\frac{3}{11} \approx 0.273$ and $0.273 > 0.26$, $\frac{3}{11} > 0.26$.

Square Root
For $x > 0$, if $a^2 = x$, then $\sqrt{x} = a$. [4.5A, p. 283]

Because $6^2 = 36$, $\sqrt{36} = 6$.

Product Property of Square Roots
If a and b are positive numbers, then $\sqrt{a \cdot b} = \sqrt{a} \cdot \sqrt{b}$. [4.5B, p. 287]

$$\sqrt{4 \cdot 25} = \sqrt{4} \cdot \sqrt{25}$$

Chapter 4 Review Exercises

1. Approximate $3\sqrt{47}$ to the nearest ten-thousandth.

2. Find the product of 0.918 and 10^5.

3. Simplify: $-\sqrt{121}$

4. Subtract: $-3.981 - 4.32$

5. Evaluate $a + b + c$ for $a = 80.59$, $b = -3.647$, and $c = 12.3$.

6. Write five and thirty-four thousandths in standard form.

7. Simplify: $\sqrt{100} - 2\sqrt{49}$

8. Find the quotient of 14.2 and 10^3.

9. Solve: $4.2z = -1.428$

10. Place the correct symbol, $<$ or $>$, between the two numbers.

    8.039   8.31

11. Evaluate $\dfrac{x}{y}$ for $x = 0.396$ and $y = 3.6$.

12. Multiply: $(9.47)(0.26)$

13. For the inequality $x \geq -1$, what numbers listed below make the inequality true?
    a. -6 **b.** -1 **c.** -0.5 **d.** $\sqrt{10}$

14. Place the correct symbol, $<$ or $>$, between the two numbers.

    $\dfrac{3}{7}$ 0.429

15. Convert 0.28 to a fraction.

16. Divide and round to the nearest tenth: $-6.8 \div 47.92$

17. *U.S. Postal Service*   The graph at the right shows U.S. Postal Service rates for Express Mail. How much would it cost to mail 25 Express Mail packages, each weighing 0.75 oz, post office to post office?

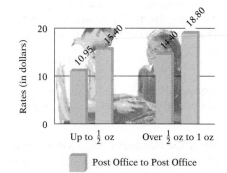

Post Office to Post Office

Post Office to Addressee

U.S. Postal Service Rates for Express Mail

18. Graph all the real numbers between -6 and -2.

19. Graph $x \geq -3$.

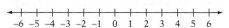

20. Find the sum of -247.8 and -193.4.

21. Find the quotient of 614.3 and 100.

22. Evaluate $a - b$ for $a = 80.32$ and $b = 29.577$.

23. Simplify: $\sqrt{90}$

24. Evaluate $60st$ for $s = 5$ and $t = -3.7$.

25. Estimate the difference between 506.81 and 64.1.

26. *Education*   A student must have a grade point average of at least 3.5 to qualify for a certain scholarship. Write an inequality for the grade point average a student must have in order to qualify for the scholarship. Does a student who has a grade point average of 3.48 qualify for the scholarship?

27. *Chemistry*   The boiling point of bromine is $58.8°C$. The melting point of bromine is $-7.2°C$. Find the difference between the boiling point and the melting point of bromine.

28. *History*   The figure at the right shows the monetary cost of four wars. **a.** What is the difference between the monetary costs of the two World Wars? **b.** How many times greater was the monetary cost of the Vietnam War than that of World War I?

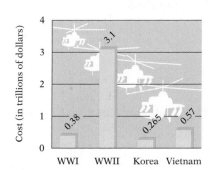

Monetary Cost of War
Source: Congressional Research Service Using Numbers from the *Statistical Abstract of the United States*

29. *Consumerism*   A 7-ounce jar of instant coffee costs $11.78. Find the cost per ounce. Round to the nearest cent.

30. *Finances*   The total of the monthly payments for a car lease is the product of the number of months of the lease and the monthly lease payment. The total of the monthly payments for a 24-month car lease is $12,371.76. Find the monthly lease payment.

31. *Business*   Use the formula $P = C + M$, where P is the price of a product to a customer, C is the cost paid by a store for the product, and M is the markup, to find the price of a treadmill that costs a business $1,124.75 and has a markup of $374.75.

32. *Physics*   The velocity of a falling object is given by the formula $v = \sqrt{64d}$, where v is the velocity in feet per second and d is the distance the object has fallen. Find the velocity of an object that has fallen a distance of 25 ft.

Chapter 4 Test

1. Write nine and thirty-three thousandths in standard form.

2. Place the correct symbol, $<$ or $>$, between the two numbers.

   4.003   4.009

3. Round 6.051367 to the nearest thousandth.

4. Find the difference between -30 and -7.247.

5. Evaluate $x - y$ for $x = 6.379$ and $y = -8.28$.

6. Estimate the difference between 92.34 and 17.95.

7. Find the total of 4.58, -3.9, and 6.017.

8. What is the product of -2.5 and 7.36?

9. Evaluate $-20cd$ for $c = 0.5$ and $d = -6.4$.

10. Solve: $5.488 = -3.92p$

11. Simplify: $\sqrt{256} - 2\sqrt{121}$

12. Find the quotient of 84.96 and 100.

13. Evaluate $\dfrac{x}{y}$ for $x = 52.7$ and $y = -6.2$.

14. Place the correct symbol, $<$ or $>$, between the two numbers.

    0.22 $\dfrac{2}{9}$

15. Approximate $2\sqrt{46}$ to the nearest ten-thousandth.

16. Simplify: $\sqrt{68}$

17. *The Film Industry*   The table at the right shows six James Bond films released between 1960 and 1970 and their gross box office incomes, in millions of dollars, in the United States. How much greater was the gross from *Thunderball* than the gross from *On Her Majesty's Secret Service*?

Film	U.S. Box Office Gross (in millions of dollars)
Dr. No	$16.1
On Her Majesty's Secret Service	$22.8
From Russia with Love	$24.8
You Only Live Twice	$43.1
Goldfinger	$51.1
Thunderball	$63.6

Source: **www.worldwideboxoffice.com**

18. Is -2.5 a solution of the equation $8.4 = 5.9 + a$?

19. Multiply: $8.973 \cdot 10^4$

20. Graph the real numbers between -2 and 2.

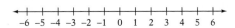

21. Graph $x \geq 3$.

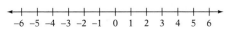

22. Evaluate $x + y$ for $x = -233.81$ and $y = 71.3$.

23. Solve: $-8v = 26$

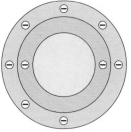

24. Chemistry   The boiling point of fluorine is $-188.14°C$. The melting point of fluorine is $-219.62°C$. Find the difference between the boiling point and the melting point of fluorine.

Flourine

25. Physics   The velocity of a falling object is given by the formula $v = \sqrt{64d}$, where v is the velocity in feet per second and d is the distance the object has fallen. Find the velocity of an object that has fallen a distance of 16 ft.

26. Accounting   The fundamental accounting equation is $A = L + S$, where A is the assets of the company, L is the liabilities of the company, and S is the stockholders' equity. Find the stockholders' equity in a company whose assets are $48.2 million and whose liabilities are $27.6 million.

27. Geometry   The lengths of the three sides of a triangle are 8.75 m, 5.25 m, and 4.5 m. Find the perimeter of the triangle. Use the formula $P = a + b + c$.

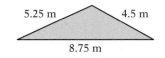

5.25 m   4.5 m

8.75 m

28. Business   Each sales representative for a company must sell at least 65,000 units per year. Write an inequality for the number of units a sales representative must sell. Has a representative who has sold 57,000 units this year met the sales goal?

29. Physics   Find the force exerted on a falling object that has a mass of 5.75 kg. Use the formula $F = ma$, where F is the force exerted by gravity on a falling object, m is the mass of the object, and a is the acceleration of gravity. The acceleration of gravity is -9.80 m/s^2. The force is measured in newtons.

30. Temperature   On January 19, 1892, the temperature in Fort Assiniboine, Montana, rose to 2.78°C from $-20.56°C$ in a period of only 15 min. Find the difference between these two temperatures.

Cumulative Review Exercises

1. Find the quotient of 387.9 and 10^4.

2. Evaluate $(x + y)^2 - 2z$ for $x = -3$, $y = 2$, and $z = -5$.

3. Solve: $-9.8 = -0.49c$

4. Write eight million seventy-two thousand ninety-two in standard form.

5. Graph all the real numbers between -4 and 1.

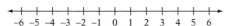

6. Graph $x \leq -2$.

7. Find the difference between -23 and -19.

8. Estimate the sum of 372, 541, 608, and 429.

9. Simplify: $\sqrt{192}$

10. Evaluate $x \div y$ for $x = 3\frac{2}{3}$ and $y = 2\frac{4}{9}$.

11. What is -36.92 increased by 18.5?

12. Simplify: $\left(\frac{5}{9}\right)\left(-\frac{3}{10}\right)\left(-\frac{6}{7}\right)$

13. Evaluate $x^4 y^2$ for $x = 2$ and $y = 10$.

14. Find the prime factorization of 260.

15. Convert $\frac{19}{25}$ to a decimal.

16. Approximate $10\sqrt{91}$ to the nearest ten-thousandth.

17. *Labor* The figure at the right shows the number of vacation days per year that are legally mandated in several countries.
   a. Which country mandates more vacation days, Ireland or Sweden?
   b. How many times more vacation days does Austria mandate than Switzerland?

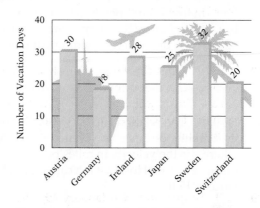

Number of Legally Mandated Vacation Days
Source: Economic Policy Institute; *World Almanac*

18. Divide: $\dfrac{-8}{0}$

19. Simplify: $-\dfrac{5}{7} + \dfrac{4}{21}$

20. Simplify: $4\sqrt{25} - \sqrt{81}$

21. Estimate the product of 62.8 and 0.47.

22. Simplify: $5(3 - 7) \div (-4) + 6(2)$

23. Evaluate $\dfrac{a}{b + c}$ for $a = \dfrac{3}{8}$, $b = \dfrac{1}{2}$, and $c = \dfrac{3}{4}$.

24. Evaluate $x - y + z$ for $x = \dfrac{5}{12}$, $y = -\dfrac{3}{8}$, and $z = -\dfrac{3}{4}$.

25. Divide and round to the nearest tenth: $2.617 \div 0.93$

26. *Pagers*   Your pager service leases alpha-numeric pagers for $5.95 per month. This fee includes 50 free pages per month. There is a charge of $.25 for each additional page after the first 50. What is your pager service bill for a month in which you sent 78 pages?

27. *Temperature*   On December 24, 1924, in Fairfield, Montana, the temperature fell from 17.22°C at noon to −29.4°C at midnight. How many degrees did the temperature fall in the 12-hour period?

28. *Consumerism*   Use the formula $C = \dfrac{M}{N}$, where C is the cost per visit at a health club, M is the membership fee, and N is the number of visits to the club, to find the cost per visit when your annual membership fee at a health club is $515 and you visit the club 125 times during the year.

29. *Business*   The figure at the right shows how the average salesperson spends the workweek. **a.** On average, how many hours per week does a salesperson work? **b.** Does the average salesperson spend more time face-to-face selling or doing both administrative work and placing service calls?

30. *Physics*   The relationship between the velocity of a car and its braking distance is given by the formula $v = \sqrt{20d}$, where v is the velocity in miles per hour and d is its braking distance in feet. How fast is a car going when its braking distance is 45 ft?

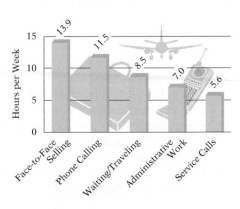

Average Salesperson's Workweek
Source: Dartnell's 28th Survey of Sales Force Compensation

Variable Expressions

Microscopes enable scientists to look at objects that are too small to be seen by the naked eye. They provide a large image of a tiny object. Scientists looking at microscopic images use scientific notation to describe the sizes of the objects they see.

Telescopes enable astronomers to look at objects that are extremely large but far away. Astronomers use scientific notation to describe distances in space.

Scientific notation replaces very large and very small numbers with more concise expressions, making these numbers easier to read and write. **Exercises 69 to 75 on page 360** provide examples of situations in which scientific notation is used.

DVD

SSM

Student Website

***Need help? For online student resources, visit* college.hmco.com/pic/aufmannPA5e.**

1. Place the correct symbol, $<$ or $>$, between the two numbers.
   54   45

For Exercises 2 to 6, add, subtract, multiply, or divide.

2. $-19 + 8$

3. $26 - 38$

4. $-2(44)$

5. $-\dfrac{3}{4}(-8)$

6. $3.97 \cdot 10^4$

7. Simplify: $(-3)^2$

8. Simplify: $(8 - 6)^2 + 12 \div 4 \cdot 3^2$

GO Figure

Luis, Kim, Reggie, and Dave are standing in line.
Dave is not first. Kim is between Luis and Reggie.
Luis is between Dave and Kim. Give the order in
which the men are standing.

5.1 Properties of Real Numbers

OBJECTIVE A Application of the Properties of Real Numbers

The Properties of Real Numbers describe the way operations on numbers can be performed. These properties have been stated in previous chapters but are restated here for review. The properties are used to rewrite variable expressions.

PROPERTIES OF REAL NUMBERS

The Commutative Property of Addition

If a and b are real numbers, then $a + b = b + a$.

$$7 + 12 = 12 + 7$$
$$19 = 19$$

The Commutative Property of Addition states that when we add two numbers, the numbers can be added in either order; the sum is the same.

The Commutative Property of Multiplication

If a and b are real numbers, then $a \cdot b = b \cdot a$.

$$7 \cdot (-2) = (-2) \cdot 7$$
$$-14 = -14$$

The Commutative Property of Multiplication states that when we multiply two numbers, the numbers can be multiplied in either order; the product is the same.

The Associative Property of Addition

If a, b, and c are real numbers, then $(a + b) + c = a + (b + c)$.

$$(7 + 3) + 8 = 7 + (3 + 8)$$
$$10 + 8 = 7 + 11$$
$$18 = 18$$

The Associative Property of Addition states that when we add three or more numbers, the numbers can be grouped in any order; the sum is the same.

The Associative Property of Multiplication

If a, b, and c are real numbers, then $(a \cdot b) \cdot c = a \cdot (b \cdot c)$.

$$(4 \cdot 5) \cdot 3 = 4 \cdot (5 \cdot 3)$$
$$20 \cdot 3 = 4 \cdot 15$$
$$60 = 60$$

The Associative Property of Multiplication states that when we multiply three or more factors, the factors can be grouped in any order; the product is the same.

The Addition Property of Zero

If a is a real number, then
$a + 0 = 0 + a = a.$

$(-7) + 0 = 0 + (-7) = -7$

The Addition Property of Zero states that the sum of a number and zero is the number.

The Multiplication Property of Zero

If a is a real number, then
$a \cdot 0 = 0 \cdot a = 0.$

$5 \cdot 0 = 0 \cdot 5 = 0$

The Multiplication Property of Zero states that the product of a number and zero is zero.

The Multiplication Property of One

If a is a real number, then
$a \cdot 1 = 1 \cdot a = a.$

$9 \cdot 1 = 1 \cdot 9 = 9$

The Multiplication Property of One states that the product of a number and 1 is the number.

The Inverse Property of Addition

If a is a real number, then
$a + (-a) = (-a) + a = 0.$

$2 + (-2) = (-2) + 2 = 0$

The sum of a number and its opposite is zero.
$-a$ is the opposite of a. $-a$ is also called the **additive inverse** of a.
a is the opposite of $-a$, or a is the *additive inverse* of $-a$.
The sum of a number and its additive inverse is zero.

The Inverse Property of Multiplication

If a is a real number and $a \neq 0$, then
$a \cdot \dfrac{1}{a} = \dfrac{1}{a} \cdot a = 1.$

$4 \cdot \dfrac{1}{4} = \dfrac{1}{4} \cdot 4 = 1$

The product of a nonzero number and its reciprocal is 1.
$\dfrac{1}{a}$ is the reciprocal of a. $\dfrac{1}{a}$ is also called the **multiplicative inverse** of a.

a is the reciprocal of $\dfrac{1}{a}$, or a is the *multiplicative inverse* of $\dfrac{1}{a}$.
The product of a nonzero number and its multiplicative inverse is 1.

The Properties of Real Numbers can be used to rewrite a variable expression in a simpler form. This process is referred to as *simplifying* the variable expression.

Simplify: $5 \cdot (4x)$

Use the Associative Property of Multiplication.	$5 \cdot (4x) = (5 \cdot 4)x$
Multiply 5 times 4.	$= 20x$

Simplify: $(6x) \cdot 2$

Use the Commutative Property of Multiplication.	$(6x) \cdot 2 = 2 \cdot (6x)$
Use the Associative Property of Multiplication.	$= (2 \cdot 6)x$
Multiply 2 times 6.	$= 12x$

Simplify: $(5y)(3y)$

Use the Commutative and Associative Properties of Multiplication.	$(5y)(3y) = 5 \cdot y \cdot 3 \cdot y$
	$= 5 \cdot 3 \cdot y \cdot y$
	$= (5 \cdot 3)(y \cdot y)$
Write $y \cdot y$ in exponential form.	$= 15y^2$
Multiply 5 times 3.	

By the Multiplication Property of One, the product of 1 and x is x.	$1 \cdot x = x$
	$1x = x$
Just as the product of 1 and x is written x, the product of -1 and x is written $-x$.	$-1 \cdot x = -x$
	$-1x = -x$

Simplify: $(-2)(-x)$

Write $-x$ as $-1x$.	$(-2)(-x) = (-2)(-1x)$
Use the Associative Property of Multiplication.	$= [(-2)(-1)]x$
Multiply -2 times -1.	$= 2x$

Simplify: $-4t + 9 + 4t$

Use the Commutative Property of Addition.	$-4t + 9 + 4t = -4t + 4t + 9$
Use the Associative Property of Addition.	$= (-4t + 4t) + 9$
Use the Inverse Property of Addition.	$= 0 + 9$
Use the Addition Property of Zero.	$= 9$

> **Take Note**
>
> Brackets, [ ], are used as a grouping symbol to group the factors -2 and -1 because parentheses have already been used in the expression to show that -2 and -1 are being multiplied. The expression $[(-2)(-1)]$ is considered easier to read than $((-2)(-1))$.

EXAMPLE 1 Simplify: $-5(7b)$

Solution $-5(7b) = (-5 \cdot 7)b$
$= -35b$

YOU TRY IT 1 Simplify: $-6(-3p)$

Your Solution

EXAMPLE 2 Simplify: $(-4r)(-9t)$

Solution $(-4r)(-9t) = [(-4)(-9)](r \cdot t)$
$= 36rt$

YOU TRY IT 2 Simplify: $(-2m)(-8n)$

Your Solution

EXAMPLE 3 Simplify: $(-8)(-z)$

Solution $(-8)(-z) = (-8)(-1z)$
$= [(-8)(-1)]z$
$= 8z$

YOU TRY IT 3 Simplify: $(-12)(-d)$

Your Solution

EXAMPLE 4 Simplify: $-5y + 5y + 7$

Solution $-5y + 5y + 7 = 0 + 7$
$= 7$

YOU TRY IT 4 Simplify: $6n + 9 + (-6n)$

Your Solution

Solutions on p. S13

<hr>

OBJECTIVE B The Distributive Property

Consider the numerical expression $6 \cdot (7 + 9)$.

This expression can be evaluated by applying the Order of Operations Agreement.

Simplify the expression inside the parentheses.
Multiply.

$6 \cdot (7 + 9) = 6 \cdot 16$
$= 96$

There is an alternative method of evaluating this expression.

Multiply each number inside the parentheses by 6 and add the products.

$6 \cdot (7 + 9) = 6 \cdot 7 + 6 \cdot 9$
$= 42 + 54$
$= 96$

Each method produces the same result. The second method uses the **Distributive Property,** which is another of the Properties of Real Numbers.

The Distributive Property

If a, b, and c are real numbers, then $a(b + c) = ab + ac$.

The Distributive Property is used to remove parentheses from a variable expression.

Simplify $3(5a + 4)$ by using the Distributive Property.

Use the Distributive Property. $3(5a + 4) = 3(5a) + 3(4)$

Simplify. $= 15a + 12$

Simplify $-4(2a + 3)$ by using the Distributive Property.

Use the Distributive Property. $-4(2a + 3) = -4(2a) + (-4)(3)$

Simplify. $= -8a + (-12)$

Rewrite addition of the opposite as
subtraction. $= -8a - 12$

The Distributive Property can also be stated in terms of subtraction.

$a(b - c) = ab - ac$

Simplify $5(2x - 4y)$ by using the Distributive Property.

Use the Distributive Property. $5(2x - 4y) = 5(2x) - 5(4y)$

Simplify. $= 10x - 20y$

Simplify $-3(2x - 8)$ by using the Distributive Property.

Use the Distributive Property. $-3(2x - 8) = -3(2x) - (-3)(8)$

Simplify. $= -6x - (-24)$

Rewrite the subtraction as addition
of the opposite. $= -6x + 24$

The Distributive Property can be extended to more than two addends inside the parentheses. For example,

$$4(2a + 3b - 5c) = 4(2a) + 4(3b) - 4(5c)$$

$$= 8a + 12b - 20c$$

The Distributive Property is used to remove the parentheses from an expression that has a negative sign in front of the parentheses. Just as $-x = -1 \cdot x$, the expression $-(x + y) = -1(x + y)$. Therefore,

$$-(x + y) = -1(x + y) = -1x - 1y = -x - y$$

When a negative sign precedes parentheses, remove the parentheses and change the sign of *each* addend inside the parentheses.

Rewrite the expression $-(4a - 3b + 7)$ without parentheses.

Remove the parentheses and change $-(4a - 3b + 7) = -4a + 3b - 7$
the sign of each addend inside the
parentheses.

EXAMPLE 5

Simplify by using the Distributive Property:
$6(5c - 12)$

Solution

$6(5c - 12) = 6(5c) - 6(12)$
$\qquad\qquad = 30c - 72$

YOU TRY IT 5

Simplify by using the Distributive Property:
$-7(2k - 5)$

Your Solution

EXAMPLE 6

Simplify by using the Distributive Property:
$-4(-2a - b)$

Solution

$-4(-2a - b) = -4(-2a) - (-4)(b)$
$\qquad\qquad\quad = 8a + 4b$

YOU TRY IT 6

Simplify by using the Distributive Property:
$-4(x - 2y)$

Your Solution

EXAMPLE 7

Simplify by using the Distributive Property:
$-2(3m - 8n + 5)$

Solution

$-2(3m - 8n + 5)$
$= -2(3m) - (-2)(8n) + (-2)(5)$
$= -6m + 16n - 10$

YOU TRY IT 7

Simplify by using the Distributive Property:
$3(-2v + 3w - 7)$

Your Solution

EXAMPLE 8

Simplify by using the Distributive Property:
$3(2a + 6b - 5c)$

Solution

$3(2a + 6b - 5c) = 3(2a) + 3(6b) - 3(5c)$
$\qquad\qquad\qquad = 6a + 18b - 15c$

YOU TRY IT 8

Simplify by using the Distributive Property:
$-4(2x - 7y - z)$

Your Solution

EXAMPLE 9

Rewrite $-(5x + 3y - 2z)$ without parentheses.

Solution

$-(5x + 3y - 2z) = -5x - 3y + 2z$

YOU TRY IT 9

Rewrite $-(c - 9d + 1)$ without parentheses.

Your Solution

Solutions on p. S13

5.1 Exercises

OBJECTIVE A **Application of the Properties of Real Numbers**

Identify the Property of Real Numbers that justifies the statement.

1. $3 \cdot (4 \cdot 7) = (3 \cdot 4) \cdot 7$ *Ass of Mul.*

2. $a + 0 = a$

3. $x + 7 = 7 + x$ *Com of Add*

4. $12 \cdot a = a \cdot 12$

5. $4r + (-4r) = 0$ *Add prop of zero*

6. $\dfrac{2}{3} \cdot \dfrac{3}{2} = 1$

7. $a(bc) = (bc)a$ *, Ass prop mult.*

8. $1 \cdot x = x$

9. $\dfrac{1}{2}(2x) = \left(\dfrac{1}{2} \cdot 2\right)x$ a. _Asso prop of x_
   $= 1 \cdot x$ b. _INVERSE "_
   $= x$ c. _X prop of ONE_

10. $(5x + 6) + (-6) = 5x + [6 + (-6)]$ a. _____
    $= 5x + 0$ b. _____
    $= 5x$ c. _____

Use the given Property of Real Numbers to complete the statement.

11. The Associative Property of Addition
    $x + (4 + y) = ?$

12. The Commutative Property of Multiplication
    $v \cdot w = ?$

13. The Inverse Property of Multiplication
    $5 \cdot ? = 1$

14. The Inverse Property of Addition
    $-7y + ? = 0$

15. The Multiplication Property of Zero
    $a \cdot ? = 0$

16. The Inverse Property of Multiplication
    For $a \neq 0, a \cdot \dfrac{1}{a} = ?$

17. The multiplicative inverse of $-\dfrac{2}{3}$ is _____ .

18. For $a \neq 0$, the multiplicative inverse of $-\dfrac{2}{a}$ is

    _____ .

19. Simplify: $3(-8x)$

    a. Use the _____ Property of Multiplication to regroup the factors.

    b. Multiply 3 times -8.

    $3(-8x)$
    $= [3(-8)]x$

    $= $ _____ x

20. Simplify: $\left(\frac{3}{5}a\right)(5)$

                                                               $\left(\frac{3}{5}a\right)(5)$

    a. Use the Commutative Property of Multiplication to change the order of the factors. $= (\underline{\hspace{1cm}})\left(\frac{3}{5}a\right)$

    b. Use the Associative Property of Multiplication to regroup the factors. $= (5 \cdot \underline{\hspace{1cm}})a$

    c. Multiply. $= \underline{\hspace{1cm}} a$

Simplify the variable expression.

21. $6(2x)$

22. $3(4y)$

23. $-5(3x)$

24. $-3(6z)$

25. $(3t) \cdot 7$

26. $(9r) \cdot 5$

27. $(-3p) \cdot 7$

28. $(-4w) \cdot 6$

29. $(-2)(-6q)$

30. $(-3)(-5m)$

31. $\frac{1}{2}(4x)$

32. $\frac{2}{3}(6n)$

33. $-\frac{5}{3}(9w)$

34. $-\frac{2}{5}(10v)$

35. $-\frac{1}{2}(-2x)$

36. $-\frac{1}{3}(-3x)$

37. $(2x)(3x)$

38. $(4k)(6k)$

39. $(-3x)(9x)$

40. $(4b)(-12b)$

41. $\left(\frac{1}{2}x\right)(2x)$

42. $\left(\frac{1}{3}h\right)(3h)$

43. $\left(-\frac{2}{3}\right)(x)\left(-\frac{3}{2}\right)$

44. $\left(-\frac{4}{3}\right)(z)\left(-\frac{3}{4}\right)$

45. $6\left(\frac{1}{6}c\right)$

46. $9\left(\frac{1}{9}v\right)$

47. $-5\left(-\frac{1}{5}a\right)$

48. $-9\left(-\frac{1}{9}s\right)$

49. $\frac{4}{5}w \cdot 15$

50. $\frac{7}{5}y \cdot 30$

51. $2v \cdot 8w$

52. $3m \cdot 7n$

53. $(-4b)(7c)$

54. $(-3k)(-6m)$

55. $3x + (-3x)$

56. $7xy + (-7xy)$

57. $-12h + 12h$

58. $5 + 8y + (-8y)$

59. $9 + 2m + (-2m)$

60. $12 - 3m + 3m$

61. $8x + 7 + (-8x)$ **62.** $13v + 12 + (-13v)$ **63.** $6t - 15 + (-6t)$ **64.** $10z - 4 + (-10z)$

65. $8 + (-8) - 5y$ **66.** $12 + (-12) - 7b$ **67.** $(-4) + 4 + 13b$ **68.** $-7 + 7 - 15t$

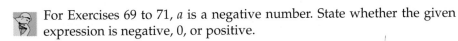 For Exercises 69 to 71, a is a negative number. State whether the given expression is negative, 0, or positive.

69. $5a(-5a)$ **70.** $5a + (-5a)$ 0 **71.** $5(5a)(-5)$

OBJECTIVE B   **The Distributive Property**

72. Simplify: $-7(4x - 3)$ $-7(4x - 3)$

   a. Use the _____ Property. $= (-7)(4x) - (-7)(3)$

   b. Simplify. $= \underline{} - (\underline{})$

   c. Rewrite subtraction as addition of the opposite. $= -28x + \underline{}$

Simplify by using the Distributive Property.

73. $2(5z + 2)$ **74.** $3(4n + 5)$ **75.** $6(2y + 5z)$

76. $4(7a + 2b)$ **77.** $3(7x - 9)$ **78.** $9(3w - 7)$

79. $-(2x - 7)$ **80.** $-(3x + 4)$ **81.** $-(-4x - 9)$

82. $-(-5y - 12)$ **83.** $-5(y + 3)$ **84.** $-4(x + 5)$

85. $-6(2x - 3)$ **86.** $-3(7y - 4)$ **87.** $-5(4n - 8)$

88. $-4(3c - 2)$

89. $-8(-6z + 3)$

90. $-2(-3k + 9)$

91. $-6(-4p - 7)$

92. $-5(-8c - 5)$

93. $5(2a + 3b + 1)$

94. $5(3x + 9y + 8)$

95. $4(3x - y - 1)$

96. $3(2x - 3y + 7)$

97. $9(4m - n + 2)$

98. $-4(3x + 2y - 5)$

99. $-6(-2v + 3w + 7)$

100. $-7(-2b - 4)$

101. $-4(-5x - 1)$

102. $-9(3x - 6y)$

103. $5(4a - 5b + c)$

104. $-4(-2m - n + 3)$

105. $-6(3p - 2r - 9)$

Rewrite without parentheses.

106. $-(4x + 6y - 8z)$ **107.** $-(5a - 9b + 7)$ **108.** $-(-6m + 3n + 1)$ **109.** $-(11p - 2q - r)$

110. Which expression is equivalent to $1 + 4(x + 3)$? (i) $5x + 15$ (ii) $4x + 13$

111. Which expression is equivalent to $3x + 3y$? (i) $7 - 4(x + y)$ (ii) $(7 - 4)(x + y)$

CRITICAL THINKING

112. Is the statement "any number divided by itself is 1" a true statement? If not, for what number or numbers is the statement not true?

113. Give examples of two operations that occur in everyday experience that are not commutative (for example, putting on socks and then shoes).

5.2 Variable Expressions in Simplest Form

OBJECTIVE A Addition of like terms

A variable expression is shown at the right. The expression can be rewritten by writing subtraction as addition of the opposite. A **term** of a variable expression is one of the addends of the expression.

$$4y^3 - 3xy + x - 9$$

$$4y^3 + (-3xy) + x + (-9)$$

The variable expression has 4 terms: $4y^3$, $-3xy$, x, and -9.

The term -9 is a **constant term,** or simply a **constant.** The terms $4y^3$, $-3xy$, and x are **variable terms.**

Each variable term consists of a **numerical coefficient** and a **variable part.** The table at the right gives the numerical coefficient and the variable part of each variable term.

Term	Numerical Coefficient	Variable Part
$4y^3$	4	y^3
$-3xy$	-3	xy
x	1	x

For an expression such as x, the numerical coefficient is 1 ($x = 1x$). The numerical coefficient for $-x$ is -1 ($-x = -1x$). The numerical coefficient of $-xy$ is -1 ($-xy = -1xy$). Usually the 1 is not written.

For the variable expression at the right, state:

$$9x^2 - x - 7yz^2 + 8$$

a.   The number of terms
b.   The coefficient of the second term
c.   The variable part of the third term
d.   The constant term

   a.   There are four terms: $9x^2$, $-x$, $-7yz^2$, and 8.
   b.   The coefficient of the second term is -1.
   c.   The variable part of the third term is yz^2.
   d.   The constant term is 8.

Like terms of a variable expression have the same variable part. Constant terms are also like terms.

> For the expression $13ab + 4 - 2ab - 10$, the terms $13ab$ and $-2ab$ are like variable terms, and 4 and -10 are like constant terms.

For the expression at the right, note that $5y^2$ and $-3y$ are not like terms because $y^2 = y \cdot y$, and $y \cdot y \neq y$. However, $6xy$ and $9yx$ are like variable terms because $xy = yx$ by the Commutative Property of Multiplication.

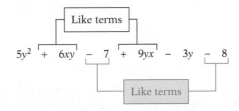

> For the variable expression $7 - 9x^2 - 8x - 9 + 4x$, state which terms are like terms.
>
> The terms $-8x$ and $4x$ are like variable terms.
>
> The terms 7 and -9 are like constant terms.

Variable expressions containing like terms are simplified by using an alternative form of the Distributive Property.

> ### Alternative Form of the Distributive Property
>
> If a, b, and c are real numbers, then $ac + bc = (a + b)c$.

Simplify: $6c + 7c$

$6c$ and $7c$ are like terms.
Use the Alternative Form of
the Distributive Property.
Then simplify.

$$6c + 7c = (6 + 7)c$$
$$= 13c$$

This example shows that to simplify a variable expression containing like terms, add the coefficients of the like terms. Adding or subtracting the like terms of a variable expression is called **combining like terms.**

Simplify: $6a + 7 - 9a + 3$

Rewrite subtraction as addition
of the opposite. Use the
Commutative Property of
Addition to rearrange terms so
that like terms are together.

$$6a + 7 - 9a + 3$$
$$= 6a + 7 + (-9a) + 3$$
$$= 6a + (-9a) + 7 + 3$$

Use the Alternative Form of the
Distributive Property to add like
variable terms. Add the like
constant terms.

$$= [6 + (-9)]a + (7 + 3)$$
$$= -3a + 10$$

Simplify: $4x^2 - 7x + x^2 - 12x$

Rewrite subtraction as addition
of the opposite. Use the
Commutative Property of
Addition to rearrange terms so
that like terms are together.

$$4x^2 - 7x + x^2 - 12x$$
$$= 4x^2 + (-7x) + x^2 + (-12x)$$
$$= 4x^2 + x^2 + (-7x) + (-12x)$$

Use the Alternative Form of the
Distributive Property to add like
terms.

$$= (4 + 1)x^2 + [-7 + (-12)]x$$
$$= 5x^2 + (-19)x$$
$$= 5x^2 - 19x$$

EXAMPLE 1 Simplify: $\dfrac{3x}{7} + \dfrac{2x}{7}$

Solution $\dfrac{3x}{7} + \dfrac{2x}{7} = \dfrac{3x + 2x}{7}$

$$= \dfrac{(3 + 2)x}{7} = \dfrac{5x}{7}$$

YOU TRY IT 1 Simplify: $\dfrac{x}{5} + \dfrac{2x}{5}$

Your Solution

Solution on p. S13

EXAMPLE 2   Simplify:
$9y - 3z - 12y + 3z + 2$

Solution $9y - 3z - 12y + 3z + 2$
$= 9y - 12y - 3z + 3z + 2$
$= -3y + 0z + 2$
$= -3y + 2$

YOU TRY IT 2   Simplify:
$12a^2 - 8a + 3 - 16a^2 + 8a$

Your Solution

EXAMPLE 3   Simplify:
$6b^2 - 9ab + 3b^2 - ab$

Solution $6b^2 - 9ab + 3b^2 - ab$
$= 6b^2 + 3b^2 - 9ab - ab$
$= 9b^2 - 10ab$

YOU TRY IT 3   Simplify:
$-7x^2 + 4xy + 8x^2 - 12xy$

Your Solution

EXAMPLE 4   Simplify:
$6u + 7v - 8 + 9u - 12v + 14$

Solution $6u + 7v - 8 + 9u - 12v + 14$
$= 6u + 9u + 7v - 12v - 8 + 14$
$= 15u - 5v + 6$

YOU TRY IT 4   Simplify:
$-2r + 7s - 12 - 8r + s + 8$

Your Solution

EXAMPLE 5   Simplify:
$5r^2t - 6rt^2 + 8rt^2 - 9r^2t$

Solution $5r^2t - 6rt^2 + 8rt^2 - 9r^2t$
$= 5r^2t - 9r^2t - 6rt^2 + 8rt^2$
$= -4r^2t + 2rt^2$

YOU TRY IT 5   Simplify:
$8x^2y - 15xy^2 + 12xy^2 - 7x^2y$

Your Solution

Solutions on p. S13

OBJECTIVE B   **General variable expressions**

General variable expressions are simplified by repeated use of the Properties of Real Numbers.

Simplify: $7(2a - 4b) - 3(4a - 2b)$

Use the Distributive Property to remove parentheses.

$7(2a - 4b) - 3(4a - 2b)$
$= 14a - 28b - 12a + 6b$

Rewrite subtraction as addition of the opposite.

$= 14a + (-28b) + (-12a) + 6b$

Use the Commutative Property of Addition to rearrange terms.

$= 14a + (-12a) + (-28b) + 6b$

Use the Alternative Form of the Distributive Property to combine like terms.

$= [14 + (-12)]a + (-28 + 6)b$
$= 2a - 22b$

To simplify variable expressions that contain grouping symbols within other grouping symbols, simplify inside the inner grouping symbols first.

Simplify: $2x - 4[3 - 2(6x + 5)]$

Use the Distributive Property to remove the parentheses.	$2x - 4[3 - 2(6x + 5)]$ $= 2x - 4[3 - 12x - 10]$
Combine like terms inside the brackets.	$= 2x - 4[-12x - 7]$
Use the Distributive Property to remove the brackets.	$= 2x + 48x + 28$
Combine like terms.	$= 50x + 28$

Simplify: $2a^2 + 3[4(2a^2 - 5) - 4(3a - 1)]$

Use the Distributive Property to remove both sets of parentheses.	$2a^2 + 3[4(2a^2 - 5) - 4(3a - 1)]$ $= 2a^2 + 3[8a^2 - 20 - 12a + 4]$
Combine like terms inside the brackets.	$= 2a^2 + 3[8a^2 - 12a - 16]$
Use the Distributive Property to remove the brackets.	$= 2a^2 + 24a^2 - 36a - 48$
Combine like terms.	$= 26a^2 - 36a - 48$

EXAMPLE 6 Simplify:
$4 - 3(2a - b) + 4(3a + 2b)$

Solution $4 - 3(2a - b) + 4(3a + 2b)$
$= 4 - 6a + 3b + 12a + 8b$
$= 6a + 11b + 4$

YOU TRY IT 6 Simplify:
$6 - 4(2x - y) + 3(x - 4y)$

Your Solution

EXAMPLE 7 Simplify:
$7y - 4(2y - 3z) - (6y - 4z)$

Solution $7y - 4(2y - 3z) - (6y - 4z)$
$= 7y - 8y + 12z - 6y + 4z$
$= -7y + 16z$

YOU TRY IT 7 Simplify:
$8c - 4(3c - 8) - 5(c + 4)$

Your Solution

EXAMPLE 8 Simplify:
$9v - 4[2(1 - 3v) - 5(2v + 4)]$

Solution $9v - 4[2(1 - 3v) - 5(2v + 4)]$
$= 9v - 4[2 - 6v - 10v - 20]$
$= 9v - 4[-16v - 18]$
$= 9v + 64v + 72$
$= 73v + 72$

YOU TRY IT 8 Simplify:
$6p + 5[3(2 - 3p) - 2(5 - 4p)]$

Your Solution

Solutions on p. S13

5.2 Exercises

OBJECTIVE A   **Addition of like terms**

1. To identify the terms of the variable expression $5x^2 - 2x - 9$, first write subtraction as addition of the opposite: $5x^2 + (\underline{\hspace{1cm}}) + (\underline{\hspace{1cm}})$. The terms of $5x^2 - 2x - 9$ are $\underline{\hspace{1cm}}$, $\underline{\hspace{1cm}}$, and $\underline{\hspace{1cm}}$.

List the terms of the variable expression. Then underline the constant term.

2. $3x^2 + 4x - 9$ 3. $-7y^2 - 2y + 6$ 4. $b + 5$ 5. $8n^2 - 1$

List the variable terms of the expression. Then underline the variable part of each term and circle the coefficient of each term.

6. $9a^2 - 12a + 4b^2$ 7. $6x^2y + 7xy^2 + 11$ 8. $3x^2 + 16$ 9. $-2n^2 + 5n - 8$

10. Simplify $5a - 7a$ by combining like terms.
    a. Rewrite subtraction as addition of the opposite.
    b. Use the Alternative Form of the Distributive Property.
    c. Add the coefficients. The variable part does not change.

$$5a - 7a$$
$$= 5a + (\underline{-7a})$$
$$= [(\underline{5}) + (\underline{-7})]a$$
$$= \underline{-2a}$$

Simplify by combining like terms.

11. $7a + 9a$ 12. $8c + 15c$ 13. $12x + 15x$ 14. $9b + 24b$

15. $9z - 6z$ 16. $12h - 4h$ 17. $9x - x$ 18. $12y - y$

19. $8z - 15z$ 20. $2p - 13p$ 21. $w - 7w$ 22. $y - 9y$

23. $12v - 12v$ 24. $11c - 11c$ 25. $9s - 8s$ 26. $6n - 5n$

27. $\dfrac{n}{5} + \dfrac{3n}{5}$ 28. $\dfrac{2n}{9} + \dfrac{5n}{9}$ 29. $\dfrac{x}{4} + \dfrac{x}{4}$ 30. $\dfrac{5x}{8} + \dfrac{3x}{8}$

31. $\dfrac{8y}{7} - \dfrac{4y}{7}$

32. $\dfrac{5y}{3} - \dfrac{y}{3}$

33. $\dfrac{5c}{6} - \dfrac{c}{6}$

34. $\dfrac{9d}{10} - \dfrac{7d}{10}$

35. $4x - 3y + 2x$

36. $3m - 6n + 4m$

37. $4r + 8p - 2r + 5p$

38. $-12t - 6s + 9t + 4s$

39. $9w - 5v - 12w + 7v$

40. $3c - 8 + 7c - 9$

41. $-4p + 9 - 5p + 2$

42. $-6y - 17 + 4y + 9$

43. $8p + 7 - 6p - 7$

44. $9m - 12 + 2m + 12$

45. $7h + 15 - 7h - 9$

46. $7v^2 - 9v + v^2 - 8v$

47. $9y^2 - 8 + 4y^2 + 9$

48. $r^2 + 4r - 8r - 5r^2$

49. $3w^2 - 7 - 9 + 9w^2$

50. $4c - 7c^2 + 8c - 8c^2$

51. $9w^2 - 15w + w - 9w^2$

52. $12v^2 + 15v - 14v - 12v^2$

53. $7a^2b + 5ab^2 - 2a^2b + 3ab^2$

54. $3xy^2 + 2x^2y - 7xy^2 - 4x^2y$

55. $8a - 9b + 2 - 8a + 9b + 3$

56. $6x^2 - 7x + 1 + 5x^2 + 5x - 1$

57. $4y^2 + 7y + 1 + y^2 - 10y + 9$

58. $-4z^2 - 6z + 1 - z^2 + 7z + 8$

59. Which numbers are zero after the following polynomial is simplified?

$-4a^2b - 4ab^2 + 4 + 4ab^2 - 4a^2b - 4$

(i) the coefficient of a^2b (ii) the coefficient of ab^2 (iii) the constant term

60. Which expressions are equivalent to $-8x - 8y - 8y - 8x$?

(i) 0 (ii) $-16x$ (iii) $-16y$ (iv) $-16x - 16y$ (v) $-16y - 16x$

Use a horizontal format to subtract $(5a^2 - a + 2) - (-2a^3 + 3a - 3)$.

Rewrite subtraction as addition of the opposite polynomial. The opposite of $-2a^2 + 3a - 3$ is $2a^3 - 3a + 3$.

$$(5a^2 - a + 2) - (-2a^3 + 3a - 3)$$
$$= (5a^2 - a + 2) + (2a^3 - 3a + 3)$$

Combine like terms.

$$= 2a^3 + 5a^2 - 4a + 5$$

Use a vertical format to subtract $(3y^3 + 4y + 9) - (2y^2 + 4y - 21)$.

The opposite of $(2y^2 + 4y - 21)$ is $(-2y^2 - 4y + 21)$.

Add the opposite of $2y^2 + 4y - 21$ to the first polynomial.

$$
\begin{array}{r}
3y^3 \qquad\;\; + 4y + \;\; 9 \\
-\, 2y^2 - 4y + 21 \\
\hline
3y^3 - 2y^2 \qquad\;\; + 30
\end{array}
$$

EXAMPLE 4

Use a horizontal format to subtract
$(7c^2 - 9c - 12) - (9c^2 + 5c - 8)$.

Solution   The opposite of $9c^2 + 5c - 8$ is $-9c^2 - 5c + 8$.

Add the opposite of $9c^2 + 5c - 8$ to the first polynomial.

$(7c^2 - 9c - 12) - (9c^2 + 5c - 8)$

$= (7c^2 - 9c - 12) + (-9c^2 - 5c + 8)$

$= -2c^2 - 14c - 4$

YOU TRY IT 4

Use a horizontal format to subtract
$(-4w^3 + 8w - 8) - (3w^3 - 4w^2 + 2w - 1)$.

Your Solution

EXAMPLE 5

Use a vertical format to subtract
$(3k^2 - 4k + 1) - (k^3 + 3k^2 - 6k - 8)$.

Solution   The opposite of $k^3 + 3k^2 - 6k - 8$ is $-k^3 - 3k^2 + 6k + 8$.

Add the opposite of $k^3 + 3k^2 - 6k - 8$ to the first polynomial.

$$
\begin{array}{r}
3k^2 - 4k + 1 \\
-k^3 - 3k^2 + 6k + 8 \\
\hline
-k^3 \qquad\;\; + 2k + 9
\end{array}
$$

YOU TRY IT 5

Use a vertical format to subtract
$(13y^3 - 6y - 7) - (4y^2 - 6y - 9)$.

Your Solution

Solutions on p. S13

EXAMPLE 6 Find the difference between $3z^2 - 4z + 1$ and $5z^2 - 8$.

Solution $(3z^2 - 4z + 1) - (5z^2 - 8)$
$= (3z^2 - 4z + 1) + (-5z^2 + 8)$
$= -2z^2 - 4z + 9$

YOU TRY IT 6 What is the difference between $-6n^4 + 5n^2 - 10$ and $4n^2 + 2$?

Your Solution

Solution on p. S13

OBJECTIVE B Applications

A company's **revenue** is the money the company earns by selling its products. A company's **cost** is the money it spends to manufacture and sell its products. A company's **profit** is the difference between its revenue and its cost. This relationship is expressed by the formula $P = R - C$, where P is the profit, R is the revenue, and C is the cost. This formula is used in the example below.

A company manufactures and sells kayaks. The total monthly cost, in dollars, to produce n kayaks is $30n + 2000$. The company's monthly revenue, in dollars, obtained from selling all n kayaks is $-0.4n^2 + 150n$. Express in terms of n the company's monthly profit.

$R = -0.4n^2 + 150n,$
$C = 30n + 2000$
Rewrite subtraction as addition of the opposite.
Simplify.

$P = R - C$
$P = (-0.4n^2 + 150n) - (30n + 2000)$

$P = (-0.4n^2 + 150n) + (-30n - 2000)$

$P = -0.4n^2 + (150n - 30n) - 2000$
$P = -0.4n^2 + 120n - 2000$

The company's monthly profit is $(-0.4n^2 + 120n - 2000)$ dollars.

EXAMPLE 7 The distance from Acton to Boyd is $(y^2 + y + 7)$ miles. The distance from Boyd to Carlyle is $(y^2 - 3)$ miles. Find the distance from Acton to Carlyle.

$y^2 + y + 7$ $y^2 - 3$
Acton   Boyd   Carlyle

Solution $(y^2 + y + 7) + (y^2 - 3)$
$= (y^2 + y^2) + y + (7 - 3)$
$= 2y^2 + y + 4$

The distance from Acton to Carlyle is $(2y^2 + y + 4)$ miles.

YOU TRY IT 7 The distance from Dover to Engel is $(5y^2 - y)$ miles. The distance from Engel to Farley is $(7y^2 + 4)$ miles. Find the distance from Dover to Farley.

$5y^2 - y$ $7y^2 + 4$
Dover   Engel   Farley

Your Solution

Solution on p. S14

5.3 Exercises

OBJECTIVE A Addition of polynomials

State whether or not the expression is a monomial.

1. 17

2. $3x^4$

3. $\dfrac{17}{x}$

4. $\sqrt{6x}$

5. $\dfrac{2}{3}y$

6. $\dfrac{2}{3y}$

7. $\dfrac{\sqrt{y}}{3}$

8. $\dfrac{y}{3}$

State whether or not the expression is a polynomial.

9. $\dfrac{1}{5}x^3 + \dfrac{1}{2}x$

10. $\dfrac{1}{5x^2} + \dfrac{1}{2x}$

11. $\sqrt{x} + 5$

12. $x + \sqrt{5}$

How many terms does the polynomial have?

13. $3x^2 - 8x + 7$

14. $5y^3 + 6$

15. $9x^2y^3z^5$

16. $n^4 + 2n^3 - n^2 + 3n + 6$

State whether the polynomial is a monomial, a binomial, or a trinomial.

17. $8x^4 - 6x^2$

18. $4a^2b^2 + 9ab + 10$

19. $7a^3bc^5$

20. $y + 1$

Write the polynomial in descending order.

21. $8x^2 - 2x + 3x^3 - 6$

22. $7y - 8 + 2y^2 + 4y^3$

23. $2a - 3a^2 + 5a^3 + 1$

24. $b - 3b^2 + b^4 - 2b^3$

25. $4 - b^2$

26. $1 - y^4$

27. Use a horizontal format to add the polynomials.

a. Use the Commutative and Associative Properties of Addition to rearrange and group like terms.

$(5y^3 + 2y - 7) + (-2y^3 - 6y + 4)$

$= [5y^3 + (\underline{\quad})] + [2y + (\underline{\quad})] + [(-7) + \underline{\quad}]$

b. Combine like terms inside each pair of brackets.

$= \underline{\quad}y^3 + (\underline{\quad})y + (\underline{\quad})$

c. Rewrite addition of a negative number as subtraction.

$= \underline{\qquad\qquad}$

Add. Use a horizontal format.

28. $(5y^2 + 3y - 7) + (6y^2 - 7y + 9)$

29. $(7m^2 - 9m - 8) + (5m^2 + 10m + 4)$

30. $(-4b^2 + 9b + 11) + (7b^2 - 12b - 13)$

31. $(-8x^2 - 11x - 15) + (4x^2 - 12x + 13)$

32. $(3w^3 + 8w^2 - 2w) + (5w^2 - 6w - 5)$

33. $(11p^3 - 9p^2 - 6p) + (10p^2 - 8p + 4)$

34. $(3a^2 - 7 + 2a - 9a^3) + (7a^3 - 12a^2 - 10a + 8)$

35. $(9x - 8x^2 - 12 + 7x^3) + (-3x^3 - 7x^2 + 5x - 9)$

36. $(7t^3 - 8t - 15) + (8t - 20 + 7t^2)$

37. $(8y^2 - 3y - 1) + (3y - 1 - 6y^3 - 8y^2)$

38. Find the sum of $6t^2 - 8t - 15$ and $7t^2 + 8t - 20$.

39. What is $8y^2 - 3y - 1$ plus $-6y^2 + 3y - 1$?

40. When using a vertical format to add polynomials, arrange the terms of each polynomial in _____ order with _____ terms in the same column.

Add. Use a vertical format.

41. $(5k^2 - 7k - 8) + (6k^2 + 9k - 10)$

42. $(8v^2 - 9v + 12) + (12v^2 - 11v - 2)$

43. $(8x^3 - 9x + 2) + (9x^3 + 9x - 7)$

44. $(13z^3 - 7z^2 + 4z) + (10z^2 + 5z - 9)$

45. $(12b^3 + 9b^2 + 5b - 10) + (4b^3 + 5b^2 - 5b + 11)$

46. $(5a^3 - a^2 + 4a - 19) + (-a^3 + a^2 - 7a + 19)$

47. $(8p^3 - 7p) + (9p^2 - 7 + p)$

48. $(12c^3 + 9c) + (-7c^2 - 8 - c)$

49. $(7a^2 - 7 - 6a) + (-6a^3 - 7a^2 + 6a - 10)$

50. $(12x^2 + 8 + 7x) + (3x^3 - 12x^2 - 7x - 11)$

51. Find the total of $9d^4 - 7d^2 + 5$ and $-6d^4 - 3d^2 - 8$.

52. What is the sum of $8z^3 + 5z^2 - 4z + 7$ and $-3z^3 - z^2 + 6z - 2$?

For Exercises 53 and 54, use the three given polynomials, in which a, b, and c are all positive numbers.

(i) $ax^2 - bx + c$ **(ii)** $-ax^2 + bx - c$ **(iii)** $ax^2 + bx + c$

53. Which two polynomials have a sum that is a binomial?

54. Which two polynomials have a sum that is zero?

OBJECTIVE B   **Subtraction of polynomials**

55. To subtract two polynomials, add the _____ of the second polynomial to the first.

56. The opposite of $-4x^2 + 9x - 2$ is $-(-4x^2 + 9x - 2) =$ _____ .

Write the opposite of the polynomial.

57. $8x^3 + 5x^2 - 3x - 6$

58. $7y^4 - 4y^2 + 10$

59. $-9a^3 + a^2 - 2a + 9$

Subtract. Use a horizontal format.

60. $(3x^2 - 2x - 5) - (x^2 + 7x - 3)$

61. $(7y^2 - 8y - 10) - (3y^2 + 2y - 9)$

62. $(11b^3 - 2b^2 + 1) - (6b^2 - 12b - 13)$

63. $(13w^3 + 3w^2 - 9) - (7w^3 - 9w + 10)$

64. $(8z^3 - 9z^2 + 4z + 12) - (10z^3 - z^2 + 4z - 9)$

65. $(15t^3 - 9t^2 + 8t + 11) - (17t^3 - 9t^2 - 8t + 6)$

66. $(9y^3 + 8y) - (-17y^2 + 5)$

67. $(8p^3 + 14p) - (9p^2 - 12)$

68. $(-6r^3 + 9r + 19) - (6r^3 + 19 - 16r)$

69. $(-4v^2 + 8v - 2) - (6v^3 + 7v + 1 - 13v^2)$

70. Find the difference between $10b^2 - 7b + 4$ and $8b^2 + 5b - 14$.

71. What is $7m^2 - 3m - 6$ minus $2m^2 - m + 5$?

Subtract. Use a vertical format.

72. $(4a^2 + 9a - 11) - (2a^2 - 3a - 9)$

73. $(8b^2 - 7b - 6) - (5b^2 + 8b + 12)$

74. $(6z^3 + 4z^2 + 1) - (3z^3 - 8z - 9)$

75. $(10y^3 - 8y - 13) - (6y^2 + 2y + 7)$

76. $(8y^2 - 9y - 16) - (3y^3 - 4y^2 + 2y + 5)$

77. $(4a^2 + 8a + 12) - (3a^3 + 4a^2 + 7a - 12)$

78. $(10b^3 - 7b) - (8b^2 + 14)$

79. $(7m - 6) - (2m^3 - m^2)$

80. $(5n^3 - 4n - 9 + 8n^2) - (2n^3 + 8n^2 + 4n - 9)$

81. $(4q^3 + 7q^2 + 8q - 9) - (-8q - 9 + 7q^2 + 14q^3)$

82. What is $8x^3 - 5x^2 + 6x$ less than $x^2 - 4x + 7$?

83. What is the difference between $7x^4 + 3x^2 - 11$ and $-5x^4 - 8x^2 + 6$?

84. True or false? The difference $(x - a) - (x + a)$ is the opposite of the difference $(x + a) - (x - a)$.

85. True or false? If a is a whole number, then the difference $(-x + a) - (-x - a)$ is a whole number.

OBJECTIVE C **Applications**

86. Read Exercise 87 and then circle the correct word to complete the following sentence. To find the distance from Ashley to Erie, add/subtract the given polynomials.

87. *Distance*   The distance from Ashley to Wyle is $(4x^2 + 3x - 5)$ kilometers. The distance from Wyle to Erie is $(6x^2 - x + 7)$ kilometers. Find the distance from Ashley to Erie.

$4x^2 + 3x - 5$ $6x^2 - x + 7$

Ashley   Wyle   Erie

88. *Distance*   The distance from Haley to Lincoln is $(2y^2 + y - 4)$ kilometers. The distance from Lincoln to Bedford is $(5y^2 - y + 3)$ kilometers. Find the distance from Haley to Bedford.

$2y^2 + y - 4$ $5y^2 - y + 3$

Haley   Lincoln   Bedford

89.   *Geometry*   Find the perimeter of the triangle shown at the right. The dimensions given are in feet. Use the formula $P = a + b + c$.

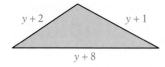

90.   *Geometry*   Find the perimeter of the triangle shown at the right. The dimensions given are in meters. Use the formula $P = a + b + c$.

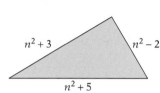

For Exercises 91 to 95, use the formula $P = R - C$, where P is the profit, R is the revenue, and C is the cost.

91.   A company's monthly cost, in dollars, is $75n + 4000$. The company's monthly revenue, in dollars, is $-0.2n^2 + 480n$. To express in terms of n the company's monthly profit, substitute _____ for R and _____ for C in the formula $P = R - C$ and then simplify.

92.   *Business*   A company's total monthly cost, in dollars, for manufacturing and selling n videotapes per month is $35n + 2000$. The company's monthly revenue, in dollars, from selling all n videotapes is $-0.2n^2 + 175n$. Express in terms of n the company's monthly profit.

93.   *Business*   A company manufactures and sells snowmobiles. The total monthly cost, in dollars, to produce n snowmobiles is $50n + 4000$. The company's revenue, in dollars, obtained from selling all n snowmobiles is $-0.6n^2 + 250n$. Express in terms of n the company's monthly profit.

94.   *Business*   A company's total monthly cost, in dollars, for manufacturing and selling n portable CD players per month is $75n + 6000$. The company's revenue, in dollars, from selling all n portable CD players is $-0.4n^2 + 800n$. Express in terms of n the company's monthly profit.

95.   *Business*   A company's total monthly cost, in dollars, for manufacturing and selling n pairs of off-road skates per month is $100n + 1500$. The company's revenue, in dollars, from selling all n pairs is $-n^2 + 800n$. Express in terms of n the company's monthly profit.

CRITICAL THINKING

96.   **a.** What polynomial must be added to $3x^2 + 6x - 9$ so that the sum is $4x^2 - 3x + 2$?

   b. What polynomial must be subtracted from $2x^2 - x - 2$ so that the difference is $5x^2 + 3x + 1$?

97.   In your own words, explain the meanings of the terms *monomial*, *binomial*, *trinomial*, and *polynomial*. Give an example of each.

5.4 Multiplication of Monomials

OBJECTIVE A Multiplication of monomials

Recall that in the exponential expression 3^4, 3 is the base and 4 is the exponent. The exponential expression 3^4 means to multiply 3, the base, 4 times. Therefore, $3^4 = 3 \cdot 3 \cdot 3 \cdot 3 = 81$.

For the variable exponential expression x^6, x is the base and 6 is the exponent. The exponent indicates the number of times the base occurs as a factor. Therefore,

$$\overbrace{x^6 = x \cdot x \cdot x \cdot x \cdot x \cdot x}^{\text{Multiply } x \text{ 6 times.}}$$

The product of exponential expressions with the *same* base can be simplified by writing each expression in factored form and then writing the result with an exponent.

$$x^3 \cdot x^2 = \overbrace{(x \cdot x \cdot x)}^{\text{3 factors}} \cdot \overbrace{(x \cdot x)}^{\text{2 factors}}$$
$$\underbrace{}_{\text{5 factors}}$$
$$= x \cdot x \cdot x \cdot x \cdot x$$
$$= x^5$$

Note that adding the exponents results in the same product.

$$x^3 \cdot x^2 = x^{3+2} = x^5$$

This suggests the following rule for multiplying exponential expressions.

Rule for Multiplying Exponential Expressions

If m and n are positive integers, then $x^m \cdot x^n = x^{m+n}$.

Simplify: $a^4 \cdot a^5$

The bases are the same.
Add the exponents.

$$a^4 \cdot a^5 = a^{4+5}$$
$$= a^9$$

Simplify: $c^3 \cdot c^4 \cdot c$

The bases are the same.
Add the exponents. Note that $c = c^1$.

$$c^3 \cdot c^4 \cdot c = c^{3+4+1}$$
$$= c^8$$

Simplify: $x^5 y^3$

The bases are *not* the same. The exponential expression is in simplest form.

$x^5 y^3$ is in simplest form.

OBJECTIVE B **Powers of monomials**

39. Use the Rule for Simplifying the Power of an Exponential Expression:
$(x^3)^4 = x^{\underline{\quad} \cdot \underline{\quad}} = \underline{\quad}$.

40. The Rule for Simplifying Powers of Products states that we raise a product to a power by multiplying each exponent _____ the parentheses by the exponent _____ the parentheses. For example,
$(x^3y)^4 = (x^{\underline{\quad} \cdot \underline{\quad}})(y^{\underline{\quad} \cdot \underline{\quad}}) = \underline{\quad\quad}$.

Simplify.

41. $(p^3)^5$ **42.** $(x^3)^5$ **43.** $(b^2)^4$ **44.** $(z^6)^3$ **45.** $(p^4)^7$

46. $(y^{10})^2$ **47.** $(c^7)^4$ **48.** $(d^9)^2$ **49.** $(3x)^2$ **50.** $(2y)^3$

51. $(x^2y^3)^6$ **52.** $(m^4n^2)^3$ **53.** $(r^3t)^4$ **54.** $(a^2b)^5$ **55.** $(-y^2)^2$

56. $(-z^3)^2$ **57.** $(2x^4)^3$ **58.** $(3n^3)^3$ **59.** $(-2a^2)^3$ **60.** $(-3b^3)^2$

61. $(3x^2y)^2$ **62.** $(4a^4b^5)^3$ **63.** $(2a^3bc^2)^3$ **64.** $(4xy^3z^2)^2$ **65.** $(-mn^5p^3)^4$

 For Exercises 66 to 69, state whether the expression can be simplified using one of the rules presented in this section.

66. $(xy)^3$ **67.** $(x + y)^3$ **68.** $(a^3 + b^4)^2$ **69.** $(a^3b^4)^2$

CRITICAL THINKING

70. *Geometry* Find the area of the rectangle shown at the right. The dimensions given are in feet. Use the formula $A = LW$.

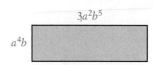

71. *Geometry* Find the area of the square shown at the right. The dimensions given are in centimeters. Use the formula $A = s^2$.

72. Evaluate $(2^3)^2$ and $2^{(3^2)}$. Are the results the same? If not, which expression has the larger value?

5.5 Multiplication of Polynomials

OBJECTIVE A Multiplication of a polynomial by a monomial

Recall that the Distributive Property states that if a, b, and c are real numbers, then $a(b + c) = ab + ac$. The Distributive Property is used to multiply a polynomial by a monomial. Each term of the polynomial is multiplied by the monomial.

Multiply: $y^2(4y^2 + 3y - 7)$

Use the Distributive Property. Multiply each term of the polynomial by y^2.

$y^2(4y^2 + 3y - 7)$

$= y^2(4y^2) + y^2(3y) - y^2(7)$

Use the Rule for Multiplying Exponential Expressions.

$= 4y^4 + 3y^3 - 7y^2$

Multiply: $3x^3(4x^4 - 2x + 5)$

Use the Distributive Property. Multiply each term of the polynomial by $3x^3$.

$3x^3(4x^4 - 2x + 5)$

$= 3x^3(4x^4) - 3x^3(2x) + 3x^3(5)$

Use the Rule for Multiplying Exponential Expressions.

$= 12x^7 - 6x^4 + 15x^3$

Multiply: $-3a(6a^4 - 3a^2)$

Use the Distributive Property. Multiply each term of the polynomial by $-3a$.

$-3a(6a^4 - 3a^2)$

$= -3a(6a^4) - (-3a)(3a^2)$

Use the Rule for Multiplying Exponential Expressions.

$= -18a^5 - (-9a^3)$

Rewrite $-(-9a^3)$ as $+ 9a^3$.

$= -18a^5 + 9a^3$

EXAMPLE 1 Multiply: $-2x(7x - 4y)$

Solution $-2x(7x - 4y)$
$= -2x(7x) - (-2x)(4y)$
$= -14x^2 + 8xy$

YOU TRY IT 1 Multiply: $-3a(-6a + 5b)$

Your Solution

EXAMPLE 2 Multiply:
$2xy(3x^2 - xy + 2y^2)$

Solution $2xy(3x^2 - xy + 2y^2)$
$= 2xy(3x^2) - 2xy(xy) + 2xy(2y^2)$
$= 6x^3y - 2x^2y^2 + 4xy^3$

YOU TRY IT 2 Multiply:
$3mn^2(2m^2 - 3mn - 1)$

Your Solution

Solutions on p. S14

OBJECTIVE B Multiplication of two binomials

In the previous objective, a monomial and a polynomial were multiplied. Using the Distributive Property, we multiplied each term of the polynomial by the monomial. Two binomials are also multiplied by using the Distributive Property. Each term of one binomial is multiplied by the other binomial.

Multiply: $(x + 2)(x + 6)$

Use the Distributive Property. Multiply each term of $(x + 6)$ by $(x + 2)$.

Use the Distributive Property again to multiply $(x + 2)x$ and $(x + 2)6$.

Simplify by combining like terms.

$(x + 2)(x + 6)$
$= (x + 2)x + (x + 2)6$
$= x(x) + 2(x) + x(6) + 2(6)$
$= x^2 + 2x + 6x + 12$
$= x^2 + 8x + 12$

Because it is frequently necessary to multiply two binomials, the terms of the binomials are labeled as shown in the diagram below and the product is computed by using a method called **FOIL.** The letters of FOIL stand for First, Outer, Inner, and Last. The FOIL method is based on the Distributive Property and involves adding the products of the first terms, the outer terms, the inner terms, and the last terms.

The product $(2x + 3)(3x + 4)$ is shown below using FOIL.

	First terms	Outer terms	Inner terms	Last terms
$(2x + 3) \cdot (3x + 4) =$	$(2x)(3x)$ +	$(2x)(4)$ +	$(3)(3x)$ +	$(3)(4)$
$=$	$6x^2$ +	$8x$ +	$9x$ +	12

$= 6x^2 + 17x + 12$

Multiply $(4x - 3)(2x + 3)$ using the FOIL method.

$(4x - 3)(2x + 3) = (4x)(2x) + (4x)(3) + (-3)(2x) + (-3)(3)$
$\qquad = 8x^2 + 12x - 6x - 9$
$\qquad = 8x^2 + 6x - 9$

EXAMPLE 3

Multiply: $(2x - 3)(x + 2)$

Solution

$(2x - 3)(x + 2)$
$= (2x)(x) + (2x)(2) + (-3)(x) + (-3)(2)$
$= 2x^2 + 4x - 3x - 6$
$= 2x^2 + x - 6$

YOU TRY IT 3

Multiply: $(3c + 7)(3c - 7)$

Your Solution

Solution on p. S14

5.5 Exercises

OBJECTIVE A Multiplication of a polynomial by a monomial

1. To multiply $-3y(y + 7)$, use the _____ Property to multiply each term of $y + 7$ by _____ .

   $-3y(y + 7) = $ _____ $(y) + ($_____$)(7) = $ _____ $-$ _____

2. To multiply $5x(x^2 + 2x - 10)$, use the Distributive Property to multiply each term of _____ by $5x$.

   $5x(x^2 + 2x - 10) = 5x($_____$) + 5x($_____$) - 5x($_____$) = $ _____ $+$ _____ $-$ _____

Multiply.

3. $x(x^2 - 3x - 4)$

4. $y(3y^2 + 4y - 8)$

5. $4a(2a^2 + 3a - 6)$

6. $3b(6b^2 - 5b - 7)$

7. $-2a(3a^2 + 9a - 7)$

8. $-4x(x^2 - 3x - 7)$

9. $m^3(4m - 9)$

10. $r^2(2r^2 + 7)$

11. $2x^3(5x^2 - 6xy + 2y^2)$

12. $4b^4(3a^2 + 4ab - b^2)$

13. $-6r^5(r^2 - 2r - 6)$

14. $-5y^4(3y^2 - 6y^3 + 7)$

15. $4a^2(3a^2 + 6a - 7)$

16. $5b^3(2b^2 - 4b - 9)$

17. $-2n^2(3 - 4n^3 - 5n^5)$

18. $-4x^3(6 - 4x^2 - 5x^4)$

19. $ab^2(3a^2 - 4ab + b^2)$

20. $x^2y^3(5y^3 - 6xy - x^3)$

21. $-x^2y^3(4x^5y^2 - 5x^3y - 7x)$

22. $-a^2b^4(3a^6b^4 + 6a^3b^2 - 5a)$

23. $6r^2t^3(1 - rt - r^3t^3)$

24. Which expression is equivalent to $b^3(b^3 - b^2)$?

   (i) $b^9 - b^6$ **(ii)** $b^{27} - b^8$ **(iii)** b^3 **(iv)** $b^6 - b^5$ **(v)** b

OBJECTIVE B Scientific notation

Very large and very small numbers are encountered in the natural sciences. For example, the mass of an electron is 0.0000000000000000000000000000000911 kg. Numbers such as this are difficult to read, so a more convenient system called **scientific notation** is used. In scientific notation, a number is expressed as the product of two factors, one a number between 1 and 10, and the other a power of 10.

To express a number in scientific notation, write it in the form $a \times 10^n$, where a is a number between 1 and 10 and n is an integer.

> **Take Note**
> There are two steps involved in writing a number in scientific notation: (1) determine the number between 1 and 10, and (2) determine the exponent on 10.

For numbers greater than 10, move the decimal point to the right of the first digit. The exponent n is positive and equal to the number of places the decimal point has been moved.

$240{,}000 = 2.4 \times 10^5$

$93{,}000{,}000 = 9.3 \times 10^7$

For numbers less than 1, move the decimal point to the right of the first nonzero digit. The exponent n is negative. The absolute value of the exponent is equal to the number of places the decimal point has been moved.

$0.0003 = 3.0 \times 10^{-4}$

$0.0000832 = 8.32 \times 10^{-5}$

Changing a number written in scientific notation to decimal notation also requires moving the decimal point.

When the exponent is positive, move the decimal point to the right the same number of places as the exponent.

$3.45 \times 10^6 = 3{,}450{,}000$

$2.3 \times 10^8 = 230{,}000{,}000$

When the exponent is negative, move the decimal point to the left the same number of places as the absolute value of the exponent.

$8.1 \times 10^{-3} = 0.0081$

$6.34 \times 10^{-7} = 0.000000634$

EXAMPLE 3 Write 824,300,000,000 in scientific notation.

Solution The number is greater than 10. Move the decimal point 11 places to the left. The exponent on 10 is 11.

$824{,}300{,}000{,}000 = 8.243 \times 10^{11}$

YOU TRY IT 3 Write 0.000000961 in scientific notation.

Your Solution

EXAMPLE 4 Write 6.8×10^{-10} in decimal notation.

Solution The exponent on 10 is negative. Move the decimal point 10 places to the left.

$6.8 \times 10^{-10} = 0.00000000068$

YOU TRY IT 4 Write 7.329×10^6 in decimal notation.

Your Solution

Solutions on p. S14

5.6 Exercises

OBJECTIVE A Division of monomials

1. As long as x is not zero, x^0 is defined to be equal to _____. Using this definition, $3^0 =$ _____ and $(7x^3)^0 =$ _____.

2. Use the Rule for Dividing Exponential Expressions:
$$\frac{p^6}{p^2} = p^{\underline{} - \underline{}} = \underline{}.$$

Simplify.

3. 27^0

4. $(3x)^0$

5. $-(17)^0$

6. $-(2a)^0$

7. 3^{-2}

8. 4^{-3}

9. 2^{-3}

10. 5^{-2}

11. x^{-5}

12. v^{-3}

13. w^{-8}

14. m^{-9}

15. y^{-1}

16. d^{-4}

17. $\dfrac{1}{a^{-5}}$

18. $\dfrac{1}{c^{-6}}$

19. $\dfrac{1}{b^{-3}}$

20. $\dfrac{1}{y^{-7}}$

21. $\dfrac{a^8}{a^2}$

22. $\dfrac{c^{12}}{c^5}$

23. $\dfrac{q^5}{q}$

24. $\dfrac{r^{10}}{r}$

25. $\dfrac{m^4 n^7}{m^3 n^5}$

26. $\dfrac{a^5 b^6}{a^3 b^2}$

27. $\dfrac{t^4 u^8}{t^2 u^5}$

28. $\dfrac{b^{11} c^4}{b^4 c}$

29. $\dfrac{x^4}{x^9}$

30. $\dfrac{r^2}{r^5}$

31. $\dfrac{b}{b^5}$

32. $\dfrac{m^5}{m^8}$

For Exercises 33 to 36, state whether the expression can be simplified using the Rule for Dividing Exponential Expressions.

33. $a^4 - a^2$

34. $\dfrac{a^4}{a^2}$

35. $\dfrac{a^4}{b^2}$

36. $\dfrac{4}{2}$

OBJECTIVE B   Scientific notation

37. A number is written in scientific notation if it is written as the product of a number between _____ and _____ and a power of _____.

38. To write the number 354,000,000 in scientific notation, move the decimal point _____ places to the _____. The exponent on 10 is _____.

39. To write the number 0.0000000086 in scientific notation, move the decimal point _____ places to the _____. The exponent on 10 is _____.

40. For the number 2.8×10^7, the exponent on 10 is _____. To write this number in decimal notation, move the decimal point _____ places to the _____.

Write the number in scientific notation.

41. 2,370,000

42. 75,000

43. 0.00045

44. 0.000076

45. 309,000

46. 819,000,000

47. 0.000000601

48. 0.00000000096

49. 57,000,000,000

50. 934,800,000,000

51. 0.000000017

52. 0.0000009217

Write the number in decimal notation.

53. 7.1×10^5

54. 2.3×10^7

55. 4.3×10^{-5}

56. 9.21×10^{-7}

57. 6.71×10^8

58. 5.75×10^9

59. 7.13×10^{-6}

60. 3.54×10^{-8}

61. 5×10^{12}

62. 1.0987×10^{11}

63. 8.01×10^{-3}

64. 4.0162×10^{-9}

 For Exercises 65 to 68, determine whether the number is written in scientific notation. If not, explain why not.

65. 84.3×10^{-3}

66. 0.97×10^4

67. $6.4 \times 10^{2.5}$

68. 4×10^{-297}

69. *Astronomy*   Astrophysicists estimate that the radius of the Milky Way galaxy is 1,000,000,000,000,000,000,000 m. Write this number in scientific notation.

70. *Geology*   The mass of Earth is 5,980,000,000,000,000,000,000,000 kg. Write this number in scientific notation.

71. *Physics*   Carbon nanotubes are cylinders of carbon atoms. Carbon nanotubes with a diameter of 0.0000000004 m have been created. Write this number in scientific notation.

72. *Biology*   The weight of a single *E. coli* bacterium is 0.000000000000665 g. Write this number in scientific notation.

73. *Archeology*   The weight of the Great Pyramid of Cheops is estimated to be 12,000,000,000 lb. Write this number in scientific notation.

74. *Physics*   The length of an infrared light wave is approximately 0.0000037 m. Write this number in scientific notation.

75. *Food Science*   The frequency (oscillations per second) of a microwave generated by a microwave oven is approximately 2,450,000,000 hertz. (One hertz is one oscillation in 1 second.) Write this number in scientific notation.

76. *Economics*   What was the U.S. trade deficit in 2002? Use the graph at the right. Write the answer in scientific notation.

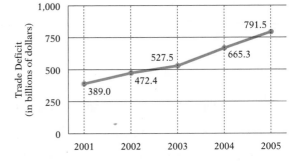

The U.S. Trade Deficit
Source: U.S. Department of Commerce, Bureau of Economic Analysis

77. $m \times 10^8$ and $n \times 10^6$ are numbers written in scientific notation. Place the correct symbol, < or >, between these two numbers.

78. $m \times 10^{-5}$ and $n \times 10^{-3}$ are numbers written in scientific notation. Place the correct symbol, < or >, between these two numbers.

CRITICAL THINKING

79. Place the correct symbol, < or >, between the two numbers.
    a. 3.45×10^{-14} ? 6.45×10^{-15} b. 5.23×10^{18} ? 5.23×10^{17}
    c. 3.12×10^{12} ? 4.23×10^{11} d. -6.81×10^{-24} ? -9.37×10^{-25}

80. a. Evaluate 3^{-x} when $x = -2, -1, 0, 1,$ and 2.
    b. Evaluate 2^{-x} when $x = -2, -1, 0, 1,$ and 2.

5.7 Verbal Expressions and Variable Expressions

OBJECTIVE A Translation of verbal expressions into variable expressions

One of the major skills required in applied mathematics is translating a verbal expression into a mathematical expression. Doing so requires recognizing the verbal phrases that translate into mathematical operations. Following is a partial list of the verbal phrases used to indicate the different mathematical operations.

Addition	more than	8 more than w	$w + 8$
	the sum of	the sum of z and 9	$z + 9$
	the total of	the total of r and s	$r + s$
	increased by	x increased by 7	$x + 7$
Subtraction	less than	12 less than b	$b - 12$
	the difference between	the difference between x and 1	$x - 1$
	decreased by	17 decreased by a	$17 - a$
Multiplication	times	negative 2 times c	$-2c$
	the product of	the product of x and y	xy
	of	three-fourths of m	$\dfrac{3}{4}m$
	twice	twice d	$2d$
Division	divided by	v divided by 15	$\dfrac{v}{15}$
	the quotient of	the quotient of y and 3	$\dfrac{y}{3}$
Power	the square of or the second power of	the square of x	x^2
	the cube of or the third power of	the cube of r	r^3
	the fifth power of	the fifth power of a	a^5

Point of Interest

The way in which expressions are symbolized has changed over time. Here are some expressions as they may have appeared in the early sixteenth century.

R p. 9 for $x + 9$. The symbol R was used for a variable to the first power. The symbol p. was used for plus.

R m. 3 for $x - 3$. The symbol R is again the variable. The symbol m. is used for minus.

The square of a variable was designated by Q, and the cube was designated by C. The expression $x^3 + x^2$ was written C p. Q.

Translating a phrase that contains the word *sum*, *difference*, *product*, or *quotient* can sometimes cause a problem. In the examples at the right, note where the operation symbol is placed.

the *sum* of x and y $x + y$

the *difference* between x and y $x - y$

the *product* of x and y $x \cdot y$

the *quotient* of x and y $\dfrac{x}{y}$

Take Note

The expression $3(c + 5)$ must have parentheses. If we write $3 \cdot c + 5$, then by the Order of Operations Agreement, only the c is multiplied by 3, but we want the 3 multiplied by the *sum* of c and 5.

Translate "three times the sum of c and five" into a variable expression.

Identify words that indicate the mathematical operations.

3 times the sum of c and 5

Use the identified words to write the variable expression. Note that the phrase times the sum of requires parentheses.

$3(c + 5)$

The sum of two numbers is thirty-seven. If x represents the smaller number, translate "twice the larger number" into a variable expression.

Write an expression for the larger number by subtracting the smaller number, x, from the sum.

larger number: $37 - x$

Identify the words that indicate the mathematical operations on the larger number.

twice the larger number

Use the identified words to write a variable expression.

$2(37 - x)$

EXAMPLE 1

Translate "the quotient of r and the sum of r and four" into a variable expression.

Solution

the quotient of r and the sum of r and four

$\dfrac{r}{r + 4}$

YOU TRY IT 1

Translate "twice x divided by the difference between x and seven" into a variable expression.

Your Solution

EXAMPLE 2

Translate "the sum of the square of y and six" into a variable expression.

Solution

the sum of the square of y and six

$y^2 + 6$

YOU TRY IT 2

Translate "the product of negative three and the square of d" into a variable expression.

Your Solution

Solutions on p. S14

OBJECTIVE B   **Translation and simplification of verbal expressions**

After a verbal expression is translated into a variable expression, it may be possible to simplify the variable expression.

Translate "a number plus five less than the product of eight and the number" into a variable expression. Then simplify.

The letter x is chosen for the unknown number. Any letter could be used.

the unknown number: x

Identify words that indicate the mathematical operations.

x plus 5 less than the product of 8 and x

Use the identified words to write the variable expression.

$x + (8x - 5)$

Simplify the expression by adding like terms.

$x + 8x - 5$
$9x - 5$

Translate "five less than twice the difference between a number and seven" into a variable expression. Then simplify.

the unknown number: x

Identify words that indicate the mathematical operations.

5 less than twice the difference between x and 7

Use the identified words to write the variable expression.

$2(x - 7) - 5$

Simplify the expression.

$2x - 14 - 5$
$2x - 19$

EXAMPLE 3

The sum of two numbers is twenty-eight. Using x to represent the smaller number, translate "the sum of the smaller number and three times the larger number" into a variable expression. Then simplify.

Solution

The smaller number is x.
The larger number is $28 - x$.

the sum of the smaller number and three times the larger number

$x + 3(28 - x)$ ▶ This is the variable expression.
$x + 84 - 3x$ ▶ Simplify.
$-2x + 84$

YOU TRY IT 3

The sum of two numbers is sixteen. Using x to represent the smaller number, translate "the difference between the larger number and twice the smaller number" into a variable expression. Then simplify.

Your Solution

Solution on p. S14

EXAMPLE 4

Translate "eight more than the product of four and the total of a number and twelve" into a variable expression. Then simplify.

Solution

Let the unknown number be x.

8 <u>more than</u> the <u>product</u> of 4 and the <u>total of</u> x and 12

$4(x + 12) + 8$ ▶ This is the variable expression.
$4x + 48 + 8$ ▶ Now simplify.
$4x + 56$

YOU TRY IT 4

Translate "the difference between fourteen and the sum of a number and seven" into a variable expression. Then simplify.

Your Solution

Solution on p. S14

OBJECTIVE C   **Applications**

Many applications of mathematics require that you identify the unknown quantity, assign a variable to that quantity, and then attempt to express other unknowns in terms of that quantity.

Ten gallons of paint were poured into two containers of different sizes. Express the amount of paint poured into the smaller container in terms of the amount poured into the larger container.

Assign a variable to the amount of paint poured into the larger container. (Any variable can be used.)

gallons of paint poured into the larger container: g

Express the amount of paint in the smaller container in terms of g. (g gallons of paint were poured into the larger container.)

The number of gallons of paint in the smaller container is $10 - g$.

EXAMPLE 5

A cyclist is riding at twice the speed of a runner. Express the speed of the cyclist in terms of the speed of the runner.

Solution

the speed of the runner: r
the speed of the cyclist is twice r: $2r$

YOU TRY IT 5

A mixture of candy contains three pounds more of milk chocolate than of caramel. Express the amount of milk chocolate in the mixture in terms of the amount of caramel in the mixture.

Your Solution

Solution on p. S14

Focus on **Problem Solving**

Look for a Pattern

A very useful problem-solving strategy is to look for a pattern. We illustrate this strategy below using a fairly old problem.

A legend says that a peasant invented the game of chess and gave it to a very rich king as a present. The king so enjoyed the game that he gave the peasant the choice of anything in the kingdom. The peasant's request was simple. "Place 1 grain of wheat on the first square, 2 grains on the second square, 4 grains on the third square, 8 on the fourth square, and continue doubling the number of grains until the last square of the chessboard is reached." How many grains of wheat must the king give the peasant?

A chessboard consists of 64 squares. To find the total number of grains of wheat on the 64 squares, we begin by looking at the amount of wheat on the first few squares.

Square 1	Square 2	Square 3	Square 4	Square 5	Square 6	Square 7	Square 8
1	2	4	8	16	32	64	128
1	3	7	15	31	63	127	255

The bottom row of numbers represents the sum of the number of grains of wheat up to and including that square. For instance, the number of grains of wheat on the first 7 squares is

$$1 + 2 + 4 + 8 + 16 + 32 + 64 = 127$$

One pattern we might observe is that the number of grains of wheat on a square can be expressed by a power of 2.

Number of grains on square $n = 2^{n-1}$.

For example, the number of grains on square $7 = 2^{7-1} = 2^6 = 64$.

A second pattern of interest is that the number *below* a square (the total number of grains up to and including that square) is one less than the number of grains of wheat *on* the next square. For example, the number *below* square 7 is one less than the number *on* square 8 ($128 - 1 = 127$). From this observation, the number of grains of wheat on the first eight squares is the number on square 8 (128) plus one less than the number on square 8 (127); the total number of grains of wheat on the first eight squares is $128 + 127 = 255$.

From this observation,

$$\frac{\text{Number of grains of}}{\text{wheat on the chessboard}} = \frac{\text{number of grains}}{\text{on square 64}} + \frac{\text{one less than the number}}{\text{of grains on square 64}}$$

$$= 2^{64-1} + 2^{64-1} - 1$$

$$= 2^{63} + 2^{63} - 1 \approx 18,000,000,000,000,000,000$$

To give you an idea of the magnitude of this number, this is more wheat than has been produced in the world since chess was invented.

Suppose that the same king decided to have a banquet in the long dining room of the palace. The king had 50 square tables and each table could seat only one person on each side. The king pushed the tables together to form one long banquet table. How many people can sit at this table? *Hint:* Try constructing a pattern by using 2 tables, 3 tables, and 4 tables.

Projects & Group Activities

Multiplication of Polynomials

Section 5.5 introduced multiplying a polynomial by a monomial and multiplying two binomials. Multiplying a binomial times a polynomial of three or more terms requires the repeated application of the Distributive Property. Each term of one polynomial is multiplied by the other polynomial. For the product $(2y - 3)(y^2 + 2y + 5)$ shown below, note that the Distributive Property is used twice. The final result is simplified by combining like terms.

$$(2y - 3)(y^2 + 2y + 5) = (2y - 3)y^2 + (2y - 3)2y + (2y - 3)5$$
$$= 2y(y^2) - 3(y^2) + 2y(2y) - 3(2y) + 2y(5) - 3(5)$$
$$= 2y^3 - 3y^2 + 4y^2 - 6y + 10y - 15$$
$$= 2y^3 + y^2 + 4y - 15$$

In Exercises 1–6, find the product of the polynomials.

1. $(y + 6)(y^2 - 3y + 4)$ 2. $(2b^2 + 4b - 5)(2b + 3)$

3. $(2a^2 + 4a - 5)(3a + 1)$ 4. $(3z^2 - 5z + 7)(z - 3)$

5. $(x - 3)(x^3 + 2x^2 - 4x - 5)$ 6. $(c^3 + 3c^2 - 4c + 5)(2c - 3)$

7. a. Multiply: $(x + 1)(x - 1)$
   b. Multiply: $(x + 1)(-x^2 + x - 1)$
   c. Multiply: $(x + 1)(x^3 - x^2 + x - 1)$
   d. Multiply: $(x + 1)(-x^4 + x^3 - x^2 + x - 1)$
   e. Use the pattern of the answers to parts a–d to multiply $(x + 1)(x^5 - x^4 + x^3 - x^2 + x - 1)$.
   f. Use the pattern of the answers to parts a–e to multiply $(x + 1)(-x^6 + x^5 - x^4 + x^3 - x^2 + x - 1)$.

Chapter 5 Summary

Key Words	Examples
The **additive inverse** of a number a is $-a$. The additive inverse of a number is also called the **opposite** number. [5.1A, p. 318]	The opposite of 15 is -15. The opposite of -24 is 24.
The **multiplicative inverse** of a nonzero number a is $\frac{1}{a}$. The multiplicative inverse of a number is also called the **reciprocal** of the number. [5.1A, p. 318]	The reciprocal of -3 is $-\frac{1}{3}$. The reciprocal of $\frac{6}{7}$ is $\frac{7}{6}$.

A **term** of a variable expression is one of the addends of the expression. A **variable term** consists of a **numerical coefficient** and a **variable part**. A **constant term** has no variable part. [5.2A, p. 327]	The variable expression $-3x^2 + 2x - 5$ has three terms: $-3x^2$, $2x$, and -5. $-3x^2$ and $2x$ are variable terms. -5 is a constant term. For the term $-3x^2$, the coefficient is -3 and the variable part is x^2.
Like terms of a variable expression have the same variable part. Constant terms are also like terms. [5.2A, p. 327]	$-6a^3b^2$ and $4a^3b^2$ are like terms.
A **monomial** is a number, a variable, or a product of numbers and variables. [5.3A, p. 335]	5 is a number, y is a variable, $8a^2b^2$ is a product of numbers and variables. 5, y, and $8a^2b^2$ are monomials.
A **polynomial** is a variable expression in which the terms are monomials. A polynomial of one term is a **monomial.** A polynomial of two terms is a **binomial.** A polynomial of three terms is a **trinomial.** [5.3A, p. 335]	5, y, and $8a^2b^2$ are monomials. $x + 9$, $y^2 - 3$, and $6a + 7b$ are binomials. $x^2 + 2x - 1$ is a trinomial.
The terms of a polynomial in one variable are usually arranged so that the exponents of the variable decrease from left to right. This is called **descending order.** [5.3A, p. 335]	$7y^4 + 5y^3 - y^2 + 6y - 8$ is written in descending order.

Essential Rules and Procedures

Properties of Addition [5.1A, pp. 317–318]

Addition Property of Zero $a + 0 = a$ or $0 + a = a$ — $-16 + 0 = -16$

Commutative Property of Addition $a + b = b + a$ — $-9 + 5 = 5 + (-9)$

Associative Property of Addition $(a + b) + c = a + (b + c)$ — $(-6 + 4) + 2 = -6 + (4 + 2)$

Inverse Property of Addition $a + (-a) = 0$ or $-a + a = 0$ — $8 + (-8) = 0$

Properties of Multiplication [5.1A, pp. 317–318]

Multiplication Property of Zero $a \cdot 0 = 0$ or $0 \cdot a = 0$ — $-3(0) = 0$

Multiplication Property of One $a \cdot 1 = a$ or $1 \cdot a = a$ — $-7(1) = -7$

Commutative Property of Multiplication $a \cdot b = b \cdot a$ — $-5(10) = 10(-5)$

Associative Property of Multiplication $(a \cdot b) \cdot c = a \cdot (b \cdot c)$ — $(-3 \cdot 4) \cdot 6 = -3 \cdot (4 \cdot 6)$

Inverse Property of Multiplication For $a \neq 0$, $a \cdot \dfrac{1}{a} = \dfrac{1}{a} \cdot a = 1$. — $8 \cdot \dfrac{1}{8} = 1$

Distributive Property [5.1B, p. 320]
$a(b + c) = ab + ac$ — $5(4x - 3) = 5(4x) - 5(3) = 20x - 15$
Alternative Form of the Distributive Property [5.2A, p. 328]
$ac + bc = (a + b)c$ — $8b + 7b = (8 + 7)b = 15b$

To add polynomials, combine like terms, which means to add the coefficients of the like terms. [5.3A, p. 335]	$(8x^2 + 2x - 9) + (-3x^2 + 5x - 7)$ $= (8x^2 - 3x^2) + (2x + 5x) + (-9 - 7)$ $= 5x^2 + 7x - 16$
To subtract two polynomials, add the opposite of the second polynomial to the first polynomial. [5.3B, pp. 336–337]	$(3y^2 - 8y + 6) - (-y^2 + 4y - 5)$ $= (3y^2 - 8y + 6) + (y^2 - 4y + 5)$ $= 4y^2 - 12y + 11$
Rule for Multiplying Exponential Expressions [5.4A, p. 344] $x^m \cdot x^n = x^{m+n}$	$b^5 \cdot b^4 = b^{5+4} = b^9$
Rule for Simplifying the Power of an Exponential Expression [5.4B, p. 346] $(x^m)^n = x^{m \cdot n}$	$(y^3)^7 = y^{3(7)} = y^{21}$
Rule for Simplifying Powers of Products [5.4B, p. 346] $(x^m y^n)^p = x^{m \cdot p} y^{n \cdot p}$	$(x^6 y^4 z^5)^2 = x^{6(2)} y^{4(2)} z^{5(2)} = x^{12} y^8 z^{10}$
The FOIL Method [5.5B, p. 351] To multiply two binomials, add the products of the First terms, the Outer terms, the Inner terms, and the Last terms.	$(4x + 3)(2x - 5)$ $= (4x)(2x) + (4x)(-5) + (3)(2x) + (3)(-5)$ $= 8x^2 - 20x + 6x - 15$ $= 8x^2 - 14x - 15$
Rule for Dividing Exponential Expressions [5.6A, p. 354] For $x \neq 0$, $\dfrac{x^m}{x^n} = x^{m-n}$.	$\dfrac{y^8}{y^3} = y^{8-3} = y^5$
Zero as an Exponent [5.6A, p. 354] For $x \neq 0$, $x^0 = 1$. The expression 0^0 is not defined.	$17^0 = 1$ $(5y)^0 = 1, y \neq 0$
Definition of Negative Exponents [5.6A, p. 355] For $x \neq 0$, $x^{-n} = \dfrac{1}{x^n}$ and $\dfrac{1}{x^{-n}} = x^n$.	$x^{-6} = \dfrac{1}{x^6}$ and $\dfrac{1}{x^{-6}} = x^6$
Scientific Notation [5.6B, p. 357] To express a number in scientific notation, write it in the form $a \times 10^n$, where a is a number between 1 and 10 and n is an integer.	
If the number is greater than 10, the exponent on 10 will be positive and equal to the number of decimal places the decimal point is moved.	$367{,}000{,}000 = 3.67 \times 10^8$
If the number is less than 1, the exponent on 10 will be negative. The absolute value of the exponent equals the number of places the decimal point has been moved.	$0.0000059 = 5.9 \times 10^{-6}$
To change a number written in scientific notation to decimal notation, move the decimal point to the right if the exponent on 10 is positive and to the left if the exponent on 10 is negative. Move the decimal point the same number of places as the absolute value of the exponent on 10.	$2.418 \times 10^7 = 24{,}180{,}000$ $9.06 \times 10^{-5} = 0.0000906$

Chapter 5 Review Exercises

1. Simplify: $4z^2 + 3z - 9z + 2z^2$

2. Multiply: $-2(9z + 1)$

3. Add: $(3z^2 + 4z - 7) + (7z^2 - 5z - 8)$

4. Multiply: $(2m^3n)(-4m^2n)$

5. Evaluate: 3^{-5}

6. Write the additive inverse of $\frac{3}{7}$.

7. Multiply: $\frac{2}{3}\left(\frac{3}{2}x\right)$

8. Simplify: $-5(2s - 5t) + 6(3t + s)$

9. Multiply: $(-5xy^4)(-3x^2y^3)$

10. Multiply: $(7a + 6)(3a - 4)$

11. Subtract:
    $(6b^3 - 7b^2 + 5b - 9) - (9b^3 - 7b^2 + b + 9)$

12. Simplify: $(2z^4)^5$

13. Multiply: $-\frac{3}{4}(-8w)$

14. Multiply: $5xyz^2(-3x^2z + 6yz^2 - x^3y^4)$

15. Write the multiplicative inverse of $-\frac{9}{4}$.

16. Multiply: $-4(3c - 8)$

17. Simplify: $2m - 6n + 7 - 4m + 6n + 9$

18. Multiply: $(4a^3b^8)(-3a^2b^7)$

19. Identify the property that justifies the statement.
    $a(b + c) = ab + ac$

20. Simplify: $(p^2q^3)^3$

21. Simplify: $\dfrac{a^4}{a^{11}}$

22. Write 0.0000397 in scientific notation.

23. Identify the property that justifies the statement.
$a + b = b + a$

24. Add: $(9y^3 + 8y^2 - 10) + (-6y^3 + 8y - 9)$

25. Simplify: $8(2c - 3d) - 4(c - 5d)$

26. Multiply: $7(2m - 6)$

27. Simplify: $\dfrac{x^3y^5}{xy}$

28. Simplify: $7a^2 + 9 - 12a^2 + 3a$

29. Multiply: $(3p - 9)(4p + 7)$

30. Multiply: $-2a^2b(4a^3 - 5ab^2 + 3b^4)$

31. Simplify: $-12x + 7y + 15x - 11y$

32. Simplify: $-7(3a - 4b) - 5(3b - 4a)$

33. Simplify: c^{-5}

34. Subtract:
$(12x^3 + 9x^2 - 5x - 1) - (6x^3 + 9x^2 + 5x - 1)$

35. Write 2.4×10^5 in decimal notation.

36. *Geometry* Find the perimeter of the triangle shown at the right. The dimensions given are in feet. Use the formula $P = a + b + c$.

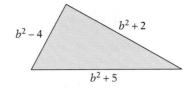

$b^2 - 4$ $b^2 + 2$ $b^2 + 5$

37. Translate "nine less than the quotient of four times a number and seven" into a variable expression.

38. Translate "the sum of three times a number and the difference between the number and seven" into a variable expression. Then simplify.

39. *Chemistry* Avogadro's number is used in chemistry, and its value is approximately 602,300,000,000,000,000,000,000. Express this number in scientific notation.

40. *Food Mixtures* Thirty pounds of a blend of coffee beans uses only mocha java and expresso beans. Express the number of pounds of expresso beans in the blend in terms of the number of pounds of mocha java beans in the blend.

Chapter 5 Test

1. Simplify: $\frac{2}{3}\left(-\frac{3}{2}r\right)$

2. Simplify: $-3(5y - 7)$

3. Simplify: $7y - 3 - 4y + 6$

4. Simplify: $4x^2 - 2z + 7z - 8x^2$

5. Simplify: $2a - 4b + 12 - 5a - 2b + 6$

6. Write the multiplicative inverse of $\frac{5}{4}$.

7. Simplify: $-2(3x - 4y) + 5(2x + y)$

8. Simplify: $9 - 2(4b - a) + 3(3b - 4a)$

9. Write 0.00000079 in scientific notation.

10. Write 4.9×10^6 in decimal notation.

11. Add: $(4x^2 - 2x - 2) + (2x^2 - 3x + 7)$

12. Simplify: $(v^2w^5)^4$

13. Simplify: $(3m^2n^3)^3$

14. Multiply: $(-5v^2z)(2v^3z^2)$

15. Multiply: $(3p - 8)(2p + 5)$

16. Multiply: $(2m^2n^2)(-4mn^3 + 2m^3 - 3n^4)$

17. Complete the statement by using the Commutative Property of Addition.
$3z + ? = 4w + 3z$

18. Simplify: $\dfrac{x^2y^5}{xy^2}$

19. Simplify: a^{-5}

20. Identify the property that justifies the statement.
$2 \cdot (c \cdot d) = (2 \cdot c) \cdot d$

21. Subtract:
$(5a^3 - 6a^2 + 4a - 8) - (8a^3 - 7a^2 + 4a + 2)$

22. Simplify: $\dfrac{1}{c^{-6}}$

23. Identify the property that justifies the statement.
$6(s + t) = 6s + 6t$

24. Complete the statement by using the Multiplication Property of Zero.
$6w \cdot 0 = ?$

25. Multiply: $(3x - 7y)(3x + 7y)$

26. Write the additive inverse of $-\dfrac{4}{7}$.

27. Write 720,000,000 in scientific notation.

28. Multiply: $(3a - 6)(4a + 2)$

29. Simplify: $2(4a - 3b) + 3(5a - 2b)$

30. Simplify: $\dfrac{m^4 n^2}{m^2 n^5}$

31. Translate "five more than three times a number" into a variable expression.

32. Translate "the sum of a number and the difference between the number and six" into a variable expression and then simplify.

33. *Food Mixtures*   A muffin batter contains 3 c more flour than sugar. Express the amount of flour in the batter in terms of the amount of sugar in the batter.

Cumulative Review Exercises

1. Find the quotient of 4.712 and −0.38.

2. Simplify: $9v - 10 + 5v + 8$

3. Multiply: $(3x - 5)(2x + 4)$

4. Evaluate $-a - b$ for $a = \dfrac{11}{24}$ and $b = -\dfrac{5}{6}$.

5. Simplify: $\sqrt{81} + 3\sqrt{25}$

6. Graph the real numbers greater than −3.

7. Simplify: $\dfrac{1}{x^{-7}}$

8. Solve: $-4t = 36$

9. Write 0.00000084 in scientific notation.

10. Add: $(5x^2 - 3x + 2) + (4x^2 + x - 6)$

11. Evaluate $-5\sqrt{x + y}$ for $x = 18$ and $y = 31$.

12. Simplify: $\dfrac{\dfrac{5}{8} + \dfrac{3}{4}}{3 - \dfrac{1}{2}}$

13. Multiply: $(-3a^2b)(4a^5b^8)$

14. Simplify: $\dfrac{x^3}{x^5}$

15. Evaluate x^3y^2 for $x = \dfrac{2}{5}$ and $y = 2\dfrac{1}{2}$.

16. Simplify: $-8p(6)$

17. Estimate the difference between 829.43 and 567.109.

18. Multiply: $-3ab^2(4a^2b + 5ab - 2ab^2)$

19. Simplify: $6(5x - 4y) - 12(x - 2y)$

20. Evaluate $\dfrac{a}{-b}$ for $a = -56$ and $b = -8$.

21. Convert 0.5625 to a fraction.

22. Simplify: $6 \cdot (-2)^3 \div 12 - (-8)$

23. Simplify: $\sqrt{300}$

24. Subtract: $(8y^2 - 7y + 4) - (3y^2 - 5y + 9)$

25. Evaluate $-6cd$ for $c = -\dfrac{2}{9}$ and $d = \dfrac{3}{4}$.

26. Simplify: $-(3a^2)^0$, $a \neq 0$

27. Simplify: $(2a^4b^3)^5$

28. Evaluate $(a - b)^2 + 5c$ for $a = -4$, $b = 6$, and $c = -2$.

29. Find the product of $2\dfrac{4}{5}$ and $\dfrac{6}{7}$.

30. Write 6.23×10^{-5} in decimal notation.

31. Translate "the quotient of ten and the difference between a number and nine" into a variable expression.

32. Translate "two less than twice the sum of a number and four" into a variable expression. Then simplify.

33. *Meteorology* The average annual precipitation in Seattle, Washington, is 38.6 in. The average annual precipitation in El Paso, Texas, is 7.82 in. Find the difference between the average annual precipitation in Seattle and the average annual precipitation in El Paso.

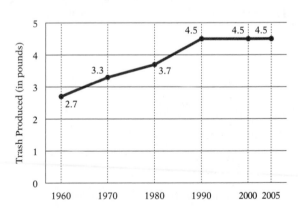

34. *Environmental Science* The graph at the right shows the amount of trash produced per person per day in the United States. On average, how much more trash did a person in the United States throw away during 2005 than during 1960?

Trash Production per Person per Day in the United States
Source: U.S. Environmental Protection Agency

35. *Astronomy* The distance from Neptune to the sun is approximately 30 times the distance from Earth to the sun. Express the distance from Neptune to the sun in terms of the distance from Earth to the sun.

36. *Investments* The cost, C, of the shares of stock in a stock purchase is equal to the cost per share, S, times the number of shares purchased, N. Use the equation $C = SN$ to find the cost of purchasing 200 shares of stock selling for $15.375 per share.

First-Degree Equations

 DVD   SSM ***Student Website***
Need help? For online student resources,
visit college.hmco.com/pic/aufmannPA5e.

This photo shows employees at a Flextronics' Xbox manufacturing facility putting the final touches on Microsoft Xbox video game systems as they roll off the assembly line for shipment across North America. The objective of Microsoft Corporation, as with any business, is to earn a profit. Profit is the difference between a company's revenue (the total amount of money the company earns by selling its products or services) and its costs (the total amount of money the company spends in doing business). Often a company wants to know its break-even point, which is the number of products that must be sold so that no profit or loss occurs. You will be calculating break-even points in **Exercises 67 to 70 on page 402.**

1. Subtract: $8 - 12$

2. Multiply: $-\dfrac{3}{4}\left(-\dfrac{4}{3}\right)$

3. Multiply: $-\dfrac{5}{8}(16)$

4. Simplify: $\dfrac{-3}{-3}$

5. Simplify: $-16 + 7y + 16$

6. Simplify: $8x - 9 - 8x$

7. Evaluate $2x + 3$ for $x = -4$.

8. Given $y = -4x + 5$, find the value of y for $x = -2$.

GO Figure

How can you cut a donut into eight equal pieces with three cuts of the knife?

6.1 Equations of the Form $x + a = b$ and $ax = b$

OBJECTIVE A   Equations of the form $x + a = b$

Recall that an **equation** expresses the equality of two mathematical expressions. The display at the right shows some examples of equations.

$$3x - 7 = 4x + 9$$
$$y = 3x - 6$$
$$2z^2 - 5z + 10 = 0$$
$$\frac{3}{x} + 7 = 9$$

The first equation in the display above is a **first-degree equation in one variable.** The equation has one variable, x, and each instance of the variable is the first power (the exponent on x is 1). First-degree equations in one variable are the topic of Sections 1 through 4 of this chapter. The second equation is a first-degree equation in two variables. These equations are discussed in Section 6.6. The remaining equations are not first-degree equations and will not be discussed in this text.

Which of the equations shown at the right are first-degree equations in one variable?

1. $5x + 4 = 9 - 3(2x + 1)$
2. $\sqrt{x} + 9 = 10$
3. $p = -14$
4. $2x - 5 = x^2 - 9$

Equation 1 is a first-degree equation in one variable.

Equation 2 is not a first-degree equation in one variable. First-degree equations do not contain square roots of variable expressions.

Equation 3 is a first-degree equation in one variable.

Equation 4 is not a first-degree equation in one variable. First-degree equations in one variable do not have exponents greater than 1 on the variable.

Recall that a **solution** of an equation is a number that, when substituted for the variable, produces a true equation.

15 is a solution of the equation $x - 5 = 10$ because $15 - 5 = 10$ is a true equation.

20 is not a solution of $x - 5 = 10$ because $20 - 5 = 10$ is a false equation.

To **solve an equation** means to determine the solutions of the equation. The simplest equation to solve is an equation of the form **variable = constant.** The constant is the solution.

Consider the equation $x = 7$, which is in the form *variable = constant.* The solution is 7 because $7 = 7$ is a true equation.

Find the solution of the equation $y = 3 + 7$.

$$y = 3 + 7$$

Simplify the right side of the equation.
$$y = 3 + 7$$
$$y = 10$$

The solution is 10.

Note that replacing x in $x + 8 = 12$ by 4 results in a true equation. The solution of the equation $x + 8 = 12$ is 4.

$$x + 8 = 12$$
$$4 + 8 = 12$$
$$12 = 12$$

If 5 is added to each side of $x + 8 = 12$, the solution is still 4.

$$x + 8 = 12$$
$$x + 8 + 5 = 12 + 5$$
$$x + 13 = 17$$

Check:
$$x + 13 = 17$$
$$\overline{4 + 13 \mid 17}$$
$$17 = 17$$

If -3 is added to each side of $x + 8 = 12$, the solution is still 4.

$$x + 8 = 12$$
$$x + 8 + (-3) = 12 + (-3)$$
$$x + 5 = 9$$

Check:
$$x + 5 = 9$$
$$\overline{4 + 5 \mid 9}$$
$$9 = 9$$

These examples suggest that adding the same number to each side of an equation does not change the solution of the equation. This is called the Addition Property of Equations.

Addition Property of Equations

The same number or variable expression can be added to each side of an equation without changing the solution of the equation.

This property is used in solving equations. Note the effect of adding, to each side of the equation $x + 8 = 12$, the opposite of the constant term 8. After simplifying, the equation is in the form *variable = constant*. The solution is the constant, 4.

$$x + 8 = 12$$
$$x + 8 + (-8) = 12 + (-8)$$
$$x + 0 = 4$$
$$x = 4$$

Check the solution.

Check:
$$x + 8 = 12$$
$$\overline{4 + 8 \mid 12}$$
$$12 = 12$$

The solution checks.

The solution is 4.

The goal in solving an equation is to rewrite it in the form variable = constant. The Addition Property of Equations is used to remove a term from one side of an equation by adding the opposite of that term to each side of the equation. The resulting equation has the same solution as the original equation.

Solve: $m - 9 = 2$

Remove the constant term -9 from the left side of the equation by adding 9, the opposite of -9, to each side of the equation. Then simplify.

$$m - 9 = 2$$
$$m - 9 + 9 = 2 + 9$$
$$m + 0 = 11$$
$$m = 11$$

You should check the solution.   The solution is 11.

In each of the equations above, the variable appeared on the left side of the equation, and the equation was rewritten in the form *variable = constant*. For some equations, it may be more practical to work toward the goal of *constant = variable*, as shown in the example below.

Solve: $12 = n - 8$

The variable is on the right side of the equation. The goal is to rewrite the equation in the form *constant = variable*.

Remove the constant term from the right side of the equation by adding 8 to each side of the equation. Then simplify.

$$12 = n - 8$$
$$12 + 8 = n - 8 + 8$$
$$20 = n + 0$$
$$20 = n$$

You should check the solution.   The solution is 20.

Because subtraction is defined in terms of addition, the Addition Property of Equations allows the same number to be subtracted from each side of an equation without changing the solution of the equation.

Solve: $z + 9 = 6$

The goal is to rewrite the equation in the form *variable = constant*.

Add the opposite of 9 to each side of the equation. This is equivalent to subtracting 9 from each side of the equation. Then simplify.

$$z + 9 = 6$$
$$z + 9 - 9 = 6 - 9$$
$$z + 0 = -3$$
$$z = -3$$

The solution checks.   The solution is -3.

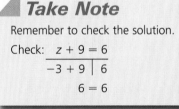

Take Note

Remember to check the solution.

Check: $z + 9 = 6$
         $-3 + 9 \mid 6$
              $6 = 6$

Solve: $5 + x - 9 = -10$

Simplify the left side of the equation by combining the constant terms.

$$5 + x - 9 = -10$$
$$x - 4 = -10$$

Add 4 to each side of the equation.

$$x - 4 + 4 = -10 + 4$$

Simplify.

$$x + 0 = -6$$
$$x = -6$$

-6 checks as a solution.   The solution is -6.

EXAMPLE 1 Solve: $6 + x = 4$

Solution
$$6 + x = 4$$
$$6 - 6 + x = 4 - 6 \quad \blacktriangleright \text{ Subtract 6.}$$
$$x = -2$$

The solution is -2.

YOU TRY IT 1 Solve: $7 + y = 12$

Your Solution

EXAMPLE 2 Solve: $a + \dfrac{3}{4} = \dfrac{1}{4}$

Solution
$$a + \frac{3}{4} = \frac{1}{4}$$
$$a + \frac{3}{4} - \frac{3}{4} = \frac{1}{4} - \frac{3}{4} \quad \blacktriangleright \text{ Subtract } \frac{3}{4}.$$
$$a = -\frac{2}{4} = -\frac{1}{2}$$

The solution is $-\dfrac{1}{2}$.

YOU TRY IT 2 Solve: $b - \dfrac{3}{8} = \dfrac{1}{2}$

Your Solution

EXAMPLE 3 Solve: $7x - 4 - 6x = 3$

Solution
$$7x - 4 - 6x = 3$$
$$x - 4 = 3 \quad \blacktriangleright \text{ Combine like terms.}$$
$$x - 4 + 4 = 3 + 4 \quad \blacktriangleright \text{ Add 4.}$$
$$x = 7$$

The solution is 7.

YOU TRY IT 3 Solve: $-5r + 3 + 6r = 1$

Your Solution

Solutions on p. S14

OBJECTIVE B **Equations of the form $ax = b$**

Note that replacing x by 3 in $4x = 12$ results in a true equation. The solution of the equation is 3.

$$4x = 12$$
$$4(3) = 12$$
$$12 = 12$$

If each side of the equation $4x = 12$ is multiplied by 2, the solution is still 3.

$$4x = 12$$
$$2(4x) = 2(12)$$
$$8x = 24$$

Check:
$$\begin{array}{c|c} 8x = 24 \\ \hline 8(3) & 24 \\ 24 = 24 \end{array}$$

If each side of the equation $4x = 12$ is multiplied by -3, the solution is still 3.

$$4x = 12$$
$$-3(4x) = -3(12)$$
$$-12x = -36$$

Check:
$$\begin{array}{c|c} -12x = -36 \\ \hline -12(3) & -36 \\ -36 = -36 \end{array}$$

These examples suggest that multiplying each side of an equation by the same nonzero number does not change the solution of the equation. This is called the Multiplication Property of Equations.

Multiplication Property of Equations

Each side of an equation can be multiplied by the same nonzero number without changing the solution of the equation.

This property is used in solving equations. Note the effect of multiplying each side of the equation $4x = 12$ by $\frac{1}{4}$, the reciprocal of the coefficient 4. After simplifying, the equation is in the form *variable* = *constant*.

$$4x = 12$$
$$\frac{1}{4} \cdot 4x = \frac{1}{4} \cdot 12$$
$$1 \cdot x = 3$$
$$x = 3$$

The solution is 3.

The Multiplication Property of Equations is used to remove a coefficient from a variable term of an equation by multiplying each side of the equation by the reciprocal of the coefficient. The resulting equation will have the same solution as the original equation.

Solve: $\frac{3}{4}x = -9$

The goal is to rewrite the equation in the form *variable* = *constant*.

Multiply each side of the equation by $\frac{4}{3}$, the reciprocal of $\frac{3}{4}$. After simplifying, the equation is in the form *variable* = *constant*.

$$\frac{3}{4}x = -9$$
$$\frac{4}{3} \cdot \frac{3}{4}x = \frac{4}{3} \cdot (-9)$$
$$1 \cdot x = -12$$
$$x = -12$$

You should check this solution.

The solution is -12.

Because division is defined in terms of multiplication, the Multiplication Property of Equations allows each side of an equation to be divided by the same nonzero number without changing the solution of the equation.

Solve: $-2x = 8$

Multiply each side of the equation by the reciprocal of -2. This is equivalent to dividing each side of the equation by -2.

$$-2x = 8$$
$$\frac{-2x}{-2} = \frac{8}{-2}$$
$$1 \cdot x = -4$$
$$x = -4$$

Check the solution.

Check:
$$\frac{-2x = 8}{-2(-4) \mid 8}$$
$$8 = 8$$

The solution checks.

The solution is -4.

When using the Multiplication Property of Equations, multiply each side of the equation by the reciprocal of the coefficient when the coefficient is a fraction. Divide each side of the equation by the coefficient when the coefficient is an integer or a decimal.

EXAMPLE 4

Solve: $48 = -12y$

Solution

$$48 = -12y$$

$$\frac{48}{-12} = \frac{-12y}{-12} \qquad \blacktriangleright \text{ Divide by } -12.$$

$$-4 = y$$

The solution is -4.

YOU TRY IT 4

Solve: $-60 = 5d$

Your Solution

EXAMPLE 5

Solve: $\frac{2x}{3} = 12$

Solution

$$\frac{2x}{3} = 12$$

$$\frac{3}{2}\left(\frac{2}{3}x\right) = \frac{3}{2}(12) \qquad \blacktriangleright \frac{2x}{3} = \frac{2}{3}x$$

$$x = 18$$

The solution is 18.

YOU TRY IT 5

Solve: $10 = \frac{-2x}{5}$

Your Solution

EXAMPLE 6

Solve and check: $3y - 7y = 8$

Solution

$$3y - 7y = 8$$

$$-4y = 8 \qquad \blacktriangleright \text{ Combine like terms.}$$

$$\frac{-4y}{-4} = \frac{8}{-4} \qquad \blacktriangleright \text{ Divide by } -4.$$

$$y = -2$$

Check:

$$\begin{array}{r|l} 3y - 7y = 8 \\ \hline 3(-2) - 7(-2) & 8 \\ -6 - (-14) & 8 \\ -6 + 14 & 8 \\ 8 = 8 \end{array}$$

-2 checks as the solution.

The solution is -2.

YOU TRY IT 6

Solve and check: $\frac{1}{3}x - \frac{5}{6}x = 4$

Your Solution

Solutions on p. S15

6.1 Exercises

OBJECTIVE A **Equations of the form $x + a = b$**

1.   What is the solution of the equation $x = 9$? Use your answer to explain why the goal in solving the equations in this section is to get the variable alone on one side of the equation.

2.   **a.** To solve $-15 = x - 8$, add _____ to each side of the equation.

   b. To solve $10 + x = 3$, subtract _____ from each side of the equation.

Solve.

3. $x + 3 = 9$	**4.** $y + 6 = 8$	**5.** $4 + x = 13$	**6.** $9 + y = 14$
7. $m - 12 = 5$	**8.** $n - 9 = 3$	**9.** $x - 3 = -2$	**10.** $y - 6 = -1$
11. $a + 5 = -2$	**12.** $b + 3 = -3$	**13.** $3 + m = -6$	**14.** $5 + n = -2$
15. $8 = x + 3$	**16.** $7 = y + 5$	**17.** $3 = w - 6$	**18.** $4 = y - 3$
19. $-7 = -7 + m$	**20.** $-9 = -9 + n$	**21.** $-3 = v + 5$	**22.** $-1 = w + 2$
23. $-5 = 1 + x$	**24.** $-3 = 4 + y$	**25.** $3 = -9 + m$	**26.** $4 = -5 + n$
27. $4 + x - 7 = 3$	**28.** $12 + y - 4 = 8$	**29.** $8t + 6 - 7t = -6$	**30.** $-5z + 5 + 6z = 12$
31. $y + \dfrac{4}{7} = \dfrac{6}{7}$	**32.** $z + \dfrac{3}{5} = \dfrac{4}{5}$	**33.** $x - \dfrac{3}{8} = \dfrac{1}{8}$	**34.** $a - \dfrac{1}{6} = \dfrac{5}{6}$

 For Exercises 35 and 36, use the following inequalities:
   (i) $n < 5$ **(ii)** $n > 5$ **(iii)** $n < -5$ **(iv)** $n > -5$

35. If solving the equation $x + 5 = n$ for x results in a positive solution, which inequality about n *must* be true?

36. If solving the equation $x - 5 = n$ for x results in a negative solution, which inequality about n *must* be true?

OBJECTIVE B **Equations of the form $ax = b$**

37.   Classify each equation as an equation of the form $x + a = b$ or $ax = b$. Explain your reasoning.

   a. $7 + p = -23$ **b.** $-16 = -2s$ **c.** $-\dfrac{7}{8}g = 49$ **d.** $2.8 = q - 9$

38. a. To solve $42 = -6x$, divide each side by _____.

b. To solve $\frac{3}{5}x = \frac{1}{2}$, multiply each side of the equation by _____.

Solve.

39. $3x = 9$

40. $8a = 16$

41. $4c = -12$

42. $5z = -25$

43. $-2r = 16$

44. $-6p = 72$

45. $-4m = -28$

46. $-12x = -36$

47. $-3y = 0$

48. $-7a = 0$

49. $12 = 2c$

50. $28 = 7x$

51. $-72 = 18v$

52. $35 = -5p$

53. $-68 = -17t$

54. $-60 = -15y$

55. $12x = 30$

56. $9v = 15$

57. $-6a = 21$

58. $-8c = 20$

59. $\frac{2}{3}x = 4$

60. $\frac{3}{4}y = 9$

61. $-\frac{4c}{7} = 16$

62. $-\frac{5n}{8} = 20$

63. $8 = \frac{4}{5}y$

64. $10 = -\frac{5}{6}c$

65. $\frac{5y}{6} = \frac{7}{12}$

66. $\frac{-3v}{4} = -\frac{7}{8}$

67. $7y - 9y = 10$

68. $8w - 5w = 9$

69. $m - 4m = 21$

70. $2a - 6a = 10$

 For Exercises 71 and 72, use the following expressions:

(i) $-\frac{3}{2}\left(\frac{3}{4}\right)$ **(ii)** $-\frac{2}{3}\left(\frac{3}{4}\right)$ **(iii)** $-\frac{3}{2}\left(\frac{4}{3}\right)$ **(iv)** $-\frac{2}{3}\left(\frac{4}{3}\right)$

71. Which expression represents the solution of the equation $\frac{3}{4} = -\frac{3}{2}x$?

72. Which expression represents the solution of the equation $-\frac{2}{3}x = \frac{4}{3}$?

CRITICAL THINKING

73. Solve the equation $ax = b$ for x. Is the solution you have written valid for all real numbers a and b?

74. Solve: $\frac{2}{\frac{1}{x}} = 8$

6.2 Equations of the Form $ax + b = c$

OBJECTIVE A **Equations of the form $ax + b = c$**

To solve an equation such as $3w - 5 = 16$, both the Addition and Multiplication Properties of Equations are used.

$$3w - 5 = 16$$

First add the opposite of the constant term -5 to each side of the equation.

$$3w - 5 + 5 = 16 + 5$$
$$3w = 21$$

Divide each side of the equation by the coefficient of w.

$$\frac{3w}{3} = \frac{21}{3}$$

The equation is in the form *variable = constant*.

$$w = 7$$

Check the solution.

Check:
$$\begin{array}{r|r} 3w - 5 & 16 \\ 3(7) - 5 & 16 \\ 21 - 5 & 16 \\ & 16 = 16 \end{array}$$

7 checks as the solution.

The solution is 7.

> **Take Note**
> Note that the Order of Operations Agreement applies to evaluating the expression $3(7) - 5$.

Solve: $8 = 4 - \frac{2}{3}x$

The variable is on the right side of the equation. Work toward the goal of *constant = variable*.

$$8 = 4 - \frac{2}{3}x$$

Subtract 4 from each side of the equation.

$$8 - 4 = 4 - 4 - \frac{2}{3}x$$
$$4 = -\frac{2}{3}x$$

Multiply each side of the equation by $-\frac{3}{2}$.

$$-\frac{3}{2} \cdot 4 = \left(-\frac{3}{2}\right)\left(-\frac{2}{3}x\right)$$

The equation is in the form *constant = variable*.

$$-6 = x$$

You should check the solution.

The solution is -6.

> **Take Note**
> Always check the solution.
> Check: $8 = 4 - \frac{2}{3}x$
> $$\begin{array}{r|r} 8 & 4 - \frac{2}{3}(-6) \\ 8 & 4 + 4 \\ 8 & = 8 \end{array}$$

EXAMPLE 1

Solve: $4m - 7 + m = 8$

Solution

$$4m - 7 + m = 8$$
$$5m - 7 = 8 \quad \blacktriangleright \text{ Combine like terms.}$$
$$5m - 7 + 7 = 8 + 7 \quad \blacktriangleright \text{ Add 7 to each side.}$$
$$5m = 15$$
$$m = 3 \quad \blacktriangleright \text{ Divide each side by 5.}$$

The solution is 3.

YOU TRY IT 1

Solve: $5v + 3 - 9v = 9$

Your Solution

Solution on p. S15

OBJECTIVE B   Applications

Some application problems can be solved by using a known formula. Here is an example.

You can afford a maximum monthly car payment of $250. Find the maximum loan amount you can afford. Use the formula $P = 0.02076L$, where P is the amount of a car payment on a 60-month loan at a 9% interest rate and L is the amount of the loan.

Strategy   To find the maximum loan amount, replace the variable P in the formula by its value (250) and solve for L.

Solution

$$P = 0.02076L$$

$$250 = 0.02076L \qquad \text{Replace } P \text{ by 250.}$$

$$\frac{250}{0.02076} = \frac{0.02076L}{0.02076} \qquad \text{Divide each side of the equation by 0.02076.}$$

$$12{,}042.39 \approx L$$

The maximum loan amount you can afford is $12,042.39.

Calculator Note

To solve for L, use your calculator: $250 \div 0.02076$. Then round the answer to the nearest cent.

EXAMPLE 2

An accountant uses the straight-line depreciation equation $V = C - 4{,}500t$ to determine the value V, after t years, of a computer that originally cost C dollars. Use this formula to determine in how many years a computer that originally cost $39,000 will be worth $25,500.

Strategy

To find the number of years, replace each of the variables by its value and solve for t.
$V = 25{,}500$, $C = 39{,}000$.

Solution

$$V = C - 4{,}500t$$
$$25{,}500 = 39{,}000 - 4{,}500t$$
$$25{,}500 - 39{,}000 = 39{,}000 - 39{,}000 - 4{,}500t$$
$$-13{,}500 = -4{,}500t$$
$$\frac{-13{,}500}{-4{,}500} = \frac{-4{,}500t}{-4{,}500}$$
$$3 = t$$

In 3 years, the computer will have a value of $25,500.

YOU TRY IT 2

The pressure P, in pounds per square inch, at a certain depth in the ocean is approximated by the equation $P = 15 + \frac{1}{2}D$, where D is the depth in feet. Use this formula to find the depth when the pressure is 45 pounds per square inch.

Your Strategy

Your Solution

Solution on p. S15

6.2 Exercises

OBJECTIVE A　**Equations of the form $ax + b = c$**

1. In your own words, state the Addition Property of Equations. Explain when this property is used.

2. In your own words, state the Multiplication Property of Equations. Explain when this property is used.

3. The first step in solving the equation $4 + 7x = 25$ is to subtract _____ from each side of the equation. The second step is to divide each side of the equation by _____.

4. Solve: $\frac{3x}{7} - 6 = -9$

   $\frac{3x}{7} - 6 = -9$

   a. Add _____ to each side.　$\frac{3x}{7} - 6 + \underline{\hspace{1cm}} = -9 + \underline{\hspace{1cm}}$

   b. Simplify.　$\frac{3x}{7} = \underline{\hspace{1cm}}$

   c. Multiply each side by _____.　$(\underline{\hspace{1cm}})\left(\frac{3x}{7}\right) = (\underline{\hspace{1cm}})(-3)$

   d. Simplify.　$x = \underline{\hspace{1cm}}$

Solve.

5. $5y + 1 = 11$

6. $3x + 5 = 26$

7. $2z - 9 = 11$

8. $7p - 2 = 26$

9. $12 = 2 + 5a$

10. $29 = 1 + 7v$

11. $-5y + 8 = 13$

12. $-7p + 6 = -8$

13. $-12a - 1 = 23$

14. $-15y - 7 = 38$

15. $10 - c = 14$

16. $3 - x = 1$

17. $4 - 3x = -5$

18. $8 - 5x = -12$

19. $-33 = 3 - 4z$

20. $-41 = 7 - 8v$

21. $-4t + 16 = 0$

22. $-6p - 72 = 0$

23. $5a + 9 = 12$

24. $7c + 5 = 20$

25. $2t - 5 = 2$

26. $3v - 1 = 4$

27. $8x + 1 = 7$

28. $6y + 5 = 8$

29. $4z - 5 = 1$

30. $8 = 5 + 6p$

31. $25 = 11 + 8v$

32. $-4 = 11 + 6z$

33. $-3 = 7 + 4y$

34. $9w - 4 = 17$

35. $8a - 5 = 31$

36. $5 - 8x = 5$

37. $7 - 12y = 7$

38. $-3 - 8z = 11$

39. $-9 - 12y = 5$

40. $5n - \dfrac{2}{9} = \dfrac{43}{9}$

41. $6z - \dfrac{1}{3} = \dfrac{5}{3}$

42. $7y - \dfrac{2}{5} = \dfrac{12}{5}$

43. $3p - \dfrac{5}{8} = \dfrac{19}{8}$

44. $\dfrac{3}{4}x - 1 = 2$

45. $\dfrac{4}{5}y + 3 = 11$

46. $\dfrac{5t}{6} + 4 = -1$

47. $\dfrac{3v}{7} - 2 = 10$

48. $\dfrac{2a}{5} - 5 = 7$

49. $\dfrac{4z}{9} + 23 = 3$

50. $\dfrac{x}{3} + 6 = 1$

51. $\dfrac{y}{4} + 5 = 2$

52. $17 = 20 + \dfrac{3}{4}x$

53. $\dfrac{2}{5}y - 3 = 1$

54. $\dfrac{7}{3}v + 2 = 8$

55. $5 - \dfrac{7}{8}y = 2$

56. $3 - \dfrac{5}{2}z = 6$

57. $\dfrac{3}{5}y + \dfrac{1}{4} = \dfrac{3}{4}$

58. $\dfrac{5}{6}x - \dfrac{2}{3} = \dfrac{5}{3}$

59. $\dfrac{3}{5} = \dfrac{2}{7}t + \dfrac{1}{5}$

60. $\dfrac{10}{3} = \dfrac{9}{5}w - \dfrac{2}{3}$

61. $\dfrac{z}{3} - \dfrac{1}{2} = \dfrac{1}{4}$

62. $\dfrac{a}{6} + \dfrac{1}{4} = \dfrac{3}{8}$

63. $5.6t - 5.1 = 1.06$

64. $7.2 + 5.2z = 8.76$

65. $6.2 - 3.3t = -12.94$

66. $2.4 - 4.8v = 13.92$

67. $6c - 2 - 3c = 10$ **68.** $12t + 6 + 3t = 16$ **69.** $4y + 5 - 12y = -3$

70. $7m - 15 - 10m = 6$ **71.** $17 = 12p - 5 - 6p$ **72.** $3 = 6n + 23 - 10n$

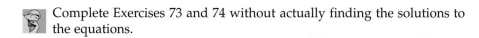 Complete Exercises 73 and 74 without actually finding the solutions to the equations.

73. Is the solution of the equation $15x + 73 = -347$ positive or negative?

74. Is the solution of the equation $17 = 25 - 40a$ positive or negative?

OBJECTIVE B **Applications**

To determine the depreciated value of an X-ray machine, an accountant uses the formula $V = C - 5{,}500t$, where V is the depreciated value of the machine in t years and C is the original cost. Use this formula for Exercises 75 to 78.

75. Find the number of years it will take for an X-ray machine that originally cost $76,000 to reach a depreciated value of $68,300.

a. Replace _____ in the given formula with 76,000 and replace _____ with 68,300. We want to solve for _____.

$V = C - 5{,}500t$

$68{,}300 = 76{,}000 - 5{,}500t$

b. Subtract _____ from each side.

$68{,}300 - \underline{\hspace{1.5cm}} = 76{,}000 - \underline{\hspace{1.5cm}} - 5{,}500t$

c. Simplify.

$\underline{\hspace{1.5cm}} = -5{,}500t$

d. Divide each side by _____.

$\dfrac{-7{,}700}{-5{,}500} = \dfrac{-5{,}500t}{-5{,}500}$

e. Simplify.

$\underline{\hspace{1cm}} = t$

f. In 1.4 _____, the X-ray machine will have a depreciated value of _____.

76. *Accounting*   An X-ray machine originally cost $70,000. In how many years will the depreciated value be $48,000?

77. *Accounting*   An X-ray machine originally cost $63,000. In how many years will the depreciated value be $47,500? Round to the nearest tenth.

78. Which of the following equations is equivalent to the formula $V = C - 5{,}500t$?

(i) $V - C = 5{,}500t$ (ii) $V - C = -5{,}500t$ (iii) $C - V = -5{,}500t$ (iv) $C + V = -5{,}500t$

Consumerism   The formula for the monthly car payment for a 60-month car loan at a 9% interest rate is $P = 0.02076L$, where P is the monthly car payment and L is the amount of the loan. Use this formula for Exercises 79 and 80.

79. If you can afford a maximum monthly car payment of $300, what is the maximum loan amount you can afford? Round to the nearest cent.

80. If you can afford a maximum of $325 for a monthly car payment, what is the largest loan amount you can afford? Round to the nearest cent.

Sports   The world record time for a 1-mile race can be approximated by the formula $t = 17.08 - 0.0067y$, where y is the year of the race between 1900 and 2000, and t is the time, in minutes, of the race. Use this formula for Exercises 81 and 82.

81. Approximate the year in which the first 4-minute mile was run. The actual year was 1954.

82. In 1985, the world record time for a 1-mile race was 3.77 min. For what year does the equation predict this record time?

Physics   Black ice is an ice covering on roads that is especially difficult to see and therefore extremely dangerous for motorists. The distance a car traveling 30 mph will slide after its brakes are applied is related to the outside air temperature by the formula $C = \frac{1}{4}D - 45$, where C is the Celsius temperature and D is the distance in feet the car will slide. Use this formula for Exercises 83 and 84.

83. Determine the distance a car will slide on black ice when the outside air temperature is $-3°C$.

84. Determine the distance a car will slide on black ice when the outside air temperature is $-11°C$.

CRITICAL THINKING

85. Solve: $x \div 28 = 1{,}481$ remainder 25

86. Make up an equation of the form $ax + b = c$ that has -3 as its solution.

6.3 General First-Degree Equations

OBJECTIVE A   **Equations of the form $ax + b = cx + d$**

An equation that contains variable terms on both the left and the right side is solved by repeated application of the Addition Property of Equations. The Multiplication Property of Equations is then used to remove the coefficient of the variable and write the equation in the form *variable = constant*.

Solve: $5z - 4 = 8z + 5$

The goal is to rewrite the equation in the form *variable = constant*.

Use the Addition Property of Equations to remove 8z from the right side by subtracting 8z from each side of the equation. After simplifying, there is only one variable term in the equation.

$$5z - 4 = 8z + 5$$
$$5z - 8z - 4 = 8z - 8z + 5$$
$$-3z - 4 = 5$$

Solve this equation by following the procedure developed in the last section. Using the Addition Property of Equations, add 4 to each side of the equation.

$$-3z - 4 + 4 = 5 + 4$$
$$-3z = 9$$

Divide each side of the equation by -3. After simplifying, the equation is in the form *variable = constant*.

$$\frac{-3z}{-3} = \frac{9}{-3}$$
$$z = -3$$

Check the solution.

$$5z - 4 = 8z + 5$$

$5(-3) - 4$	$8(-3) + 5$
$-15 - 4$	$-24 + 5$
$-19 =$	-19

-3 checks as a solution.

The solution is -3.

Point of Interest

Evariste Galois (1812–1832), even though he was killed in a duel at the age of 21, made significant contributions to solving equations. There is a branch of mathematics called Galois Theory that shows what kinds of equations can and cannot be solved. In fact, Galois, fearing he would be killed the next morning, stayed up all night before the duel, frantically writing notes pertaining to this new branch of mathematics.

EXAMPLE 1

Solve: $2c + 5 = 8c + 2$

Solution

$$2c + 5 = 8c + 2$$
$$2c - 8c + 5 = 8c - 8c + 2 \quad \blacktriangleright \text{Subtract } 8c.$$
$$-6c + 5 = 2$$
$$-6c + 5 - 5 = 2 - 5 \quad \blacktriangleright \text{Subtract } 5.$$
$$-6c = -3$$
$$\frac{-6c}{-6} = \frac{-3}{-6} \quad \blacktriangleright \text{Divide by } -6.$$
$$c = \frac{1}{2}$$

The solution is $\frac{1}{2}$.

YOU TRY IT 1

Solve: $r - 7 = 5 - 3r$

Your Solution

Solution on p. S15

EXAMPLE 2

Solve: $6a + 3 - 9a = 3a + 7$

Solution

$$6a + 3 - 9a = 3a + 7$$
$$-3a + 3 = 3a + 7$$
$$-3a - 3a + 3 = 3a - 3a + 7$$
$$-6a + 3 = 7$$
$$-6a + 3 - 3 = 7 - 3$$
$$-6a = 4$$
$$\frac{-6a}{-6} = \frac{4}{-6}$$

$$a = -\frac{2}{3}$$

The solution is $-\frac{2}{3}$.

▶ Combine like terms.

▶ $\frac{4}{-6} = -\frac{2}{3}$

▶ Remember to check the solution.

YOU TRY IT 2

Solve: $4a - 2 + 5a = 2a - 2 + 3a$

Your Solution

Solution on p. S15

OBJECTIVE B   **Equations with parentheses**

When an equation contains parentheses, one of the steps in solving the equation requires the use of the Distributive Property. The Distributive Property is used to remove parentheses from a variable expression.

Solve: $6 - 2(3x - 1) = 3(3 - x) + 5$

The goal is to rewrite the equation in the form *variable = constant*.

Use the Distributive Property to remove parentheses. Then combine like terms on each side of the equation.

$$6 - 2(3x - 1) = 3(3 - x) + 5$$
$$6 - 6x + 2 = 9 - 3x + 5$$
$$8 - 6x = 14 - 3x$$

Using the Addition Property of Equations, add $3x$ to each side of the equation. After simplifying, there is only one variable term in the equation.

$$8 - 6x + 3x = 14 - 3x + 3x$$
$$8 - 3x = 14$$

Subtract 8 from each side of the equation. After simplifying, there is only one constant term in the equation.

$$8 - 8 - 3x = 14 - 8$$
$$-3x = 6$$

Divide each side of the equation by -3, the coefficient of x. The equation is in the form *variable = constant*.

$$\frac{-3x}{-3} = \frac{6}{-3}$$
$$x = -2$$

-2 checks as a solution.

The solution is -2.

Calculator Note

A calculator can be used to check the solution to the equation at the right. First evaluate the left side of the equation for $x = -2$. Enter

$6\ -\ 2\ (\ 3\ \times\ 2\ +/-\ -\ 1\ )\ =$

The display reads 20. Then evaluate the right side of the equation for $x = -2$. Enter

$3\ (\ 3\ -\ 2\ +/-\ )\ +\ 5$

The display reads 20, the same value as the left side of the equation. The solution checks.

The solution shown above illustrates the steps involved in solving first-degree equations.

Steps in Solving General First-Degree Equations

1. Use the Distributive Property to remove parentheses.

2. Combine like terms on each side of the equation.

3. Rewrite the equation with only one variable term.

4. Rewrite the equation with only one constant term.

5. Rewrite the equation so that the coefficient of the variable is 1.

EXAMPLE 3

Solve: $4 - 3(2t + 1) = 15$

Solution

$$4 - 3(2t + 1) = 15$$
$$4 - 6t - 3 = 15 \quad \blacktriangleright \text{ Distributive Property}$$
$$-6t + 1 = 15$$
$$-6t + 1 - 1 = 15 - 1 \quad \blacktriangleright \text{ Subtract 1.}$$
$$-6t = 14$$
$$\frac{-6t}{-6} = \frac{14}{-6} \quad \blacktriangleright \quad \frac{14}{-6} = -\frac{7}{3}$$
$$t = -\frac{7}{3} \quad \blacktriangleright \text{ The solution checks.}$$

The solution is $-\frac{7}{3}$.

YOU TRY IT 3

Solve: $6 - 5(3y + 2) = 26$

Your Solution

EXAMPLE 4

Solve: $5x - 3(2x - 3) = 4(x - 2)$

Solution

$$5x - 3(2x - 3) = 4(x - 2)$$
$$5x - 6x + 9 = 4x - 8 \quad \blacktriangleright \text{ Distributive Property}$$
$$-x + 9 = 4x - 8 \quad \blacktriangleright \text{ Combine like terms.}$$
$$-x - 4x + 9 = 4x - 4x - 8 \quad \blacktriangleright \text{ Subtract } 4x.$$
$$-5x + 9 = -8$$
$$-5x + 9 - 9 = -8 - 9 \quad \blacktriangleright \text{ Subtract 9.}$$
$$-5x = -17$$
$$\frac{-5x}{-5} = \frac{-17}{-5} \quad \blacktriangleright \text{ Divide by } -5.$$
$$x = \frac{17}{5} \quad \blacktriangleright \text{ The solution checks.}$$

The solution is $\frac{17}{5}$.

YOU TRY IT 4

Solve: $2w - 7(3w + 1) = 5(5 - 3w)$

Your Solution

Solutions on p. S15

Take Note

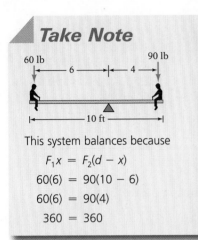

This system balances because

$$F_1 x = F_2(d - x)$$
$$60(6) = 90(10 - 6)$$
$$60(6) = 90(4)$$
$$360 = 360$$

OBJECTIVE C   **Applications**

A lever system is shown at the right. It consists of a lever, or bar; a fulcrum; and two forces, F_1 and F_2. The distance d represents the length of the lever, x represents the distance from F_1 to the fulcrum, and $d - x$ represents the distance from F_2 to the fulcrum.

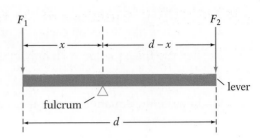

When a lever system balances, $F_1 x = F_2(d - x)$. This is known as Archimedes' Principle of Levers.

EXAMPLE 5

A lever 10 ft long is used to move a 100-pound rock. The fulcrum is placed 2 ft from the rock. What minimum force must be applied to the other end of the lever to move the rock?

Strategy

To find the minimum force needed, replace the variables F_1, d, and x by the given values and solve for F_2.
$F_1 = 100$, $d = 10$, $x = 2$

Solution

$$F_1 x = F_2(d - x)$$
$$100 \cdot 2 = F_2(10 - 2)$$
$$200 = 8F_2$$
$$\frac{200}{8} = \frac{8F_2}{8}$$
$$25 = F_2$$

Check:

$$\begin{array}{c|c} F_1 x = F_2(d - x) \\ \hline 100 \cdot 2 & 25(10 - 2) \\ 200 & 25(8) \\ 200 = 200 \end{array}$$

25 checks as the solution.

The minimum force required is 25 lb.

YOU TRY IT 5

A lever is 25 ft long. A force of 45 lb is applied to one end of the lever, and a force of 80 lb is applied to the other end. What is the location of the fulcrum when the system balances?

Your Strategy

Your Solution

Solution on p. S16

6.3 Exercises

OBJECTIVE A **Equations of the form** *ax* + *b* = *cx* + *d*

1. **a.** To rewrite the equation $8x + 3 = 2x + 21$ with all the variable terms on the left side, subtract _____ from each side.

   b. To rewrite the equation $8x + 3 = 2x + 21$ with all the constant terms on the right side, subtract _____ from each side.

2. Solve: $5n - 7 = 28 - 2n$ $5n - 7 = 28 - 2n$

   a. To rewrite the equation with all the variable terms on the left side, add _____ to each side. $5n + \_\_\_\_ - 7 = 28 - 2n + \_\_\_\_$

   b. Simplify. $\_\_\_\_ - 7 = 28$

   c. Add _____ to each side. $7n - 7 + \_\_\_\_ = 28 + \_\_\_\_$

   d. Simplify. $\_\_\_\_ = \_\_\_\_$

   e. Divide each side by _____. $\dfrac{7n}{} = \dfrac{35}{}$

   f. Simplify. $n = \_\_\_\_$

Solve.

3. $4x + 3 = 2x + 9$

4. $6z + 5 = 3z + 20$

5. $7y - 6 = 3y + 6$

6. $8w - 5 = 5w + 10$

7. $12m + 11 = 5m + 4$

8. $8a + 9 = 2a - 9$

9. $7c - 5 = 2c - 25$

10. $7r - 1 = 5r - 13$

11. $2n - 3 = 5n - 18$

12. $4t - 7 = 10t - 25$

13. $3z + 5 = 19 - 4z$

14. $2m + 3 = 23 - 8m$

15. $5v - 3 = 4 - 2v$

16. $3r - 8 = 2 - 2r$

17. $7 - 4a = 2a$

18. $5 - 3x = 5x$

19. $12 - 5y = 3y - 12$

20. $8 - 3m = 8m - 14$

21. $7r = 8 + 2r$

22. $-2w = 4 - 5w$

23. $5a + 3 = 3a + 10$

24. $7y + 3 = 5y + 12$

25. $x - 7 = 5x - 21$

26. $3y - 4 = 9y - 24$

27. $5n - 1 + 2n = 4n + 8$

28. $3y + 1 + y = 2y + 11$

29. $3z - 2 - 7z = 4z + 6$

30. $2a + 3 - 9a = 3a + 33$

31. $4t - 8 + 12t = 3 - 4t - 11$

32. $6x - 5 + 9x = 7 - 4x - 12$

33. Suppose a is a positive number. Will solving the equation $2x - a = 4x$ for x result in a positive solution or a negative solution?

34. Suppose a is a negative number. Will solving the equation $-5x = a - 3x$ for x result in a positive solution or a negative solution?

OBJECTIVE B **Equations with parentheses**

35. Use the Distributive Property to remove the parentheses from the equation $3(x - 6) + 13 = 31$: _____ $+ 13 = 31$.

36. Use the Distributive Property to remove the parentheses from the equation $-2(4x + 1) - 5 = 11$: _____ $- 5 = 11$.

37. Which of the following equations is equivalent to $7 - 4(3x - 2) = 9$?

(i) $3(3x - 2) = 9$ **(ii)** $7 - 12x - 2 = 9$ **(iii)** $7 - 12x - 8 = 9$ **(iv)** $7 - 12x + 8 = 9$

38. Which of the following equations are equivalent to $-3(6x + 1) = 18$?

(i) $-18x + 1 = 18$ **(ii)** $-18x + 3 = 18$ **(iii)** $6x + 1 = -6$ **(iv)** $-18x - 3 = 18$

39. $3(4y + 5) = 25$

40. $5(3z - 2) = 8$

41. $-2(4x + 1) = 22$

42. $-3(2x - 5) = 30$

43. $5(2k + 1) - 7 = 28$

44. $7(3t - 4) + 8 = -6$

45. $3(3v - 4) + 2v = 10$

46. $4(3x + 1) - 5x = 25$

47. $3y + 2(y + 1) = 12$

48. $7x + 3(x + 2) = 33$

49. $7v - 3(v - 4) = 20$

50. $15m - 4(2m - 5) = 34$

51. $6 + 3(3x - 3) = 24$

52. $9 + 2(4p - 3) = 24$

53. $9 - 3(4a - 2) = 9$

54. $17 - 8(x - 3) = 1$

55. $3(2z - 5) = 4z + 1$

56. $4(3z - 1) = 5z + 17$

57. $2 - 3(5x + 2) = 2(3 - 5x)$

58. $5 - 2(3y + 1) = 3(2 - 3y)$

59. $4r + 11 = 5 - 2(3r + 3)$

60. $3v + 6 = 9 - 4(2v - 2)$

OBJECTIVE C   Applications

For Exercises 61 to 66, use the lever system equation $F_1 x = F_2(d - x)$.

61. Two children sit on a seesaw that is 12 ft long. One child weighs 84 lb and the other child weighs 60 lb. To determine how far from the 60-pound child the fulcrum should be placed so that the seesaw balances, use the lever system equation and replace the variables with the following values: $F_1 = $ _____, $F_2 = $ _____, and $d = $ _____. Then solve for _____.

62. When two people sit on the ends of a seesaw that is 10 ft long and has its fulcrum in the middle, the seesaw is not balanced. When one person moves 1 ft toward the center of the seesaw, the seesaw is balanced. Which person is heavier, the person who moved in toward the fulcrum or the other person?

63. *Physics*   Two people are sitting 15 ft apart on a seesaw. One person weighs 180 lb; the second person weighs 120 lb. How far from the 180-pound person should the fulcrum be placed so that the seesaw balances?

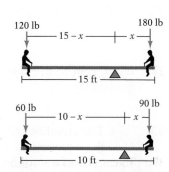

64. *Physics*   Two children are sitting on a seesaw that is 10 ft long. One child weighs 60 lb; the second child weighs 90 lb. How far from the 90-pound child should the fulcrum be placed so that the seesaw balances?

65. *Physics* A metal bar 8 ft long is used to move a 150-pound rock. The fulcrum is placed 1.5 ft from the rock. What minimum force must be applied to the other end of the bar to move the rock? Round to the nearest tenth.

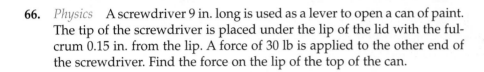

66. *Physics* A screwdriver 9 in. long is used as a lever to open a can of paint. The tip of the screwdriver is placed under the lip of the lid with the fulcrum 0.15 in. from the lip. A force of 30 lb is applied to the other end of the screwdriver. Find the force on the lip of the top of the can.

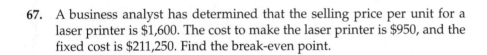

Business To determine the break-even point, or the number of units that must be sold so that no profit or loss occurs, an economist uses the formula $Px = Cx + F$, where P is the selling price per unit, x is the number of units that must be sold to break even, C is the cost to make each unit, and F is the fixed cost. Use this formula for Exercises 67 to 70.

67. A business analyst has determined that the selling price per unit for a laser printer is $1,600. The cost to make the laser printer is $950, and the fixed cost is $211,250. Find the break-even point.

68. An economist has determined that the selling price per unit for a gas barbecue is $325. The cost to make one gas barbecue is $175, and the fixed cost is $39,000. Find the break-even point.

69. A manufacturer of thermostats determines that the cost per unit for a programmable thermostat is $38 and that the fixed cost is $24,400. The selling price for the thermostat is $99. Find the break-even point.

70. A manufacturing engineer determines that the cost per unit for a computer mouse is $12 and that the fixed cost is $19,240. The selling price for the computer mouse is $49. Find the break-even point.

CRITICAL THINKING

71. If $5a - 4 = 3a + 2$, what is the value of $4a^3$?

72. If $3 + 2(4a - 3) = 5$ and $4 - 3(2 - 3b) = 11$, which is larger, a or b?

73. The equation $x = x + 1$ has no solution, whereas the solution of the equation $2x + 3 = 3$ is zero. Is there a difference between no solution and a solution of zero? Explain your answer.

6.4 Translating Sentences into Equations

OBJECTIVE A   **Translate a sentence into an equation and solve**

An equation states that two mathematical expressions are equal. Therefore, to translate a sentence into an equation, you must recognize the words or phrases that mean "equals." Some of these words and phrases are listed below.

equals	is	represents
amounts to	totals	is the same as

Translate "five less than four times a number is four more than the number" into an equation and solve.

Assign a variable to the unknown number.   the unknown number: n

Find two verbal expressions for the same value.

Five less than four times a number	is	four more than the number

Write an equation. $4n - 5 = n + 4$

Solve the equation.
Subtract n from each side. $3n - 5 = 4$

Add 5 to each side. $3n = 9$

Divide each side by 3. $n = 3$

The solution checks.   The number is 3.

Take Note

You can check the solution to a translation problem.

Check:

5 less than 4 times 3	4 more than 3
$4 \cdot 3 - 5$	$3 + 4$
$12 - 5$	7
$7 = 7$	

EXAMPLE 1

Translate "eight less than three times a number equals five times the number" into an equation and solve.

Solution

the unknown number: x

Eight less than three times a number	equals	five times the number

$$3x - 8 = 5x$$
$$3x - 3x - 8 = 5x - 3x$$
$$-8 = 2x$$
$$\frac{-8}{2} = \frac{2x}{2}$$
$$-4 = x$$

-4 checks as the solution.

The number is -4.

YOU TRY IT 1

Translate "six more than one-half a number is the total of the number and nine" into an equation and solve.

Your Solution

Solution on p. S16

EXAMPLE 2

Translate "four more than five times a number is six less than three times the number" into an equation and solve.

Solution

the unknown number: m

Four more than five times a number	is	six less than three times the number

$$5m + 4 = 3m - 6$$
$$5m - 3m + 4 = 3m - 3m - 6$$
$$2m + 4 = -6$$
$$2m + 4 - 4 = -6 - 4$$
$$2m = -10$$
$$\frac{2m}{2} = \frac{-10}{2}$$
$$m = -5$$

-5 checks as the solution.

The number is -5.

YOU TRY IT 2

Translate "seven less than a number is equal to five more than three times the number" into an equation and solve.

Your Solution

EXAMPLE 3

The sum of two numbers is nine. Eight times the smaller number is five less than three times the larger number. Find the numbers.

Solution

the smaller number: p
the larger number: $9 - p$

Eight times the smaller number	is	five less than three times the larger number

$$8p = 3(9 - p) - 5$$
$$8p = 27 - 3p - 5$$
$$8p = 22 - 3p$$
$$8p + 3p = 22 - 3p + 3p$$
$$11p = 22$$
$$\frac{11p}{11} = \frac{22}{11}$$
$$p = 2$$

$9 - p = 9 - 2 = 7$

These numbers check as solutions.

The smaller number is 2.
The larger number is 7.

YOU TRY IT 3

The sum of two numbers is fourteen. One more than three times the smaller number equals the sum of the larger number and three. Find the two numbers.

Your Solution

Solutions on p. S16

OBJECTIVE B **Applications**

EXAMPLE 4

In the 2006 national election, $40 million was spent for online political ads. This was two-fifths of the amount spent for newspaper political ads. (*Source:* PQ Media) How much was spent for political ads in newspapers?

Strategy

To find the amount spent for political ads in newspapers, write and solve an equation using n to represent the amount spent for political ads in newspapers.

Solution

| $40 million | is | $\frac{2}{5}$ of the amount spent for newspaper ads |

$$40 = \frac{2}{5}n$$
$$\frac{5}{2} \cdot 40 = \frac{5}{2} \cdot \frac{2}{5}n$$
$$100 = n$$

The amount spent for political ads in newspapers was $100 million.

YOU TRY IT 4

In 2006, the Internal Revenue Service's tax code was 66,498 pages long. This was 20,836 pages longer than the 2001 tax code. (*Source:* Office of Management and Budget, CCH) Find the number of pages in the Internal Revenue Service's tax code in 2001.

Your Strategy

Your Solution

EXAMPLE 5

A wallpaper hanger charges a fee of $50 plus $28 for each roll of wallpaper used in a room. If the total charge for hanging wallpaper is $218, how many rolls of wallpaper were used?

Strategy

To find the number of rolls of wallpaper used, write and solve an equation using n to represent the number of rolls of wallpaper.

Solution

| $50 plus $28 for each roll of wallpaper | is | $218 |

$$50 + 28n = 218$$
$$50 - 50 + 28n = 218 - 50$$
$$28n = 168$$
$$\frac{28n}{28} = \frac{168}{28}$$
$$n = 6$$

The wallpaper hanger used 6 rolls.

YOU TRY IT 5

The fee charged by a ticketing agency for a concert is $9.50 plus $57.50 for each ticket purchased. If your total charge for tickets is $527, how many tickets are you purchasing?

Your Strategy

Your Solution

Solutions on p. S16

EXAMPLE 6

A bank charges a checking account customer a fee of $12 per month plus $1.50 for each use of an ATM. For the month of July, the customer was charged $24. How many times did this customer use an ATM during the month of July?

Strategy

To find the number of times an ATM was used, write and solve an equation using n for the number of times an ATM was used.

Solution

| $12.00 plus $1.50 per ATM use | is | $24 |

$$12 + 1.50n = 24$$
$$12 - 12 + 1.50n = 24 - 12$$
$$1.50n = 12$$
$$\frac{1.50n}{1.50} = \frac{12}{1.50}$$
$$n = 8$$

The customer used the ATM 8 times in July.

EXAMPLE 7

A guitar wire 22 in. long is cut into two pieces. The length of the longer piece is 4 in. more than twice the length of the shorter piece. Find the length of the shorter piece.

Strategy

To find the length, write and solve an equation using x to represent the length of the shorter piece and $22 - x$ to represent the length of the longer piece.

Solution

| The longer piece | is | 4 in. more than twice the shorter piece |

$$22 - x = 2x + 4$$
$$22 - x - 2x = 2x - 2x + 4$$
$$22 - 3x = 4$$
$$22 - 22 - 3x = 4 - 22$$
$$-3x = -18$$
$$\frac{-3x}{-3} = \frac{-18}{-3}$$
$$x = 6$$

The shorter piece is 6 in. long.

YOU TRY IT 6

An auction website charges a customer a fee of $10 to place an ad plus $5.50 for each day the ad is posted on the website. A customer is charged $43 for advertising a used trumpet on this website. For how many days did the customer have the ad for the trumpet posted on the website?

Your Strategy

Your Solution

YOU TRY IT 7

A board 18 ft long is cut into two pieces. One foot more than twice the length of the shorter piece is 2 ft less than the length of the longer piece. Find the length of each piece.

Your Strategy

Your Solution

Solutions on pp. S16–S17

6.4 Exercises

OBJECTIVE A Translate a sentence into an equation and solve

For Exercises 1 and 2, translate the sentence into an equation. Use x to represent the unknown number.

1. Three times a number   is   negative thirty.
   $\downarrow$ $\downarrow$ $\downarrow$
   _____   _____   _____

2. Ten plus a number   is equal to   negative eight.
   $\downarrow$ $\downarrow$ $\downarrow$
   _____   _____   _____

3. The sentence "The sum of a number and twelve equals three" can be translated as _____ + _____ = _____.

4. The sentence "The difference between nine and a number is negative six" can be translated as _____ − _____ = _____.

Translate into an equation and solve.

5. Seven plus a number is forty. Find the number.

6. Six less than a number is five. Find the number.

7. Four more than a number is negative two. Find the number.

8. The product of a number and eight is equal to negative forty. Find the number.

9. The sum of a number and twelve is twenty. Find the number.

10. The difference between nine and a number is seven. Find the number.

11. Three-fifths of a number is negative thirty. Find the number.

12. The quotient of a number and six is twelve. Find the number.

13. Four more than three times a number is thirteen. Find the number.

14. The sum of twice a number and five is fifteen. Find the number.

15. The difference between nine times a number and six is twelve. Find the number.

16. Six less than four times a number is twenty-two. Find the number.

17. Seventeen less than the product of five and a number is three. Find the number.

18. Eight less than the product of eleven and a number is negative nineteen. Find the number.

19. Forty equals nine less than the product of seven and a number. Find the number.

20. Twenty-three equals the difference between eight and the product of five and a number. Find the number.

21. Twice the difference between a number and twenty-five is three times the number. Find the number.

22. Four times a number is three times the difference between thirty-five and the number. Find the number.

23. The sum of two numbers is twenty. Three times the smaller is equal to two times the larger. Find the two numbers.

24. The sum of two numbers is fifteen. One less than three times the smaller is equal to the larger. Find the two numbers.

25. The sum of two numbers is twenty-one. Twice the smaller number is three more than the larger number. Find the two numbers.

26. The sum of two numbers is thirty. Three times the smaller number is twice the larger number. Find the two numbers.

27. The sum of two numbers is eighteen. Three times one number is two less than the other number. Which of the following equations do *not* represent this relationship?

(i) $3(18 - n) = n - 2$ **(ii)** $3n = 2 - (18 - n)$

(iii) $3n = (18 - n) - 2$ **(iv)** $3(n - 18) = n - 2$

OBJECTIVE B **Applications**

28. Running at 5 mph burns 472 calories per hour. This is 118 calories less than the number of calories burned per hour when running at 6 mph. Find the number of calories burned per hour when running at 6 mph.

  a. Let n represent the number of calories burned per hour running at a rate of _____ mph, and let $n - 118$ represent the number of calories burned per hour running at a rate of _____ mph.

  b. The equation that can be used to find n is _____ $= n - 118$.

Write an equation and solve.

29. *Airports* In a recent year, the number of passengers traveling through Atlanta's Hartsfield-Jackson International Airport was 42 million. This represents twice the number of passengers traveling through Las Vegas McCarran International Airport in the same year. (*Source:* Bureau of Transportation Statistics) Find the number of passengers traveling through Las Vegas McCarran International Airport that year.

Hartsfield-Jackson International Airport

30. *Health* In 1985, 595.4 billion cigarettes were smoked. This is 202.3 billion more cigarettes than were smoked in 2005. (*Source:* Orzechowski & Walker) Find the number of cigarettes smoked in 2005.

31. *Advertising*   In 2005, advertisers spent $6.3 billion on outdoor advertising. This is $3.7 billion more than advertisers spent on outdoor advertising in 1990. (*Source:* Outdoor Advertising Association of America) Find the amount that advertisers spent on outdoor advertising in 1990.

32. *Pets*   In 2001, pet owners in the United States spent $6.6 million on dog food, excluding treats. This is $1.3 million less than they spent on dog food in 2006. (*Source:* Euromonitor International) Find the amount pet owners spent on dog food in 2006.

33. *The Military*   According to the Census Bureau, there were 3,400,000 U.S. military personnel on active duty in 1967. This is 20 times the number of U.S. military personnel on active duty in 1915. In 2006, there were 1,400,000 U.S. military personnel on active duty. (*Source:* Census Bureau) Find the number of U.S. military personnel on active duty in 1915.

34. *Taxes*   According to the Census Bureau, per capita state taxes collected in a recent year averaged $2190. This represents two and one-half times the average per capita income tax collected that year. (*Source:* Orzechowski & Walker) Find the average per capita income tax collected that year.

35. *Recycling*   According to the Environmental Protection Agency, 58 million tons of waste was collected for recycling in 2005. This is 2 tons less than twice the amount of waste collected for recycling in 1990. Find the amount of waste collected for recycling in 1990.

36. *Banking*   According to the American Banking Association, the number of ATMs in the United States in 2005 was 396,000. This is 27 more than three times the number of ATMs in the United States in 1995. Find the number of ATMs in the United States in 1995.

37. *Consumerism*   A technical information hotline charges a customer $18 plus $1.50 per minute to answer questions about software. For how many minutes did a customer who received a bill for $34.50 use this service?

38. *Consumerism*   The total cost to paint the inside of a house was $2,692. This cost included $250 for materials and $66 per hour for labor. How many hours of labor were required?

39. *Carpentry*   A 12-foot board is cut into two pieces. Twice the length of the shorter piece is 3 feet less than the length of the longer piece. Find the length of each piece.

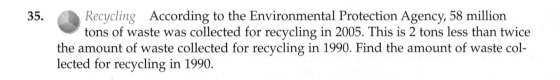

40. *Sports*   A 14-yard fishing line is cut into two pieces. Three times the length of the longer piece is four times the length of the shorter piece. Find the length of each piece.

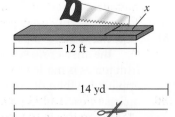

41. *Financial Aid*   Seven thousand dollars is divided into two scholarships. Twice the amount of the smaller scholarship is $1,000 less than the amount of the larger scholarship. What is the amount of the larger scholarship?

42. *Investments*   An investment of $10,000 is divided into two accounts, one for stocks and one for mutual funds. The value of the stock account is $2,000 less than twice the value of the mutual fund account. Find the amount in each account.

43. *Food Mixtures*   A 10-pound blend of coffee contains Colombian coffee, French Roast, and Java. There is 1 lb more of French Roast than of Colombian and 2 lb more of Java than of French Roast. How many pounds of each are in the mixture?

44. *Agriculture*   A 60-pound soil supplement contains nitrogen, iron, and potassium. There is twice as much potassium as iron and three times as much nitrogen as iron. How many pounds of each element are in the soil supplement?

CRITICAL THINKING

An equation that is never true is called a **contradiction.** For example, the equation $x = x + 1$ is a contradiction. There is no value of x that will make the equation true. An equation that is true for all real numbers is called an **identity.** The equation $x + x = 2x$ is an identity. This equation is true for any real number. A **conditional equation** is one that is true for some real numbers and false for some real numbers. The equation $2x = 4$ is a conditional equation. This equation is true when x is 2 and false for any other real number. Determine whether each equation below is a contradiction, an identity, or a conditional equation. If it is a conditional equation, find the solution.

45. $6x + 2 = 5 + 3(2x - 1)$

46. $3 - 2(4x + 1) = 5 + 8(1 - x)$

47. $6 + 4(2y + 1) = 5 - 8y$

48. $3t - 5(t + 1) = 2(2 - t) - 9$

49. $3v - 2 = 5v - 2(2 + v)$

50. $9z = 15z$

51. It is always important to check the answer to an application problem to be sure the answer makes sense. Consider the following problem. A 4-quart mixture of fruit juices is made from apple juice and cranberry juice. There are 6 more quarts of apple juice than of cranberry juice. Write and solve an equation for the number of quarts of each juice used. Does the answer to this question make sense? Explain.

6.5 The Rectangular Coordinate System

OBJECTIVE A   The rectangular coordinate system

Before the fifteenth century, geometry and algebra were considered separate branches of mathematics. That all changed when René Descartes, a French mathematician who lived from 1596 to 1650, founded analytic geometry. In this geometry, a **coordinate system** is used to study relationships between variables.

A **rectangular coordinate system** is formed by two number lines, one horizontal and one vertical, that intersect at the zero point of each line. The point of intersection is called the **origin.** The two lines are called **coordinate axes,** or simply **axes.**

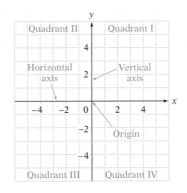

The axes determine a **plane,** which can be thought of as a large, flat sheet of paper. The two axes divide the plane into four regions called **quadrants,** which are numbered counterclockwise from I to IV.

Point of Interest

Although Descartes is given credit for introducing analytic geometry, others, notably Pierre Fermat, were working on the same concept. Nowhere in Descartes's work is there a coordinate system as we draw it with two axes. Descartes did not use the word *coordinate* in his work. This word was introduced by Gottfried Leibnitz, who also first used the words *abscissa* and *ordinate.*

Each point in the plane can be identified by a pair of numbers called an **ordered pair.** The first number of the pair measures a horizontal distance and is called the **abscissa,** or **x-coordinate.** The second number of the pair measures a vertical distance and is called the **ordinate,** or **y-coordinate.** The ordered pair (x, y) associated with a point is also called the **coordinates** of the point.

```
Horizontal distance ─────┐   ┌───── Vertical distance
                         ↓   ↓
Ordered pair ────────→ (2, 3)
                         ↑   ↑
x-coordinate ─────────────┘   └───── y-coordinate
```

To **graph,** or **plot,** a point in the plane, place a dot at the location given by the ordered pair. The **graph of an ordered pair** is the dot drawn at the coordinates of the point in the plane. The points whose coordinates are $(3, 4)$ and $(-2.5, -3)$ are graphed in the figures below.

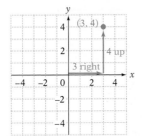

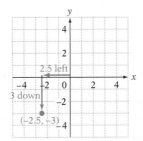

The points whose coordinates are (3, −1) and (−1, 3) are shown graphed at the right. Note that the graphed points are in different locations. The order of the coordinates of an ordered pair is important.

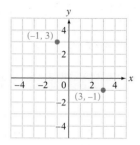

Each point in the plane is associated with an ordered pair, and each ordered pair is associated with a point in the plane. Although only the labels for integers are given on a coordinate grid, the graph of any ordered pair can be approximated. For example, the points whose coordinates are (−2.3, 4.1) and ($\sqrt{2}$, −$\sqrt{3}$) are shown in the graph at the right.

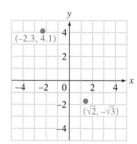

EXAMPLE 1 Graph the ordered pairs (−2, −3), (3, −2), (1, 3), and (4, 1).

Solution

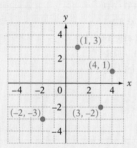

YOU TRY IT 1 Graph the ordered pairs (−1, 3), (1, 4), (−4, 0), and (−2, −1).

Your Solution

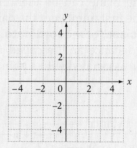

EXAMPLE 2 Find the coordinates of each point.

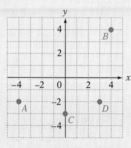

Solution $A(-4, -2)$ $C(0, -3)$
$B(4, 4)$ $D(3, -2)$

YOU TRY IT 2 Find the coordinates of each point.

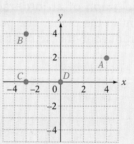

Your Solution

Solutions on p. S17

OBJECTIVE B   **Scatter diagrams**

Discovering a relationship between two variables is an important task in the study of mathematics. These relationships occur in many forms and in a wide variety of applications. Here are some examples:

A botanist wants to know the relationship between the number of bushels of wheat yielded per acre and the amount of watering per acre.

An environmental scientist wants to know the relationship between the incidence of skin cancer and the amount of ozone in the atmosphere.

A business analyst wants to know the relationship between the price of a product and the number of products that are sold at that price.

A researcher may investigate the relationship between two variables by means of *regression analysis*, which is a branch of statistics. The study of the relationship between two variables may begin with a **scatter diagram,** which is a graph of the ordered pairs of the known data.

The following table gives data collected by a university registrar comparing the grade point averages of graduating high school seniors and their scores on a national test.

GPA, x	3.50	3.50	3.25	3.00	3.00	2.75	2.50	2.50	2.00	2.00	1.50
Test, y	1,500	1,100	1,200	1,200	1,000	1,000	1,000	900	800	900	700

The scatter diagram for these data is shown below.

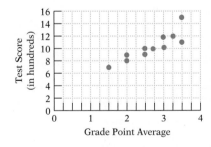

Each ordered pair represents the GPA and test score for a student. For example, the ordered pair (2.75, 1,000) indicates a student with a GPA of 2.75 who had a test score of 1,000.

The dot on the scatter diagram at (3, 12) represents the student with a GPA of 3.00 and a test score of 1,200.

EXAMPLE 3

A nutritionist collected data on the number of grams of sugar and grams of fiber in 1-ounce servings of six brands of cereal. The data are recorded in the following table. Graph the scatter diagram for the data.

Sugar, x	6	8	6	5	7	5
Fiber, y	2	1	4	4	2	3

Solution

Graph the ordered pairs on the rectangular coordinate system. The horizontal axis represents the grams of sugar. The vertical axis represents the grams of fiber.

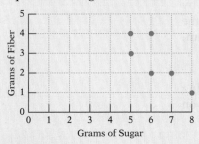

EXAMPLE 4

To test a heart medicine, a doctor measured the heart rates, in beats per minute, of five patients before and after they took the medication. The results are recorded in the scatter diagram. One patient's heart rate before taking the medication was 75 beats per minute. What was this patient's heart rate after taking the medication?

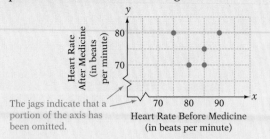

The jags indicate that a portion of the axis has been omitted.

Solution

Locate 75 beats per minute on the x-axis. Follow the vertical line from 75 to a point plotted in the diagram. Follow a horizontal line from that point to the y-axis. Read the number where that line intersects the y-axis.

The ordered pair is (75, 80), which indicates that the patient's heart rate before taking the medication was 75 and the heart rate after taking the medication was 80.

YOU TRY IT 3

A sports statistician collected data on the total number of yards gained by a college football team and the number of points scored by the team. The data are recorded in the following table. Graph the scatter diagram for the data.

Yards, x	300	400	350	400	300	450
Points, y	18	24	14	21	21	30

Your Solution

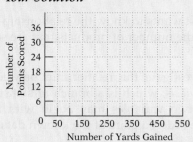

YOU TRY IT 4

A study by the FAA showed that narrow, over-the-wing emergency exit rows slow passenger evacuation. The scatter diagram below shows the space between seats, in inches, and the evacuation time, in seconds, for a group of 35 passengers. What was the evacuation time when the space between seats was 20 in.?

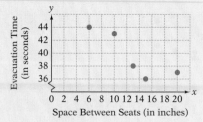

Your Solution

Solutions on p. S17

6.5 Exercises

OBJECTIVE A **The rectangular coordinate system**

For Exercises 1 and 2, fill in each blank with *left, right, up,* or *down.*

1. To graph the point $(5, -1)$, start at the origin and move 5 units _____ and 1 unit _____ .

2. To graph the point $(-6, 7)$, start at the origin and move 6 units _____ and 7 units _____ .

3. [pencil icon] Explain how to locate the point $(-4, 3)$ in a rectangular coordinate system.

4. [pencil icon] Explain how to locate the point $(2, -5)$ in a rectangular coordinate system.

In which quadrant does the given point lie?

5. $(5, 4)$ 6. $(3, -2)$ 7. $(-8, 1)$ 8. $(-7, -6)$

On which axis does the given point lie?

9. $(0, -6)$ 10. $(8, 0)$

11. Describe the signs of the coordinates of a point plotted in **a.** Quadrant I and **b.** Quadrant III.

12. Describe the signs of the coordinates of a point plotted in **a.** Quadrant II and **b.** Quadrant IV.

For Exercises 13 to 21, graph the ordered pairs.

13. $(5, 2), (3, -5), (-2, 1),$ and $(0, 3)$

14. $(-3, -3), (5, -1), (-2, 4),$ and $(0, -5)$

15. $(-2, -3), (1, -1), (-4, 5),$ and $(-1, 0)$

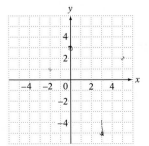

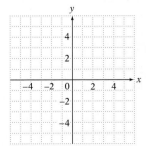

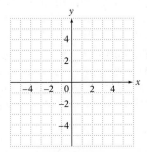

16. $(2, 5)$, $(0, 0)$, $(3, -4)$, and $(-1, 4)$

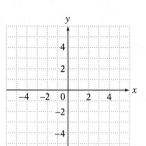

17. $(2, -5)$, $(-4, -1)$, $(-3, 1)$, and $(0, 2)$

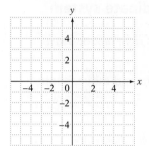

18. $(3, 1)$, $(4, -3)$, $(-2, 5)$, and $(-4, -2)$

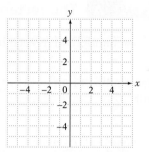

19. $(-1, -5)$, $(-2, 3)$, $(4, -1)$, and $(-3, 0)$

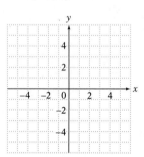

20. $(4, -5)$, $(-3, 2)$, $(5, 0)$, and $(-5, -1)$

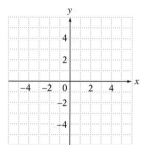

21. $(4, 1)$, $(-3, 5)$, $(4, 0)$, and $(-1, -2)$

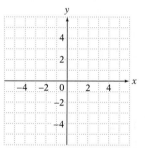

For Exercises 22 to 30, find the coordinates of each point.

22.

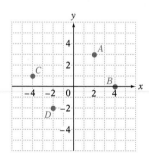

23.

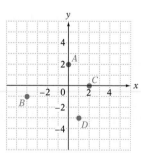

24.

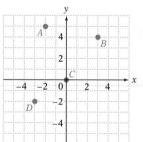

25.

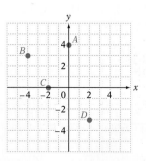

26.

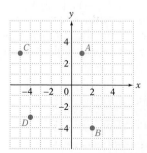

27.

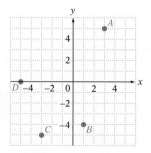

Find the ordered-pair solution of $y = \frac{2}{3}x - 3$ that corresponds to $x = 6$.

$$y = \frac{2}{3}x - 3$$

Replace x by 6. $\quad y = \frac{2}{3}(6) - 3$

Solve for y. $\quad y = 4 - 3$

$\qquad\qquad\qquad y = 1$

The ordered-pair solution is (6, 1).

EXAMPLE 1

Is $(-3, 2)$ a solution of the equation $y = -2x - 5$?

Solution

$$y = -2x - 5$$

2	$-2(-3) - 5$
2	$6 - 5$

$2 \neq 1$

▸ Replace x by -3 and y by 2.

No, $(-3, 2)$ is not a solution of the equation $y = -2x - 5$.

YOU TRY IT 1

Is $(-2, 4)$ a solution of the equation $y = -\frac{1}{2}x + 3$?

Your Solution

EXAMPLE 2

Find the ordered-pair solution of the equation $y = -3x + 1$ corresponding to $x = 2$.

Solution

$$y = -3x + 1$$
$$y = -3(2) + 1 \quad \text{▸ Replace } x \text{ by 2.}$$
$$y = -6 + 1$$
$$y = -5$$

The ordered-pair solution is $(2, -5)$.

YOU TRY IT 2

Find the ordered-pair solution of the equation $y = 2x - 3$ corresponding to $x = 0$.

Your Solution

Solutions on p. S17

OBJECTIVE B **Equations of the form $y = mx + b$**

The **graph of an equation in two variables** is a graph of the ordered-pair solutions of the equation.

Consider $y = 2x + 1$. Choosing $x = -2, -1, 0, 1,$ and 2 and determining the corresponding values of y produces some of the ordered-pair solutions of the equation. These are recorded in the table at the right. The graph of the ordered pairs is shown in Figure 6.1.

x	$2x + 1$	y	(x, y)
-2	$2(-2) + 1$	-3	$(-2, -3)$
-1	$2(-1) + 1$	-1	$(-1, -1)$
0	$2(0) + 1$	1	$(0, 1)$
1	$2(1) + 1$	3	$(1, 3)$
2	$2(2) + 1$	5	$(2, 5)$

Choosing values of x that are not integers produces more ordered pairs to graph, such as $\left(-\frac{5}{2}, -4\right)$ and $\left(\frac{3}{2}, 4\right)$, as shown in Figure 6.2. Choosing still other values of x would result in more and more ordered pairs being graphed. The result would be so many dots that the graph would appear as the straight line shown in Figure 6.3, which is the graph of $y = 2x + 1$.

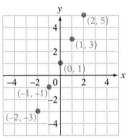

Figure 6.1

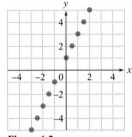

Figure 6.2

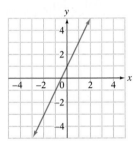

Figure 6.3

Equations in two variables have characteristic graphs. The equation $y = 2x + 1$ is an example of a *linear equation* because its graph is a straight line.

> ## Linear Equation in Two Variables
>
> Any equation of the form $y = mx + b$, where m is the coefficient of x and b is a constant, is a **linear equation in two variables.** The graph of a linear equation in two variables is a straight line.

To graph a linear equation, choose some values of x and then find the corresponding values of y. Because a straight line is determined by two points, it is sufficient to find only two ordered-pair solutions. However, it is recommended that at least three ordered-pair solutions be used to ensure accuracy.

Take Note

If the three points you graph do not lie on a straight line, you have made an arithmetic error in calculating a point or you have plotted a point incorrectly.

Graph $y = -\frac{3}{2}x + 2$.

This is a linear equation with $m = -\frac{3}{2}$ and $b = 2$. Find at least three solutions. Because m is a fraction, choose values of x that will simplify the calculations. We have chosen -2, 0, and 4 for x. (Any values of x could have been selected.)

x	$y = -\dfrac{3}{2}x + 2$	y	(x, y)
-2	$-\dfrac{3}{2}(-2) + 2$	5	$(-2, 5)$
0	$-\dfrac{3}{2}(0) + 2$	2	$(0, 2)$
4	$-\dfrac{3}{2}(4) + 2$	-4	$(4, -4)$

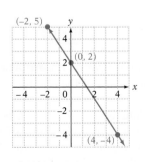

Graph the ordered pairs and then draw a line through the points.

Remember that a graph is a drawing of the ordered-pair solutions of the equation. Therefore, every point on the graph is a solution of the equation and every solution of the equation is a point on the graph.

The graph at the right is the graph of $y = x + 2$. Note that $(-4, -2)$ and $(1, 3)$ are points on the graph and that these points are solutions of $y = x + 2$. The point whose coordinates are $(4, 1)$ is not a point on the graph and is not a solution of the equation.

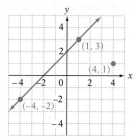

EXAMPLE 3   Graph $y = 3x - 2$.

Solution

x	y
0	-2
-1	-5
2	4

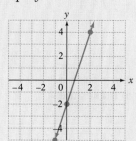

YOU TRY IT 3   Graph $y = 3x + 1$.

Your Solution

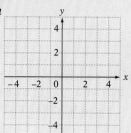

EXAMPLE 4   For the graph shown below, what is the y value when $x = 1$?

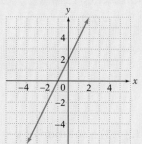

Solution   Locate 1 on the x-axis. Follow the vertical line from 1 to a point on the graph. Follow a horizontal line from that point to the y-axis. Read the number where that line intersects the y-axis.

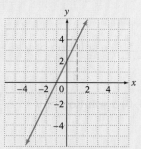

The y value is 4 when $x = 1$.

YOU TRY IT 4   For the graph shown below, what is the x value when $y = 5$?

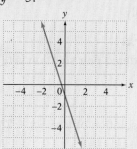

Your Solution

Solutions on p. S17

6.6 Exercises

OBJECTIVE A Solutions of linear equations in two variables

For Exercises 1 to 6, is the equation a linear equation in two variables?

1. $y = -x^2 - 3x + 4$

2. $y = \dfrac{1}{z + 5}$

3. $y = 6x - 3$

4. $y = -7x - 1$

5. $x = 2x - 8$

6. $y = -\dfrac{2}{3}y + 1$

7. To determine if the ordered pair $(1, 8)$ is a solution of the equation $y = 5x + 3$, replace _____ by 1 and _____ by 8 to see if the ordered pair $(1, 8)$ makes the equation $y = 5x + 3$ a true statement.

8. Find the ordered-pair solution of $y = -2x - 7$ corresponding to $x = -6$.
   a. Replace _____ by _____. $y = -2(-6) - 7$
   b. Multiply. $y = $ _____ $- 7$
   c. Subtract. $y = $ _____
   d. The ordered pair solution is (_____ , _____).

9. Is $(3, 4)$ a solution of $y = -x + 7$?

10. Is $(2, -3)$ a solution of $y = x + 5$?

11. Is $(-1, 2)$ a solution of $y = \dfrac{1}{2}x - 1$?

12. Is $(1, -3)$ a solution of $y = -2x - 1$?

13. Is $(4, 1)$ a solution of $y = \dfrac{1}{4}x + 1$?

14. Is $(-5, 3)$ a solution of $y = -\dfrac{2}{5}x + 1$?

15. Is $(0, 4)$ a solution of $y = \dfrac{3}{4}x + 4$?

16. Is $(-2, 0)$ a solution of $y = -\dfrac{1}{2}x - 1$?

17. Is $(0, 0)$ a solution of $y = 3x + 2$?

18. Is $(0, 0)$ a solution of $y = -\dfrac{3}{4}x$?

19. Find the ordered-pair solution of $y = 3x - 2$ corresponding to $x = 3$.

20. Find the ordered-pair solution of $y = 4x + 1$ corresponding to $x = -1$.

21. Find the ordered-pair solution of $y = \frac{2}{3}x - 1$ corresponding to $x = 6$.

22. Find the ordered-pair solution of $y = \frac{3}{4}x - 2$ corresponding to $x = 4$.

23. Find the ordered-pair solution of $y = -3x + 1$ corresponding to $x = 0$.

24. Find the ordered-pair solution of $y = \frac{2}{5}x - 5$ corresponding to $x = 0$.

25. Find the ordered-pair solution of $y = \frac{2}{5}x + 2$ corresponding to $x = -5$.

26. Find the ordered-pair solution of $y = -\frac{1}{6}x - 2$ corresponding to $x = 12$.

 For Exercises 27 and 28, use the linear equation $y = -3x + 6$.

27. If x is negative, which of the following inequalities about y *must* be true?

   (i) $y < 0$ (ii) $y = 0$ (iii) $y < 6$ (iv) $y > 6$

28. If x is positive, which of the following inequalities about y *must* be true?

   (i) $y > 0$ (ii) $y = 6$ (iii) $y < 6$ (iv) $y > 6$

OBJECTIVE B   **Equations of the form $y = mx + b$**

For Exercises 29 to 32, is the graph of the equation a straight line?

29. $y = -\frac{1}{2}x + 5$

30. $y = \frac{1}{x} + 5$

31. $y = 2x^2 + 5$

32. $y = -2x - 5$

33. Find three points on the graph of $y = 6x - 5$ by finding the y values that correspond to x values of $-1, 0$, and 1.

   a. When $x = -1$, $y = 6(_____) - 5 = _____$. A point on the graph is
      (_____, _____).

   b. When $x = 0$, $y = 6(_____) - 5 = _____$. A point on the graph is
      (_____, _____).

   c. When $x = 1$, $y = 6(_____) - 5 = _____$. A point on the graph is
      (_____, _____).

34. To find points on the graph of $y = \frac{2}{5}x - 3$, it is helpful to choose x values that are divisible by _____.

For Exercises 35 to 64, graph the equation.

35. $y = 2x - 4$

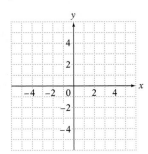

36. $y = x - 1$

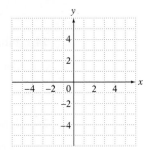

37. $y = -x + 2$

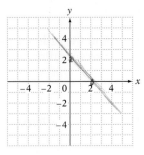

38. $y = x + 3$

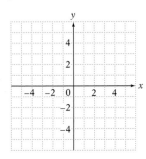

39. $y = x - 3$

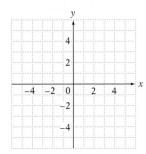

40. $y = -2x + 1$

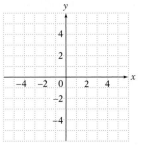

41. $y = -2x + 3$

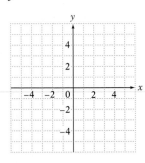

42. $y = -4x + 1$

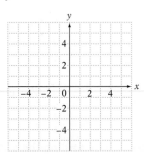

43. $y = -3x + 4$

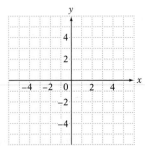

44. $y = 4x - 5$

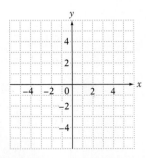

45. $y = 2x - 1$

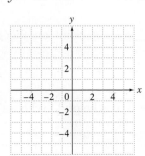

46. $y = 2x$

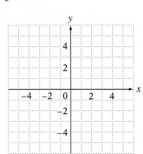

47. $y = 3x$

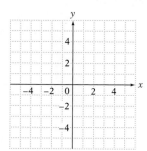

48. $y = \dfrac{3}{2}x$

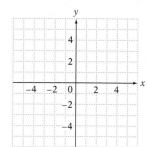

49. $y = \dfrac{1}{3}x$

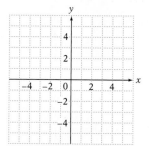

50. $y = -\dfrac{5}{2}x$

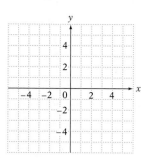

51. $y = -\dfrac{4}{3}x$

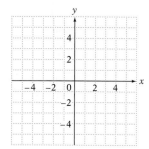

52. $y = \dfrac{2}{3}x + 1$

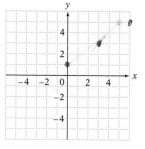

53. $y = \dfrac{3}{2}x - 1$

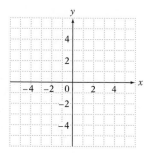

54. $y = \dfrac{1}{4}x + 2$

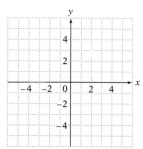

55. $y = \dfrac{2}{5}x - 1$

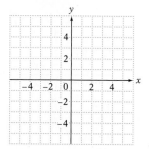

56. $y = -\dfrac{1}{2}x + 3$

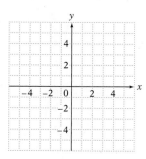

57. $y = -\dfrac{2}{3}x + 1$

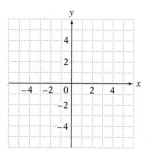

58. $y = -\dfrac{3}{4}x - 3$

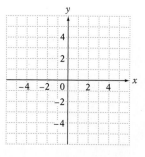

59. $y = -\dfrac{5}{3}x - 2$

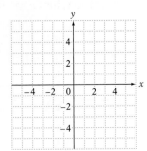

60. $y = \dfrac{1}{2}x - 1$

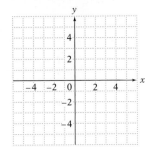

61. $y = \dfrac{5}{2}x - 1$

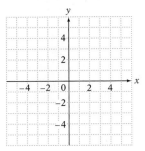

62. $y = -\dfrac{1}{4}x + 1$

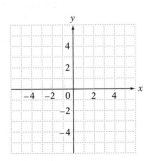

63. $y = x$

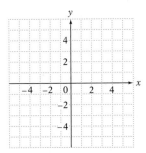

64. $y = -x$

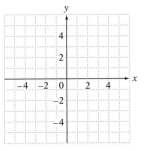

65. For the graph shown below, what is the y value when $x = 3$?

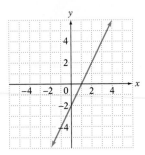

66. For the graph shown below, what is the y value when $x = -2$?

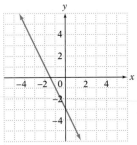

67. For the graph shown below, what is the y value when $x = 4$?

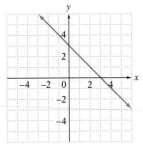

68. For the graph shown below, what is the y value when $x = 2$?

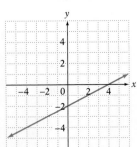

69. For the graph shown below, what is the y value when $x = 3$?

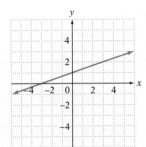

70. For the graph shown below, what is the y value when $x = -1$?

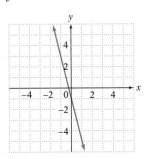

71. For the graph shown below, what is the x value when $y = -2$?

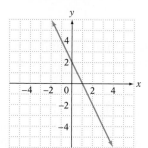

72. For the graph shown below, what is the x value when $y = -5$?

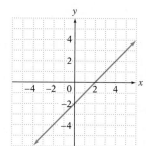

73. For the graph shown below, what is the x value when $y = -2$?

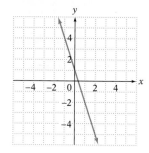

74. For the graph shown below, what is the x value when $y = 2$?

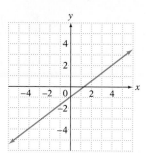

75. For the graph shown below, what is the x value when $y = -1$?

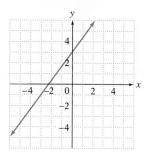

76. For the graph shown below, what is the x value when $y = 3$?

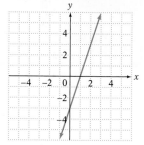

77. If (a, b) is a point on the line $y = mx$, where m is a positive constant, do a and b have the same sign or opposite signs?

78. If (a, b) is a point on the line $y = mx$, where m is negative constant, do a and b have the same sign or opposite signs?

CRITICAL THINKING

79. What are the coordinates of the point at which the graph of $y = 2x + 1$ crosses the y-axis?

80. Select the correct word and fill in the blank.
   a. If $y = 3x - 4$ and the value of x changes from 3 to 4, then the value of y increases/decreases by _____.
   b. If $y = -2x + 1$ and the value of x changes from 3 to 4, then the value of y increases/decreases by _____.

81. Suppose you are checking whether an ordered pair is a solution of an equation and the result is $4 = -1$. What does this mean?

Focus on Problem Solving

Making a Table

Sometimes a table can be used to organize information so that it is in a useful form. For example, in this chapter, we used tables to organize ordered-pair solutions of equations.

A basketball player scored 11 points in a game. The player can score 1 point for making a free throw, 2 points for making a field goal from within the three-point line, and 3 points for making a field goal from outside the three-point line. Find the number of possible combinations in which the player could have scored 11 points.

The following table lists the possible combinations for scoring 11 points.

Points																
Free throws	0	2	1	3	5	0	2	4	6	8	1	3	5	7	9	11
2-point field goals	1	0	2	1	0	4	3	2	1	0	5	4	3	2	1	0
3-point field goals	3	3	2	2	2	1	1	1	1	1	0	0	0	0	0	0
Total points	11	11	11	11	11	11	11	11	11	11	11	11	11	11	11	11

There are 16 possible ways in which the basketball player could have scored 11 points.

1. A football team scores 17 points. A touchdown counts as 6 points, an extra point as 1 point, a field goal as 3 points, and a safety as 2 points. Find the number of possible combinations in which the team can score 17 points. Remember that the number of extra points cannot exceed the number of touchdowns scored.

2. Repeat Exercise 1. Assume that no safety was scored.

3. Repeat Exercise 1. Assume that no safety was scored and that the team scored two field goals.

4. Find the number of possible combinations of nickels, dimes, and quarters that one might get when receiving $.85 in change.

5. Repeat Exercise 4. Assume that no combination contains coins that could be exchanged for a larger coin. That is, the combination of three quarters and two nickels would not be allowed because the two nickels could be exchanged for a dime.

6. Find the number of possible combinations of $1, $5, $10, and $20 bills that one might get when receiving $33.

Projects & Group Activities

Collecting, Organizing, and Analyzing Data

Decide on two quantities that may be related, and collect at least 10 pairs of values. Here are some examples:

- The heights and weights of the students in an elementary school class
- The time spent studying for a test and the grade earned on the test
- The distance a student commutes to class and the number of miles on the odometer of the car used for commuting
- The number of credit hours a student is taking this semester and the amount the student spent on textbooks this term

Draw a scatter diagram of the data. Is there a trend? That is, as you move from left to right on the graph, do the points tend to rise or fall?

Chapter 6 Summary

Key Words	Examples
An **equation** expresses the equality of two mathematical expressions. [6.1A, p. 381]	$5x + 6 = 7x - 3$ $y = 4x - 10$ $3a^2 - 6a + 4 = 0$
In a **first-degree equation in one variable,** the equation has only one variable, and each instance of the variable is the first power (the exponent on the variable is 1). [6.1A, p. 381]	$3x - 8 = 4$ $z = -11$ $6(x + 7) = 2 - (x - 9)$
A **solution** of an equation is a number that, when substituted for the variable, produces a true equation. [6.1A, p. 381]	6 is a solution of $x - 4 = 2$ because $6 - 4 = 2$ is a true equation.
To **solve an equation** means to find the solutions of the equation. The goal is to rewrite the equation in the form **variable = constant.** [6.1A, p. 381]	$x = 5$ is in the form *variable = constant*. The solution of the equation $x = 5$ is the constant 5 because $5 = 5$ is a true equation.
Some of the words and phrases that translate to "equals" are **equals, is, is the same as, amounts to, totals,** and **represents.** [6.4A, p. 403]	"Eight plus a number is ten" translates to $8 + x = 10$.
A **rectangular coordinate system** is formed by two number lines, one horizontal and one vertical, that intersect at the zero point of each line. The point of intersection is called the **origin.** The number lines that make up a rectangular coordinate system are called **coordinate axes.** A rectangular coordinate system divides the plane into four regions called **quadrants.** [6.5A, p. 411]	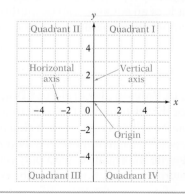

An **ordered pair** (x, y) is used to locate a point in a rectangular coordinate system. The first number of the pair is the **abscissa.** The second number is the **ordinate.** The **coordinates** of the point are the numbers in the ordered pair associated with the point. To **graph,** or **plot,** a point in the plane, place a dot at the location given by the ordered pair. The **graph of an ordered pair** is the dot drawn at the coordinates of the point in the plane. [6.5A, p. 411]

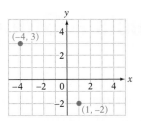

A **scatter diagram** is a graph of a set of ordered pairs of data. [6.5B, p. 413]

The distance, in miles, a house is from a fire station and the amount, in thousands of dollars, of fire damage that the house sustained in a fire are given in the scatter diagram.

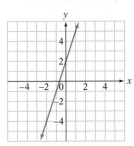

An equation of the form $y = mx + b$, where m is the coefficient of x and b is a constant, is a **linear equation in two variables.** A **solution of a linear equation in two variables** is an ordered pair (x, y) whose coordinates make the equation a true statement. The **graph of an equation in two variables** is a graph of the ordered-pair solutions of the equation. The graph of a linear equation in two variables is a straight line. [6.6A, p. 420; 6.6B, pp. 421–422]

$y = 3x + 2$ is a linear equation in two variables; $m = 3$ and $b = 2$. Ordered-pair solutions of $y = 3x + 2$ are shown below, along with the graph of the equation.

x	y
1	5
0	2
−1	−1

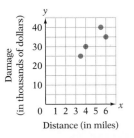

Essential Rules and Procedures

Addition Property of Equations [6.1A, p. 382]
The same number or variable expression can be added to each side of an equation without changing the solution of the equation.

$$x + 7 = 20$$
$$x + 7 + (-7) = 20 + (-7)$$
$$x = 13$$

Multiplication Property of Equations [6.1B, p. 385]
Each side of an equation can be multiplied by the same nonzero number without changing the solution of the equation.

$$\frac{3}{4}x = 24$$
$$\frac{4}{3} \cdot \frac{3}{4}x = \frac{4}{3} \cdot 24$$
$$x = 32$$

Steps in Solving General First-Degree Equations [6.3B, p. 397]
1. Use the Distributive Property to remove parentheses.
2. Combine like terms on each side of the equation.
3. Rewrite the equation with only one variable term.
4. Rewrite the equation with only one constant term.
5. Rewrite the equation so that the coefficient of the variable is 1.

$$8 - 4(2x + 3) = 2(1 - x)$$
$$8 - 8x - 12 = 2 - 2x$$
$$-8x - 4 = 2 - 2x$$
$$-6x - 4 = 2$$
$$-6x = 6$$
$$x = -1$$

Cumulative Review Exercises

1. Evaluate $-3ab$ for $a = -2$ and $b = 3$.

2. Simplify: $-3(4p - 7)$

3. Simplify: $\left(\dfrac{2}{3}\right)\left(-\dfrac{9}{8}\right) + \dfrac{3}{4}$

4. Solve: $-\dfrac{2}{3}y = 12$

5. Evaluate $(-b)^3$ for $b = -2$.

6. Evaluate $4xy^2 - 2xy$ for $x = -2$ and $y = 3$.

7. Simplify: $\sqrt{121}$

8. Simplify: $\sqrt{48}$

9. Simplify: $4(3v - 2) - 5(2v - 3)$

10. Simplify: $-4(-3m)$

11. Is -9 a solution of the equation $-5d = -45$?

12. Solve: $5 - 7a = 3 - 5a$

13. Simplify: $6 - 2(7z - 3) + 4z$

14. Evaluate $\dfrac{a^2 + b^2}{2ab}$ for $a = -2$ and $b = -1$.

15. Solve: $8z - 9 = 3$

16. Simplify: $(2m^2n^5)^5$

17. Multiply: $-3a^3(2a^2 + 3ab - 4b^2)$

18. Multiply: $(2x - 3)(3x + 1)$

19. Simplify: 2^{-4}

20. Simplify: $\dfrac{x^8}{x^2}$

21. Simplify: $(-5x^3y)(-3x^5y^2)$

22. Solve: $5 - 3(2x - 8) = -2(1 - x)$

23. Graph $y = \dfrac{5}{3}x + 1$.

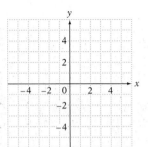

24. Graph $y = -\dfrac{2}{5}x$.

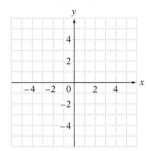

25. Write 3.5×10^{-8} in decimal notation.

26. Translate "the product of five and the sum of a number and two" into a variable expression. Then simplify the variable expression.

27. *Physics* Find the time it takes a falling object to increase its speed from 50 ft/s to 98 ft/s. Use the equation $v = v_0 + 32t$, where v is the final velocity, v_0 is the initial velocity, and t is the time it takes for the object to fall.

28. *Zoology* The number of dogs in the world is 1,000 times the number of wolves in the world. Express the number of dogs in the world in terms of the number of wolves in the world.

29. *The Film Industry* The figure at the right shows the top-grossing movies in the United States in the 1970s. Find the total box office gross for these four films.

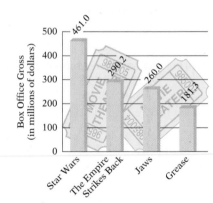

Top-Grossing Movies in the 1970s
Source: www.worldwideboxoffice.com

30. *Finances* A homeowner's mortgage payment for one month for principal and interest was $949. The principal payment was $204 less than the interest payment. Find the amount of the interest payment.

31. *Geography* The Aleutian Trench in the Pacific Ocean is 8,100 m deep. Each story of an average skyscraper is 4.2 m tall. How many stories, to the nearest whole number, would a skyscraper as tall as the Aleutian Trench have?

32. *Charities* A donation of $12,000 is given to two charities. One charity received twice as much as the other charity. How much did each charity receive?

Measurement and Proportion

Anthony Reyes of the St. Louis Cardinals throws a pitch to one of the Detroit Tigers during game one of the 2006 World Series at Comerica Park in Detroit, Michigan. If the Tigers score a run during this play, Reyes' earned run average (ERA) would suffer. His ERA is the number of earned runs that have been scored for every nine innings he has pitched. A pitcher's ERA can be calculated by setting up a proportion, as seen in the **Project on page 481.**

DVD

SSM

Student Website

Need help? For online student resources, visit college.hmco.com/pic/aufmannPA5e.

1. Simplify: $\dfrac{8}{10}$

2. Write as a decimal: $\dfrac{372}{15}$

For Exercises 3 to 14, add, subtract, multiply, or divide.

3. $36 \times \dfrac{1}{9}$

4. $\dfrac{5}{3} \times 6$

5. $5\dfrac{3}{4} \times 8$

6. $3\overline{)714}$

7. $3.732 \times 10,000$

8. $41.07 \div 1,000$

9. $6 - 0.875$

10. $5 + 0.96$

11. 3.25×0.04

12. $35 \times \dfrac{1.61}{1}$

13. $1.67 \times \dfrac{1}{3.34}$

14. $315 \div 84$

GO Figure

Suppose you threw six darts and all six hit the target shown. Which of the following could be your score?

4   15   58   28   29   31

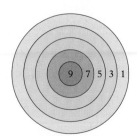

7.1 Exercises

OBJECTIVE A   **The metric system**

1. In the metric system, what is the basic unit of length? Of liquid measure? Of weight?

2. **a.** Explain how to convert meters to centimeters.
   b. Explain how to convert milliliters to liters.

3. **a.** Complete the table.

Metric System Prefix	Symbol	Magnitude	Means Multiply the Basic Unit By:
tera-	T	10^{12}	1 000 000 000 000
giga-	G	___	1 000 000 000
mega-	M	10^{6}	_____
kilo-	___	___	1 000
hecto-	h	___	100
deca-	da	10^{1}	_____
deci-	d	$\frac{1}{10}$	_____
centi-	___	$\frac{1}{10^{2}}$	_____
milli-	___	___	0.001
micro-	μ	$\frac{1}{10^{6}}$	_____
nano-	n	$\frac{1}{10^{9}}$	_____
pico-	p	___	0.000 000 000 001

   b. How can the magnitude column in the table above be used to determine how many places to move the decimal point when converting to the basic unit in the metric system?

Name the unit in the metric system that would be used to measure each of the following.

4. the distance from New York to London

5. the weight of a truck

6. a person's waist

7. the amount of coffee in a mug

8. the weight of a thumbtack

9. the amount of water in a swimming pool

10. the distance a baseball player hits a baseball

11. a person's hat size

12. the amount of fat in a slice of cheddar cheese

13. a person's weight

14. the amount of maple syrup served with pancakes

15. the amount of water in a watercooler

16. the amount of vitamin C in a vitamin tablet

17. a serving of cereal

18. the width of a hair

19. a person's height

20. the amount of medication in an aspirin

21. the weight of a lawn mower

22. the weight of a slice of bread

23. the contents of a bottle of salad dressing

24. the amount of water a family uses monthly

25. the weight of newspapers collected at a recycling center

26. the amount of liquid in a bowl of soup

27. the distance to the bank

Convert.

28. 42 cm = _____ mm

29. 91 cm = _____ mm

30. 360 g = _____ kg

31. 1 856 g = _____ kg

32. 5 194 ml = _____ L

33. 7 285 ml = _____ L

34. 2 m = _____ mm

35. 8 m = _____ mm

36. 217 mg = _____ g

37. 34 mg = _____ g

38. 4.52 L = _____ ml

39. 0.029 7 L = _____ ml

40. 8 406 m = _____ km

41. 7 530 m = _____ km

42. 2.4 kg = _____ g

43. 9.2 kg = _____ g

44. 6.18 kl = _____ L

45. 0.036 kl = _____ L

46. 9.612 km = _____ m

47. 2.35 km = _____ m

48. 0.24 g = _____ mg

49. 0.083 g = _____ mg

50. 298 cm = _____ m

51. 71.6 cm = _____ m

52. 4 231 L = _____ kl

53. 3 206 L = _____ kl

54. 5.84 m = _____ cm

55. 0.99 m = _____ cm

56. 87 mm = _____ cm

57. 605 mm = _____ cm

For Exercises 58 to 63, fill in the blank with the correct unit of measurement.

58. 8 750 cg = 87.5_____

59. 0.05 m = 5_____

60. 1.78 kl = 1 780_____

61. 28 300 mg = 2 830_____

62. 5 ml = 0.005_____

63. 0.7 km = 70 000_____

For Exercises 64 and 65, fill in the blank and circle the correct word to complete each setence.

64. One box of pasta weighs 454 g. To determine the weight in kilograms of 10 boxes of pasta, first <u>multiply/divide</u> the weight of one box of pasta by 10. Then convert the weight of 10 boxes of pasta to _____ .

65. To determine how many pieces of wood, each 70 cm in length, can be cut from a board 3.5 m long, first convert the length of the board to _____ . Then <u>multiply/divide</u> the length of the board by 70.

Solve.

66. *The Olympics*   **a.** One of the events in the summer Olympics is the 50 000-meter walk. How many kilometers do the entrants in this event walk? **b.** One of the events in the winter Olympic Games is the 10 000-meter speed skating race. How many kilometers do the entrants in this event skate?

67. *Gemology*   A carat is a unit of weight equal to 200 mg. Find the weight in grams of a 15-carat precious stone.

68. *Crafts*   How many pieces of material, each 75 cm long, can be cut from a bolt of fabric that is 9 m long?

69. *Health Clubs*   An athletic club uses 800 ml of chlorine each day for its swimming pool. How many liters of chlorine are used in a month of 30 days?

70. *Carpentry*   Each of the four shelves in a bookcase measures 175 cm. Find the cost of the shelves when the price of lumber is $16.50 per meter.

71. *Consumerism*   The printed label from a container of milk is shown at the right. To the nearest whole number, how many 230-milliliter servings are in the container?

Chad Hedrick

72. *Consumerism*   A 1.19-kilogram container of Quaker Oats contains 30 servings. Find the number of grams in one serving of the oatmeal. Round to the nearest gram.

73. *Nutrition Labels*   The nutrition label for a corn bread mix is shown at the right. **a.** How many kilograms of mix are in the package? **b.** How many grams of sodium are contained in two servings of the corn bread?

Nutrition Facts
Serving Size ⅙ pkg. (31 g mix)
Servings Per Container 6

Amount Per Serving	Mix	Prepared
Calories	110	160
Calories from Fat	10	50

	% Daily Value*	
Total Fat 1g	1%	9%
Saturated Fat 0g	0%	7%
Cholesterol 0mg	0%	12%
Sodium 210mg	9%	11%
Total Carbohydrate 24g	8%	8%
Sugars 6g		
Protein 2g		

74. *Health*   A patient is advised to supplement her diet with 3 g of calcium per day. The calcium tablets she purchases contain 500 mg of calcium per tablet. How many tablets per day should the patient take?

75. *Chemistry*   A laboratory assistant is in charge of ordering acid for three chemistry classes of 40 students each. Each student requires 80 ml of acid. How many liters of acid should be ordered? The assistant must order by the whole liter.

76. *Consumerism*   A case of 12 one-liter bottles of apple juice costs $19.80. A case of 24 cans, each can containing 340 ml of apple juice, costs $14.50. Which case of apple juice costs less per milliliter?

77. *Light*   The distance between Earth and the sun is 150 000 000 km. Light travels 300 000 000 m in 1 s. How long does it take for light to reach Earth from the sun?

78. *Business*   A health food store buys nuts in 10-kilogram containers and repackages the nuts for resale. The store packages the nuts in 200-gram bags, costing $.06 each, and sells them for $2.89 per bag. Find the profit on a 10-kilogram container of nuts costing $75.

79. *Business*   For $149.50, a cosmetician buys 5 L of moisturizer and repackages it in 125-milliliter jars. Each jar costs the cosmetician $.55. Each jar of moisturizer is sold for $8.95. Find the profit on the 5 L of moisturizer.

80. *Business*   A service station operator bought 85 kl of gasoline for $38,500. The gasoline was sold for $.658 per liter. Find the profit on the 85 kl of gasoline.

CRITICAL THINKING

81. A 280-milliliter serving is taken from a 3-liter bottle of water. How much water remains in the container? Write the answer in two different ways.

82. Why is it necessary to have internationally standardized units of measurement?

7.2 Ratios and Rates

OBJECTIVE A   Ratios and rates

In previous work, we have used quantities with units, such as 12 ft, 3 h, 2¢, and 15 acres. In these examples, the units are feet, hours, cents, and acres.

A **ratio** is the quotient or comparison of two quantities with the *same* unit. We can compare the measure of 3 ft to the measure of 8 ft by writing a quotient.

$$\frac{3\ \text{ft}}{8\ \text{ft}} = \frac{3}{8} \qquad 3\ \text{ft is}\ \frac{3}{8}\ \text{of 8 ft.}$$

A ratio can be written in three ways:

1. As a fraction $\dfrac{3}{8}$

2. As two numbers separated by a colon $3:8$

3. As two numbers separated by the word *to*   3 to 8

The ratio of 15 mi to 45 mi is written as

$$\frac{15\ \text{mi}}{45\ \text{mi}} = \frac{15}{45} = \frac{1}{3}\ \text{or}\ 1:3\ \text{or}\ 1\ \text{to}\ 3$$

A ratio is in **simplest form** when the two numbers do not have a common factor. The units are not written in a ratio.

A **rate** is the comparison of two quantities with *different* units.

A catering company prepares 9 gal of coffee for every 50 people at a reception. This rate is written

$$\frac{9\ \text{gal}}{50\ \text{people}}$$

You traveled 200 mi in 6 h. The rate is written

$$\frac{200\ \text{mi}}{6\ \text{h}} = \frac{100\ \text{mi}}{3\ \text{h}}$$

A rate is in **simplest form** when the numbers have no common factors. The units are written as part of the rate.

Many rates are written as unit rates. A **unit rate** is a rate in which the number in the denominator is 1. The word *per* generally indicates a unit rate. It means "for each" or "for every." For example,

23 mi per gallon   ▶ The unit rate is $\dfrac{23\ \text{mi}}{1\ \text{gal}}$.

65 mi per hour   ▶ The unit rate is $\dfrac{65\ \text{mi}}{1\ \text{h}}$.

$4.78 per pound   ▶ The unit rate is $\dfrac{\$4.78}{1\ \text{lb}}$.

Point of Interest

It is believed that billiards was invented in France during the reign of Louis XI (1423–1483). In the United States, the standard billiard table is 4 ft 6 in. by 9 ft. This is a ratio of 1:2. The same ratio holds for carom and snooker tables, which are 5 ft by 10 ft.

Unit rates make comparisons easier. For example, if you travel 37 mph and I travel 43 mph, we know that I am traveling faster than you are. It is more difficult to compare speeds if we are told that you are traveling $\frac{111 \text{ mi}}{3 \text{ h}}$ and I am traveling $\frac{172 \text{ mi}}{4 \text{ h}}$.

To find a unit rate, divide the number in the numerator of the rate by the number in the denominator of the rate. A unit rate is often written in decimal form.

A student received $57 for working 6 h at the bookstore. Find the wage per hour (the unit rate).

Write the rate as a fraction. $\frac{\$57}{6 \text{ h}}$

Divide the number in the numerator of the rate (57) by the number in the denominator (6). $57 \div 6 = 9.5$

The unit rate is $\frac{\$9.50}{1 \text{ h}} = \$9.50/\text{h}$. This is read "$9.50 per hour."

EXAMPLE 1

Write the comparison of 12 to 8 as a ratio in simplest form using a fraction, a colon, and the word *to*.

Solution

$\frac{12}{8} = \frac{3}{2}$

$12:8 = 3:2$

$12 \text{ to } 8 = 3 \text{ to } 2$

YOU TRY IT 1

Write the comparison of 12 to 20 as a ratio in simplest form using a fraction, a colon, and the word *to*.

Your Solution

EXAMPLE 2

Write "12 hits in 26 times at bat" as a rate in simplest form.

Solution

$\frac{12 \text{ hits}}{26 \text{ at-bats}} = \frac{6 \text{ hits}}{13 \text{ at-bats}}$

YOU TRY IT 2

Write "20 bags of grass seed for 8 acres" as a rate in simplest form.

Your Solution

EXAMPLE 3

Write "285 mi in 5 h" as a unit rate.

Solution

$\frac{285 \text{ mi}}{5 \text{ h}}$

$285 \div 5 = 57$

The unit rate is 57 mph.

YOU TRY IT 3

Write "$8.96 for 3.5 lb" as a unit rate.

Your Solution

Solutions on p. S18

7.2 Exercises

OBJECTIVE A   **Ratios and rates**

For Exercises 1 and 2, fill in the blank or circle the correct words to complete each sentence.

1. In a ratio, the units <u>are/are not</u> written. In a rate, the units <u>are/are not</u> written.

2. A unit rate is a rate in which the number in the denominator is _____ . To write a rate as a unit rate, divide the number in the _____ by the number in the _____ .

Write the comparison as a ratio in simplest form using a fraction, a colon, and the word *to.*

3. 16 in. to 24 in.

4. 8 lb to 60 lb

5. 9 h to 24 h

6. $55 to $150

7. 9 ft to 2 ft

8. 50 min to 6 min

9. 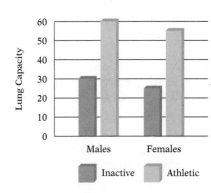 *Physical Fitness*   The figure at the right shows the lung capacities of inactive versus athletic 45-year-olds. Write the comparison of the lung capacity of an inactive male to that of an athletic male as a ratio in simplest form using a fraction, a colon, and the word *to.*

Lung Capacity (in milliliters of oxygen per kilogram of body weight per minute)

Write as a ratio in simplest form using a fraction.

10. *Sports*   A baseball player had 3 errors in 42 fielding attempts. What is the ratio of the number of times the player did not make an error to the total number of attempts?

11. *Sports*   A basketball team won 18 games and lost 8 games during the season. What is the ratio of the number of games won to the total number of games?

12. *Mechanics*   Find the ratio of two meshed gears if one gear has 24 teeth and the other gear has 36 teeth.

Write as a rate in simplest form.

13. 7 children in 5 families

14. 243 apple trees on 6 acres

15. 160 mi in 6 h

16. $28 for 8 T-shirts   **17.** $81 for 9 h of work   **18.** 87 students in 6 classes

For Exercises 19 to 24, write as a unit rate.

19. $460 earned for 40 h of work   **20.** $38,700 earned in 12 months

21. 387.8 mi in 7 h   **22.** 364.8 mi on 9.5 gal of gas

23. $19.08 for 4.5 lb   **24.** $20.16 for 15 oz

25. *Sports* NCAA statistics show that for every 2,800 college seniors playing college basketball, only 50 will play as rookies in the National Basketball Association. Write the ratio of the number of National Basketball Association rookies to the number of college seniors playing basketball.

26. *Travel* An airplane flew 1,155 mi in 2.5 h. Find the rate of travel.

27. *Population Density* The table at the right shows the population and the area of three countries. Find the population density (people per square mile) for each country. Round to the nearest tenth.

Country	Population	Area (in square miles)
Australia	20,264,000	2,968,000
India	1,095,352,000	1,269,000
United States	300,219,000	3,718,000

28. *Investments* An investor purchased 100 shares of stock for $2,500. One year later the investor sold the stock for $3,200. What was the investor's profit per share?

29. True or false? The ratio of 5 m to 5 cm is 1 m per centimeter.

30. True or false? The ratio of 10 g to 10 kg is 1 to 1000.

CRITICAL THINKING

31. *Social Security* According to the Social Security Administration, the number of workers per retiree is expected to be as given in the table below.

Year	2010	2020	2030	2040
Number of workers per retiree	3.1	2.5	2.1	2.0

Why is the shrinking number of workers per retiree important to the Social Security Administration?

7.3 The U.S. Customary System of Measurement

OBJECTIVE A The U.S. Customary System of Measurement

The standard U.S. Customary System units of length are **inch, foot, yard,** and **mile.** The abbreviations for these units of length are in., ft, yd, and mi. Equivalences between units of length in the U.S. Customary System are

1 ft = 12 in.
1 yd = 3 ft
1 yd = 36 in.
1 mi = 5,280 ft

Weight is a measure of how strongly Earth is pulling on an object. The U.S. Customary System units of weight are **ounce, pound,** and **ton.** The abbreviation for ounces is oz, and the abbreviation for pounds is lb. Equivalences between units of weight in the U.S. Customary System are

1 lb = 16 oz
1 ton = 2,000 lb

Liquids are measured in units of **capacity.** The standard U.S. Customary System units of capacity (and their abbreviations) are the **fluid ounce** (fl oz), **cup** (c), **pint** (pt), **quart** (qt), and **gallon** (gal). Equivalences between units of capacity in the U.S. Customary System are

1 c = 8 fl oz
1 pt = 2 c
1 qt = 4 c
1 gal = 4 qt

Area is a measure of the amount of surface in a region. The standard U.S. Customary System units of area are **square inch** (in^2), **square foot** (ft^2), **square yard** (yd^2), **square mile** (mi^2), and **acre.** Equivalences between units of area in the U.S. Customary System are

$1 ft^2$ = $144 in^2$
$1 yd^2$ = $9 ft^2$
$1 acre$ = $43,560 ft^2$
$1 mi^2$ = $640 acres$

In solving application problems, scientists, engineers, and other professionals find it useful to include the units as they work through the solutions to problems so that the answers are in the proper units. Using units to organize and check the correctness of an application is called **dimensional analysis.** Applying dimensional analysis to application problems requires converting units as well as multiplying and dividing units.

Point of Interest

The ancient Greeks devised the foot measurement, which they usually divided into 16 fingers. It was the Romans who subdivided the foot into 12 units called inches. The word *inch* is derived from the Latin word *uncia,* meaning "a twelfth part."

The Romans also used a unit called pace, which equaled two steps. One thousand paces equaled 1 mi. The word *mile* is derived from the Latin word *mille,* which means "one thousand."

The word *quart* has its root in the Latin word *quartus,* which means "one-fourth"; a quart is one-fourth of a gallon. The same Latin word is the root of other English words such as *quartet, quadrilateral,* and *quarter.*

The equivalent measures listed on page 453 can be used to form conversion rates to change one unit of measurement to another. For example, the equivalent measures **1 mi** and **5,280 ft** are used to form the following conversion rates:

$$\frac{1 \text{ mi}}{5{,}280 \text{ ft}} \qquad \frac{5{,}280 \text{ ft}}{1 \text{ mi}}$$

Because **1 mi = 5,280 ft,** both of the conversion rates $\frac{1 \text{ mi}}{5{,}280 \text{ ft}}$ and $\frac{5{,}280 \text{ ft}}{1 \text{ mi}}$ are equal to 1.

To convert 3 mi to feet, multiply 3 mi by the conversion rate $\frac{5{,}280 \text{ ft}}{1 \text{ mi}}$.

$$3 \text{ mi} = 3 \text{ mi} \cdot 1 = \frac{3 \text{ mi}}{1} \cdot \frac{5{,}280 \text{ ft}}{1 \text{ mi}} = \frac{3 \text{ mi} \cdot 5{,}280 \text{ ft}}{1 \text{ mi}} = 3 \cdot 5{,}280 \text{ ft} = 15{,}840 \text{ ft}$$

There are three important points to notice in the above example. First, you can think of dividing the numerator and denominator by the common unit "mile" just as you would divide the numerator and denominator of a fraction by a common factor. Second, the conversion rate $\frac{5{,}280 \text{ ft}}{1 \text{ mi}}$ is equal to 1, and multiplying an expression by 1 does not change the value of the expression.

In the above example, we had the choice of two conversion rates, $\frac{1 \text{ mi}}{5{,}280 \text{ ft}}$ or $\frac{5{,}280 \text{ ft}}{1 \text{ mi}}$. In the conversion rate chosen, the unit in the numerator is the same as the unit desired in the answer (ft). The unit in the denominator is the same as the unit in the given measurement (mi).

EXAMPLE 1

Convert 36 fl oz to cups.

Solution

The equivalence is 1 c = 8 fl oz.

The conversion rate must have cups in the numerator and fluid ounces in the denominator: $\frac{1 \text{ c}}{8 \text{ fl oz}}$

$$36 \text{ fl oz} = 36 \text{ fl oz} \cdot 1 = \frac{36 \text{ fl oz}}{1} \cdot \frac{1 \text{ c}}{8 \text{ fl oz}}$$

$$= \frac{36 \text{ fl oz} \cdot 1 \text{ c}}{8 \text{ fl oz}}$$

$$= \frac{36 \text{ c}}{8} = 4\frac{1}{2} \text{ c}$$

YOU TRY IT 1

Convert 40 in. to feet.

Your Solution

Solution on p. S18

EXAMPLE 2

Convert $4\frac{1}{2}$ tons to pounds.

Solution

The equivalence is 1 ton = 2,000 lb.

The conversion rate must have pounds in the numerator and tons in the denominator:

$\dfrac{2,000 \text{ lb}}{1 \text{ ton}}$.

$4\frac{1}{2}$ tons $= 4\frac{1}{2}$ tons $\cdot 1 = \dfrac{9}{2}$ tons $\cdot \dfrac{2,000 \text{ lb}}{1 \text{ ton}}$

$= \dfrac{9 \text{ tons}}{2} \cdot \dfrac{2,000 \text{ lb}}{1 \text{ ton}}$

$= \dfrac{9 \text{ tons} \cdot 2,000 \text{ lb}}{2 \cdot 1 \text{ ton}}$

$= \dfrac{9 \cdot 2,000 \text{ lb}}{2}$

$= \dfrac{18,000 \text{ lb}}{2}$

$= 9,000 \text{ lb}$

EXAMPLE 3

How many seconds are in 1 h?

Solution

We need to convert hours to minutes and minutes to seconds. The equivalences are
1 h = 60 min and 1 min = 60 s.

Choose the conversion rates so that we can divide by the unit "hours" and by the unit "minutes."

$1 \text{ h} = 1 \text{ h} \cdot 1 \cdot 1 = \dfrac{1 \text{ h}}{1} \cdot \dfrac{60 \text{ min}}{1 \text{ h}} \cdot \dfrac{60 \text{ s}}{1 \text{ min}}$

$= \dfrac{1 \text{ h} \cdot 60 \text{ min} \cdot 60 \text{ s}}{1 \cdot 1 \text{ h} \cdot 1 \text{ min}}$

$= \dfrac{60 \cdot 60 \text{ s}}{1} = 3,600 \text{ s}$

There are 3,600 s in 1 h.

YOU TRY IT 2

Convert $2\frac{1}{2}$ yd to inches.

Your Solution

YOU TRY IT 3

How many yards are in 1 mi?

Your Solution

Solutions on p. S18

OBJECTIVE B Applications

In solving these application problems, we will write the units throughout the solution as we work through the arithmetic. Note that, just as in the conversions in Objective A, conversion rates are used to set up the units before the arithmetic is performed.

Barbaro

In 2006, a horse named Barbaro won the Kentucky Derby, covering the 1.25-mile course in 2.0227 min. Find Barbaro's average speed for the race in miles per hour. Round to the nearest tenth.

Strategy　To find the average speed in miles per hour:

▶ Write Barbaro's speed as a rate in fraction form. The distance (1.25 mi) is in the numerator, and the time (2.0227 min) is in the denominator.

▶ Multiply the fraction by the conversion rate $\frac{60 \text{ min}}{1 \text{ h}}$.

Solution　Barbaro's rate is $\frac{1.25 \text{ mi}}{2.0227 \text{ min}}$.

$$\frac{1.25 \text{ mi}}{2.0227 \text{ min}} = \frac{1.25 \text{ mi}}{2.0227 \text{ min}} \cdot \frac{60 \text{ min}}{1 \text{ h}}$$

$$= \frac{75 \text{ mi}}{2.0227 \text{ h}} \approx 37.1 \text{ mph}$$

Barbaro's average speed was 37.1 mph.

EXAMPLE 4

A carpet is to be placed in a room that is 20 ft wide and 30 ft long. At $28.50 per square yard, how much will it cost to carpet the room? Use the formula $A = LW$, where A is the area, L is the length, and W is the width, to find the area of the room.

Strategy

To find the cost of the carpet:

▶ Use the formula $A = LW$ to find the area.

▶ Use the conversion rate $\frac{1 \text{ yd}^2}{9 \text{ ft}^2}$ to find the area in square yards.

▶ Multiply by $\frac{\$28.50}{1 \text{ yd}^2}$ to find the cost.

Solution

$A = LW = 30 \text{ ft} \cdot 20 \text{ ft} = 600 \text{ ft}^2$　▶ ft · ft = ft²

$600 \text{ ft}^2 = \frac{600 \text{ ft}^2}{1} \cdot \frac{1 \text{ yd}^2}{9 \text{ ft}^2} = \frac{600 \text{ yd}^2}{9} = \frac{200 \text{ yd}^2}{3}$

$\text{Cost} = \frac{200 \text{ yd}^2}{3} \cdot \frac{\$28.50}{1 \text{ yd}^2} = \$1,900$

The cost of the carpet is $1,900.

YOU TRY IT 4

Find the number of gallons of water in a fish tank that is 36 in. long and 24 in. wide and is filled to a depth of 16 in. Use the formula $V = LWH$, where V is the volume, L is the length, W is the width, and H is the depth of the water. (1 gal = 231 in³) Round to the nearest tenth.

Your Strategy

Your Solution

Solution on p. S18

OBJECTIVE C **Conversion between the U.S. Customary System and the metric system**

Because more than 90% of the world's population uses the metric system of measurement, converting U.S. Customary units to metric units is essential in trade and commerce—for example, in importing foreign goods and exporting domestic goods. Also, metric units are being used throughout the United States today. Cereal is packaged by the gram, 35-millimeter film is available, and soda is sold by the liter.

Approximate equivalences between the U.S. Customary System and the metric system are shown below.

Units of Length	*Units of Weight*	*Units of Capacity*
1 in. = 2.54 cm	1 oz ≈ 28.35 g	1 L ≈ 1.06 qt
1 m ≈ 3.28 ft	1 lb ≈ 454 g	1 gal ≈ 3.79 L
1 m ≈ 1.09 yd	1 kg ≈ 2.2 lb	
1 mi ≈ 1.61 km		

Point of Interest

The definition of 1 in. has been changed as a consequence of the wide acceptance of the metric system. One inch is now exactly 25.4 mm.

These equivalences can be used to form conversion rates to change one measurement to another. For example, because $1 \text{ mi} \approx 1.61 \text{ km}$, the conversion rates $\dfrac{1 \text{ mi}}{1.61 \text{ km}}$ and $\dfrac{1.61 \text{ km}}{1 \text{ mi}}$ are each approximately equal to 1.

The procedure used to convert from one system to the other is identical to the procedure used to perform conversions in the U.S. Customary System in Objective A.

Convert 55 mi to kilometers.

The equivalence is $1 \text{ mi} \approx 1.61 \text{ km}$. The conversion rate must have kilometers in the numerator and miles in the denominator: $\dfrac{1.61 \text{ km}}{1 \text{ mi}}$.

$$55 \text{ mi} = \frac{55 \text{ mi}}{1}$$

Note that we are multiplying 55 mi by 1, so we are not changing its value.

$$\approx \frac{55 \text{ mi}}{1} \cdot \boxed{\frac{1.61 \text{ km}}{1 \text{ mi}}}$$

Divide the numerator and denominator by the common unit "mile."

$$\approx \frac{55 \text{ mi}}{1} \cdot \frac{1.61 \text{ km}}{1 \text{ mi}}$$

Multiply 55 times 1.61.

$$\approx \frac{88.55 \text{ km}}{1}$$

$55 \text{ mi} \approx 88.55 \text{ km}$

EXAMPLE 5

Convert 200 m to feet.

Solution

$$200 \text{ m} = \frac{200 \text{ m}}{1} \approx \frac{200 \text{ m}}{1} \cdot \frac{3.28 \text{ ft}}{1 \text{ m}}$$

$$\approx 656 \text{ ft}$$

YOU TRY IT 5

Convert 45 cm to inches. Round to the nearest hundredth.

Your Solution

EXAMPLE 6

Convert 45 mph to kilometers per hour.

Solution

$$45 \text{ mph} = \frac{45 \text{ mi}}{1 \text{ h}} \approx \frac{45 \text{ mi}}{1 \text{ h}} \cdot \frac{1.61 \text{ km}}{1 \text{ mi}}$$

$$\approx 72.45 \text{ km per hour}$$

YOU TRY IT 6

Convert 75 km per hour to miles per hour. Round to the nearest hundredth.

Your Solution

EXAMPLE 7

The price of gasoline is $2.99 per gallon. Find the price per liter. Round to the nearest tenth of a cent.

Solution

$$\$2.99 \text{ per gallon} = \frac{\$2.99}{\text{gal}} \approx \frac{\$2.99}{\text{gal}} \cdot \frac{1 \text{ gal}}{3.79 \text{ L}}$$

$$\approx \$.789 \text{ per liter}$$

The price is approximately $.789 per liter.

YOU TRY IT 7

The price of milk is $3.59 per gallon. Find the price per liter. Round to the nearest cent.

Your Solution

EXAMPLE 8

The price of gasoline is $.699 per liter. Find the price per gallon. Round to the nearest cent.

Solution

$$\$.699 \text{ per liter} = \frac{\$.699}{1 \text{ L}} \approx \frac{\$.699}{1 \text{ L}} \cdot \frac{3.79 \text{ L}}{1 \text{ gal}}$$

$$\approx \$2.65 \text{ per gallon}$$

The price is approximately $2.65 per gallon.

YOU TRY IT 8

The price of ice cream is $2.65 per liter. Find the price per gallon. Round to the nearest cent.

Your Solution

Solutions on pp. S18–S19

7.3 Exercises

OBJECTIVE A **The U.S. Customary System of Measurement**

1. To write a conversion rate, place the unit in the given measurement in the _____ of the rate, and place the desired unit in the _____ of the rate.

2. Convert 102 ft to yards.
   a. The equivalence is 1 _____ = 3 _____ . The conversion rate must have _____ in the numerator and _____ in the denominator. The

   conversion rate is ——— .

   b. Multiply 102 ft by 1 in the form of the conversion rate.

   $$102 \text{ ft} \cdot 1 = \frac{102 \text{ ft}}{1} \cdot \underline{}$$

   $$= \frac{102 \text{ ft} \cdot 1 \text{ yd}}{\underline{}}$$

   c. Divide the numerator and denominator by the common unit, feet. Then divide the number in the numerator by the number in the denominator.

   $$= \frac{102 \text{ ft} \cdot 1 \text{ yd}}{3 \text{ ft}}$$

   $$= \frac{102 \text{ yd}}{3} = \underline{} \text{ yd}$$

3. Convert 64 in. to feet.

4. Convert 14 ft to yards.

5. Convert 42 oz to pounds.

6. Convert 4,400 lb to tons.

7. Convert 7,920 ft to miles.

8. Convert 42 c to quarts.

9. Convert 500 lb to tons.

10. Convert 90 oz to pounds.

11. Convert 10 qt to gallons.

12. How many pounds are in $1\frac{1}{4}$ tons?

13. How many fluid ounces are in $2\frac{1}{2}$ c?

14. How many ounces are in $2\frac{5}{8}$ lb?

15. Convert $2\frac{1}{4}$ mi to feet.

16. Convert 17 c to quarts.

17. Convert $7\frac{1}{2}$ in. to feet.

18. Convert $2\frac{1}{4}$ gal to quarts.

19. Convert 60 fl oz to cups.

20. Convert $1\frac{1}{2}$ qt to cups.

21. Convert $7\frac{1}{2}$ pt to quarts.

22. Convert 20 fl oz to pints.

23. Write a conversion rate for changing days into seconds.

24. Write a conversion rate for changing miles into inches.

OBJECTIVE B **Applications**

For Exercises 25 and 26, use the following information: One bottle of vitamin water contains 20 fl oz. The unit cost of the vitamin water is $3.81 per quart.

25. Complete the strategy to find the cost per bottle of vitamin water.

a. Find the number of quarts per bottle by multiplying $\dfrac{20 \text{ fl oz}}{1 \text{ bottle}}$ by two conversion factors: $\dfrac{20 \text{ fl oz}}{1 \text{ bottle}} \cdot \dfrac{\boxed{\ } \text{ c}}{\boxed{\ } \text{ fl oz}} \cdot \dfrac{\boxed{\ } \text{ qt}}{\boxed{\ } \text{ c}}$

b. Find the cost per bottle by multiplying the number of quarts per bottle by the unit cost, $\dfrac{\$\boxed{\ }}{1 \text{ qt}}$.

26. Complete the strategy to find the number of gallons of vitamin water in a case of 24 bottles.

a. Find the number of fluid ounces in a case:

$\dfrac{20 \text{ fl oz}}{1 \text{ bottle}} \cdot \dfrac{\boxed{\ } \text{ bottles}}{1 \text{ case}} = \dfrac{480 \text{ fl oz}}{1 \text{ case}}$

b. Convert the number of fluid ounces in a case to gallons by multiplying by three conversion factors:

$\dfrac{480 \text{ fl oz}}{1 \text{ case}} \cdot \dfrac{\boxed{\ } \text{ c}}{\boxed{\ } \text{ fl oz}} \cdot \dfrac{\boxed{\ } \text{ qt}}{\boxed{\ } \text{ c}} \cdot \dfrac{\boxed{\ } \text{ gal}}{\boxed{\ } \text{ qt}}$

27. *Time* When a person reaches the age of 35, for how many seconds has that person lived?

28. *Interior Decorating* Fifty-eight feet of material are purchased for making pleated draperies. Find the total cost of the material if the price is $39 per yard.

When three terms of a proportion are given, the fourth term can be found. To solve a proportion for an unknown term, use the fact that the product of the means equals the product of the extremes.

Solve: $\dfrac{n}{5} = \dfrac{9}{16}$

$$\frac{n}{5} = \frac{9}{16}$$

Find the number (n) that will make the proportion true.

The product of the means equals the product of the extremes. Solve for n.

$$5 \cdot 9 = n \cdot 16$$

$$45 = 16n$$

$$\frac{45}{16} = \frac{16n}{16}$$

$$2.8125 = n$$

Calculator Note

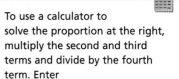

To use a calculator to solve the proportion at the right, multiply the second and third terms and divide by the fourth term. Enter

$$5 \times 9 \div 16 =$$

The display reads 2.8125.

EXAMPLE 1 Determine whether $\dfrac{15}{3} = \dfrac{90}{18}$ is a true proportion.

Solution $\dfrac{15}{3} \diagdown \dfrac{90}{18} \longrightarrow 3 \cdot 90 = 270$
$\phantom{\dfrac{15}{3}} \nearrow \phantom{\dfrac{90}{18}} \longrightarrow 15 \cdot 18 = 270$

The product of the means equals the product of the extremes.

The proportion is true.

YOU TRY IT 1 Is $\dfrac{50 \text{ mi}}{3 \text{ gal}} = \dfrac{250 \text{ mi}}{12 \text{ gal}}$ a true proportion?

Your Solution

EXAMPLE 2 Solve: $\dfrac{5}{9} = \dfrac{x}{45}$

Solution $\dfrac{5}{9} = \dfrac{x}{45}$

$$9 \cdot x = 5 \cdot 45$$

$$9x = 225$$

$$\frac{9x}{9} = \frac{225}{9}$$

$$x = 25$$

YOU TRY IT 2 Solve: $\dfrac{7}{12} = \dfrac{42}{x}$

Your Solution

EXAMPLE 3 Solve: $\dfrac{6}{n} = \dfrac{45}{124}$. Round to the nearest tenth.

Solution $\dfrac{6}{n} = \dfrac{45}{124}$

$$n \cdot 45 = 6 \cdot 124$$

$$45n = 744$$

$$\frac{45n}{45} = \frac{744}{45}$$

$$n \approx 16.5$$

YOU TRY IT 3 Solve: $\dfrac{5}{n} = \dfrac{3}{322}$. Round to the nearest hundredth.

Your Solution

Solutions on p. S19

EXAMPLE 4   Solve: $\dfrac{x+2}{3} = \dfrac{7}{8}$

Solution $\dfrac{x+2}{3} = \dfrac{7}{8}$

$$3 \cdot 7 = (x+2)8$$

$$21 = 8x + 16$$

$$5 = 8x$$

$$0.625 = x$$

YOU TRY IT 4   Solve: $\dfrac{4}{5} = \dfrac{3}{x-3}$

Your Solution

Solution on p. S19

OBJECTIVE B   Applications

Proportions are useful in many types of application problems. In cooking, proportions are used when a larger batch of ingredients is used than the recipe calls for. In mixing cement, the amounts of cement, sand, and rock are mixed in the same ratio. A map is drawn on a proportional basis, such as 1 in. representing 50 mi.

In setting up a proportion, keep the same units in the numerators and the same units in the denominators. For example, if *feet* is in the numerator on one side of the proportion, then *feet* must be in the numerator on the other side of the proportion.

A customer sees an ad in a newspaper advertising 2 tires for $162.50. The customer wants to buy 5 tires and use one for the spare. How much will the 5 tires cost?

Write a proportion.
Let c = the cost of the 5 tires.

$$\dfrac{2 \text{ tires}}{\$162.50} = \dfrac{5 \text{ tires}}{c}$$

$$162.50 \cdot 5 = 2 \cdot c$$

$$812.50 = 2c$$

$$\dfrac{812.50}{2} = \dfrac{2c}{2}$$

$$406.25 = c$$

Take Note

It is also correct to write the proportion with the costs in the numerators and the number of tires in the denominators:

$\dfrac{\$162.50}{2 \text{ tires}} = \dfrac{c}{5 \text{ tires}}$. The solution

will be the same.

The 5 tires will cost $406.25.

EXAMPLE 5

During a Friday, the ratio of stocks declining in price to those advancing was 5 to 3. If 450,000 shares advanced, how many shares declined on that day?

Strategy

To find the number of shares declining in price, write and solve a proportion using n to represent the number of shares declining in price.

Solution

$$\frac{5 \text{ (declining)}}{3 \text{ (advancing)}} = \frac{n \text{ shares declining}}{450,000 \text{ shares advancing}}$$

$$3n = 5 \cdot 450,000$$

$$3n = 2,250,000$$

$$\frac{3n}{3} = \frac{2,250,000}{3}$$

$$n = 750,000$$

750,000 shares declined in price.

YOU TRY IT 5

An automobile can travel 396 mi on 11 gal of gas. At the same rate, how many gallons of gas would be necessary to travel 832 mi? Round to the nearest tenth.

Your Strategy

Your Solution

EXAMPLE 6

From previous experience, a manufacturer knows that in an average production run of 5,000 calculators, 40 will be defective. What number of defective calculators can be expected in a run of 45,000 calculators?

Strategy

To find the number of defective calculators, write and solve a proportion using n to represent the number of defective calculators.

Solution

$$\frac{40 \text{ defective calculators}}{5,000 \text{ calculators}} = \frac{n \text{ defective calculators}}{45,000 \text{ calculators}}$$

$$5,000 \cdot n = 40 \cdot 45,000$$

$$5,000n = 1,800,000$$

$$\frac{5,000n}{5,000} = \frac{1,800,000}{5,000}$$

$$n = 360$$

The manufacturer can expect 360 defective calculators.

YOU TRY IT 6

An automobile recall was based on tests that showed 15 transmission defects in 1,200 cars. At this rate, how many defective transmissions will be found in 120,000 cars?

Your Strategy

Your Solution

Solutions on p. S19

7.4 Exercises

OBJECTIVE A **Proportion**

1. An equation of the form $\frac{a}{b} = \frac{c}{d}$ that states that two ratios or rates are equal is called
   a _____ . The extremes are the terms _____ and _____ , and the means
   are the terms _____ and _____ . In a true proportion, the product of the means
   and the product of the extremes are _____ .

2. The first step in solving the proportion $\frac{x}{85} = \frac{5}{17}$ is to write the equation that states
   that the product of the means equals the product of the extremes:
   $85 \cdot$ _____ $=$ _____ $\cdot\, 17$.

Determine whether the proportion is true or not true.

3. $\dfrac{27}{8} = \dfrac{9}{4}$

4. $\dfrac{3}{18} = \dfrac{4}{19}$

5. $\dfrac{45}{135} = \dfrac{3}{9}$

6. $\dfrac{3}{4} = \dfrac{54}{72}$

7. $\dfrac{6 \text{ min}}{5 \text{ cents}} = \dfrac{30 \text{ min}}{25 \text{ cents}}$

8. $\dfrac{7 \text{ tiles}}{4 \text{ ft}} = \dfrac{42 \text{ tiles}}{20 \text{ ft}}$

9. $\dfrac{300 \text{ ft}}{4 \text{ rolls}} = \dfrac{450 \text{ ft}}{7 \text{ rolls}}$

10. $\dfrac{\$65}{5 \text{ days}} = \dfrac{\$26}{2 \text{ days}}$

Solve. Round to the nearest hundredth.

11. $\dfrac{2}{3} = \dfrac{n}{15}$

12. $\dfrac{7}{15} = \dfrac{n}{15}$

13. $\dfrac{n}{5} = \dfrac{12}{25}$

14. $\dfrac{n}{8} = \dfrac{7}{8}$

15. $\dfrac{3}{8} = \dfrac{n}{12}$

16. $\dfrac{5}{8} = \dfrac{40}{n}$

17. $\dfrac{3}{n} = \dfrac{7}{40}$

18. $\dfrac{7}{12} = \dfrac{25}{n}$

19. $\dfrac{16}{n} = \dfrac{25}{40}$

20. $\dfrac{15}{45} = \dfrac{72}{n}$

21. $\dfrac{120}{n} = \dfrac{144}{25}$

22. $\dfrac{65}{20} = \dfrac{14}{n}$

23. $\dfrac{0.5}{2.3} = \dfrac{n}{20}$

24. $\dfrac{1.2}{2.8} = \dfrac{n}{32}$

25. $\dfrac{0.7}{1.2} = \dfrac{6.4}{n}$

26. $\dfrac{2.5}{0.6} = \dfrac{165}{n}$

27. $\dfrac{x}{6.25} = \dfrac{16}{87}$

28. $\dfrac{x}{2.54} = \dfrac{132}{640}$

29. $\dfrac{1.2}{0.44} = \dfrac{y}{14.2}$

30. $\dfrac{12.5}{y} = \dfrac{102}{55}$

31. $\dfrac{n+2}{5} = \dfrac{1}{2}$ **32.** $\dfrac{5+n}{8} = \dfrac{3}{4}$ **33.** $\dfrac{4}{3} = \dfrac{n-2}{6}$ **34.** $\dfrac{3}{5} = \dfrac{n-7}{8}$

35. $\dfrac{2}{n+3} = \dfrac{7}{12}$ **36.** $\dfrac{5}{n+1} = \dfrac{7}{3}$ **37.** $\dfrac{7}{10} = \dfrac{3+n}{2}$ **38.** $\dfrac{3}{2} = \dfrac{5+n}{4}$

39. $\dfrac{x-4}{3} = \dfrac{3}{4}$ **40.** $\dfrac{x-1}{8} = \dfrac{5}{2}$ **41.** $\dfrac{6}{1} = \dfrac{x-2}{5}$ **42.** $\dfrac{7}{3} = \dfrac{x-4}{8}$

43. $\dfrac{5}{8} = \dfrac{2}{x-3}$ **44.** $\dfrac{5}{2} = \dfrac{1}{x-6}$ **45.** $\dfrac{3}{x-4} = \dfrac{5}{3}$ **46.** $\dfrac{8}{x-6} = \dfrac{5}{4}$

47. **a.** Write a true proportion using the numbers 9, 4, 2, and 18.
        b. Write a true proportion using only the numbers 8, 2, and 4.

48. **a.** Write a proportion in which the product of the means and the product of the extremes is 60.
        b. Using different numbers than you used in part (a), write another proportion in which the product of the means and the product of the extremes is 60.

OBJECTIVE B   **Applications**

For Exercises 49 and 50, use the information that Jane ran 4 mi in 50 min. Let n be the number of miles Jane can run in 30 min at the same rate.

49. To determine how many miles Jane can run in 30 min, solve the proportion $\dfrac{4 \text{ mi}}{50 \text{ min}} = \dfrac{\boxed{} \text{ mi}}{\boxed{} \text{ min}}$.

50. To determine how many miles Jane can run in 30 min, one student used the proportion $\dfrac{50}{30} = \dfrac{4}{n}$ and a second student used the proportion $\dfrac{30}{n} = \dfrac{50}{4}$. Can either of these proportions be used to solve this problem?

Solve.

51. *Biology*   In a drawing, the length of an amoeba is 2.6 in. The scale of the drawing is 1 in. on the drawing equals 0.002 in. on the amoeba. Find the actual length of the amoeba.

52. *Insurance* A life insurance policy costs $15.22 for every $1,000 of insurance. At this rate, what is the cost of $75,000 of insurance?

53. *Sewing* Six children's robes can be made from 6.5 yd of material. How many robes can be made from 26 yd of material?

54. *Computers* A computer manufacturer finds that an average of 3 defective hard drives are found in every 100 drives manufactured. How many defective drives are expected to be found in the production of 1,200 hard drives?

55. *Taxes* The property tax on a $180,000 home is $4,320. At this rate, what is the property tax on a home appraised at $280,000?

56. *Medicine* The dosage of a certain medication is 2 mg for every 80 lb of body weight. How many milligrams of this medication are required for a person who weighs 220 lb?

57. *Travel* An automobile was driven 84 mi and used 3 gal of gasoline. At the same rate of consumption, how far would the car travel on 14.5 gal of gasoline?

58. *Nutrition* If a 56-gram serving of pasta contains 7 g of protein, how many grams of protein are in a 454-gram box of the pasta?

59. *Construction* A building contractor estimates that 5 overhead lights are needed for every 400 ft² of office space. Using this estimate, how many light fixtures are necessary for an office building of 35,000 ft²?

60. *Sports* A softball player has hit 9 home runs in 32 games. At the same rate, how many home runs will the player hit in a 160-game schedule?

61. *Health* A dieter has lost 3 lb in 5 weeks. At this rate, how long will it take the dieter to lose 36 lb?

62. *Business* An automobile recall was based on engineering tests that showed 22 defects in 1,000 cars. At this rate, how many defects would be found in 125,000 cars?

63. *Health*   Walking 5 mi in 2 h will burn 650 calories. Walking at the same rate, how many miles would a person need to walk to lose 1 lb? (The burning of 3,500 calories is equivalent to the loss of 1 lb.) Round to the nearest hundredth.

64. *Travel*   An account executive bought a new car and drove 22,000 mi in the first 4 months. At the same rate, how many miles will the account executive drive in 3 years?

65. *Investments*   An investment of $1,500 earns $120 each year. At the same rate, how much additional money must be invested to earn $300 each year?

$1,500	$1,500 + x
earns	earns
$120	$300

66. *Investments*   A stock investment of $3,500 earns a dividend of $280. At the same rate, how much additional money would have to be invested so that the total dividend is $400?

67. *Cartography*   The scale on a map is $\frac{1}{2}$ in. equals 8 mi. What is the actual distance between two points that are $1\frac{1}{4}$ in. apart on the map?

68. *Energy*   A slow-burning candle will burn 1.5 in. in 40 min. How many inches of the candle will burn in 4 h?

69. *Mixtures*   A saltwater solution is made by dissolving $\frac{2}{3}$ lb of salt in 5 gal of water. At this rate, how many pounds of salt are required for 12 gal of water?

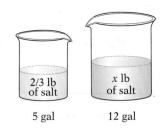

2/3 lb of salt   x lb of salt

5 gal   12 gal

70. *Business*   A management consulting firm recommends that the ratio of midmanagement salaries to junior management salaries be 7:5. Using this recommendation, find the yearly midmanagement salary when the junior management salary is $90,000.

CRITICAL THINKING

71. *Elections*   A survey of voters in a city claimed that 2 people of every 5 who voted cast a ballot in favor of city amendment A, and 3 people of every 4 who voted cast a ballot against amendment A. Is this possible? Explain your answer.

72. Determine whether the statement is true or false.

a. A quotient $(a \div b)$ is a ratio.

b. If $\frac{a}{b} = \frac{c}{d}$, then $\frac{b}{a} = \frac{d}{c}$.

c. If $\frac{a}{b} = \frac{c}{d}$, then $\frac{a}{c} = \frac{b}{d}$.

d. If $\frac{a}{b} = \frac{c}{d}$, then $\frac{a}{d} = \frac{c}{b}$.

7.5 Direct and Inverse Variation

OBJECTIVE A   Direct variation

An equation of the form $y = kx$ can be used to describe many important relationships in business, science, and engineering. The equation $y = kx$, where k is a constant, is an example of a **direct variation.** The constant k is called the **constant of variation** or the **constant of proportionality.** The equation $y = kx$ is read "y varies directly as x."

For example, the distance traveled by a car traveling at a constant rate of 55 mph is represented by $y = 55x$, where x is the number of hours and y is the total distance traveled. The number 55 is the constant of proportionality.

The distance (d) sound travels varies directly as the time (t) it travels. If sound travels 8,920 ft in 8 s, find the distance sound travels in 3 s.

This is a direct variation. k is the constant of proportionality.

$$d = kt$$

Substitute 8,920 for d and 8 for t.

$$8,920 = k \cdot 8$$

Solve for k.

$$\frac{8,920}{8} = \frac{k \cdot 8}{8}$$
$$1,115 = k$$

Write the direct variation equation for d, substituting 1,115 for k.
Find d when $t = 3$.

$$d = 1,115t$$

$$d = 1,115 \cdot 3$$
$$d = 3,345$$

Sound travels 3,345 ft in 3 s.

A direct variation equation can be written in the form $y = kx^n$, where n is a positive integer. For example, the equation $y = kx^2$ is read "y varies directly as the square of x."

The load (L) that a horizontal beam can safely support is directly proportional to the square of the depth (d) of the beam. A beam with a depth of 8 in. can support 800 lb. Find the load that a beam with a depth of 6 in. can support.

This is a direct variation. k is the constant of proportionality.

$$L = kd^2$$

Substitute 800 for L and 8 for d.

$$800 = k \cdot 8^2$$
$$800 = k \cdot 64$$

Solve for k.

$$\frac{800}{64} = \frac{k \cdot 64}{64}$$
$$12.5 = k$$

Write the direct variation equation for L, substituting 12.5 for k.
Find L when $d = 6$.

$$L = 12.5d^2$$

$$L = 12.5 \cdot 6^2$$
$$L = 12.5 \cdot 36$$
$$L = 450$$

The beam can support a load of 450 lb.

EXAMPLE 1

Find the constant of variation if y varies directly as x, and $y = 5$ when $x = 35$.

Strategy

To find the constant of variation, substitute 5 for y and 35 for x in the direct variation equation $y = kx$ and solve for k.

Solution

$y = kx$

$5 = k \cdot 35$ ▸ $y = 5$ when $x = 35$.

$\dfrac{5}{35} = \dfrac{k \cdot 35}{35}$ ▸ Solve for k.

$\dfrac{1}{7} = k$

The constant of variation is $\dfrac{1}{7}$.

YOU TRY IT 1

Find the constant of variation if y varies directly as x, and $y = 120$ when $x = 8$.

Your Strategy

Your Solution

EXAMPLE 2

Given that L varies directly as P, and $L = 24$ when $P = 16$, find P when $L = 80$. Round to the nearest tenth.

Strategy

To find P when $L = 80$:

▸ Write the basic direct variation equation, replace the variables by the given values, and solve for k.

▸ Write the direct variation equation, replacing k by its value. Substitute 80 for L and solve for P.

Solution

$L = kP$

$24 = k \cdot 16$ ▸ $L = 24$ when $P = 16$.

$\dfrac{24}{16} = \dfrac{k \cdot 16}{16}$ ▸ Solve for k.

$1.5 = k$

$L = 1.5P$ ▸ $k = 1.5$

$80 = 1.5P$ ▸ $L = 80$

$\dfrac{80}{1.5} = \dfrac{1.5P}{1.5}$ ▸ Solve for P.

$53.3 \approx P$

The value of P is approximately 53.3 when $L = 80$.

YOU TRY IT 2

Given that S varies directly as R, and $S = 8$ when $R = 30$, find S when $R = 200$. Round to the nearest tenth.

Your Strategy

Your Solution

Solutions on p. S19

EXAMPLE 3

The distance (d) required for a car to stop varies directly as the square of the velocity (v) of the car. If a car traveling at 40 mph requires 130 ft to stop, find the stopping distance for a car traveling at 60 mph.

Strategy

To find the stopping distance:

▶ Write the basic direct variation equation, replace the variables by the given values, and solve for k.
▶ Write the direct variation equation, replacing k by its value. Substitute 60 for v and solve for d.

Solution

$$d = kv^2$$ ▶ $d = 130$ when $v = 40$.
$$130 = k \cdot 40^2$$ ▶ Solve for k.
$$130 = k \cdot 1{,}600$$
$$0.08125 = k$$

$$d = 0.08125 \cdot v^2$$
$$= 0.08125 \cdot 60^2$$ ▶ $v = 60$
$$= 0.08125 \cdot 3{,}600 = 292.5$$ ▶ Solve for d.

The stopping distance is 292.5 ft.

YOU TRY IT 3

The distance (d) a body falls from rest varies directly as the square of the time (t) of the fall. An object falls 64 ft in 2 s. How far will the object fall in 9 s?

Your Strategy

Your Solution

Solution on p. S19

OBJECTIVE B Inverse variation

The equation $y = \dfrac{k}{x}$, where k is a constant, is an example of an **inverse variation**. The equation $y = \dfrac{k}{x}$ is read "y varies inversely as x" or "y is inversely proportional to x." In general, an inverse variation equation can be written $y = \dfrac{k}{x^n}$, where n is a positive integer. For example, the equation $y = \dfrac{k}{x^2}$ is read "y varies inversely as the square of x."

The volume (V) of a gas at a fixed temperature varies inversely as the pressure (P). The inverse variation equation would be written as

$$V = \frac{k}{P}$$

The gravitational force (F) between two planets is inversely proportional to the square of the distance (d) between the planets. This inverse variation would be written as

$$F = \frac{k}{d^2}$$

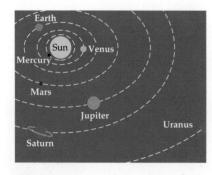

Given that y varies inversely as the square of x, and $y = 5$ when $x = 2$, find y when $x = 40$.

Write the basic inverse variation equation, where y varies inversely as the square of x.	$y = \dfrac{k}{x^2}$
Replace x and y by the given values.	$5 = \dfrac{k}{2^2}$
Solve for the constant of variation.	$20 = k$
Write the inverse variation equation by substituting the value of k into the basic inverse variation equation.	$y = \dfrac{20}{x^2}$
To find y when $x = 40$, substitute 40 for x in the equation and solve for y.	$y = \dfrac{20}{40^2}$
	$y = 0.0125$

EXAMPLE 4

A company that produces personal computers has determined that the number of computers it can sell (S) is inversely proportional to the price (P) of the computer. Two thousand computers can be sold when the price is $2,500. How many computers can be sold when the price of a computer is $2,000?

Strategy

To find the number of computers:

▸ Write the basic inverse variation equation, replace the variables by the given values, and solve for k.
▸ Write the inverse variation equation, replacing k by its value. Substitute 2,000 for the price and solve for the number sold.

Solution

$$S = \frac{k}{P}$$

$$2{,}000 = \frac{k}{2{,}500} \qquad ▸ S = 2{,}000 \text{ when } P = 2{,}500.$$

$$5{,}000{,}000 = k \qquad ▸ \text{ Solve for } k.$$

$$S = \frac{5{,}000{,}000}{P}$$

$$= \frac{5{,}000{,}000}{2{,}000} = 2{,}500 \qquad ▸ P = 2{,}000$$

When the price is $2,000, 2,500 computers can be sold.

YOU TRY IT 4

The resistance (R) to the flow of electric current in a wire of fixed length is inversely proportional to the square of the diameter (d) of the wire. If a wire of diameter 0.01 cm has a resistance of 0.5 ohm, what is the resistance in a wire that is 0.02 cm in diameter?

Your Strategy

Your Solution

Solution on p. S20

7.5 Exercises

OBJECTIVE A **Direct variation**

1. Which of the following are direct variations? Why?

   a. $y = kx$ **b.** $y = \dfrac{k}{x}$ **c.** $y = k + x$ **d.** $y = \dfrac{k}{x^2}$

2. To find the constant of variation given that y varies directly as x and $y = 14$ when $x = 3$, first replace x by _____ and y by _____ in the equation $y = kx$. Then solve for _____, the constant of variation.

3. Given that n varies directly as m, and $n = 1.8$ when $m = 1.5$, find n when $m = 3.5$.

   a. Write the direct variation equation. Replace the variables m and n by the given values.

   $n = km$

   _____ $= k \cdot$ _____

   b. To solve for k, _____ each side by _____ .

   _____ $= k$

   c. Write the direct variation equation, replacing k by the value from part (b).

   $n =$ _____m

   d. To find n when $m = 3.5$, use the direct variation equation from part (c). Replace m by _____ and solve for n.

   $n = 1.2 \cdot$ _____

   $n =$ _____

   e. The value of n when m is 3.5 is _____ .

4. Find the constant of variation when y varies directly as x, and $y = 12$ when $x = 2$.

5. Find the constant of variation when t varies directly as s, and $t = 25$ when $s = 150$.

6. Find the constant of variation when n varies directly as the square of m, and $n = 64$ when $m = 2$.

7. Find the constant of variation when z varies directly as the square of y, and $z = 54$ when $y = 3$.

8. Given that P varies directly as R, and $P = 20$ when $R = 5$, find P when $R = 8$.

9. Given that T varies directly as S, and $T = 36$ when $S = 9$, find T when $S = 7$.

10. Given that M is directly proportional to P, and $M = 15$ when $P = 30$, find M when $P = 10$.

11. Given that A is directly proportional to B, and $A = 6$ when $B = 18$, find A when $B = 30$.

12. Given that y is directly proportional to the square of x, and $y = 10$ when $x = 2$, find y when $x = 1$.

13. Given that W is directly proportional to the square of V, and $W = 50$ when $V = 5$, find W when $V = 9$.

14. 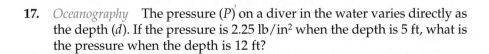 If $y = kx$ is a direct variation equation in which $0 < k < 1$ and $y = b$ when $x = a$, is $a > b$ or is $b > a$?

Solve.

15. *Compensation*   A worker's wage (w) is directly proportional to the number of hours (h) worked. If \$82 is earned for working 8 h, how much is earned for working 30 h?

16. *Mechanics*   The distance (d) a spring will stretch varies directly as the force (F) applied to the spring. If a force of 12 lb is required to stretch a spring 3 in., what force is required to stretch the spring 5 in.?

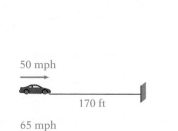

17. *Oceanography*   The pressure (P) on a diver in the water varies directly as the depth (d). If the pressure is 2.25 lb/in² when the depth is 5 ft, what is the pressure when the depth is 12 ft?

18. *Computers*   The number of words typed (w) is directly proportional to the time (t) spent typing. A typist can type 260 words in 4 min. Find the number of words typed in 15 min.

19. *Travel*   The stopping distance (s) of a car varies directly as the square of its speed (v). If a car traveling at 50 mph requires 170 ft to stop, find the stopping distance for a car traveling at 65 mph.

20. *Physics*   The distance (d) an object falls is directly proportional to the square of the time (t) of the fall. If an object falls a distance of 8 ft in 0.5 s, how far will the object fall in 5 s?

21. *Energy*   The current (I) varies directly as the voltage (V) in an electric circuit. If the current is 4 amps when the voltage is 100 volts, find the current when the voltage is 75 volts.

22. *Travel*   The distance traveled (d) varies directly as the time (t) of travel, assuming that the speed is constant. If it takes 45 min to travel 50 mi, how many hours would it take to travel 180 mi?

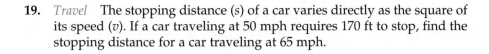

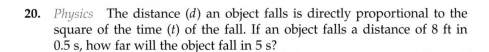

OBJECTIVE B   **Inverse variation**

23. 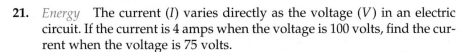 Which of the following equations represents "d varies inversely as t"? Explain your answer.

    a. $d = 400t$ **b.** $d = \dfrac{16}{t}$ **c.** $d = \dfrac{t}{25}$ **d.** $d = t - 50$

24. Find the constant of variation when B varies inversely as A^2, and $B = 8$ when $A = 2$.

    a. Write the inverse variation equation. Replace the variables A and B by the given values.

$$B = \frac{k}{A^2}$$

$$\underline{\hspace{2cm}} = \frac{k}{\Box}$$

$$8 = \frac{k}{\Box}$$

    b. Simplify the denominator.

    c. To solve for k, _____ each side by _____ .

$$\underline{\hspace{2cm}} = k$$

    d. The constant of variation is _____ .

25. Find the constant of variation when y varies inversely as x, and $y = 8$ when $x = 5$.

26. Find the constant of proportionality when t varies inversely as s, and $t = 0.5$ when $s = 10$.

27. Find the constant of variation when P varies inversely as the square of Q, and $P = 6$ when $Q = 2$.

28. Find the constant of variation when w varies inversely as the square of v, and $w = 5$ when $v = 3$.

29. If y varies inversely as x, and $y = 50$ when $x = 4$, find y when $x = 20$.

30. If L varies inversely as W, and $L = 12$ when $W = 5$, find L when $W = 2$.

31. If y varies inversely as the square of x, and $y = 40$ when $x = 2$, find y when $x = 4$.

32. If L varies inversely as the square of d, and $L = 25$ when $d = 2$, find L when $d = 5$.

Solve.

33. *Geometry* The length (L) of a rectangle of fixed area varies inversely as the width (W). If the length of the rectangle is 8 ft when the width is 5 ft, find the length of the rectangle when the width is 4 ft.

34. *Travel* The time (t) of travel of an automobile trip varies inversely as the speed (v). At an average speed of 65 mph, a trip took 4 h. The return trip took 5 h. Find the average speed of the return trip.

35. *Energy* The current (I) in an electric circuit is inversely proportional to the resistance (R). If the current is 0.25 amp when the resistance is 8 ohms, find the resistance when the current is 1.2 amps.

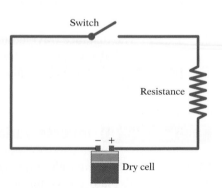

36. *Physics* The volume (V) of a gas varies inversely as the pressure (P) on the gas. If the volume of the gas is 12 ft³ when the pressure is 15 lb/ft², find the volume of the gas when the pressure is 4 lb/ft².

Percent

The young woman in this photo is interested in purchasing a car. She must make decisions about the make and model car she will buy, as well as the price she can afford to pay for it. She knows that the price of the car determines the amount of sales tax she will pay.

After she purchases the car, this woman will have ongoing expenses. These include the monthly car payment, auto insurance, gasoline, and expenses for upkeep; for example, paying for tune-ups and new tires. The **Project on page 526** illustrates how to determine monthly expenses for maintaining a car.

DVD

SSM

Student Website

Need help? For online student resources,
visit **college.hmco.com/pic/aufmannPA5e.**

For Exercises 1 to 6, multiply or divide.

1. $19 \times \dfrac{1}{100}$

2. 23×0.01

3. 0.47×100

4. $0.06 \times 47{,}500$

5. $60 \div 0.015$

6. $8 \div \dfrac{1}{4}$

7. Multiply $\dfrac{5}{8} \times 100$. Write the answer as a decimal.

8. Write $\dfrac{200}{3}$ as a mixed number.

9. Divide $28 \div 16$. Write the answer as a decimal.

GO Figure

I have two brothers and one sister. My father's parents have 10 grandchildren. My mother's parents have 11 grandchildren. If no divorces or remarriages occurred, how many first cousins do I have?

8.1 Percent

OBJECTIVE A   Percents as decimals or fractions

Percent means "parts of 100." The figure at the right has 100 parts. Because 19 of the 100 parts are shaded, 19% of the figure is shaded.

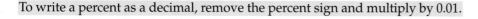

19 parts to 100 parts can be expressed as the ratio $\frac{19}{100}$. One percent can be expressed as 1 part to 100, or $\frac{1}{100}$. Thus 1% is $\frac{1}{100}$ or 0.01.

"A population growth rate of 5%," "a manufacturer's discount of 40%," and "an 8% increase in pay" are typical examples of the many ways in which percent is used in applied problems. When solving problems involving a percent, it is usually necessary either to rewrite the percent as a fraction or a decimal, or to rewrite a fraction or a decimal as a percent.

To write a percent as a fraction, remove the percent sign and multiply by $\frac{1}{100}$.

Write 67% as a fraction.

Remove the percent sign and multiply by $\frac{1}{100}$. $67\% = 67\left(\frac{1}{100}\right) = \frac{67}{100}$

To write a percent as a decimal, remove the percent sign and multiply by 0.01.

Write 19% as a decimal.

Remove the percent sign and multiply by 0.01. This is the same as moving the decimal point two places to the left.

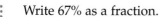

$$19\% \quad = \quad 19(0.01) \quad = \quad 0.19$$

Move the decimal point two places to the left. Then remove the percent sign.

EXAMPLE 1   Write 150% as a fraction and as a decimal.

Solution $150\% = 150\left(\frac{1}{100}\right) = \frac{150}{100} = 1\frac{1}{2}$
$150\% = 150(0.01) = 1.50$

YOU TRY IT 1   Write 110% as a fraction and as a decimal.

Your Solution

EXAMPLE 2   Write $66\frac{2}{3}\%$ as a fraction.

Solution $66\frac{2}{3}\% = 66\frac{2}{3}\left(\frac{1}{100}\right)$

$= \frac{200}{3}\left(\frac{1}{100}\right) = \frac{2}{3}$

YOU TRY IT 2   Write $16\frac{3}{8}\%$ as a fraction.

Your Solution

Solutions on p. S20

EXAMPLE 3 Write 0.35% as a decimal.

Solution 0.35% = 0.35(0.01) = 0.0035

YOU TRY IT 3 Write 0.8% as a decimal.

Your Solution

Solution on p. S20

OBJECTIVE B Fractions and decimals as percents

A fraction or decimal can be written as a percent by multiplying by 100%. Since 100% is $\frac{100}{100} = 1$, multiplying by 100% is the same as multiplying by 1.

Write $\frac{7}{8}$ as a percent.

Multiply $\frac{7}{8}$ by 100%.

$$\frac{7}{8} = \frac{7}{8}(100\%) = \frac{700}{8}\% = 87.5\%$$

Write 0.64 as a percent.

Multiply by 100%. This is the same as moving the decimal point two places to the right.

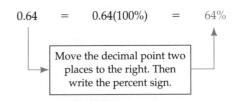

0.64 = 0.64(100%) = 64%

Move the decimal point two places to the right. Then write the percent sign.

EXAMPLE 4 Write 1.78 as a percent.

Solution 1.78 = 1.78(100%) = 178%

YOU TRY IT 4 Write 0.038 as a percent.

Your Solution

EXAMPLE 5 Write $\frac{3}{11}$ as a percent. Write the remainder in fractional form.

Solution $\frac{3}{11} = \frac{3}{11}(100\%) = \frac{300}{11}\%$

$= 27\frac{3}{11}\%$

YOU TRY IT 5 Write $\frac{9}{7}$ as a percent. Write the remainder in fractional form.

Your Solution

EXAMPLE 6 Write $1\frac{1}{7}$ as a percent. Round to the nearest tenth of a percent.

Solution $1\frac{1}{7} = \frac{8}{7} = \frac{8}{7}(100\%)$

$= \frac{800}{7}\% \approx 114.3\%$

YOU TRY IT 6 Write $1\frac{5}{9}$ as a percent. Round to the nearest tenth of a percent.

Your Solution

Solutions on p. S20

8.1 Exercises

OBJECTIVE A   **Percents as decimals or fractions**

1. **a.** Explain how to convert a percent to a fraction.
   b. Explain how to convert a percent to a decimal.

2. Explain why multiplying a number by 100% does not change the value of the number.

3. To write 53% as a decimal, remove the percent sign and multiply by _____:
   $53\% = 53 \cdot$ _____ $=$ _____ .

4. To write 6.7% as a decimal, remove the percent sign and multiply by _____:
   $6.7\% = 6.7 \cdot$ _____ $=$ _____ .

5. To write 80% as a fraction, remove the percent sign and multiply by _____ :
   $80\% = 80 \cdot$ _____ $=$ _____ .

6. To write $4\frac{1}{3}\%$ as a fraction, remove the percent sign and multiply by _____ :
   $4\frac{1}{3}\% = 4\frac{1}{3} \cdot$ _____ $= \dfrac{}{3} \cdot \dfrac{1}{100} =$ _____ .

Write as a fraction and as a decimal.

7. 5%

8. 60%

9. 30%

10. 90%

11. 250%

12. 140%

13. 28%

14. 66%

15. 35%

16. 8%

17. 29%

18. 83%

Write as a fraction.

19. $11\frac{1}{9}\%$

20. $12\frac{1}{2}\%$

21. $37\frac{1}{2}\%$

22. $31\frac{1}{4}\%$

23. $66\frac{2}{3}\%$

24. $45\frac{5}{11}\%$

25. $6\frac{2}{3}\%$

26. $68\frac{3}{4}\%$

27. $\frac{1}{2}\%$

28. $83\frac{1}{3}\%$

29. $6\frac{1}{4}\%$

30. $3\frac{1}{3}\%$

Write as a decimal.

31. 7.3%   **32.** 9.1%   **33.** 15.8%   **34.** 16.7%

35. 0.3%   **36.** 0.9%   **37.** 121.2%   **38.** 18.23%

39. 62.14%   **40.** 0.15%   **41.** 8.25%   **42.** 5.05%

43. When a certain percent is written as a fraction, the result is a proper fraction. Is the percent less than, equal to, or greater than 100%?

44. When a certain percent is written as a fraction, the result is an improper fraction. Is the percent less than, equal to, or greater than 100%?

45. *Pets* The figure at the right shows some ways in which owners pamper their dogs. What fraction of the owners surveyed would buy a house or a car with their dog in mind?

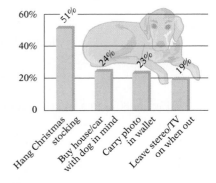

How Owners Pamper Their Dogs
Source: Purina Survey

OBJECTIVE B Fractions and decimals as percents

46. To write 0.46 as a percent, multiply by _____:
$0.46 = 0.46 \cdot$ _____ = _____.

47. To write 1.25 as a percent, multiply by _____:
$1.25 = 1.25 \cdot$ _____ = _____.

48. To write $\frac{3}{10}$ as a percent, multiply by _____:
$\frac{3}{10} = \frac{3}{10} \cdot$ _____ = _____.

49. To write $\frac{7}{5}$ as a percent, multiply by _____:
$\frac{7}{5} = \frac{7}{5} \cdot$ _____ = $\dfrac{\rule{2cm}{0.4pt}}{5}$% = _____.

Write as a percent.

50. 0.15   **51.** 0.37   **52.** 0.05   **53.** 0.02

54. 0.175   **55.** 0.125   **56.** 1.15   **57.** 1.36

58. 0.62   **59.** 0.96   **60.** 2.09   **61.** 0.07

Write as a percent. Round to the nearest tenth of a percent.

62. $\dfrac{27}{50}$

63. $\dfrac{83}{100}$

64. $\dfrac{37}{200}$

65. $\dfrac{1}{3}$

66. $\dfrac{5}{11}$

67. $\dfrac{4}{9}$

68. $\dfrac{7}{8}$

69. $\dfrac{9}{20}$

70. $1\dfrac{2}{3}$

71. $2\dfrac{1}{2}$

72. $\dfrac{2}{5}$

73. $\dfrac{1}{6}$

Write as a percent. Write the remainder in fractional form.

74. $\dfrac{17}{50}$

75. $\dfrac{17}{25}$

76. $\dfrac{3}{8}$

77. $\dfrac{9}{16}$

78. $1\dfrac{1}{4}$

79. $2\dfrac{5}{8}$

80. $1\dfrac{5}{9}$

81. $2\dfrac{5}{6}$

82. $\dfrac{12}{25}$

83. $\dfrac{7}{30}$

84. $\dfrac{3}{7}$

85. $\dfrac{2}{9}$

Complete Exercises 86 and 87 without actually finding the percents.

86. Does $\dfrac{4}{3}$ represent a percent greater than 100% or less than 100%?

87. Does 0.055 represent a percent greater than 1% or less than 1%?

CRITICAL THINKING

88. Determine whether the statement is true or false. If the statement is false, give an example to show that it is false.
   a. Multiplying a number by a percent always decreases the number.
   b. Dividing by a percent always increases the number.
   c. The word *percent* means "per hundred."
   d. A percent is always less than 1.

89. *Compensation* Employee A had an annual salary of $42,000, Employee B had an annual salary of $48,000, and Employee C had an annual salary of $46,000 before each employee was given a 5% raise. Which of the three employee's annual salary is now the highest? Explain how you arrived at your answer.

90. *Compensation* Each of three employees earned an annual salary of $45,000 before Employee A was given a 3% raise, Employee B was given a 6% raise, and Employee C was given a 4.5% raise. Which of the three employees now has the highest annual salary? Explain how you arrived at your answer.

8.2 The Basic Percent Equation

OBJECTIVE A The basic percent equation

A real estate broker receives a payment that is 6% of a $275,000 sale. To find the amount the broker receives, we must answer the question, "6% of $275,000 is what?" This sentence can be written using mathematical symbols and then solved for the unknown number. Recall that *of* is written as · (times), *is* is written as = (equals), and *what* is written as *n* (the unknown number).

6%	of	$275,000	is	what?
↓	↓	↓	↓	↓
percent	·	base	=	amount
6%		$275,000		*n*

$$0.06 \cdot \$275{,}000 = n$$
$$\$16{,}500 = n$$

The broker receives a payment of $16,500.

The solution was found by solving the basic percent equation for amount.

The Basic Percent Equation

Percent · base = amount

Calculator Note

The percent key % on a scientific calculator moves the decimal point to the left two places when pressed after a multiplication or division computation. For the example at the left, enter

800 ⊠ 2 · 5 % =

The display reads 20.

Find 2.5% of 800.

Use the basic percent equation.
Percent = 2.5% = 0.025,
base = 800, amount = *n*

Percent · base = amount
$$0.025 \cdot 800 = n$$
$$20 = n$$

2.5% of 800 is 20.

A recent promotional game at a grocery store listed the probability of winning a prize as "1 chance in 2." A percent can be used to describe the chance of winning. This requires answering the question, "What percent of 2 is 1?"

The chance of winning can be found by solving the basic percent equation for percent.

What	percent	of	2	is	1?
	↓	↓	↓	↓	↓
	percent	·	base	=	amount
	n		2		1

$$n \cdot 2 = 1$$
$$n = \frac{1}{2}$$
$$n = \frac{1}{2}(100\%) = 50\%$$ ▶ Write the fraction as a percent.

There is a 50% chance of winning a prize.

32 is what percent of 20?

Use the basic percent equation.
Percent = n, base = 20, amount = 32

Write 1.6 as a percent.

32 is 160% of 20.

$$\text{Percent} \cdot \text{base} = \text{amount}$$
$$n \cdot 20 = 32$$
$$\frac{20n}{20} = \frac{32}{20}$$
$$n = 1.6$$
$$n = 160\%$$

> ## Take Note
> We have written $n \cdot 20 = 32$ because that is the form of the basic percent equation. We could have written $20n = 32$. The important point is that each side of the equation is divided by 20, the coefficient of n.

Each year an investor receives a payment that equals 8% of the value of an investment. This year that payment amounted to $640. To find the value of the investment this year, we must answer the question, "8% of what value is $640?"

The value of the investment can be found by solving the basic percent equation for the base.

8% of what is $640?

Percent · base = amount
8% n 640

$$0.08 \cdot n = 640$$
$$\frac{0.08n}{0.08} = \frac{640}{0.08}$$
$$n = 8{,}000$$

This year the investment is worth $8,000.

62% of what is 800? Round to the nearest tenth.

Use the basic percent equation.
Percent = 62% = 0.62, base = n,
amount = 800

$$\text{Percent} \cdot \text{base} = \text{amount}$$
$$0.62 \cdot n = 800$$
$$\frac{0.62n}{0.62} = \frac{800}{0.62}$$
$$n \approx 1{,}290.3$$

62% of 1,290.3 is approximately 800.

> ## Take Note
> The base in the basic percent equation will usually follow the phrase *percent of*. Some percent problems may use the word *find*. In this case, we can substitute *what is* for *find*. See Example 1 below.

Note from the previous problems that if any two parts of the basic percent equation are given, the third part can be found.

EXAMPLE 1 Find 9.4% of 240.

Strategy To find the amount, solve the basic percent equation.
Percent = 9.4% = 0.094, base = 240, amount = n

Solution Percent · base = amount
$$0.094 \cdot 240 = n$$
$$22.56 = n$$

22.56 is 9.4% of 240.

YOU TRY IT 1 Find $33\frac{1}{3}\%$ of 45.

Your Strategy

Your Solution

Solution on p. S20

EXAMPLE 2   What percent of 30 is 12?

Strategy   To find the percent, solve the basic percent equation.
Percent = n, base = 30, amount = 12

Solution   Percent · base = amount
$$n \cdot 30 = 12$$
$$\frac{30n}{30} = \frac{12}{30}$$
$$n = 0.4$$
$$n = 40\%$$

12 is 40% of 30.

YOU TRY IT 2   25 is what percent of 40?

Your Strategy

Your Solution

EXAMPLE 3   60 is 2.5% of what?

Strategy   To find the base, solve the basic percent equation.
Percent = 2.5% = 0.025, base = n, amount = 60

Solution   Percent · base = amount
$$0.025 \cdot n = 60$$
$$\frac{0.025n}{0.025} = \frac{60}{0.025}$$
$$n = 2{,}400$$

60 is 2.5% of 2,400.

YOU TRY IT 3 $16\frac{2}{3}$% of what is 15?

Your Strategy

Your Solution

Solutions on p. S20

OBJECTIVE B   **Percent problems using proportions**

Percent problems can also be solved by using proportions. The proportion method is based on writing two ratios with quantities that can be found in the basic percent equation. One ratio is the percent ratio, written as $\frac{\text{percent}}{100}$. The second ratio is the amount-to-base ratio, written as $\frac{\text{amount}}{\text{base}}$. These two ratios form the proportion

$$\frac{\textbf{percent}}{\textbf{100}} = \frac{\textbf{amount}}{\textbf{base}}$$

The proportion method can be illustrated by a diagram. The rectangle at the left is divided into two parts. The whole rectangle is represented by 100 and the part by percent. On the other side, the whole rectangle is represented by the base and the part by amount. The ratio of the percent to 100 is equal to the ratio of the *amount* to the *base*.

What is 32% of 85?

Sketch a diagram.

Percent = 32,
base = 85,
amount = n

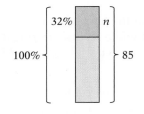

$$\frac{percent}{100} = \frac{amount}{base}$$

$$\frac{32}{100} = \frac{n}{85}$$

$$100 \cdot n = 32 \cdot 85$$

$$100n = 2{,}720$$

$$\frac{100n}{100} = \frac{2{,}720}{100}$$

$$n = 27.2$$

32% of 85 is 27.2.

EXAMPLE 4   24% of what is 16? Round to the nearest hundredth.

Solution $$\frac{percent}{100} = \frac{amount}{base}$$

$$\frac{24}{100} = \frac{16}{n}$$

$$24 \cdot n = 100 \cdot 16$$

$$24n = 1{,}600$$

$$n = \frac{1{,}600}{24} \approx 66.67$$

16 is approximately 24% of 66.67.

YOU TRY IT 4   8 is 25% of what?

Your Solution

EXAMPLE 5   Find 1.2% of 42.

Solution $$\frac{percent}{100} = \frac{amount}{base}$$

$$\frac{1.2}{100} = \frac{n}{42}$$

$$1.2 \cdot 42 = 100 \cdot n$$

$$50.4 = 100n$$

$$\frac{50.4}{100} = \frac{100n}{100}$$

$$0.504 = n$$

1.2% of 42 is 0.504.

YOU TRY IT 5   Find 0.74% of 1,200.

Your Solution

EXAMPLE 6   What percent of 52 is 13?

Solution $$\frac{percent}{100} = \frac{amount}{base}$$

$$\frac{n}{100} = \frac{13}{52}$$

$$n \cdot 52 = 100 \cdot 13$$

$$52n = 1{,}300$$

$$\frac{52n}{52} = \frac{1{,}300}{52}$$

$$n = 25$$

25% of 52 is 13.

YOU TRY IT 6   What percent of 180 is 54?

Your Solution

Solutions on p. S20

Figure 8.1 Causes of Death for
Police Officers Killed
in the Line of Duty
Source: International Union of
Police Associations

OBJECTIVE C Applications

The circle graph at the left shows the causes of death for all police
officers who died while on duty during a recent year. What percent of
the deaths were due to traffic accidents? Round to the nearest tenth of
a percent.

Strategy To find the percent:
▶ Find the total number of officers who died in the line of
duty.
▶ Use the basic percent equation.
Percent = n, base = total number killed,
amount = number of deaths due to traffic accidents = 73

Solution $58 + 73 + 6 + 19 = 156$

Percent · base = amount

$$n \cdot 156 = 73$$

$$\frac{156n}{156} = \frac{73}{156}$$

$$n \approx 0.468$$

$$n \approx 46.8\%$$

46.8% of the deaths were due to traffic accidents.

EXAMPLE 7

During a recent year, 276 billion product
coupons were issued by manufacturers.
Shoppers redeemed 4.8 billion of these
coupons. (*Source:* NCH NuWorld Consumer
Behavior Study, America Coupon Council)
What percent of the coupons issued were
redeemed by customers? Round to the nearest
tenth of a percent.

Strategy

To find the percent, use the basic percent
equation.
Percent = n, base = number of coupons
issued = 276 billion, amount = number of
coupons redeemed = 4.8 billion

Solution

Percent · base = amount

$$n \cdot 276 = 4.8$$

$$\frac{276n}{276} = \frac{4.8}{276}$$

$$n \approx 0.017$$

$$n \approx 1.7\%$$

Of the product coupons issued, 1.7% were
redeemed by customers.

YOU TRY IT 7

An instructor receives a monthly
salary of $4,330, and $649.50 is
deducted for income tax. Find the
percent of the instructor's salary
deducted for income tax.

Your Strategy

Your Solution

Solution on p. S21

EXAMPLE 8

A taxpayer pays a tax rate of 35% for state and federal taxes. The taxpayer has an income of $47,500. Find the amount of state and federal taxes paid by the taxpayer.

Strategy

To find the amount, solve the basic percent equation.
Percent = 35% = 0.35, base = 47,500, amount = n

Solution

Percent · base = amount
$$0.35 \cdot 47{,}500 = n$$
$$16{,}625 = n$$

The amount of taxes paid is $16,625.

YOU TRY IT 8

According to Board-Trac, approximately 19% of the country's 2.4 million surfers are women. Estimate the number of female surfers in this country. Write the number in standard form.

Your Strategy

Your Solution

EXAMPLE 9

A department store has a blue blazer on sale for $114, which is 60% of the original price. What is the difference between the original price and the sale price?

Strategy

To find the difference between the original price and the sale price:

▶ Find the original price. Solve the basic percent equation.
  Percent = 60% = 0.60,
  amount = 114, base = n
▶ Subtract the sale price from the original price.

Solution

Percent · base = amount
$$0.60 \cdot n = 114$$
$$\frac{0.60n}{0.60} = \frac{114}{0.60}$$
$$n = 190$$

$$190 - 114 = 76$$

The difference in price is $76.

YOU TRY IT 9

An electrician's wage this year is $30.13 per hour, which is 115% of last year's hourly wage. What was the increase in the hourly wage over the past year?

Your Strategy

Your Solution

Solutions on p. S21

8.2 Exercises

OBJECTIVE A The basic percent equation

For Exercises 1 to 4, state the number or variable that will replace each word in the basic percent equation, percent · base = amount. If the percent is known, write the percent as a decimal.

1. 12% of what is 68?

    percent = _____   base = _____   amount = _____

2. What percent of 64 is 16?

    percent = _____   base = _____   amount = _____

3. What is 8% of 450?

    percent = _____   base = _____   amount = _____

4. 32 is what percent of 96?

    percent = _____   base = _____   amount = _____

Solve. Use the basic percent equation.

5. 8% of 100 is what?   6. 16% of 50 is what?

7. 0.05% of 150 is what?   8. 0.075% of 625 is what?

9. 15 is what percent of 90?   10. 24 is what percent of 60?

11. What percent of 16 is 6?   12. What percent of 24 is 18?

13. 10 is 10% of what?   14. 37 is 37% of what?

15. 2.5% of what is 30?   16. 10.4% of what is 52?

17. Find 10.7% of 485.   18. Find 12.8% of 625.

19. 80% of 16.25 is what?   20. 26% of 19.5 is what?

21. 54 is what percent of 2,000?   22. 8 is what percent of 2,500?

23. 16.4 is what percent of 4.1?

24. 5.3 is what percent of 50?

25. 18 is 240% of what?

26. 24 is 320% of what?

27. Given that 25% of x equals y, is $x < y$ or is $x > y$?

28. Given that 200% of x equals y, is $x < y$ or is $x > y$?

OBJECTIVE B **Percent problems using proportions**

For Exercises 29 and 30, state the number or variable that will replace each word in the proportion $\frac{percent}{100} = \frac{amount}{base}$.

29. What is 36% of 25?

percent = _____   amount = _____   base = _____

30. 89% of what is 1,780?

percent = _____   amount = _____   base = _____

Solve. Use the proportion method.

31. 26% of 250 is what?

32. Find 18% of 150.

33. 37 is what percent of 148?

34. What percent of 150 is 33?

35. 68% of what is 51?

36. 126 is 84% of what?

37. What percent of 344 is 43?

38. 750 is what percent of 50?

39. 82 is 20.5% of what?

40. 2.4% of what is 21?

41. What is 6.5% of 300?

42. Find 96% of 75.

43. 7.4 is what percent of 50?

44. What percent of 1,500 is 693?

45. Find 50.5% of 124.

46. What is 87.4% of 225?

47. 120% of what is 6?

48. 14 is 175% of what?

49. What is 250% of 18?

50. 325% of 4.4 is what?

51. 87 is what percent of 29?

52. What percent of 38 is 95?

53. For $\frac{1}{4}$%, the percent ratio is the ratio $\frac{\frac{1}{4}}{100}$. Which of the following fractions is equivalent to this ratio?

(i) $\frac{1}{25}$ (ii) $\frac{1}{400}$ (iii) $\frac{25}{1}$

54. For 0.75%, the percent ratio is the ratio $\frac{0.75}{100}$. Which of the following fractions is equivalent to this ratio?

(i) $\frac{3}{4}$ (ii) $\frac{3}{400}$ (iii) $\frac{75}{10,000}$

55. True or false? For all positive values of N, 20% of N is equal to N% of 20.

OBJECTIVE C   Applications

56. Read Exercise 58 and state the number or variable that will replace each word in the basic percent equation. If the percent is known, write the percent as a decimal.

   percent = _____   base = _____   amount = _____

57. Read Exercise 59 and state the number or variable that will replace each word in the basic percent equation. If the percent is known, write the percent as a decimal.

   percent = _____   base = _____   amount = _____

Solve.

58. *Automotive Technology*   A mechanic estimates that the brakes of an RV still have 6,000 mi of wear. This amount is 12% of the estimated safe-life use of the brakes. What is the estimated safe-life use of the brakes?

59. *Fire Science*   A fire department received 24 false alarms out of a total of 200 alarms received. What percent of the alarms received were false alarms?

60. *Fireworks*   The value of the fireworks imported to the United States in a recent year was $163.1 million. During that year, the value of the fireworks imported from China was $157.2 million. (*Source:* **www.census.gov**) What percent of the value of the fireworks imported to the United States was the value of the fireworks imported from China? Round to the nearest tenth of a percent.

61. *Charitable Giving*   In 2005, Americans gave $260.28 billion to charities. Of that amount, individuals contributed $199 billion and corporations donated $13.77 billion. (*Source:* Giving USA) What percent of charitable giving in 2005 came from individual contributions? Round to the nearest tenth of a percent. Is this more or less than three-quarters of the total donations that year?

62. *Fire Science*   The graph at the right shows firefighter deaths by type of duty for a recent year. What percent of the deaths occurred during training? Round to the nearest tenth of a percent.

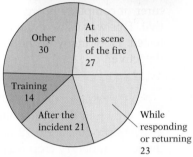

Firefighters Deaths
Source: U.S. Fire Administration

63. *The Labor Force*   In 2006, the number of working Americans age 55 or over was 24.6 million. Twenty-five percent of these workers were 65 or older. (*Source:* Challenger, Gray, & Christmas, Inc.) In 2006, how many workers were age 65 or older?

64. *Lobsters*   Each year, 183 million pounds of lobster are caught in the United States and Canada. Twenty-five percent of this amount is sold live. (*Source:* Lobster Institute at the University of Maine) How many pounds of lobster are sold live each year in the United States and Canada?

65. *Organ Donation*   The graph at the right shows the number of living kidney donors in 2005, by age. What percent of the donors are aged 18 to 34? Round to the nearest tenth of a percent.

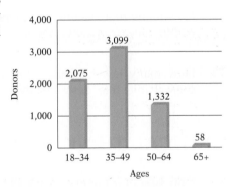

Number of Living Kidney Donors
Source: Organ Procurement and Transplantation Network

66. *Amusement Parks*   In 2005, 253 million people visited amusement parks worldwide. This was 102.2% of the number of visitors in 2004. (*Source:* Amusement Business; Economics Research Associates) How many people visited amusement parks in 2004? Round to the nearest million.

67. *Home Schooling*   In a recent year, 1.1 million students were home-schooled. This was 2.2% of all students in the United States. (*Source:* Home Schooling in the United States; U.S. Department of Education) Find the number of students in the United States that year.

68. *Business*   An antiques shop owner expects to receive $16\frac{2}{3}\%$ of the shop's sales as profit. What is the expected profit in a month when the total sales are $24,000?

69. *Financing* A used car is sold for $18,900. The buyer of the car makes a down payment of $3,780. What percent of the selling price is the down payment?

70. 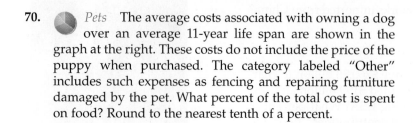 *Pets* The average costs associated with owning a dog over an average 11-year life span are shown in the graph at the right. These costs do not include the price of the puppy when purchased. The category labeled "Other" includes such expenses as fencing and repairing furniture damaged by the pet. What percent of the total cost is spent on food? Round to the nearest tenth of a percent.

71. *Manufacturing* During a quality control test, a manufacturer of computer boards found that 56 boards were defective. This was 0.7% of the total number of computer boards tested. How many of the tested computer boards were not defective?

72. *Agriculture* Of the 572 million pounds of cranberries grown in the United States in a recent year, Wisconsin growers produced 291.72 million pounds. What percent of the total cranberry crop was produced in Wisconsin?

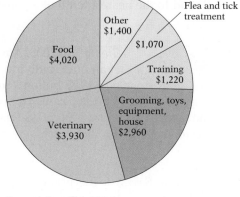

Cost of Owning a Dog
Source: American Kennel Club, *USA Today* research

Politics The results of a survey in which 32,840 full-time college and university faculty members were asked to describe their political views are shown at the right. Use these data for Exercises 73 and 74.

73. How many more faculty members described their political views as liberal than described their views as far-left?

74. How many fewer faculty members described their political views as conservative than described their views as middle-of-the-road?

Political View	Percent of Faculty Members Responding
Far left	5.3%
Liberal	42.3%
Middle of the road	34.3%
Conservative	17.7%
Far right	0.3%

Source: Higher Education Research Institute, UCLA

CRITICAL THINKING

75. Find 10% of a number and subtract it from the original number. Now take 10% of the new number and subtract it from the new number. Is this the same as taking 20% of the original number? Explain.

76. Increase a number by 10%. Now decrease the new number by 10%. Is the result the original number? Explain.

77. *Compensation* Your employer agrees to give you a 5% raise after one year on the job, a 6% raise the next year, and a 7% raise the following year. Is your salary after the third year greater than, less than, or the same as it would be if you had received a 6% raise each year?

78. Find five different uses of percents and explain why percent was used in those instances.

8.3 Percent Increase and Percent Decrease

OBJECTIVE A   Percent increase

Percent increase is used to show how much a quantity has increased over its original value. The statements "sales volume increased by 11% over last year's sales volume" and "employees received an 8% pay increase" are illustrations of the use of percent increase.

The population of the world is expected to increase from 6.6 billion people in 2007 to 9.4 billion people in 2050. (*Source:* U.S. Census Bureau) Find the expected percent increase in the world population from 2007 to 2050. Round to the nearest tenth of a percent.

Strategy   To find the percent increase:

> ▸ Find the expected increase in the population from 2007 to 2050.
> ▸ Use the basic percent equation.
>   Percent = n, base = 6.6, amount = the amount of increase

Solution $9.4 - 6.6 = 2.8$

Percent · base = amount

$$n \cdot 6.6 = 2.8$$

$$\frac{6.6n}{6.6} = \frac{2.8}{6.6}$$

$$n \approx 0.424 = 42.4\%$$

> ▸ The base is the *original value*. The amount is the *amount of increase.*

The population is expected to increase 42.4% from 2007 to 2050.

Point of Interest

The largest 1-day percent increase in the Dow Jones Industrial Average occurred on October 6, 1931. The Dow gained approximately 15% of its value.

EXAMPLE 1

A sales associate was earning $11.60 per hour before an 8% increase in pay. What is the new hourly wage? Round to the nearest cent.

Strategy

To find the new hourly wage:

> ▸ Use the basic percent equation to find the increase in pay.
>   Percent = 8% = 0.08, base = 11.60, amount = n
> ▸ Add the amount of increase to the original wage.

Solution

Percent · base = amount

$$0.08 \cdot 11.60 = n$$

$$0.93 \approx n$$

$\$11.60 + \$.93 = \$12.53$

The new hourly wage is $12.53.

YOU TRY IT 1

An automobile manufacturer increased the average mileage on a car from 17.5 mi/gal to 18.2 mi/gal. Find the percent increase in mileage.

Your Strategy

Your Solution

Solution on p. S21

Point of Interest

OBJECTIVE B Percent decrease

Percent decrease is used to show how much a quantity has decreased from its original value. The statements "the president's approval rating has decreased 9% over the last month" and "there has been a 15% decrease in the number of industrial accidents" are illustrations of the use of percent decrease.

The population of Philadelphia, Pennsylvania, in 2000 was 1.52 million. In 2005, the city's population was 1.46 million. (*Source*: U.S. Census Bureau) Find the percent decrease in the population of Philadelphia from 2000 to 2005. Round to the nearest tenth of a percent.

Strategy   To find the percent decrease:
▶ Find the decrease in the population from 2000 to 2005.
▶ Use the basic percent equation.
   Percent = n, base = 1.52, amount = the amount of decrease

Solution $1.52 - 1.46 = 0.06$

Percent · base = amount
$$n \cdot 1.52 = 0.06$$
$$\frac{1.52n}{1.52} = \frac{0.06}{1.52}$$
$$n \approx 0.039 = 3.9\%$$

▶ The base is the *original value*. The amount is the *amount of decrease.*

The population decreased by 3.9% from 2000 to 2005.

EXAMPLE 2

During a recent year, violent crime in a small city decreased from 27 crimes per 1,000 people to 24 crimes per 1,000 people. Find the percent decrease in violent crime. Round to the nearest tenth of a percent.

Strategy

To find the percent decrease in crime:
▶ Find the decrease in the number of crimes.
▶ Use the basic percent equation to find the percent decrease in crime.

Solution

$27 - 24 = 3$
Percent · base = amount
$$n \cdot 27 = 3$$
$$n = \frac{3}{27} \approx 0.111$$

▶ The amount is the amount of decrease.

Violent crime decreased by approximately 11.1% during the year.

YOU TRY IT 2

The market value of a luxury car decreased by 24% during the past year. Find the value of a luxury car that cost $47,000 last year. Round to the nearest dollar.

Your Strategy

Your Solution

Solution on p. S21

8.3 Exercises

OBJECTIVE A   **Percent increase**

For Exercises 1 and 2, suppose the price of an item increases from $15 to $18.

1. Fill in each blank with *original* or *new:* The amount of the increase is found by subtracting the _____ price from the _____ price.

   The amount of increase is $_____ − $_____ = $_____.

2. To find the percent increase in price, use the basic percent equation. Let *n* represent the unknown percent.

   base = the original price = $_____   amount = amount of increase = $_____

Solve. Round percents to the nearest tenth of a percent.

3. *Poverty*   According to the Census Bureau, there were 32.9 million Americans living in poverty in 2001. By 2005, the number had increased to 37.0 million. Find the percent increase in the number of Americans living in poverty from 2001 to 2005.

4. *Passports*   The graph at the right shows the number of passports, in millions, issued to U.S. citizens each year from 2003 to 2006. Find the percent increase in the number of passports issued from 2003 to 2006.

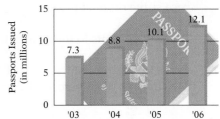

Passports Issued to U.S. Citizens
Source: State Department

5. *Education*   In the 1990–91 school year, the average cost for tuition, fees, and room and board at a private college was $13,476. In the 2005–06 school year, the cost had risen to $29,026. (*Source:* The College Board) Find the percent increase in the average cost of attending a private college.

6. *Education*   In the 1990–91 school year, the average cost for tuition, fees, and room and board at a public college was $5,074. In the 2005–06 school year, the cost had risen to $12,127. (*Source:* The College Board) Find the percent increase in the average cost of attending a public college.

7. *Population*   In 2006, the population of Clark County, Nevada, which contains Las Vegas, was 2 million. The population is projected to increase to 3 million by 2020. (*Source:* Nevada State Demographer's Office) Find the projected percent increase in the population of Clark County from 2006 to 2020.

8. *Health*   From 1997 to 2004, the number of new diabetes diagnoses grew from 0.88 million to 1.36 million. (*Source:* Centers for Disease Control) Find the percent increase in the number of new diabetes diagnoses from 1997 to 2004.

9.   *Volunteers* According to AmeriCorps, the number of college graduates join-ing federal service organizations grew from 11,825 applicants in 2001 to 19,979 applicants in 2006. Find the percent increase in the number of applicants from 2001 to 2006.

10.   *Wealth* The table at the right shows the number of millionaire households in the United States for selected years. Find the percent increase in the number of millionaire households from 1975 to 2005.

Year	Number of Households Containing Millionaires
1975	350,000
1997	3,500,000
2005	8,900,000

Source: Affluent Market Institute;
CNNMoney.com

OBJECTIVE B Percent decrease

For Exercises 11 and 12, suppose the price of an item decreases from $48 to $38.

11.   Fill in each blank with *original* or *new*: The amount of the decrease is found by sub-tracting the _____ price from the _____ price.

The amount of decrease is $_____ – $_____ = $_____ .

12.   To find the percent decrease in price, use the basic percent equation. Let n represent the unknown percent.

base = the original price = $_____   amount = amount of decrease = $_____

Solve. Round percents to the nearest tenth of a percent.

13.   *Consumerism* A family reduced its normal monthly food bill of $320 by $50. What percent decrease does this represent?

14.   *Adolescent Inmates* The graph at the right shows the num-ber of inmates, in thousands, under the age of 18 held in states prisons for various years. Find the percent decrease in the number of inmates younger than 18 held in state prisons from 2001 to 2005.

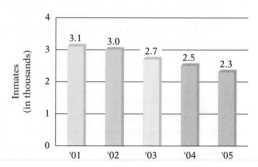

Inmates Younger Than 18 in State Prisons
Source: U.S. Department of Justice

15.   *Manufacturing* A new production method reduced the time needed to clean a piece of metal from 8 min to 5 min. What per-cent decrease does this represent?

16.   *Marketing* As a result of an increased number of service lines at a grocery store, the average amount of time a customer waits in line has decreased from 3.8 min to 2.5 min. Find the percent decrease.

17.   *Consumerism* It is estimated that the value of a new car is reduced 30% after 1 year of ownership. Find the value of a $21,900 new car after 1 year.

18.   *Business* A sales manager's average monthly expense for gasoline was $92. After joining a car pool, the manager was able to decrease gasoline expenses by 22%. What is the average monthly gasoline bill now?

19. *Airline Industry*   In 2000, there were 67 near collisions of aircraft in the United States. This number decreased to 31 in 2006. (*Source*: FAA; NTSB) Find the percent decrease in the number of near collisions from 2000 to 2006.

20. *Automobile Sales*   The number of minivans sold during the first nine months of 2005 was 874,949. During the first nine months of 2006, 766,461 minivans were sold. (*Source*: Autodata) Find the percent decrease in the number of minivans sold for these time periods.

21. *Federal Funding*   Federal funding for the National Center for Missing Adults declined from $1.5 million in 2002 to $.1 million in 2006. (*Source*: National Center for Missing Adults) Find the percent decrease in federal funding from 2002 to 2006.

22. *The Military*   The graph at the right shows the number of active-duty U.S. military personnel, in thousands, in 1990 and in 2005. Which branch of the military had the greatest percent decrease in personnel from 1990 to 2005? What was the percent decrease for this branch of the service?

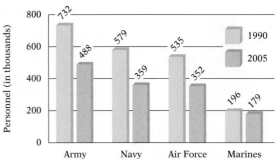

Number of Active-Duty U.S. Military Personnel
Source: Department of Defense

23. *Internet Providers*   The graph at the right shows the number of AOL subscribers, in millions, from 2001 to 2006. **a.** For which pairs of years can a percent increase in subscribers be calculated? **b.** For which pairs of years can a percent decrease in subscribers be calculated?

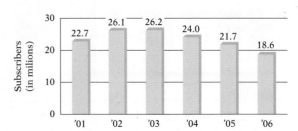

AOL Subscribers
Source: Time Warner, Inc.

CRITICAL THINKING

24. *Consumerism*   A department store gives you three discount coupons: a 10% discount, a 20% discount, and a 30% discount off any item. You have decided to purchase a CD player costing $225 and use all three coupons. Is there a particular order in which you should ask to have the discount coupons applied to your purchase so that your purchase price is as small as possible? Explain.

25. *Consumerism*   A wide-screen TV costing $3,000 was on sale for 30% off. An additional 10% off the sale price was offered to customers who paid by check. Calculate the sales price after the two discounts. Is this the same as a discount of 40%? Find the equivalent discount of the successive discounts.

26. Define *per millage*. Explain its relation to percent.

8.4 Markup and Discount

OBJECTIVE A   Markup

Cost is the price a merchandising business or retailer pays for a product. **Selling price,** or **retail price,** is the price for which a merchandising business or retailer sells a product to a customer.

The difference between selling price and cost is called **markup.** Markup is added to cost to cover the expenses of operating a business and provide a profit to the owners.

Markup can be expressed as a percent of the cost, or it can be expressed as a percent of the selling price. Here we present markup as a percent of the cost, which is the most common practice.

The percent markup is called the **markup rate,** and it is expressed as the markup based on the cost.

A diagram is useful when expressing the markup equations. In the diagram at the right, the total length is the selling price. One part of the diagram is the cost, and the other part is the markup.

Markup

Selling Price

Cost

Take Note

If C is added to both sides of the first equation, $M = P - C$, the result is the second equation listed, $P = C + M$.

The Markup Equations

$$M = P - C$$
$$P = C + M$$
$$M = r \cdot C$$

where M = markup
P = selling price
C = cost
r = markup rate

The manager of a clothing store buys a sports jacket for $80 and sells the jacket for $116. Find the markup rate.

Strategy   To find the markup rate:

▸ Find the markup by solving the formula $M = P - C$ for M. $P = 116, C = 80$
▸ Solve the formula $M = r \cdot C$ for r.
M = the markup, $C = 80$

Solution $M = P - C$
$M = 116 - 80$
$M = 36$

$M = r \cdot C$
$36 = r \cdot 80$
$\dfrac{36}{80} = \dfrac{80r}{80}$
$0.45 = r$

M

116

80

The markup rate is 45%.

EXAMPLE 1

A soft top for a convertible that costs a dealer $250 has a markup rate of 35%. Find the markup.

Strategy

To find the markup, solve the formula $M = r \cdot C$ for M.
$r = 35\% = 0.35$, $C = 250$

Solution

$M = r \cdot C$

$M = 0.35 \cdot 250$

$M = 87.50$

The markup is $87.50.

YOU TRY IT 1

An outboard motor costing $650 has a markup rate of 45%. Find the markup.

Your Strategy

Your Solution

EXAMPLE 2

A graphing calculator costing $45 is sold for $80. Find the markup rate. Round to the nearest tenth of a percent.

Strategy

To find the markup rate:

▶ Find the markup by solving the formula $M = P - C$ for M.
$P = 80$, $C = 45$
▶ Solve the formula $M = r \cdot C$ for r.
M = the markup, $C = 45$

Solution

$M = P - C$

$M = 80 - 45$

$M = 35$

$M = r \cdot C$

$35 = r \cdot 45$

$\dfrac{35}{45} = \dfrac{45r}{45}$

$0.778 \approx r$

The markup rate is 77.8%.

YOU TRY IT 2

A laser printer costing $950 is sold for $1,450. Find the markup rate. Round to the nearest tenth of a percent.

Your Strategy

Your Solution

Solutions on p. S21

EXAMPLE 3

A fishing reel with a cost of $50 has a markup rate of 22%. Find the selling price.

Strategy

To find the selling price:

▸ Find the markup by solving the equation $M = r \cdot C$ for M.
$r = 22\% = 0.22, C = 50$
▸ Solve the formula $P = C + M$ for P.
$C = 50, M = $ the markup

Solution

$M = r \cdot C$
$M = 0.22 \cdot 50$
$M = 11$

$P = C + M$
$P = 50 + 11$
$P = 61$

The selling price is $61.

YOU TRY IT 3

A basketball with a cost of $82.50 has a markup rate of 42%. Find the selling price.

Your Strategy

Your Solution

Solution on p. S21

OBJECTIVE B **Discount**

A retailer may reduce the regular price of a product for a promotional sale because the product is damaged, an odd size or color, or a discontinued item. The **discount,** or **markdown,** is the amount by which a retailer reduces the regular price of a product.

The percent discount is called the **discount rate** and is usually expressed as a percent of the original price (the regular selling price).

Discount or markdown
Sale price
Regular price

The Discount Equations
....

$D = R - S$ where $D = $ discount or markdown
$D = r \cdot R$ $S = $ sale price
$S = (1 - r)R$ $R = $ regular price
                                  $r = $ discount rate

8.5 Simple Interest

OBJECTIVE A Simple interest

If you deposit money in a savings account at a bank, the bank will pay you for the privilege of using that money. The amount you deposit in the savings account is called the **principal.** The amount the bank pays you for the privilege of using the money is called **interest.**

If you borrow money from the bank in order to buy a car, the amount you borrow is called the **principal.** The additional amount of money you must pay the bank, above and beyond the amount borrowed, is called **interest.**

Whether you deposit money or borrow it, the amount of interest paid is usually computed as a percent of the principal. The percent used to determine the amount of interest to be paid is the **interest rate.** Interest rates are given for specific periods of time, such as months or years.

Interest computed on the original principal is called **simple interest.** Simple interest is the cost of a loan that is for a period of about 1 year or less.

> ### The Simple Interest Formula
>
> $I = Prt$, where I = simple interest earned, P = principal,
> r = annual simple interest rate, t = time (in years)

In the simple interest formula, t is the time in years. If a time period is given in days or months, it must be converted to years and then substituted in the formula for t. For example,

120 days is $\dfrac{120}{365}$ of a year. 6 months is $\dfrac{6}{12}$ of a year.

Shannon O'Hara borrowed $5,000 for 90 days at an annual simple interest rate of 7.5%. Find the simple interest due on the loan.

Strategy To find the simple interest owed, use the simple interest formula.

$P = 5,000, r = 7.5\% = 0.075, t = \dfrac{90}{365}$

Solution $I = Prt$

$I = 5,000(0.075)\left(\dfrac{90}{365}\right)$

$I \approx 92.47$

The simple interest due on the loan is $92.47.

In the example above, we calculated that the simple interest due on Shannon O'Hara's 90-day, $5,000 loan was $92.47. This means that at the end of the 90 days, Shannon owes $5,000 + $92.47 = $5,092.47. The principal plus the interest owed on a loan is called the **maturity value.**

Formula for the Maturity Value of a Simple Interest Loan

$M = P + I,$ where M = the maturity value, P = the principal, I = the simple interest

The example below illustrates solving the simple interest formula for the interest rate. The solution requires the Multiplication Property of Equations.

Ed Pabas took out a 45-day, $12,000 loan. The simple interest on the loan was $168. To the nearest tenth of a percent, what is the simple interest rate?

Strategy   To find the simple interest rate, use the simple interest formula. $P = 12,000$, $t = \dfrac{45}{365}$, $I = 168$

Solution
$$I = Prt$$
$$168 = 12,000r\left(\frac{45}{365}\right)$$
$$168 = \frac{540,000}{365}r$$
$$\frac{365}{540,000}(168) = \frac{365}{540,000} \cdot \frac{540,000}{365}r$$
$$0.114 \approx r$$

The simple interest rate on the loan is 11.4%.

EXAMPLE 1

You arrange for a 9-month bank loan of $9,000 at an annual simple interest rate of 8.5%. Find the total amount you must repay to the bank.

Strategy

To calculate the maturity value:

▶ Find the simple interest due on the loan by solving the simple interest formula for I.
$t = \dfrac{9}{12}$, $P = 9,000$, $r = 8.5\% = 0.085$

▶ Use the formula for the maturity value of a simple interest loan, $M = P + I$.

Solution
$$I = Prt$$
$$I = 9,000(0.085)\left(\frac{9}{12}\right)$$
$$I = 573.75$$

$$M = P + I$$
$$M = 9,000 + 573.75$$
$$M = 9,573.75$$

The total amount owed to the bank is $9,573.75.

YOU TRY IT 1

William Carey borrowed $12,500 for 8 months at an annual simple interest rate of 9.5%. Find the total amount due on the loan.

Your Strategy

Your Solution

Solution on p. S22

8.5 Exercises

OBJECTIVE A **Simple interest**

1. In the table below, the interest rate is an annual simple interest rate. Complete the table by calculating the simple interest due on the loan.

Loan Amount	Interest Rate	Period	Prt	=	Interest, I
$5,000	6%	1 month	$ _____ · _____ · _____	=	_____
$5,000	6%	2 months	$ _____ · _____ · _____	=	_____
$5,000	6%	3 months	$ _____ · _____ · _____	=	_____
$5,000	6%	4 months	$ _____ · _____ · _____	=	_____
$5,000	6%	5 months	$ _____ · _____ · _____	=	_____

2. Use the pattern of the amounts of interest in the table in Exercise 1 to find the simple interest due on a $5,000 loan that has an annual simple interest rate of 6% for a period of **a.** 6 months, **b.** 7 months, **c.** 8 months, and **d.** 9 months.

 For Exercises 3 and 4, refer to your answers to Exercises 1 and 2.

3. If you know the simple interest due on a 1-month loan, explain how to use that amount to calculate the simple interest due on a 7-month loan for the same principal and the same interest rate.

4. If the time period of a loan is doubled but the principal and interest rate remain the same, how many times greater is the simple interest due on the loan?

Solve.

5. Hector Elizondo took out a 75-day loan of $7,500 at an annual interest rate of 9.5%. Find the simple interest due on the loan.

6. Kristi Yang borrowed $15,000. The term of the loan was 90 days, and the annual simple interest rate was 7.4%. Find the simple interest due on the loan.

7. A home builder obtained a preconstruction loan of $50,000 for 8 months at an annual interest rate of 9.5%. What is the simple interest due on the loan?

8. To finance the purchase of 15 new cars, the Lincoln Car Rental Agency borrowed $100,000 for 9 months at an annual interest rate of 9%. What is the simple interest due on the loan?

9. The Mission Valley Credit Union charges its customers an interest rate of 2% per month on money that is transferred into an account that is overdrawn. Find the interest owed to the credit union for 1 month when $800 is transferred into an overdrawn account.

10. Assume that Visa charges Francesca 1.6% per month on her unpaid balance. Find the interest owed to Visa when her unpaid balance for the month is $1,250.

11. Find the simple interest Kara Tanamachi owes on a $1\frac{1}{2}$-year loan of $1,500 at an annual interest rate of 7.5%.

12. Find the simple interest Jacob Zucker owes on a 2-year loan of $8,000 at an annual interest rate of 9%.

13. A corporate executive took out a $25,000 loan at an 8.2% annual simple interest rate for 1 year. Find the maturity value of the loan.

14. An auto parts dealer borrowed $150,000 at a 9.5% annual simple interest rate for 1 year. Find the maturity value of the loan.

15. A credit union loans a member $5,000 for the purchase of a used car. The loan is made for 18 months at an annual simple interest rate of 6.9%. What is the maturity value of the car loan?

16. Capitol City Bank approves a home-improvement loan application for $14,000 at an annual simple interest rate of 10.25% for 270 days. What is the maturity value of the loan?

17. Michele Gabrielle borrowed $3,000 for 9 months and paid $168.75 in simple interest on the loan. Find the annual simple interest rate that Michele paid on the loan.

18. A $12,000 investment earned $462 in interest in 6 months. Find the annual simple interest rate on the loan.

19. Don Glover borrowed $18,000 for 210 days and paid $604.80 in simple interest on the loan. What annual simple interest rate did Don pay on the loan?

20. An investor earned $937.50 on an investment of $50,000 in 75 days. Find the annual simple interest rate earned on the investment.

CRITICAL THINKING

21. Visit a savings and loan officer to collect information about the different kinds of home loans. Write a short essay describing the different kinds of loans available.

Focus on **Problem Solving**

Using a Calculator as a Problem-Solving Tool

A calculator is an important tool for problem solving. Here are a few problems you can solve with a calculator. You may need to research some of the questions to find information you do not know.

1. Choose any single-digit positive number. Multiply the number by 1,507 and 7,373. What is the answer? Choose another positive single-digit number and again multiply by 1,507 and 7,373. What is the answer? What pattern do you see? Why does this work?

2. The U.S. gross domestic product in 2005 was $12,455,825,000,000. Is this more or less than the amount of money that would be placed on the last square of a standard checkerboard if 1 cent were placed on the first square, 2 cents were placed on the second square, 4 cents were placed on the third square, 8 cents were placed on the fourth square, and so on until the 64th square was reached?

3. Which of the reciprocals of the first 16 natural numbers have a terminating decimal representation, and which have a repeating decimal representation?

4. What is the largest natural number n for which
$4^n > 1 \cdot 2 \cdot 3 \cdot 4 \cdot 5 \cdot \cdots \cdot n$?

5. If $1,000 bills are stacked one on top of another, is the height of $1 billion less than or greater than the height of the Washington Monument?

6. What is the value of $1 + \cfrac{1}{1 + \cfrac{1}{1 + \cfrac{1}{1 + \cfrac{1}{1 + 1}}}}$?

7. Calculate 15^2, 35^2, 65^2, and 85^2. Study the results. Make a conjecture about a relationship between a number ending in 5 and its square. Use your conjecture to find 75^2 and 95^2. Does your conjecture work for 125^2?

8. Find the sum of the first 1,000 natural numbers. (*Hint:* You could just start adding $1 + 2 + 3 + \cdots$, but even if you performed one operation every 3 seconds, it would take you an hour to find the sum. Instead, try pairing the numbers and then adding the pairs. Pair 1 and 1,000, 2 and 999, 3 and 998, and so on. What is the sum of each pair? How many pairs are there? Use this information to answer the original question.)

9. For a borrower to qualify for a home loan, a bank requires that the monthly mortgage payment be less than 25% of the borrower's monthly take-home income. A laboratory technician has deductions for taxes, insurance, and retirement that amount to 25% of the technician's monthly gross income. What minimum monthly income must this technician earn to receive a bank loan that has a mortgage payment of $1,200 per month?

Projects & Group Activities

Buying a Car

Suppose a student has an after-school job to earn money to buy and maintain a car. We will make assumptions about the monthly costs in several categories in order to determine how many hours per week the student must work to support the car. Assume that the student earns $9.50 per hour.

1. Monthly payment

   Assume that the car cost $8,500 with a down payment of $1,020. The remainder is financed for 3 years at an annual simple interest rate of 9%.

   Monthly payment = _____

2. Insurance

   Assume that insurance costs $1,500 per year.

   Monthly insurance payment = _____

3. Gasoline

   Assume that the student travels 750 miles per month, that the car travels 25 miles per gallon of gasoline, and that gasoline costs $2.50 per gallon.

   Number of gallons of gasoline purchased per month = _____

   Monthly cost for gasoline = _____

4. Miscellaneous

   Assume $.55 per mile for upkeep.

   Monthly expense for upkeep = _____

5. Total monthly expenses for the monthly payment, insurance, gasoline, and miscellaneous = _____

6. To find the number of hours per month that the student must work to finance the car, divide the total monthly expenses by the hourly rate.

   Number of hours per month ≈ _____

7. To find the number of hours per week that the student must work, divide the number of hours per month by 4.

   Number of hours per week = _____

   The student has to work _____ hours per week to pay the monthly car expenses.

If you own a car, make out your own expense record. If you do not own a car, make assumptions about the kind of car that you would like to purchase, and calculate the total monthly expenses that you would have. An insurance company will give you rates on different kinds of insurance. An automobile club can give you approximations of miscellaneous expenses.

Chapter 8 Summary

Key Words	Examples
Percent means "parts of 100." [8.1A, p. 493]	23% means 23 of 100 equal parts.
Percent increase is used to show how much a quantity has increased over its original value. [8.3A, p. 509]	The city's population increased 5%, from 10,000 people to 10,500 people.
Percent decrease is used to show how much a quantity has decreased from its original value. [8.3B, p. 510]	Sales decreased 10%, from 10,000 units in the third quarter to 9,000 units in the fourth quarter.
Cost is the price a business pays for a product. **Selling price,** or **retail price,** is the price for which a business sells a product to a customer. **Markup** is the difference between selling price and cost. The percent markup is called the **markup rate.** In this text, the markup rate is expressed as the markup rate based on cost. [8.4A, p. 514]	A business pays $90 for a pair of cross trainers; the cost is $90. The business sells the cross trainers for $135; the selling price is $135. The markup is $135 − $90 = $45. The markup rate is 45 ÷ 90 = 0.5 = 50%.
Discount or **markdown** is the difference between the regular price and the discount price. The discount is frequently stated as a percent, called the **discount rate.** [8.4B, p. 516]	A DVD box set that regularly sells for $50 is on sale for $40. The discount is $50 − $40 = $10. The discount rate is 10 ÷ 50 = 0.2 = 20%.
Principal is the amount of money originally deposited or borrowed. **Interest** is the amount paid for the privilege of using someone else's money. The percent used to determine the amount of interest is the **interest rate.** Interest computed on the original amount is called **simple interest.** The principal plus the interest owed on a loan is called the **maturity value.** [8.5A, p. 521]	Consider a 1-year loan of $5,000 at an annual simple interest rate of 8%. The principal is $5,000. The interest rate is 8%. The interest paid on the loan is $400. The maturity value is $5,000 + $400 = $5,400.

Essential Rules and Procedures

To write a percent as a fraction, drop the percent sign and multiply by $\frac{1}{100}$. [8.1A, p. 493]	$56\% = 56\left(\frac{1}{100}\right) = \frac{56}{100} = \frac{14}{25}$
To write a percent as a decimal, drop the percent sign and multiply by 0.01. [8.1A, p. 493]	$87\% = 87(0.01) = 0.87$
To write a fraction as a percent, multiply by 100%. [8.1B, p. 494]	$\frac{7}{20} = \frac{7}{20}(100\%) = \frac{700}{20}\% = 35\%$
To write a decimal as a percent, multiply by 100%. [8.1B, p. 494]	$0.325 = 0.325(100\%) = 32.5\%$

The Basic Percent Equation [8.2A, p. 498]
Percent · base = amount

8% of 250 is what number?
Percent · base = amount
$$0.08 \cdot 250 = n$$
$$20 = n$$

Proportion Method of Solving a Percent Problem [8.2B, p. 500]
$$\frac{\text{percent}}{100} = \frac{\text{amount}}{\text{base}}$$

8% of 250 is what number?
$$\frac{\text{percent}}{100} = \frac{\text{amount}}{\text{base}}$$
$$\frac{8}{100} = \frac{n}{250}$$
$$100 \cdot n = 8 \cdot 250$$
$$100n = 2{,}000$$
$$n = 20$$

Markup Equations: [8.4A, p. 514]
M = markup, P = selling price, C = cost, r = markup rate:
$$M = P - C$$
$$P = C + M$$
$$M = r \cdot C$$

The manager of a sporting goods store buys a golf club for $275 and sells the golf club for $343.75. Find the markup rate.
$$M = P - C$$
$$M = 343.75 - 275 = 68.75$$

$$M = r \cdot C$$
$$68.75 = r \cdot 275$$
$$0.25 = r$$
The markup rate is 25%.

Discount Equations: [8.4B, p. 516]
D = discount or markdown, S = sale price, R = regular price, r = discount rate:
$$D = R - S$$
$$D = r \cdot R$$
$$S = (1 - r)R$$

A golf club that regularly sells for $343.75 is on sale for $275. Find the discount rate.
$$D = R - S$$
$$D = 343.75 - 275 = 68.75$$

$$D = r \cdot R$$
$$68.75 = r \cdot 343.75$$
$$0.2 = r$$
The discount rate is 20%.

Simple Interest Formula [8.5A, p. 521]
I = simple interest earned, P = principal, r = annual simple interest rate, t = time (in years):
$$I = Prt$$

You borrow $10,000 for 180 days at an annual interest rate of 8%. Find the simple interest due on the loan.
$$I = Prt$$
$$I = 10{,}000(0.08)\left(\frac{180}{365}\right)$$
$$I \approx 394.52$$

Formula for the Maturity Value of a Simple Interest Loan [8.5A, p. 522]
M = maturity value, P = principal, I = simple interest:
$$M = P + I$$

Suppose you paid $400 in interest on a 1-year loan of $5,000. The maturity value of the loan is $5,000 + $400 = $5,400.

Chapter 8 Review Exercises

1. Write 32% as a fraction.

2. Write 22% as a decimal.

3. Write 25% as a fraction and as a decimal.

4. Write $3\frac{2}{5}\%$ as a fraction.

5. Write $\frac{7}{40}$ as a percent.

6. Write $1\frac{2}{7}$ as a percent. Round to the nearest tenth of a percent.

7. Write 2.8 as a percent.

8. 42% of 50 is what?

9. What percent of 3 is 15?

10. 12 is what percent of 18? Round to the nearest tenth of a percent.

11. 150% of 20 is what number?

12. Find 18% of 85.

13. 32% of what number is 180?

14. 4.5 is what percent of 80?

15. Find 0.58% of 2.54.

16. 0.0048 is 0.05% of what number?

17. *Tourism*   The table at the right shows the countries with the highest projected numbers of tourists visiting in 2020. What percent of the tourists projected to visit these countries will be visiting China? Round to the nearest tenth of a percent.

Country	Projected Number of Tourists in 2020
China	137 million
France	93 million
Spain	71 million
USA	102 million

Source: The State of the World Atlas by Dan Smith

18. *Business*   A company spent 7% of its $120,000 budget for advertising. How much did the company spend for advertising?

19. *Manufacturing*   A quality control inspector found that 1.2% of 4,000 cellular telephones were defective. How many of the phones were not defective?

20. *Television*   According to the Cabletelevision Advertising Bureau, cable households watch an average of 61.35 h of television per week. On average, what percent of the week do cable households spend watching TV? Round to the nearest tenth of a percent.

21. *Business*   A resort lodge expects to make a profit of 22% of total income. What is the expected profit on $750,000 of income?

22. *Sports*   A basketball auditorium increased its 9,000 seating capacity by 18%. How many seats were added to the auditorium?

23. *Travel*   An airline knowingly overbooks flights by selling 12% more tickets than there are seats available. How many tickets would this airline sell for an airplane that has 175 seats?

24. *Elections*   In a recent city election, 25,400 out of 112,000 registered voters voted. What percent of the registered voters voted in the election? Round to the nearest tenth of a percent.

25. *Compensation*   A sales clerk was earning $10.50 an hour before an 8% increase in pay. What is the clerk's new hourly wage?

26. *Computers*   A computer system that sold for $2,400 one year ago can now be bought for $1,800. What percent decrease does this represent?

27. *Business*   A car dealer advertises a 6% markup rate on a car that cost the dealer $18,500. Find the selling price of the car.

28. *Business*   A parka costing $110 is sold for $181.50. Find the markup rate.

29. *Business*   A tennis racket that regularly sells for $80 is on sale for 30% off the regular price. Find the sale price.

30. *Travel*   An airline is offering a 40% discount on round-trip air fares. Find the sale price of a round-trip ticket that normally sells for $650.

31. *Finance*   Find the simple interest on a 45-day loan of $3,000 at an annual simple interest rate of 8.6%.

32. *Finance*   A corporation borrowed $500,000 for 60 days and paid $7,397.26 in simple interest. What annual simple interest rate did the corporation pay on the loan? Round to the nearest hundredth of a percent.

33. *Finance*   A realtor took out a $10,000 loan at an 8.4% annual simple interest rate for 9 months. Find the maturity value of the loan.

Chapter 8 Test

1. Write 86.4% as a decimal.

2. Write 0.4 as a percent.

3. Write $\frac{5}{4}$ as a percent.

4. Write $83\frac{1}{3}\%$ as a fraction.

5. Write 32% as a fraction.

6. Write 1.18 as a percent.

7. 18 is 20% of what number?

8. What is 68% of 73?

9. What percent of 320 is 180?

10. 28 is 14% of what number?

11. *Insurance*   An insurance company expects that 2.2% of a company's employees will have an industrial accident. How many accidents are expected for a company that employs 1,500 people?

12. *Education*   A student missed 16 questions on a history exam of 90 questions. What percent of the questions did the student answer correctly? Round to the nearest tenth.

13. *Compensation*   An administrative assistant has a wage of $480 per week. This is 120% of last year's wage. What is the dollar increase in the assistant's weekly wage over last year?

14. *Education*   The table at the right shows the average cost of tuition, room, and board at both public and private colleges in the United States. What is the percent increase in cost for a student who goes from public college to private college? Round to the nearest tenth of a percent.

Average Tuition, Room, and Board	
Public college	$ 12,127
Private college	$ 29,026

Source: The College Board

15. *Business*   The number of management trainees working for a company has increased from 36 to 42. What percent increase does this represent?

16. *Nutrition* The table at the right shows the fat, saturated fat, cholesterol, and calorie content in a 90-gram ground beef burger and in a 90-gram soy burger.
    a. As compared with the beef burger, by what percent is the fat content decreased in the soy burger?
    b. What is the percent decrease in cholesterol in the soy burger as compared with the beef burger?
    c. Calculate the percent decrease in calories in the soy burger as compared with the beef burger.

	Beef Burger	Soy Burger
Fat	24 g	4 g
Saturated Fat	10 g	1.5 g
Cholesterol	75 mg	0 mg
Calories	280	140

17. *Business* Last year a company's travel expenses totaled $25,000. This year the travel expenses totaled $23,000. What percent decrease does this represent?

18. *Art* A painting has a value of $1,500. This is 125% of the painting's value last year. What is the dollar increase in the value of the painting?

19. *Business* The manager of a stationery store uses a markup rate of 60%. What is the markup on a box of notepaper that costs the store $21?

20. *Business* An electric keyboard costing $225 is sold for $349. Find the markup rate on the keyboard. Round to the nearest tenth of a percent.

21. *Business* A telescope is on sale for $180 after a markdown of 40% off the regular price. Find the regular price.

22. *Business* The regular price of a 10-foot by 10-foot dome tent is $370. The tent is now marked down $51.80. Find the discount rate on the tent.

23. *Finance* Find the simple interest on a 9-month loan of $5,000 when the annual interest rate is 8.4%.

24. *Finance* Maribeth Bakke took out a 150-day, $40,000 business loan that had an annual simple interest rate of 9.25%. Find the maturity value of the loan.

25. *Finance* Gene Connery paid $672 in simple interest on an 8-month loan for $12,000. Find the simple interest rate on the loan.

Cumulative Review Exercises

1. Evaluate $a - b$ for $a = 102.5$ and $b = 77.546$.

2. Evaluate 5^4.

3. Find the product of 4.67 and 3.007.

4. Multiply: $(2x - 3)(2x - 5)$

5. Divide: $3\frac{5}{8} \div 2\frac{7}{12}$

6. Multiply: $-2a^2b(-3ab^2 + 4a^2b^3 - ab^3)$

7. 120% of 35 is what?

8. Solve: $x - 2 = -5$

9. Find the product of 1.005 and 10^5.

10. Simplify: $-\frac{5}{8} - \left(-\frac{3}{4}\right) + \frac{5}{6}$

11. Simplify: $\dfrac{3 - \frac{7}{8}}{\frac{11}{12} + \frac{1}{4}}$

12. Multiply: $(-3a^2b)(4a^5b^4)$

13. Graph $y = -2x + 5$.

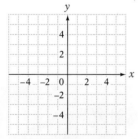

14. Graph $y = \frac{5}{3}x - 2$.

15. Find the quotient of $\frac{7}{8}$ and $\frac{5}{16}$.

16. Simplify: $4 - (-3) + 5 - 8$

17. Solve: $\frac{3}{4}x = -9$

18. Solve: $6x - 9 = -3x + 36$

19. Write 322.4 mi in 5 h as a unit rate.

20. Solve: $\dfrac{32}{n} = \dfrac{5}{7}$

21. 2.5 is what percent of 30? Round to the nearest tenth of a percent.

22. Find 42% of 160.

23. Simplify: $44 - (-6)^2 \div (-3) + 2$

24. Solve: $3(x - 2) + 2 = 11$

Solve.

25. *Health* According to the table at the right, what fraction of the population aged 75–84 are affected by Alzheimer's disease?

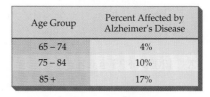

Age Group	Percent Affected by Alzheimer's Disease
65 – 74	4%
75 – 84	10%
85 +	17%

Source: Mayo Clinic Family Health Book, Encyclopedia Americana, Associated Press

26. *Business* A suit that regularly sells for $202.50 is on sale for 36% off the regular price. Find the sale price.

27. *Business* A graphing calculator with a selling price of $67.20 has a markup of 60%. Find the cost of the graphing calculator.

28. *Sports* A baseball team has won 13 out of the first 18 games played. At this rate, how many games will the team win in a 162-game season?

29. *Sports* A wrestler needs to lose 8 lb in three days in order to make the proper weight class. The wrestler loses $3\frac{1}{2}$ lb the first day and $2\frac{1}{4}$ lb the second day. How many pounds must the wrestler lose the third day in order to make the weight class?

30. *Physics* The speed of a falling object is given by the formula $v = \sqrt{64d}$, where v is the speed of the falling object in feet per second and d is the distance in feet that the object has fallen. Find the speed of an object that has fallen 81 ft.

31. *Sports* In the luge event in the 2006 Olympics, sliders reached a top speed of 130 km/h. Convert this speed to meters per second. Round to the nearest hundredth.

32. *Contractors* A plumber charged $1,632 for work done on a medical building. This charge included $192 for materials and $40 per hour for labor. Find the number of hours the plumber worked on the medical building.

33. *Physics* The current (I) in an electric circuit is inversely proportional to the resistance (R). If the current is 2 amperes when the resistance is 20 ohms, find the resistance when the current is 8 amperes.

Geometry

Hyde Park in London, England, is one of London's finest historic landscapes, covering approximately 350 acres. The best way to appreciate the design of Hyde Park is to view the garden from over-head. Each geometric shape, having its own set of dimensions, combines to form the entire park. **Example 4 on page 562** illustrates how to use a geometric formula to determine the size of an area and to calculate how much grass seed is needed for an area that size.

Student Website
Need help? For online student resources,
visit **college.hmco.com/pic/aufmannPA5e.**

DVD   SSM

1. Simplify: $2(18) + 2(10)$

2. Evaluate abc for $a = 2$, $b = 3.14$, and $c = 9$.

3. Evaluate xyz^3 for $x = \frac{4}{3}$, $y = 3.14$, and $z = 3$.

4. Solve: $x + 47 = 90$

5. Solve: $32 + 97 + x = 180$

6. Solve: $\frac{5}{12} = \frac{6}{x}$

GO Figure

Draw the figure that would come next.

9.1 Introduction to Geometry

OBJECTIVE A   Problems involving lines and angles

The word *geometry* comes from the Greek words for *earth* and *measure*. In ancient Egypt, geometry was used by the Egyptians to measure land and to build structures such as the pyramids. Today geometry is used in many fields, such as physics, medicine, and geology. Geometry is also used in applied fields such as mechanical drawing and astronomy. Geometric forms are used in art and design.

Point of Interest

Geometry is one of the oldest branches of mathematics. Around 350 B.C., Euclid of Alexandria wrote *Elements*, which contained all of the known concepts of geometry. Euclid's contribution was to unify various concepts into a single deductive system that was based on a set of axioms.

Three basic concepts of geometry are point, line, and plane. A **point** is symbolized by drawing a dot. A **line** is determined by two distinct points and extends indefinitely in both directions, as the arrows on the line shown at the right indicate. This line contains points A and B and is represented by $\overleftrightarrow{AB}$. A line can also be represented by a single letter, such as ℓ.

A **ray** starts at a point and extends indefinitely in *one* direction. The point at which a ray starts is called the **endpoint** of the ray. The ray shown at the right is denoted by $\overrightarrow{AB}$. Point A is the endpoint of the ray.

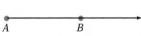

A **line segment** is part of a line and has two endpoints. The line segment shown at the right is denoted by $\overline{AB}$.

The distance between the endpoints of $\overline{AC}$ is denoted by AC. If B is a point on $\overline{AC}$, then AC (the distance from A to C) is the sum of AB (the distance from A to B) and BC (the distance from B to C).

$$AC = AB + BC$$

Given the figure above and the fact that $AB = 22$ cm and $AC = 31$ cm, find BC.

Write an equation for the distances between points on the line segment.

$$AC = AB + BC$$

Substitute the given distances for AB and AC into the equation.

$$31 = 22 + BC$$

Solve for BC.

$$9 = BC$$

$BC = 9$ cm

In this section we will be discussing figures that lie in a plane. A **plane** is a flat surface and can be pictured as a table top or blackboard that extends in all directions. Figures that lie in a plane are called **plane figures.**

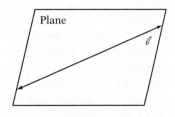

Plane

Lines in a plane can be intersecting or parallel. **Intersecting lines** cross at a point in the plane. **Parallel lines** never intersect. The distance between them is always the same.

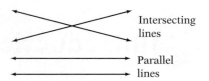

Intersecting lines

Parallel lines

The symbol ∥ means "is parallel to." In the figure at the right, $j \parallel k$ and $\overline{AB} \parallel \overline{CD}$. Note that j contains $\overline{AB}$ and k contains $\overline{CD}$. Parallel lines contain parallel line segments.

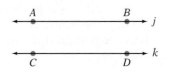

An **angle** is formed by two rays with the same endpoint. The **vertex** of the angle is the point at which the two rays meet. The rays are called the **sides** of the angle.

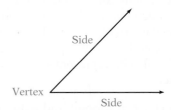

Side

Vertex

Side

If A and C are points on rays r_1 and r_2, and B is the vertex, then the angle is called $\angle B$ or $\angle ABC$, where $\angle$ is the symbol for angle. Note that the angle is named by the vertex, or the vertex is the second point listed when the angle is named by giving three points. $\angle ABC$ could also be called $\angle CBA$.

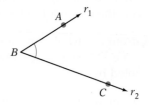

An angle can also be named by a variable written between the rays close to the vertex. In the figure at the right, $\angle x = \angle QRS$ and $\angle y = \angle SRT$. Note that in this figure, more than two rays meet at R. In this case, the vertex cannot be used to name an angle.

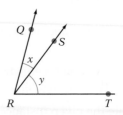

Point of Interest

The Babylonians knew that Earth is in approximately the same position in the sky every 365 days. Historians suggest that one complete revolution of a circle is called 360° because 360 is the closest number to 365 that is divisible by many natural numbers.

An angle is measured in **degrees.** The symbol for degrees is a small raised circle, °. Probably because early Babylonians believed that Earth revolves around the sun in approximately 360 days, the angle formed by a ray rotating through a circle has a measure of 360° (360 degrees).

360°

A **protractor** is used to measure an angle. Place the center of the protractor at the vertex of the angle with the edge of the protractor along a side of the angle. The angle shown in the figure below measures 58°.

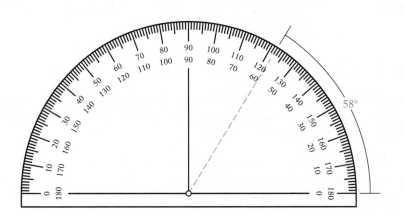

The Leaning Tower of Pisa is the bell tower of the Cathedral in Pisa, Italy. Its construction began on August 9, 1173, and continued for about 200 years. The tower was designed to be vertical, but it started to lean during its construction. By 1350 it was 2.5° off from the vertical; by 1817, it was 5.1° off; and by 1990, it was 5.5° off. In 2001, work on the structure that returned its list to 5° was completed. (*Source: Time* magazine, June 25, 2001, pp. 34–35)

A 90° angle is called a **right angle.** The symbol ∟ represents a right angle.

Perpendicular lines are intersecting lines that form right angles.

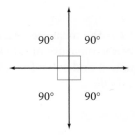

The symbol ⊥ means "is perpendicular to." In the figure at the right, $p \perp q$ and $\overline{AB} \perp \overline{CD}$. Note that line p contains $\overline{AB}$ and line q contains $\overline{CD}$. Perpendicular lines contain perpendicular line segments.

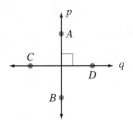

Complementary angles are two angles whose measures have the sum 90°.

$$\angle A + \angle B = 70° + 20° = 90°$$

$\angle A$ and $\angle B$ are complementary angles.

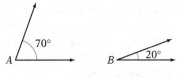

A 180° angle is called a **straight angle.**

$\angle AOB$ is a straight angle.

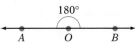

Supplementary angles are two angles whose measures have the sum 180°.

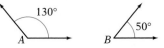

$$\angle A + \angle B = 130° + 50° = 180°$$

$\angle A$ and $\angle B$ are supplementary angles.

An **acute angle** is an angle whose measure is between 0° and 90°. $\angle B$ above is an acute angle. An **obtuse angle** is an angle whose measure is between 90° and 180°. $\angle A$ above is an obtuse angle.

Two angles that share a common side are **adjacent angles.** In the figure at the right, $\angle DAC$ and $\angle CAB$ are adjacent angles. $\angle DAC = 45°$ and $\angle CAB = 55°$.

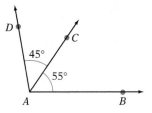

$$\angle DAB = \angle DAC + \angle CAB$$
$$= 45° + 55° = 100°$$

In the figure at the right, $\angle EDG = 80°$. $\angle FDG$ is three times the measure of $\angle EDF$. Find the measure of $\angle EDF$.

Let x = the measure of $\angle EDF$. Then $3x$ = the measure of $\angle FDG$. Write an equation and solve for x, the measure of $\angle EDF$.

$$\angle EDF + \angle FDG = \angle EDG$$
$$x + 3x = 80$$
$$4x = 80$$
$$x = 20$$

$\angle EDF = 20°$

EXAMPLE 1

Given $MN = 15$ mm, $NO = 18$ mm, and $MP = 48$ mm, find OP.

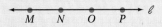

Solution

$MN + NO + OP = MP$ ▸ $MN = 15, NO = 18,$
$15 + 18 + OP = 48$ $MP = 48$
$33 + OP = 48$ ▸ Add 15 and 18.
$OP = 15$ ▸ Subtract 33 from each side.

$OP = 15$ mm

YOU TRY IT 1

Given $QR = 24$ cm, $ST = 17$ cm, and $QT = 62$ cm, find RS.

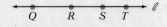

Your Solution

Solution on p. S22

EXAMPLE 2

Given $XY = 9$ m and YZ is twice XY, find XZ.

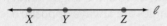

Solution

$XZ = XY + YZ$

$XZ = XY + 2(XY)$ ▸ YZ is twice XY.

$XZ = 9 + 2(9)$ ▸ $XY = 9$

$XZ = 9 + 18$

$XZ = 27$

$XZ = 27$ m

YOU TRY IT 2

Given $BC = 16$ ft and $AB = \frac{1}{4}(BC)$, find AC.

```
•————•————————•———→ ℓ
A    B         C
```

Your Solution

EXAMPLE 3

Find the complement of a 38° angle.

Strategy

Complementary angles are two angles whose sum is 90°. To find the complement, let x represent the complement of a 38° angle. Write an equation and solve for x.

Solution

$x + 38° = 90°$

$x = 52°$

The complement of a 38° angle is a 52° angle.

YOU TRY IT 3

Find the supplement of a 129° angle.

Your Strategy

Your Solution

EXAMPLE 4

Find the measure of ∠x.

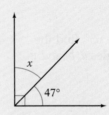

Strategy

To find the measure of ∠x, write an equation using the fact that the sum of the measure of ∠x and 47° is 90°. Solve for ∠x.

Solution

∠$x + 47° = 90°$

∠$x = 43°$

The measure of ∠x is 43°.

YOU TRY IT 4

Find the measure of ∠a.

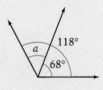

Your Strategy

Your Solution

Solutions on p. S22

OBJECTIVE B Problems involving angles formed by intersecting lines

Point of Interest

Many cities in the New World, unlike those in Europe, were designed using rectangular street grids. Washington, D.C. was planned that way except that diagonal avenues were added, primarily for the purpose of enabling quick troop movement in the event that the city required defense. As an added precaution, monuments were constructed at major intersections so that attackers would not have a straight shot down a boulevard.

Four angles are formed by the intersection of two lines. If the two lines are perpendicular, each of the four angles is a right angle. If the two lines are not perpendicular, then two of the angles formed are acute angles and two of the angles are obtuse angles. The two acute angles are always opposite each other, and the two obtuse angles are always opposite each other.

In the figure at the right, $\angle w$ and $\angle y$ are acute angles. $\angle x$ and $\angle z$ are obtuse angles.

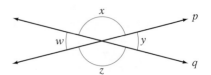

Two angles that are on opposite sides of the intersection of two lines are called **vertical angles.** Vertical angles have the same measure. $\angle w$ and $\angle y$ are vertical angles. $\angle x$ and $\angle z$ are vertical angles.

Vertical angles have the same measure.

$$\angle w = \angle y$$
$$\angle x = \angle z$$

Recall that two angles that share a common side are called **adjacent angles.** For the figure shown above, $\angle x$ and $\angle y$ are adjacent angles, as are $\angle y$ and $\angle z$, $\angle z$ and $\angle w$, and $\angle w$ and $\angle x$. Adjacent angles of intersecting lines are supplementary angles.

Adjacent angles of intersecting lines are supplementary angles.

$$\angle x + \angle y = 180°$$
$$\angle y + \angle z = 180°$$
$$\angle z + \angle w = 180°$$
$$\angle w + \angle x = 180°$$

Given that $\angle c = 65°$, find the measures of angles a, b, and d.

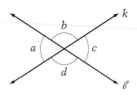

$\angle a = \angle c$ because $\angle a$ and $\angle c$ are vertical angles.

$$\angle a = 65°$$

$\angle b$ is supplementary to $\angle c$ because $\angle b$ and $\angle c$ are adjacent angles of intersecting lines.

$$\angle b + \angle c = 180°$$
$$\angle b + 65° = 180°$$
$$\angle b = 115°$$

$\angle d = \angle b$ because $\angle d$ and $\angle b$ are vertical angles.

$$\angle d = 115°$$

A line that intersects two other lines at different points is called a **transversal.**

If the lines cut by a transversal t are parallel lines and the transversal is perpendicular to the parallel lines, all eight angles formed are right angles.

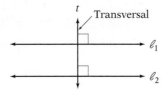

If the lines cut by a transversal t are parallel lines and the transversal is not perpendicular to the parallel lines, all four acute angles have the same measure and all four obtuse angles have the same measure. For the figure at the right,

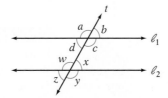

$$\angle b = \angle d = \angle x = \angle z$$

$$\angle a = \angle c = \angle w = \angle y$$

Alternate interior angles are two non-adjacent angles that are on opposite sides of the transversal and between the parallel lines. In the figure above, $\angle c$ and $\angle w$ are alternate interior angles; $\angle d$ and $\angle x$ are alternate interior angles. Alternate interior angles have the same measure.

Alternate interior angles have the same measure.

$$\angle c = \angle w$$
$$\angle d = \angle x$$

Alternate exterior angles are two non-adjacent angles that are on opposite sides of the transversal and outside the parallel lines. In the figure above, $\angle a$ and $\angle y$ are alternate exterior angles; $\angle b$ and $\angle z$ are alternate exterior angles. Alternate exterior angles have the same measure.

Alternate exterior angles have the same measure.

$$\angle a = \angle y$$
$$\angle b = \angle z$$

Corresponding angles are two angles that are on the same side of the transversal and are both acute angles or are both obtuse angles. For the figure above, the following pairs of angles are corresponding angles: $\angle a$ and $\angle w$, $\angle d$ and $\angle z$, $\angle b$ and $\angle x$, $\angle c$ and $\angle y$. Corresponding angles have the same measure.

Corresponding angles have the same measure.

$$\angle a = \angle w$$
$$\angle d = \angle z$$
$$\angle b = \angle x$$
$$\angle c = \angle y$$

Given that $\ell_1 \parallel \ell_2$ and $\angle c = 58°$, find the measures of $\angle f$, $\angle h$, and $\angle g$.

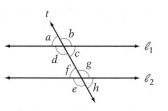

$\angle c$ and $\angle f$ are alternate interior angles.

$\angle c$ and $\angle h$ are corresponding angles.

$\angle g$ is supplementary to $\angle h$.

$\angle f = \angle c = 58°$

$\angle h = \angle c = 58°$

$\angle g + \angle h = 180°$
$\angle g + 58° = 180°$
$\angle g = 122°$

EXAMPLE 5

Find x.

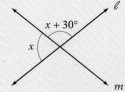

Strategy

The angles labeled are adjacent angles of intersecting lines and are therefore supplementary angles. To find x, write an equation and solve for x.

Solution

$x + (x + 30°) = 180°$
$2x + 30° = 180°$
$2x = 150°$
$x = 75°$

YOU TRY IT 5

Find x.

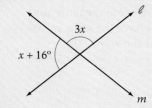

Your Strategy

Your Solution

EXAMPLE 6

Given $\ell_1 \parallel \ell_2$, find x.

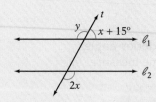

Strategy

$2x = y$ because alternate exterior angles have the same measure. $(x + 15°) + y = 180°$ because adjacent angles of intersecting lines are supplementary angles. Substitute $2x$ for y and solve for x.

Solution

$(x + 15°) + 2x = 180°$
$3x + 15° = 180°$
$3x = 165°$
$x = 55°$

YOU TRY IT 6

Given $\ell_1 \parallel \ell_2$, find x.

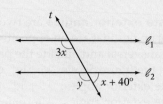

Your Strategy

Your Solution

Solutions on p. S22

OBJECTIVE C   **Problems involving the angles of a triangle**

If the lines cut by a transversal are not parallel lines, the three lines will intersect at three points. In the figure at the right, the transversal t intersects lines p and q. The three lines intersect at points A, B, and C. These three points define three line segments, $\overline{AB}$, $\overline{BC}$, and $\overline{AC}$. The plane figure formed by these three line segments is called a **triangle.**

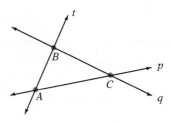

Each of the three points of intersection is the vertex of four angles. The angles within the region enclosed by the triangle are called **interior angles.** In the figure at the right, angles a, b, and c are interior angles. The sum of the measures of the interior angles of a triangle is 180°.

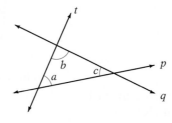

$$\angle a + \angle b + \angle c = 180°$$

> ### The Sum of the Measures of the Interior Angles of a Triangle
>
> The sum of the measures of the interior angles of a triangle is 180°.

An angle adjacent to an interior angle is an **exterior angle.** In the figure at the right, angles m and n are exterior angles for angle a. The sum of the measures of an interior and an exterior angle is 180°.

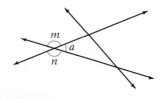

$$\angle a + \angle m = 180°$$
$$\angle a + \angle n = 180°$$

Given that $\angle c = 40°$ and $\angle d = 100°$, find the measure of $\angle e$.

$\angle d$ and $\angle b$ are supplementary angles.

$$\angle d + \angle b = 180°$$
$$100° + \angle b = 180°$$
$$\angle b = 80°$$

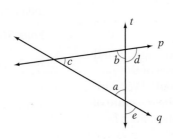

The sum of the interior angles is 180°.

$$\angle c + \angle b + \angle a = 180°$$
$$40° + 80° + \angle a = 180°$$
$$120° + \angle a = 180°$$
$$\angle a = 60°$$

$\angle a$ and $\angle e$ are vertical angles.

$$\angle e = \angle a = 60°$$

EXAMPLE 7

Given that $\angle y = 55°$, find the measures of angles a, b, and d.

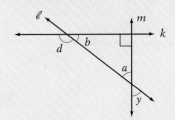

Strategy

▶ To find the measure of angle a, use the fact that $\angle a$ and $\angle y$ are vertical angles.
▶ To find the measure of angle b, use the fact that the sum of the measures of the interior angles of a triangle is 180°.
▶ To find the measure of angle d, use the fact that the sum of an interior and an exterior angle is 180°.

Solution

$\angle a = \angle y = 55°$

$\angle a + \angle b + 90° = 180°$
$55° + \angle b + 90° = 180°$
$\angle b + 145° = 180°$
$\angle b = 35°$

$\angle d + \angle b = 180°$
$\angle d + 35° = 180°$
$\angle d = 145°$

EXAMPLE 8

Two angles of a triangle measure 53° and 78°. Find the measure of the third angle.

Strategy

To find the measure of the third angle, use the fact that the sum of the measures of the interior angles of a triangle is 180°. Write an equation using x to represent the measure of the third angle. Solve the equation for x.

Solution

$x + 53° + 78° = 180°$
$x + 131° = 180°$
$x = 49°$

The measure of the third angle is 49°.

YOU TRY IT 7

Given that $\angle a = 45°$ and $\angle x = 100°$, find the measures of angles b, c, and y.

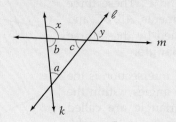

Your Strategy

Your Solution

YOU TRY IT 8

One angle in a triangle is a right angle, and one angle measures 34°. Find the measure of the third angle.

Your Strategy

Your Solution

Solutions on p. S23

Area of a Trapezoid

Let b_1 and b_2 represent the lengths of the bases and h the height of a trapezoid. The area, A, of the trapezoid is given by

$A = \frac{1}{2}h(b_1 + b_2)$.

Find the area of a trapezoid that has bases measuring 15 in. and 5 in. and a height of 8 in.

$A = \frac{1}{2}h(b_1 + b_2)$

$= \frac{1}{2} \cdot 8(15 + 5) = 4(20) = 80$

The area is 80 in².

15 in.

8 in.

5 in.

The area of a circle is equal to the product of π and the square of the radius.

r

$A = \pi r^2$

The Area of a Circle

The area, A, of a circle with radius r is given by $A = \pi r^2$.

Find the area of a circle that has a radius of 6 cm.

Use the formula for the area of a circle.
$r = 6$.

$A = \pi r^2$
$A = \pi(6)^2$
$A = \pi(36)$

The exact area of the circle is 36π cm².

$A = 36\pi$

An approximate measure is found by using the π key on a calculator.

$A \approx 113.10$

The approximate area of the circle is 113.10 cm².

Calculator Note

To approximate 36π on your calculator, enter 36 ✕ π = .

For your reference, all of the formulas for the perimeter and area of the geometric figures presented in this section are listed in the Chapter Summary, which begins on page 596.

EXAMPLE 4

The Parks and Recreation Department of a city plans to plant grass seed in a playground that has the shape of a trapezoid, as shown below. Each bag of grass seed will seed 1,500 ft². How many bags of grass seed should the department purchase?

Strategy

To find the number of bags to be purchased:

▶ Use the formula for the area of a trapezoid to find the area of the playground.
▶ Divide the area of the playground by the area one bag will seed (1,500).

Solution

$$A = \frac{1}{2}h(b_1 + b_2)$$

$$A = \frac{1}{2} \cdot 64(80 + 115)$$

$A = 6,240$ ▶ The area of the playground is 6,240 ft².

$6,240 \div 1,500 = 4.16$

Because a portion of a fifth bag is needed, 5 bags of grass seed should be purchased.

EXAMPLE 5

Find the area of a circle with a diameter of 5 ft. Give the exact measure.

Strategy

To find the area:

▶ Find the radius of the circle.
▶ Use the formula for the area of a circle. Leave the answer in terms of π.

Solution

$$r = \frac{1}{2}d = \frac{1}{2}(5) = 2.5$$

$$A = \pi r^2 = \pi(2.5)^2 = \pi(6.25) = 6.25\pi$$

The area of the circle is 6.25π ft².

YOU TRY IT 4

An interior designer decides to wallpaper two walls of a room. Each roll of wallpaper will cover 30 ft². Each wall measures 8 ft by 12 ft. How many rolls of wallpaper should be purchased?

Your Strategy

Your Solution

YOU TRY IT 5

Find the area of a circle with a radius of 11 cm. Round to the nearest hundredth.

Your Strategy

Your Solution

Solutions on p. S23

9.2 Exercises

OBJECTIVE A **Perimeter of a plane geometric figure**

Name each polygon.

1.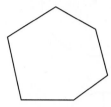

_____

2.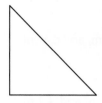

_____

3.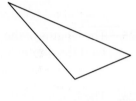

_____

4.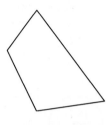

_____

Classify the triangle as isosceles, equilateral, or scalene.

5.

_____

6.

_____

7.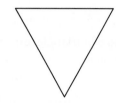

_____

8.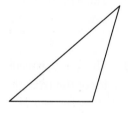

_____

Classify the triangle as acute, obtuse, or right.

9.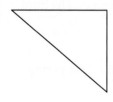

_____

10.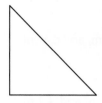

_____

11.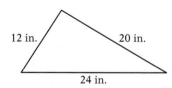

_____

12.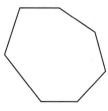

_____

Find the perimeter of the figure.

13.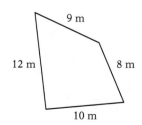
12 in. 20 in. 24 in.

14.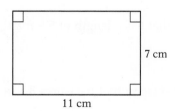
7 cm 11 cm

15.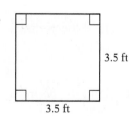
3.5 ft 3.5 ft

16.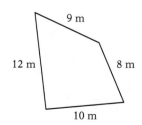
9 m 12 m 8 m 10 m

17.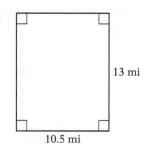
13 mi 10.5 mi

18.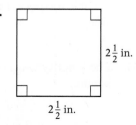
$2\frac{1}{2}$ in. $2\frac{1}{2}$ in.

For Exercises 19 to 24, find the circumference of the figure.
Give both the exact value and an approximation to the nearest hundredth.

19.

20.

21.

22.

23.

24.

25. The lengths of the three sides of a triangle are 3.8 cm, 5.2 cm, and 8.4 cm. Find the perimeter of the triangle.

26. The lengths of the three sides of a triangle are 7.5 m, 6.1 m, and 4.9 m. Find the perimeter of the triangle.

27. The length of each of two sides of an isosceles triangle is $2\frac{1}{2}$ cm. The third side measures 3 cm. Find the perimeter of the triangle.

28. The length of each side of an equilateral triangle is $4\frac{1}{2}$ in. Find the perimeter of the triangle.

29. A rectangle has a length of 8.5 m and a width of 3.5 m. Find the perimeter of the rectangle.

30. Find the perimeter of a rectangle that has a length of $5\frac{1}{2}$ ft and a width of 4 ft.

31. Find the perimeter of a regular pentagon that measures 3.5 in. on each side.

32. What is the perimeter of a regular hexagon that measures 8.5 cm on each side?

33. The length of each side of a square is 12.2 cm. Find the perimeter of the square.

34. Find the perimeter of a square that is 0.5 m on each side.

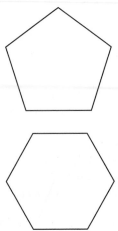

35. Find the circumference of a circle that has a diameter of 1.5 in. Give the exact value.

36. The diameter of a circle is 4.2 ft. Find the circumference of the circle. Round to the nearest hundredth.

37. The radius of a circle is 36 cm. Find the circumference of the circle. Round to the nearest hundredth.

38. Find the circumference of a circle that has a radius of 2.5 m. Give the exact value.

39. How many feet of fencing should be purchased for a rectangular garden that is 18 ft long and 12 ft wide?

40. How many meters of binding are required to bind the edge of a rectangular quilt that measures 3.5 m by 8.5 m?

41. Wall-to-wall carpeting is installed in a room that is 12 ft long and 10 ft wide. The edges of the carpet are nailed to the floor. Along how many feet must the carpet be nailed down?

42. The length of a rectangular park is 55 yd. The width is 47 yd. How many yards of fencing are needed to surround the park?

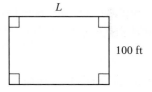

43. The perimeter of a rectangular playground is 440 ft. If the width is 100 ft, what is the length of the playground?

44. A rectangular vegetable garden has a perimeter of 64 ft. The length of the garden is 20 ft. What is the width of the garden?

45. Each of two sides of a triangular banner measures 18 in. If the perimeter of the banner is 46 in., what is the length of the third side of the banner?

46. The perimeter of an equilateral triangle is 13.2 cm. What is the length of each side of the triangle?

47. The perimeter of a square picture frame is 48 in. Find the length of each side of the frame.

48. A square rug has a perimeter of 32 ft. Find the length of each edge of the rug.

Solve. For Exercises 49 to 55, round to the nearest hundredth.

49. The circumference of a circle is 8 cm. Find the length of a diameter of the circle.

50. The circumference of a circle is 15 in. Find the length of a radius of the circle.

51. Find the length of molding needed to put around a circular table that is 4.2 ft in diameter.

52. How much binding is needed to bind the edge of a circular rug that is 3 m in diameter?

53. A bicycle tire has a diameter of 24 in. How many feet does the bicycle travel when the wheel makes eight revolutions?

54. A tricycle tire has a diameter of 12 in. How many feet does the tricycle travel when the wheel makes 12 revolutions?

55. The distance from the surface of Earth to its center is 6,356 km. What is the circumference of Earth?

56. Bias binding is to be sewed around the edge of a rectangular tablecloth measuring 72 in. by 45 in. If the bias binding comes in packages containing 15 ft of binding, how many packages of bias binding are needed for the tablecloth?

57. Which has the greater perimeter, a square whose side measures 1 ft or a rectangle that has a length of 2 in. and a width of 1 in.?

58. The perimeter of an isosceles triangle is 54 ft. Let s be the length of one of the two equal sides. Is it possible for s to be 30 ft?

OBJECTIVE B   **Area of a plane geometric figure**

Find the area of the figure.

59.
5 ft
12 ft

60.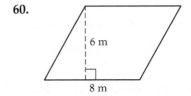
6 m
8 m

61.
4.5 in.
4.5 in.

62.

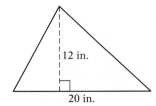

12 in.

20 in.

63.

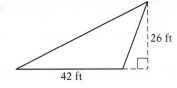

26 ft

42 ft

64.

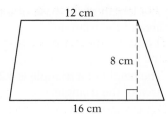

12 cm

8 cm

16 cm

For Exercises 65 to 70, find the area of the figure.
Give both the exact value and an approximation to the nearest hundredth.

65.

4 cm

66.

12 m

67.

5.5 mi

68.

18 in.

69.

17 ft

70.

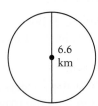

6.6 km

Solve.

71. The length of a side of a square is 12.5 cm. Find the area of the square.

72. Each side of a square measures $3\frac{1}{2}$ in. Find the area of the square.

73. The length of a rectangle is 38 in., and the width is 15 in. Find the area of the rectangle.

74. Find the area of a rectangle that has a length of 6.5 m and a width of 3.8 m.

75. The length of the base of a parallelogram is 16 in., and the height is 12 in. Find the area of the parallelogram.

76. The height of a parallelogram is 3.4 m, and the length of the base is 5.2 m. Find the area of the parallelogram.

77. The length of the base of a triangle is 6 ft. The height is 4.5 ft. Find the area of the triangle.

78. The height of a triangle is 4.2 cm. The length of the base is 5 cm. Find the area of the triangle.

79. The length of one base of a trapezoid is 35 cm, and the length of the other base is 20 cm. If the height is 12 cm, what is the area of the trapezoid?

80. The height of a trapezoid is 5 in. The bases measure 16 in. and 18 in. Find the area of the trapezoid.

81. The radius of a circle is 5 in. Find the area of the circle. Give the exact value.

82. The diameter of a circle is 6.5 m. Find the area of the circle. Give the exact value.

83. The lens on the Hale telescope at Mount Palomar, California, has a diameter of 200 in. Find its area. Give the exact value.

84. An irrigation system waters a circular field that has a 50-foot radius. Find the area watered by the irrigation system. Give the exact value.

85. Find the area of a rectangular flower garden that measures 14 ft by 9 ft.

86. What is the area of a square patio that measures 8.5 m on each side?

87. Artificial turf is being used to cover a playing field. If the field is rectangular with a length of 100 yd and a width of 75 yd, how much artificial turf must be purchased to cover the field?

88. A fabric wall hanging is to fill a space that measures 5 m by 3.5 m. Allowing for 0.1 m of the fabric to be folded back along each edge, how much fabric must be purchased for the wall hanging?

89. The area of a rectangle is 80 in². The length of the rectangle is 16 in. Find the width.

    a. Use the formula for the area of a rectangle. $A = $ _____

    b. Replace A by _____ and L by _____ .   _____ $= 16W$

    c. Divide both sides of the equation by _____ .   _____ $= W$

    d. The width of the rectangle is _____ in.

90. The area of a parallelogram is 96 m². The height of the parallelogram is 8 m. Find the length of the base.

   a. Use the formula for the area of a parallelogram. $A = $ _____
   b. Replace A by _____ and h by _____ .   _____ $= b \cdot 8$
   c. Divide both sides of the equation by _____ .   _____ $= b$
   d. The length of the base is _____ m.

91. The area of a rectangle is 300 in². If the length of the rectangle is 30 in., what is the width?

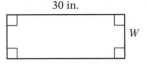

92. The width of a rectangle is 12 ft. If the area is 312 ft², what is the length of the rectangle?

93. The height of a triangle is 5 m. The area of the triangle is 50 m². Find the length of the base of the triangle.

94. The area of a parallelogram is 42 m². If the height of the parallelogram is 7 m, what is the length of the base?

95. You plan to stain the wooden deck attached to your house. The deck measures 10 ft by 8 ft. If a quart of stain will cover 50 ft², how many quarts of stain should you buy?

96. You want to tile your kitchen floor. The floor measures 12 ft by 9 ft. How many tiles, each a square with side $1\frac{1}{2}$ ft, should you purchase for the job?

97. You are wallpapering two walls of a child's room, one measuring 9 ft by 8 ft and the other measuring 11 ft by 8 ft. The wallpaper costs $18.50 per roll, and each roll of wallpaper will cover 40 ft². What is the cost to wallpaper the two walls?

98. An urban renewal project involves reseeding a park that is in the shape of a square, 60 ft on each side. Each bag of grass seed costs $5.75 and will seed 1,200 ft². How much money should be budgeted for buying grass seed for the park?

99. A circle has a radius of 8 in. Find the increase in area when the radius is increased by 2 in. Round to the nearest hundredth.

100. A circle has a radius of 6 cm. Find the increase in area when the radius is doubled. Round to the nearest hundredth.

101. You want to install wall-to-wall carpeting in your living room, which measures 15 ft by 24 ft. If the cost of the carpet you would like to purchase is $15.95 per square yard, what is the cost of the carpeting for your living room? (*Hint:* $9 \text{ ft}^2 = 1 \text{ yd}^2$)

102. You want to paint the walls of your bedroom. Two walls measure 15 ft by 9 ft, and the other two walls measure 12 ft by 9 ft. The paint you wish to purchase costs $19.98 per gallon, and each gallon will cover 400 ft^2 of wall. Find the total amount you will spend on paint.

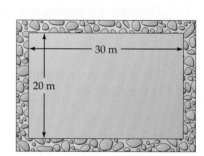

103. A walkway 2 m wide surrounds a rectangular plot of grass. The plot is 30 m long and 20 m wide. What is the area of the walkway?

104. Pleated draperies for a window must be twice as wide as the width of the window. Draperies are being made for four windows, each 2 ft wide and 4 ft high. Since the drapes will fall slightly below the window sill and extra fabric will be needed for hemming the drapes, 1 ft must be added to the height of the window. How much material must be purchased to make the drapes?

105. A circle has a radius of 5 cm.
   a. Can the exact area of the circle be A cm, where A is a decimal approximation?
   b. Can the exact area of the circle be A cm^2, where A is a whole number?

106. A rectangle has a perimeter of 20 units. What dimensions will result in a rectangle with the greatest possible area? Consider only whole-number dimensions.

107. If both the length and the width of a rectangle are doubled, how many times larger is the area of the resulting rectangle?

CRITICAL THINKING

108. Find the ratio of the areas of two squares if the ratio of the lengths of their sides is 2:3.

109. Suppose a circle is cut into 16 equal pieces, which are then arranged as shown at the right. The figure formed resembles a parallelogram. What variable expression could describe the base of the parallelogram? What variable could describe its height? Explain how the formula for the area of a circle is derived from this approach.

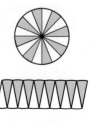

110. Prepare a report on the history of quilts in the United States. Find examples of quilt patterns that incorporate regular polygons.

9.3 Triangles

OBJECTIVE A   The Pythagorean Theorem

A **right triangle** contains one right angle. The side opposite the right angle is called the **hypotenuse.** The other two sides are called **legs.**

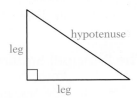

The angles in a right triangle are usually labeled with the capital letters A, B, and C, with C reserved for the right angle. The side opposite angle A is side a, the side opposite angle B is side b, and c is the hypotenuse.

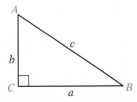

The Greek mathematician Pythagoras is generally credited with the discovery that the square of the hypotenuse of a right triangle is equal to the sum of the squares of the two legs. This is called the **Pythagorean Theorem.**

The figure at the right is a right triangle with legs measuring 3 units and 4 units and a hypotenuse measuring 5 units. Each side of the triangle is also the side of a square. The number of square units in the area of the largest square is equal to the sum of the numbers of square units in the areas of the smaller squares.

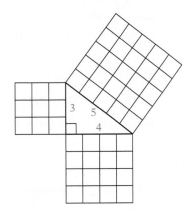

Square of the hypotenuse	=	sum of the squares of the two legs

$$5^2 = 3^2 + 4^2$$
$$25 = 9 + 16$$
$$25 = 25$$

Point of Interest

The first known proof of the Pythagorean Theorem is in a Chinese textbook that dates from 150 B.C. The book is called *Nine Chapters on the Mathematical Art.* The diagram below is from that book and was used in the proof of the theorem.

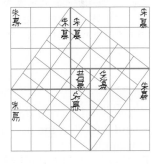

Pythagorean Theorem

If a and b are the lengths of the legs of a right triangle and c is the length of the hypotenuse, then $c^2 = a^2 + b^2$.

If the lengths of two sides of a right triangle are known, the Pythagorean Theorem can be used to find the length of the third side.

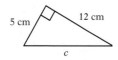

Consider a right triangle with legs that measure 5 cm and 12 cm. Use the Pythagorean Theorem, with $a = 5$ and $b = 12$, to find the length of the hypotenuse. (If you let $a = 12$ and $b = 5$, the result is the same.)

$$c^2 = a^2 + b^2$$
$$c^2 = 5^2 + 12^2$$
$$c^2 = 25 + 144$$
$$c^2 = 169$$

This equation states that the square of c is 169. Since $13^2 = 169$, $c = 13$, and the length of the hypotenuse is 13 cm. We can find c by taking the square root of 169: $\sqrt{169} = 13$. This suggests the following property.

Calculator Note

The way in which you evaluate the square root of a number depends on the type of calculator you have. Here are two possible keystrokes to find $\sqrt{35}$:

35   √   =

or

√   35   ENTER

The first method is used on many scientific calculators. The second method is used on many graphing calculators.

The Principal Square Root Property

If $r^2 = s$, then $r = \sqrt{s}$, and r is called the **square root** of s.

The Principal Square Root Property and its application can be illustrated as follows: Because $5^2 = 25$, $5 = \sqrt{25}$. Therefore, if $c^2 = 25$, $c = \sqrt{25} = 5$.

Recall that numbers whose square roots are integers, such as 25, are perfect squares. If a number is not a perfect square, a calculator can be used to find an approximate square root when a decimal approximation is required.

The length of one leg of a right triangle is 8 in. The hypotenuse is 12 in. Find the length of the other leg. Round to the nearest hundredth.

Use the Pythagorean Theorem. $a^2 + b^2 = c^2$
$a = 8, c = 12$ $8^2 + b^2 = 12^2$
Solve for b^2. $64 + b^2 = 144$
(If you let $b = 8$ and solve for a^2, the result is the same.) $b^2 = 80$

Use the Principal Square Root Property.
Since $b^2 = 80$, b is the square root of 80. $b = \sqrt{80}$

Use a calculator to approximate $\sqrt{80}$. $b \approx 8.94$

The length of the other leg is approximately 8.94 in.

EXAMPLE 1

The two legs of a right triangle measure 12 ft and 9 ft. Find the hypotenuse of the right triangle.

Strategy

To find the hypotenuse, use the Pythagorean Theorem. $a = 12, b = 9$

Solution

$$c^2 = a^2 + b^2$$
$$c^2 = 12^2 + 9^2$$
$$c^2 = 144 + 81$$
$$c^2 = 225$$
$$c = \sqrt{225} \quad \blacktriangleright \text{ The Principal Square Root}$$
$$c = 15 \qquad\qquad \text{Property}$$

The length of the hypotenuse is 15 ft.

YOU TRY IT 1

The hypotenuse of a right triangle measures 6 m, and one leg measures 2 m. Find the measure of the other leg. Round to the nearest hundredth.

Your Strategy

Your Solution

Solution on p. S23

OBJECTIVE B Similar triangles

Similar objects have the same shape but not necessarily the same size. A tennis ball is similar to a basketball. A model ship is similar to an actual ship.

Similar objects have corresponding parts; for example, the rudder on the model ship corresponds to the rudder on the actual ship. The relationship between the sizes of each of the corresponding parts can be written as a ratio, and each ratio will be the same. If the rudder on the model ship is $\frac{1}{100}$ the size of the rudder on the actual ship, then the model wheelhouse is $\frac{1}{100}$ the size of the actual wheelhouse, the width of the model is $\frac{1}{100}$ the width of the actual ship, and so on.

The two triangles ABC and DEF shown at the right are similar. Side $\overline{AB}$ corresponds to side $\overline{DE}$, side $\overline{BC}$ corresponds to side $\overline{EF}$, and side $\overline{AC}$ corresponds to side $\overline{DF}$. The ratios of corresponding sides are equal.

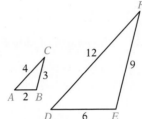

$$\frac{AB}{DE} = \frac{2}{6} = \frac{1}{3}, \frac{BC}{EF} = \frac{3}{9} = \frac{1}{3}, \text{ and } \frac{AC}{DF} = \frac{4}{12} = \frac{1}{3}.$$

Since the ratios of corresponding sides are equal, three proportions can be formed.

$$\frac{AB}{DE} = \frac{BC}{EF}, \frac{AB}{DE} = \frac{AC}{DF}, \text{ and } \frac{BC}{EF} = \frac{AC}{DF}.$$

The corresponding angles in similar triangles are equal. Therefore,

$\angle A = \angle D$, $\angle B = \angle E$, and $\angle C = \angle F$.

Triangles ABC and DEF at the right are similar triangles. AH and DK are the heights of the triangles. The ratio of heights of similar triangles equals the ratio of corresponding sides.

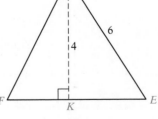

Ratio of corresponding sides $= \dfrac{1.5}{6} = \dfrac{1}{4}$

Ratio of heights $= \dfrac{1}{4}$

> ## Properties of Similar Triangles
>
> For similar triangles, the ratios of corresponding sides are equal. The ratio of corresponding heights is equal to the ratio of corresponding sides.

Point of Interest

Many mathematicians have studied similar objects. Thales of Miletus (c. 624 B.C.–547 B.C.) discovered that he could determine the heights of pyramids and other objects by measuring a small object and the length of its shadow and then making use of similar triangles.

The two triangles at the right are similar triangles. Find the length of side $\overline{EF}$. Round to the nearest tenth.

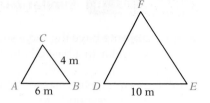

The triangles are similar, so the ratios of corresponding sides are equal.

$$\frac{EF}{BC} = \frac{DE}{AB}$$

$$\frac{EF}{4} = \frac{10}{6}$$

$$6(EF) = 4(10)$$
$$6(EF) = 40$$
$$EF \approx 6.7$$

The length of side EF is approximately 6.7 m.

EXAMPLE 2

Triangles ABC and DEF are similar. Find FG, the height of triangle DEF.

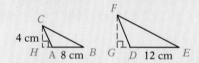

Strategy

To find FG, write a proportion using the fact that, in similar triangles, the ratio of corresponding sides equals the ratio of corresponding heights. Solve the proportion for FG.

Solution

$$\frac{AB}{DE} = \frac{CH}{FG}$$

$$\frac{8}{12} = \frac{4}{FG}$$

$$8(FG) = 12(4)$$
$$8(FG) = 48$$
$$FG = 6$$

The height FG of triangle DEF is 6 cm.

YOU TRY IT 2

Triangles ABC and DEF are similar. Find FG, the height of triangle DEF.

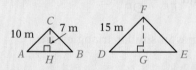

Your Strategy

Your Solution

Solution on pp. S23–S24

OBJECTIVE C   **Congruent triangles**

Congruent objects have the same shape *and* the same size.

The two triangles at the right are congruent. They have the same size.

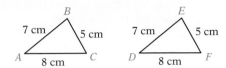

Congruent and similar triangles differ in that the corresponding sides and angles of congruent triangles must be equal, whereas for similar triangles, corresponding angles are equal, but corresponding sides are not necessarily the same length.

The three major rules used to determine whether two triangles are congruent are given below.

> ### Side-Side-Side Rule (SSS)
> ....
>
> Two triangles are congruent if the three sides of one triangle equal the corresponding three sides of a second triangle.

In the triangles at the right, $AC = DE$, $AB = EF$, and $BC = DF$. The corresponding sides of triangles ABC and DEF are equal. The triangles are congruent by the SSS Rule.

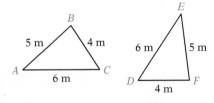

> ### Side-Angle-Side Rule (SAS)
> ....
>
> If two sides and the included angle of one triangle equal two sides and the included angle of a second triangle, the two triangles are congruent.

In the two triangles at the right, $AB = EF$, $AC = DE$, and $\angle BAC = \angle DEF$. The triangles are congruent by the SAS Rule.

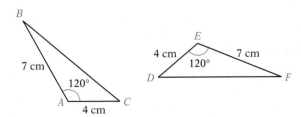

Angle-Side-Angle Rule (ASA)

If two angles and the included side of one triangle equal two angles and the included side of a second triangle, the two triangles are congruent.

For triangles *ABC* and *DEF* at the right, $\angle A = \angle F$, $\angle C = \angle E$, and $AC = EF$. The triangles are congruent by the ASA Rule.

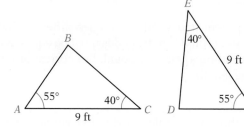

Given triangle *PQR* and triangle *MNO*, do the conditions $\angle P = \angle O$, $\angle Q = \angle M$, and $PQ = MO$ guarantee that triangle *PQR* is congruent to triangle *MNO*?

Draw a sketch of the two triangles and determine whether one of the rules for congruence is satisfied.

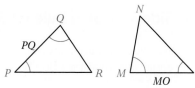

Because two angles and the included side of one triangle equal two angles and the included side of the second triangle, the triangles are congruent by the ASA Rule.

EXAMPLE 3

In the figure below, is triangle *ABC* congruent to triangle *DEF*?

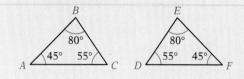

Strategy

To determine whether the triangles are congruent, determine whether one of the rules for congruence is satisfied.

Solution

The triangles do not satisfy the SSS Rule, the SAS Rule, or the ASA Rule. The triangles are not necessarily congruent.

YOU TRY IT 3

In the figure below, is triangle *PQR* congruent to triangle *MNO*?

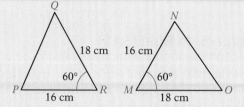

Your Strategy

Your Solution

Solution on p. S24

47.

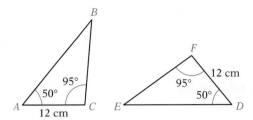

48.

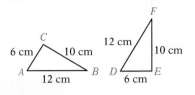

49. Given triangle *ABC* and triangle *DEF*, do the conditions ∠*C* = ∠*E*, *AC* = *EF*, and *BC* = *DE* guarantee that triangle *ABC* is congruent to triangle *DEF*? If they are congruent, by what rule are they congruent?

50. Given triangle *PQR* and triangle *MNO*, do the conditions *PR* = *NO*, *PQ* = *MO*, and *QR* = *MN* guarantee that triangle *PQR* is congruent to triangle *MNO*? If they are congruent, by what rule are they congruent?

51. Given triangle *LMN* and triangle *QRS*, do the conditions ∠*M* = ∠*S*, ∠*N* = ∠*Q*, and ∠*L* = ∠*R* guarantee that triangle *LMN* is congruent to triangle *QRS*? If they are congruent, by what rule are they congruent?

52. Given triangle *DEF* and triangle *JKL*, do the conditions ∠*D* = ∠*K*, ∠*E* = ∠*L*, and *DE* = *KL* guarantee that triangle *DEF* is congruent to triangle *JKL*? If they are congruent, by what rule are they congruent?

53. Given triangle *ABC* and triangle *PQR*, do the conditions ∠*B* = ∠*P*, *BC* = *PQ*, and *AC* = *QR* guarantee that triangle *ABC* is congruent to triangle *PQR*? If they are congruent, by what rule are they congruent?

For Exercises 54 and 55, determine whether the given conditions guarantee that the triangles are congruent.

54. The ratio of the corresponding sides of two similar triangles is $\frac{1}{1}$.

55. One right triangle has a hypotenuse of length 10 in. and an acute angle of 40°. A second right triangle has a hypotenuse of length 10 in. and an acute angle of 50°.

CRITICAL THINKING

56. *Home Maintenance* You need to clean the gutters of your home. The gutters are 24 ft above the ground. For safety, the distance a ladder reaches up a wall should be four times the distance from the bottom of the ladder to the base of the side of the house. Therefore, the ladder must be 6 ft from the base of the house. Will a 25-foot ladder be long enough to reach the gutters? Explain how you determined your answer.

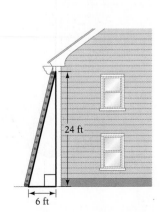

9.4 Solids

OBJECTIVE A   Volume of a solid

Geometric solids are figures in space. Five common geometric solids are the rectangular solid, the sphere, the cylinder, the cone, and the pyramid.

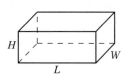

A **rectangular solid** is one in which all six sides, called **faces,** are rectangles. The variable L is used to represent the length of a rectangular solid, W its width, and H its height.

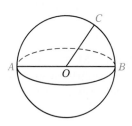

A **sphere** is a solid in which all points are the same distance from point O, called the **center** of the sphere. The **diameter,** d, of a sphere is a line across the sphere going through point O. The **radius,** r, is a line from the center to a point on the sphere. AB is a diameter and OC is a radius of the sphere shown at the right.

$$d = 2r \quad \text{or} \quad r = \frac{1}{2}d$$

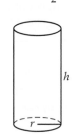

The most common cylinder, called a **right circular cylinder,** is one in which the bases are circles and are perpendicular to the height of the cylinder. The variable r is used to represent the radius of a base of a cylinder, and h represents the height. In this text, only right circular cylinders are discussed.

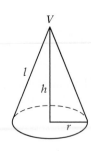

A **right circular cone** is obtained when one base of a right circular cylinder is shrunk to a point, called the **vertex,** V. The variable r is used to represent the radius of the base of the cone, and h represents the height. The variable l is used to represent the **slant height,** which is the distance from a point on the circumference of the base to the vertex. In this text, only right circular cones are discussed.

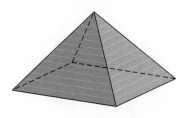

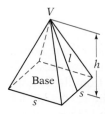

The base of a **regular pyramid** is a regular polygon, and the sides are isosceles triangles. The height, h, is the distance from the vertex, V, to the base and is perpendicular to the base. The variable l is used to represent the **slant height,** which is the height of one of the isosceles triangles on the face of the pyramid. The regular square pyramid at the right has a square base. This is the only type of pyramid discussed in this text.

A **cube** is a special type of rectangular solid. Each of the six faces of a cube is a square. The variable *s* is used to represent the length of one side of a cube.

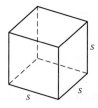

Volume is a measure of the amount of space occupied by a geometric solid. Volume can be used to describe the amount of trash in a landfill, the amount of concrete poured for the foundation of a house, or the amount of water in a town's reservoir.

A cube that is 1 ft on each side has a volume of 1 cubic foot, which is written 1 ft³. A cube that measures 1 cm on each side has a volume of 1 cubic centimeter, written 1 cm³.

The volume of a solid is the number of cubes that are necessary to exactly fill the solid. The volume of the rectangular solid at the right is 24 cm³ because it will hold exactly 24 cubes, each 1 cm on a side. Note that the volume can be found by multiplying the length times the width times the height.

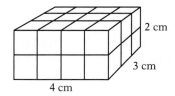

Point of Interest

Originally, the human body was used as the standard of measure. A mouthful was used as a unit of measure in ancient Egypt; it was later referred to as a *half jigger*. In French, the word for *inch* is *pouce*, which means "thumb." A *span* was the distance from the tip of the outstretched thumb to the tip of the little finger. The *cubit* referred to the distance from the elbow to the end of the fingers. A *fathom* was the distance from the tip of the fingers on one hand to the tip of the fingers on the other hand when standing with arms fully extended out from the sides. The *hand*, where 1 hand = 4 inches, is still used today to measure horses.

The formulas for the volumes of the geometric solids described above are given below.

Volumes of Geometric Solids

The volume, *V*, of a **rectangular solid** with length *L*, width *W*, and height *H* is given by $V = LWH$.

The volume, *V*, of a **cube** with side *s* is given by $V = s^3$.

The volume, *V*, of a **sphere** with radius *r* is given by $V = \frac{4}{3}\pi r^3$.

The volume, *V*, of a **right circular cylinder** is given by $V = \pi r^2 h$, where *r* is the radius of the base and *h* is the height.

The volume, *V*, of a **right circular cone** is given by $V = \frac{1}{3}\pi r^2 h$, where *r* is the radius of the circular base and *h* is the height.

The volume, *V*, of a **regular square pyramid** is given by $V = \frac{1}{3}s^2 h$, where *s* is the length of a side of the base and *h* is the height.

Find the volume of a sphere with a diameter of 6 in.

First find the radius of the sphere. $r = \dfrac{1}{2}d = \dfrac{1}{2}(6) = 3$

Use the formula for the volume of a sphere. $V = \dfrac{4}{3}\pi r^3$

$$V = \dfrac{4}{3}\pi(3)^3$$

$$V = \dfrac{4}{3}\pi(27)$$

The exact volume of the sphere is 36π in³. $V = 36\pi$

An approximate measure can be found by using the π key on a calculator. $V \approx 113.10$

The approximate volume is 113.10 in³.

Calculator Note

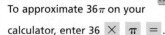

To approximate 36π on your calculator, enter 36 $\times$ π $=$.

EXAMPLE 1

The length of a rectangular solid is 5 m, the width is 3.2 m, and the height is 4 m. Find the volume of the solid.

Strategy

To find the volume, use the formula for the volume of a rectangular solid. $L = 5$, $W = 3.2$, $H = 4$

Solution

$V = LWH = 5(3.2)(4) = 64$

The volume of the rectangular solid is 64 m³.

YOU TRY IT 1

Find the volume of a cube that measures 2.5 m on a side.

Your Strategy

Your Solution

EXAMPLE 2

The radius of the base of a cone is 8 cm. The height is 12 cm. Find the volume of the cone. Round to the nearest hundredth.

Strategy

To find the volume, use the formula for the volume of a cone. An approximation is asked for; use the π key on a calculator. $r = 8$, $h = 12$

Solution

$V = \dfrac{1}{3}\pi r^2 h$

$V = \dfrac{1}{3}\pi(8)^2(12) = \dfrac{1}{3}\pi(64)(12) = 256\pi \approx 804.25$

The volume is approximately 804.25 cm³.

YOU TRY IT 2

The diameter of the base of a cylinder is 8 ft. The height of the cylinder is 22 ft. Find the exact volume of the cylinder.

Your Strategy

Your Solution

Solutions on p. S24

OBJECTIVE B   **Surface area of a solid**

The **surface area** of a solid is the total area on the surface of the solid. Suppose you want to cover a geometric solid with wallpaper. The amount of wallpaper needed is equal to the surface area of the figure.

When a rectangular solid is cut open and flattened out, each face is a rectangle. The surface area, SA, of the rectangular solid is the sum of the areas of the six rectangles:

$$SA = LW + LH + WH + LW + WH + LH$$

which simplifies to

$$SA = 2LW + 2LH + 2WH$$

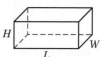

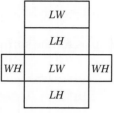

The surface area of a cube is the sum of the areas of the six faces of the cube. The area of each face is s^2. Therefore, the surface area, SA, of a cube is given by the formula $SA = 6s^2$.

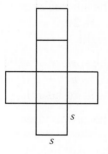

When a cylinder is cut open and flattened out, the top and bottom of the cylinder are circles. The side of the cylinder flattens out to a rectangle. The length of the rectangle is the circumference of the base, which is $2\pi r$; the width is h, the height of the cylinder. Therefore, the area of the rectangle is $2\pi rh$. The area of each circle is πr^2. The surface area, SA, of the cylinder is

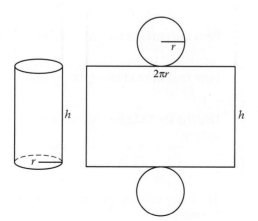

$$SA = \pi r^2 + 2\pi rh + \pi r^2$$

which simplifies to

$$SA = 2\pi r^2 + 2\pi rh$$

The surface area of a pyramid is the area of the base plus the area of the four isosceles triangles. A side of the square base is s; therefore, the area of the base is s^2. The slant height, l, is the height of each triangle, and s is the base of each triangle. The surface area, SA, of a pyramid is

$$SA = s^2 + 4\left(\frac{1}{2}sl\right)$$

which simplifies to

$$SA = s^2 + 2sl$$

Formulas for the surface areas of geometric solids are given below.

> ## Surface Areas of Geometric Solids
>
> The surface area, SA, of a **rectangular solid** with length L, width W, and height H is given by $SA = 2LW + 2LH + 2WH$.
>
> The surface area, SA, of a **cube** with side s is given by $SA = 6s^2$.
>
> The surface area, SA, of a **sphere** with radius r is given by $SA = 4\pi r^2$.
>
> The surface area, SA, of a **right circular cylinder** is given by $SA = 2\pi r^2 + 2\pi rh$, where r is the radius of the base and h is the height.
>
> The surface area, SA, of a **right circular cone** is given by $SA = \pi r^2 + \pi rl$, where r is the radius of the circular base and l is the slant height.
>
> The surface area, SA, of a **regular pyramid** is given by $SA = s^2 + 2sl$, where s is the length of a side of the base and l is the slant height.

Find the surface area of a sphere with a diameter of 18 cm.

First find the radius of the sphere.

$$r = \frac{1}{2}d = \frac{1}{2}(18) = 9$$

Use the formula for the surface area of a sphere.

$$SA = 4\pi r^2$$
$$SA = 4\pi (9)^2$$
$$SA = 4\pi (81)$$
$$SA = 324\pi$$

The exact surface area of the sphere is 324π cm^2.

An approximate measure can be found by using the π key on a calculator.

$$SA \approx 1,017.88$$

The approximate surface area is 1,017.88 cm^2.

Calculator Note

To approximate 324π on your calculator, enter 324 × π = .

EXAMPLE 3

The diameter of the base of a cone is 5 m, and the slant height is 4 m. Find the surface area of the cone. Give the exact measure.

Strategy

To find the surface area of the cone:

▸ Find the radius of the base of the cone.
▸ Use the formula for the surface area of a cone. Leave the answer in terms of π.

Solution

$$r = \frac{1}{2}d = \frac{1}{2}(5) = 2.5$$

$SA = \pi r^2 + \pi rl$
$SA = \pi(2.5)^2 + \pi(2.5)(4)$
$SA = \pi(6.25) + \pi(2.5)(4)$
$SA = 6.25\pi + 10\pi$
$SA = 16.25\pi$

The surface area of the cone is 16.25π m².

YOU TRY IT 3

The diameter of the base of a cylinder is 6 ft, and the height is 8 ft. Find the surface area of the cylinder. Round to the nearest hundredth.

Your Strategy

Your Solution

EXAMPLE 4

Find the area of a label used to cover a soup can that has a radius of 4 cm and a height of 12 cm. Round to the nearest hundredth.

Strategy

To find the area of the label, use the fact that the surface area of the side of a cylinder is given by $2\pi rh$. An approximation is asked for; use the π key on a calculator. $r = 4$, $h = 12$

Solution

Area of the label = $2\pi rh$
Area of the label = $2\pi(4)(12) = 96\pi \approx 301.59$

The area is approximately 301.59 cm².

YOU TRY IT 4

Which has a larger surface area, a cube with a side measuring 10 cm or a sphere with a diameter measuring 8 cm?

Your Strategy

Your Solution

Solutions on p. S24

9.4 Exercises

OBJECTIVE A   Volume of a solid

1. Refer to Exercises 3 to 8 below. Fill in each blank with the name of the solid shown in the given exercise.

   a. Exercise 4 _____   b. Exercise 6 _____

   c. Exercise 7 _____   d. Exercise 8 _____

2. Find the volume of the solid shown at the right.

   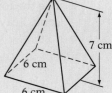

   a. Use the formula for the volume of a _____ . $V = \frac{1}{3}s^2h$

   b. Replace s by _____ and h by _____ . $V = \frac{1}{3}(\underline{\quad})^2(\underline{\quad})$

   c. Multiply. $V = \frac{1}{3}(\underline{\quad})(7) = (\underline{\quad})(7) = \underline{\quad}$

   d. Fill in the blank with the correct unit: The volume of the pyramid is 84 _____ .

For Exercises 3 to 8, find the volume of the figure. For calculations involving π, give both the exact value and an approximation to the nearest hundredth.

3.
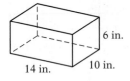
6 in.
14 in.   10 in.

4.
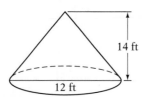
14 ft
12 ft

5.
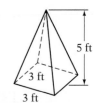
5 ft
3 ft
3 ft

6.
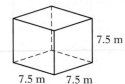
7.5 m
7.5 m   7.5 m

7.

3 cm

8.
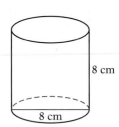
8 cm
8 cm

Solve.

9. A rectangular solid has a length of 6.8 m, a width of 2.5 m, and a height of 2 m. Find the volume of the solid.

10. Find the volume of a rectangular solid that has a length of 4.5 ft, a width of 3 ft, and a height of 1.5 ft.

11. Find the volume of a cube whose side measures 2.5 in.

Focus on **Problem Solving**

Trial and Error

S ome problems in mathematics are solved by using **trial and error**. The trial-and-error method of arriving at a solution to a problem involves repeated tests or experiments until a satisfactory conclusion is reached.

Many of the Critical Thinking exercises in this text require a trial-and-error method of solution. For example, an exercise on page 592 reads as follows:

Explain how you could cut through a cube so that the face of the resulting solid is **a.** a square, **b.** an equilateral triangle, **c.** a trapezoid, **d.** a hexagon.

There is no formula to apply to this problem; there is no computation to perform. This problem requires picturing a cube and the results after it is cut through at different places on its surface and at different angles. For part (a), cutting perpendicular to the top and bottom of the cube and parallel to two of its sides will result in a square. The other shapes may prove more difficult.

When solving problems of this type, keep an open mind. Sometimes when using the trial-and-error method, we are hampered by narrowness of vision; we cannot expand our thinking to include other possibilities. Then, when we see someone else's solution, it appears so obvious to us! For example, for the question above, it is necessary to conceive of cutting through the cube at places other than the top surface; we need to be open to the idea of beginning the cut at one of the corner points of the cube.

A topic of the Projects and Group Activities in this chapter is symmetry. Here again, the trial-and-error method is used to determine the lines of symmetry inherent in an object. For example, in determining lines of symmetry for a square, begin by drawing a square. The horizontal line of symmetry and the vertical line of symmetry may be immediately obvious to you.

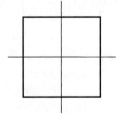

But there are two others. Do you see that a line drawn through opposite corners of the square is also a line of symmetry?

Many of the questions in this text that require an answer of "always true," "sometimes true," or "never true" are best solved by the trial-and-error method. For example, consider the following statement:

Two rectangles that have the same area have the same perimeter.

Try some numbers. Each of two rectangles, one measuring 6 units by 2 units and another measuring 4 units by 3 units, has an area of 12 square units, but the perimeter of the first is 16 units and the perimeter of the second is 14 units, so the answer "always true" has been eliminated. We still need to determine whether there is a case for which it *is* true. After experimenting with a lot of numbers, you may come to realize that we are trying to determine whether it is possible for two different pairs of factors of a number to have the same sum. Is it?

Don't be afraid to make many experiments, and remember that *errors*, or tests that "don't work," are a part of the trial-and-*error* process.

Projects & Group Activities

Lines of Symmetry

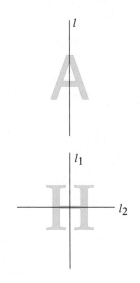

Look at the letter A printed at the left. If the letter were folded along line *l*, the two sides of the letter would match exactly. This letter has **symmetry** with respect to line *l*. Line *l* is called the **axis of symmetry.**

Now consider the letter H printed below at the left. Both lines l_1 and l_2 are axes of symmetry for this letter; the letter could be folded along either line and the two sides would match exactly.

1. Does the letter A have more than one axis of symmetry?

2. Find axes of symmetry for other capital letters of the alphabet.

3. Which lower-case letters have one axis of symmetry?

4. Do any of the lower-case letters have more than one axis of symmetry?

5. Find the numbers of axes of symmetry for the plane geometric figures presented in this chapter.

6. There are other types of symmetry. Look up the meanings of *point symmetry* and *rotational symmetry*. Which plane geometric figures provide examples of these types of symmetry?

7. Find examples of symmetry in nature, art, and architecture.

Preparing a Circle Graph

In Section 1 of this chapter, a protractor was used to measure angles. Preparing a circle graph requires the ability to use a protractor to draw angles.

To draw an angle of 142°, first draw a ray. Place a dot at the endpoint of the ray. This dot will be the vertex of the angle.

Place the straight bottom edge of the protractor on the ray as shown in the figure at the right. Make sure the center of the bottom edge of the protractor is located directly over the vertex point. Locate the position of the 142° mark. Place a dot next to the mark.

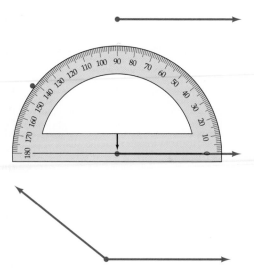

Remove the protractor and draw a ray from the vertex to the dot at the 142° mark.

An example of preparing a circle graph is given on the next page.

The revenues (in thousands of dollars) from four segments of a car dealership for the first quarter of a recent year were

New car sales:	$2,100	Used car/truck sales:	$1,500
New truck sales:	$1,200	Parts/service:	$700

To draw a circle graph to represent the percent that each segment contributed to the total revenue from all four segments, proceed as follows:

Find the total revenue from all four segments.

$$2,100 + 1,200 + 1,500 + 700 = 5,500$$

Find what percent each segment is of the total revenue of $5,500.

New car sales: $\dfrac{2,100}{5,500} \approx 38.2\%$

New truck sales: $\dfrac{1,200}{5,500} \approx 21.8\%$

Used car/truck sales: $\dfrac{1,500}{5,500} \approx 27.3\%$

Parts/service: $\dfrac{700}{5,500} \approx 12.7\%$

Each percent represents the part of the circle for that sector. Because the circle contains 360°, multiply each percent by 360° to find the measure of the angle for each sector. Round to the nearest whole number.

New car sales:

$$0.382 \times 360° \approx 138°$$

New truck sales:

$$0.218 \times 360° \approx 78°$$

Used car/truck sales:

$$0.273 \times 360° \approx 98°$$

Parts/service:

$$0.127 \times 360° \approx 46°$$

Draw a circle and use a protractor to draw the sectors representing the percents that each segment contributed to the total revenue from all four segments.

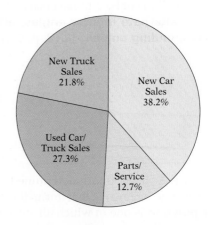

Collect data appropriate for display in a circle graph. [Some possibilities are last year's sales for the top three car manufacturers in the United States, votes cast in the last election for your state governor, the majors of the students in your math class, and the number of students enrolled in each class (senior, junior, etc.) at your college.] Then prepare the circle graph.

Chapter 9 Summary

Key Words	Examples

A **line** is determined by two distinct points and extends indefinitely in both directions. A **line segment** is part of a line and has two endpoints. **Parallel lines** never meet; the distance between them is always the same. **Perpendicular lines** are intersecting lines that form right angles. [9.1A, pp. 537–539]

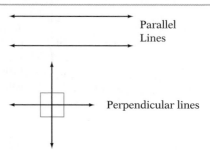

A **ray** starts at a point and extends indefinitely in one direction. The point at which a ray starts is the **endpoint** of the ray. An **angle** is formed by two rays with the same endpoint. The **vertex** of an angle is the point at which the two rays meet. An angle is measured in **degrees**. A 90° angle is a **right angle**. A 180° angle is a **straight angle**. An **acute angle** is an angle whose measure is between 0° and 90°. An **obtuse angle** is an angle whose measure is between 90° and 180°. **Complementary angles** are two angles whose measures have the sum 90°. **Supplementary angles** are two angles whose measures have the sum 180°. [9.1A, pp. 537–540]

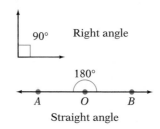

Two angles that are on opposite sides of the intersection of two lines are **vertical angles**; vertical angles have the same measure. Two angles that share a common side are **adjacent angles**; adjacent angles of intersecting lines are supplementary angles. [9.1A, p. 540; 9.1B, p. 542]

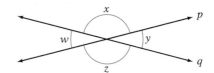

Angles *w* and *y* are vertical angles.
Angles *x* and *y* are adjacent angles.

A line that intersects two other lines at two different points is a **transversal**. If the lines cut by a transversal are parallel lines, equal angles are formed: **alternate interior angles, alternate exterior angles,** and **corresponding angles.** [9.1B, p. 543]

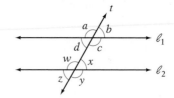

Parallel lines ℓ_1 and ℓ_2 are cut by transversal *t*. All four acute angles have the same measure. All four obtuse angles have the same measure.

A **polygon** is a closed figure determined by three or more line segments. The line segments that form the polygon are its **sides.** A **regular polygon** is one in which all sides have the same length and all angles have the same measure. Polygons are classified by the number of sides. A **quadrilateral** is a four-sided polygon. A parallelogram, a rectangle, a square, a rhombus, and a trapezoid are all quadrilaterals. [9.2A, pp. 553–554]

Number of Sides	Name of the Polygon
3	Triangle
4	Quadrilateral
5	Pentagon
6	Hexagon
7	Heptagon
8	Octagon
9	Nonagon
10	Decagon

A **triangle** is a plane figure formed by three line segments. An **isosceles triangle** has two sides of equal length. The three sides of an **equilateral triangle** are of equal length. A **scalene triangle** has no two sides of equal length. An **acute triangle** has three acute angles. An **obtuse triangle** has one obtuse angle. A **right triangle** has a right angle. [9.1C, p. 545; 9.2A, pp. 553–554]

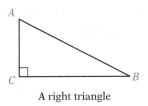

A right triangle

A **circle** is a plane figure in which all points are the same distance from the center of the circle. A **diameter** of a circle is a line segment across the circle through the center. A **radius** of a circle is a line segment from the center of the circle to a point on the circle. [9.2A, p. 556]

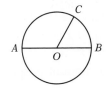

AB is a diameter of the circle.
OC is a radius.

Similar triangles have the same shape but not necessarily the same size. The ratios of corresponding sides are equal. The ratio of corresponding heights is equal to the ratio of corresponding sides. **Congruent triangles** have the same shape and the same size. [9.3B, p. 573; 9.3C, p. 575]

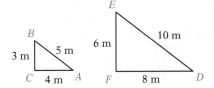

Triangles ABC and DEF are similar triangles. The ratio of corresponding sides is $\frac{1}{2}$.

Essential Rules and Procedures

Triangles [9.1C, p. 545, 9.3C, pp. 575–576]
Sum of the measures of the interior angles = 180°
Sum of an interior and the corresponding exterior angle = 180°
Rules to determine congruence: SSS Rule, SAS Rule, ASA Rule

In a right triangle, the measure of one acute angle is 12°. Find the measure of the other acute angle.

$$x + 12° + 90° = 180°$$
$$x + 102° = 180°$$
$$x = 78°$$

Formulas for Perimeter (the distance around a figure) [9.2A, pp. 554–556]
Triangle: $P = a + b + c$
Rectangle: $P = 2L + 2W$
Square: $P = 4s$
Circumference of a circle: $C = \pi d$ or $C = 2\pi r$

The length of a rectangle is 8 m. The width is 5.5 m. Find the perimeter of the rectangle.

$$P = 2L + 2W$$
$$P = 2(8) + 2(5.5)$$
$$P = 16 + 11$$
$$P = 27$$

The perimeter is 27 m.

Formulas for Area (the amount of surface in a region) [9.2B, pp. 558–561]

Triangle: $A = \frac{1}{2}bh$

Rectangle: $A = LW$
Square: $A = s^2$
Circle: $A = \pi r^2$
Parallelogram: $A = bh$

Trapezoid: $A = \frac{1}{2}h(b_1 + b_2)$

The length of the base of a parallelogram is 12 cm, and the height is 4 cm. Find the area of the parallelogram.

$A = bh$

$A = 12(4)$

$A = 48$

The area is 48 cm².

Formulas for Volume (the amount of space inside a figure in space) [9.4A, p. 583]
Rectangular solid: $V = LWH$
Cube: $V = s^3$

Sphere: $V = \frac{4}{3}\pi r^3$

Right circular cylinder: $V = \pi r^2 h$

Right circular cone: $V = \frac{1}{3}\pi r^2 h$

Regular pyramid: $V = \frac{1}{3}s^2 h$

Find the volume of a cube that measures 3 in. on a side.

$V = s^3$

$V = 3^3$

$V = 27$

The volume is 27 in³.

Formulas for Surface Area (the total area on the surface of a solid) [9.4B, p. 586]
Rectangular solid: $SA = 2LW + 2LH + 2WH$
Cube: $SA = 6s^2$
Sphere: $SA = 4\pi r^2$
Right circular cylinder: $SA = 2\pi r^2 + 2\pi rh$
Right circular cone: $SA = \pi r^2 + \pi rl$
Regular pyramid: $SA = s^2 + 2sl$

Find the surface area of a sphere with a diameter of 10 cm. Give the exact value.

$r = \frac{1}{2}d = \frac{1}{2}(10) = 5$

$SA = 4\pi r^2$

$SA = 4\pi(5^2)$

$SA = 4\pi(25)$

$SA = 100\pi$

The surface area is 100π cm².

Pythagorean Theorem [9.3A, p. 571]
If a and b are the legs of a right triangle and c is the length of the hypotenuse, then $c^2 = a^2 + b^2$.

Two legs of a right triangle measure 6 ft and 8 ft. Find the hypotenuse of the right triangle.

$c^2 = a^2 + b^2$

$c^2 = 6^2 + 8^2$

$c^2 = 36 + 64$

$c^2 = 100$

$c = \sqrt{100}$

$c = 10$

The length of the hypotenuse is 10 ft.

Principal Square Root Property [9.3A, p. 572]
If $r^2 = s$, then $r = \sqrt{s}$, and r is called the **square root** of s.

If $c^2 = 16$, then $c = \sqrt{16} = 4$.

Chapter 9 Review Exercises

1. Given that $\angle a = 74°$ and $\angle b = 52°$, find the measures of angles x and y.

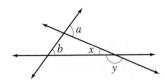

2. Triangles ABC and DEF are similar. Find the perimeter of triangle ABC.

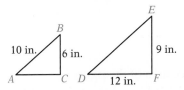

3. Find the volume of the figure.

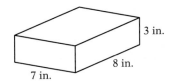

4. Find the measure of $\angle x$.

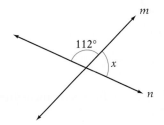

5. Determine whether the two triangles are congruent. If they are congruent, state by what rule they are congruent.

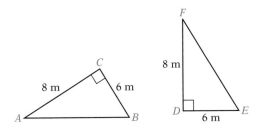

6. Find the surface area of the figure. Round to the nearest hundredth.

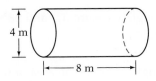

7. Given that $BC = 11$ cm and AB is three times the length of BC, find the length of AC.

8. Find x.

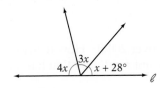

9. Find the area of the figure.

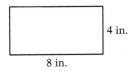

10. Find the volume of the figure.

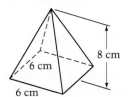

11. Find the perimeter of the figure.

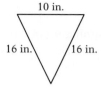

10 in.

16 in.   16 in.

12. Given that $\ell_1 \parallel \ell_2$, find the measures of angles a and b.

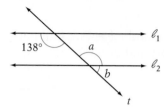

138° a ℓ_1

b ℓ_2

t

13. Find the surface area of the figure.

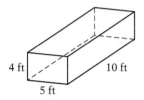

4 ft   10 ft

5 ft

14. Find the unknown side of the triangle. Round to the nearest hundredth.

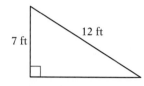

7 ft   12 ft

15. Find the volume of a cube whose side measures 3.5 in.

16. Find the supplement of a 32° angle.

17. Find the volume of a rectangular solid with a length of 6.5 ft, a width of 2 ft, and a height of 3 ft.

18. Two angles of a triangle measure 37° and 48°. Find the measure of the third angle.

19. The height of a triangle is 7 cm. The area of the triangle is 28 cm². Find the length of the base of the triangle.

20. Find the volume of a sphere that has a diameter of 12 mm. Give the exact value.

21. The perimeter of a square picture frame is 86 cm. Find the length of each side of the frame.

22. A can of paint will cover 200 ft². How many cans of paint should be purchased in order to paint a cylinder that has a height of 15 ft and a radius of 6 ft?

23. The length of a rectangular park is 56 yd. The width is 48 yd. How many yards of fencing are needed to surround the park?

24. What is the area of a square patio that measures 9.5 m on each side?

25. A walkway 2 m wide surrounds a rectangular plot of grass. The plot is 40 m long and 25 m wide. What is the area of the walkway?

Chapter 9 Test

1. For the right triangle shown below, determine the length of side *BC*. Round to the nearest hundredth.

2. Determine whether the two triangles are congruent. If they are congruent, state by what rule they are congruent.

    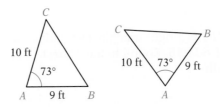

3. Determine the area of a rectangle with a length of 15 m and a width of 7.4 m.

4. Determine the area of a triangle whose base is 7 ft and whose height is 12 ft.

5. Determine the exact volume of a right circular cone whose radius is 7 cm and whose height is 16 cm.

6. Determine the exact surface area of a pyramid whose square base is 3 m on each side and whose slant height is 11 m.

7. Determine the volume of the solid shown below. Round to the nearest hundredth.

    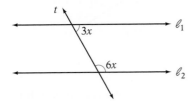

8. Determine the area of the trapezoid shown below.

    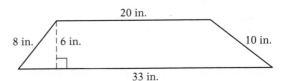

9. Given that $\ell_1 \parallel \ell_2$, find *x*.

    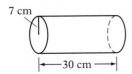

10. Determine the surface area of the figure shown below.

11. Find *x*.

    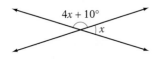

12. Name the figure shown below.

13. Determine whether the two triangles are congruent. If they are congruent, state by what rule they are congruent.

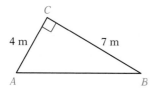

14. Determine the volume of the rectangular solid shown below.

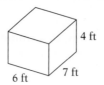

15. Figure *ABC* is a right triangle. Determine the length of side *AB*. Round to the nearest hundredth.

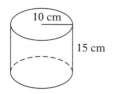

16. Given that ℓ_1 and ℓ_2 are parallel lines, determine the measure of angle *a*.

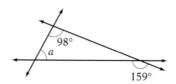

17. Determine the exact surface area of the right circular cylinder shown below.

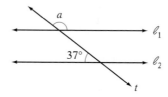

18. Determine the measure of angle *a*.

19. Triangles *ABC* and *DEF* are similar triangles. Determine the length of line segment *FG*. Round to the nearest hundredth.

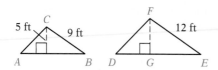

20. Triangles *ABC* and *DEF* are similar triangles. Determine the length of side *BC*. Round to the nearest hundredth.

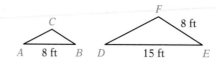

21. Determine the perimeter of a square whose side is 5 m.

22. Determine the perimeter of a rectangle whose length is 8 cm and whose width is 5 cm.

23. Find the perimeter of a right triangle with legs that measure 12 ft and 18 ft. Round to the nearest tenth.

24. Two angles of a triangle measure 41° and 37°. Find the measure of the third angle.

25. Find the supplement of a 41° angle.

Cumulative Review Exercises

1. Find 8.5% of 2,400.

2. Find all the factors of 78.

3. Divide: $4\frac{2}{3} \div 5\frac{3}{5}$

4. Add: $(3x^2 + 5x - 2) + (4x^2 - x + 7)$

5. Divide and round to the nearest tenth: $82.93 \div 6.5$

6. Write 0.000029 in scientific notation.

7. Find the measure of $\angle x$.

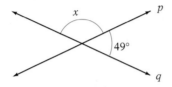

8. Find the unknown side of the triangle.

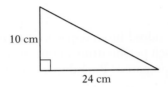

9. Find the area of the triangle.

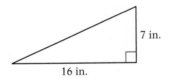

10. Given that $\ell_1 \parallel \ell_2$, find x.

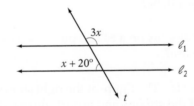

11. Multiply: $(4x^2y^2)(-3x^3y)$

12. Solve: $3(2x + 5) = 18$

13. A triangle has a right angle and a 32° angle. Find the measure of the third angle.

14. Graph $x > -3$.

$$\begin{array}{c} \longleftrightarrow \\ -6 \; -5 \; -4 \; -3 \; -2 \; -1 \; 0 \; 1 \; 2 \; 3 \; 4 \; 5 \; 6 \end{array}$$

15. Simplify: $5(2x + 4) - (3x + 2)$

16. Evaluate $2x + 3y^2z$ for $x = 5$, $y = -1$, and $z = -4$.

17. Evaluate $x^2y - 2z$ for $x = \frac{1}{2}$, $y = \frac{4}{5}$, and $z = -\frac{3}{10}$.

18. Convert 60 mph to kilometers per hour. Round to the nearest tenth. (1 mi $\approx$ 1.61 km)

19. Solve: $4x + 2 = 6x - 8$

20. Graph $y = -\dfrac{3}{2}x + 3$.

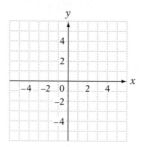

21. Convert 3 482 m to kilometers.

22. Write $\dfrac{3}{8}$ as a percent.

23. Find the simple interest on a 270-day loan of $20,000 at an annual interest rate of 8.875%.

24. *Catering* Two hundred fifty people are expected to attend a reception. Assuming that each person drinks 12 oz of coffee, how many gallons of coffee should be prepared? Round to the nearest whole number.

25. *Business* The charge for cellular phone service for a business executive is $22 per month plus $.25 per minute of phone use. In a month when the executive's phone bill was $43.75, how many minutes did the executive use the cellular phone?

26. *Taxes* If the sales tax on a $12.50 purchase is $.75, what is the sales tax on a $75 purchase?

27. *Foreign Trade* The figure at the right shows the value, in trillions of dollars, of the imports and exports during the first and second quarters of a recent year. Find the percent increase in the value of the imports from the first quarter to the second quarter. Round to the nearest tenth of a percent.

28. *Geometry* The volume of a box is 144 ft³. The length of the box is 12 ft, and the width is 4 ft. Find the height of the box.

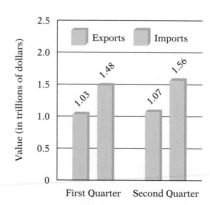

Value of Imports and Exports
Source: Bureau of Economic Analysis

29. *Oceanography* The pressure, P, in pounds per square inch, at a certain depth in the ocean can be approximated by the equation $P = 15 + \dfrac{1}{2}D$, where D is the depth in feet. Use this equation to find the depth when the pressure is 35 lb/in².

30. *Astronautics* The weight of an object is related to the distance the object is above the surface of Earth. A formula for this relationship is $d = 4{,}000\sqrt{\dfrac{E}{S}} - 4{,}000$, where E is the object's weight on the surface of Earth and S is the object's weight at a distance of d miles above Earth's surface. A space explorer who weighs 196 lb on the surface of Earth weighs 49 lb in space. How far above Earth's surface is the space explorer?

Statistics and Probability

Jockey Thierry Thulliez uses his crop on Domedriver of Ireland as they near the finish line in the fifth race at the NetJets Breeders' Cup Mile race at Arlington Park in Arlington Heights, Illinois. Domedriver won the race, but he was considered a "longshot," meaning the odds were against him winning. Rock of Gibraltar was considered the "favorite," meaning the odds were in his favor. *Odds in favor* and *odds against* are closely linked to probability. **Example 3 on page 632** shows how to calculate the probability of a horse winning a race given the odds against that horse winning.

DVD

SSM

Student Website
Need help? For online student resources,
visit **college.hmco.com/pic/aufmannPA5e.**

Prep TEST

1. Simplify: $\dfrac{3}{2+7}$

2. Approximate $\sqrt{13}$ to the nearest thousandth.

3. Bill-related mail accounted for 49 billion of the 102 billion pieces of first-class mail handled by the U.S. Postal Service during a recent year. (Source: USPS) What percent of the pieces of first-class mail handled by the U.S. Postal Service was bill-related mail? Round to the nearest tenth of a percent.

4. The table at the right shows the estimated costs of funding an education at a public college. Between which two enrollment years is the estimated increase in cost greatest? What is the increase between these two years?

Enrollment Year	Cost of Public College
2005	$70,206
2006	$74,418
2007	$78,883
2008	$83,616
2009	$88,633
2010	$93,951

Source: The College Board's Annual Survey of Colleges

5. During the 1924 Summer Olympics in Paris, France, the United States won 45 gold medals, 27 silver medals, and 27 bronze medals. (*Source: The Ultimate Book of Sports List*)
   a. Find the ratio of gold medals won by the United States to silver medals won by the United States during the 1924 Summer Olympics. Write the ratio as a fraction in simplest form.

   b. Find the ratio of silver medals won by the United States to bronze medals won by the United States during the 1924 Summer Olympics. Write the ratio using a colon.

6. The table (below right) shows the number of television viewers, in millions, who watch pay-cable channels, such as HBO and Showtime, each night of the week. (*Source:* Neilsen Media Research analyzed by Initiative Media North America)
   a. Arrange the numbers in the table from smallest to largest.

Mon	Tue	Wed	Thu	Fri	Sat	Sun
3.9	4.5	4.2	3.9	5.2	7.1	5.5

   b. Find the average number of viewers per night.

7. Approximately 15% of the nation's 1.4 million service members on active military duty are women. (*Source:* Women in Military Service for America Memorial Foundation)
   a. Approximately how many women are on active military duty?

   b. What fraction of the service members on active military duty are women?

GO Figure

In the addition at the right, each letter stands for a different digit. If N = 1, I = 5, U = 7, F = 2, and T = 3, what is the value of S?

```
  FUN
   IN
+ THE
-----
  SUN
```

10.1 Organizing Data

OBJECTIVE A   Frequency distributions

Statistics is the study of collecting, organizing, and interpreting data. Data are collected from a **population,** which is the set of all observations of interest. Here are two examples of populations.

> A medical researcher wants to determine the effectiveness of a new drug to control blood pressure. The population for the researcher is the amount of change in blood pressure for each patient receiving the medication.

> The quality control inspector of a precision instrument company wants to determine the diameters of ball bearings. The population for the inspector is the measure of the diameter of each ball bearing.

A **frequency distribution** is one method of organizing the data collected from a population. A frequency distribution is constructed by dividing the data gathered from the population into **classes.** Here is an example:

A ski association surveys 40 of its members, asking them to report the percent of their ski terrain that is rated expert. The results of the survey follow.

Percent of Expert Terrain at 40 Ski Resorts

14	24	8	31	27	9	12	32	24	27
12	21	24	23	12	31	30	31	26	34
13	18	29	33	34	21	28	23	11	10
25	20	14	18	15	11	17	29	21	25

To organize these data into a frequency distribution:

1. Find the smallest number (8) and the largest number (34) in the table. The difference between these two numbers is the **range** of the data.

   Range $= 34 - 8 = 26$

2. Decide how many classes the frequency distribution will contain. Usually frequency distributions have from 6 to 12 classes. The frequency distribution for this example will contain 6 classes.

3. Divide the range by the number of classes. If necessary, round the quotient to a whole number. This number is called the **class width.**

   $\frac{26}{6} \approx 4$. The class width is 4.

4. Form the classes of the frequency distribution.

   Classes
8–12	Add 4 to the smallest number.
13–17	Add 4 again.
18–22	Continue until a class contains
23–27	the largest number in the set
28–32	of data.
33–37	

   These are the **lower class limits.** ⬆ ⬆ These are the **upper class limits.**

Calculator Note

Recall that $\frac{26}{6}$ can be read $26 \div 6$.

To calculate the class width, press

$26 \div 6 =$.

The display reads 4.3333333.

Take Note

For your convenience, the data presented on page 607 are repeated below.

14	12	13	25
24	21	18	20
8	24	29	14
31	23	33	18
27	12	34	15
9	31	21	11
12	30	28	17
32	31	23	29
24	26	11	21
27	34	10	25

5. Complete the table by tabulating the data for each class. For each number from the data, place a slash next to the class that contains the number. Count the number of tallies in each class. This is the **class frequency.**

Frequency Distribution for Ski Resort Data

Classes	Tally	Frequency
8–12	////////	8
13–17	/////	5
18–22	National	6
23–27	//////////	10
28–32	////////	8
33–37	///	3

Organizing data into a frequency distribution enables us to make statements about the data. For example, 27 (6 + 10 + 8 + 3) of the ski resorts reported that 18% or more of their terrain was rated expert.

An insurance adjuster tabulated the dollar amounts of 50 auto accident claims. The results are given in the following table. Use these data for Example 1 and You Try It 1.

Dollar Amount of 50 Auto Insurance Claims

475	224	722	721	815	351	596	625	981	748
993	881	361	560	574	742	703	998	435	873
882	278	455	803	985	305	522	900	638	810
677	688	410	505	890	186	829	631	882	991
484	339	950	579	539	422	326	793	453	118

EXAMPLE 1

For the table of auto insurance claims, make a frequency distribution that has 6 classes.

Strategy

To make the frequency distribution:

▶ Find the range.
▶ Divide the range by 6, the number of classes. Round the quotient to the nearest whole number. This is the class width.
▶ Tabulate the data for each class.

Solution

Range = 998 − 118 = 880

Class width $= \dfrac{880}{6} \approx 147$

Dollar Amount of Insurance Claims

Classes	Tally	Frequency
118–265	///	3
266–413	////////	7
414–561	//////////	10
562–709	/////////	9
710–857	/////////	9
858–1,005	////////////	12

YOU TRY IT 1

For the table of auto insurance claims, make a frequency distribution that has 8 classes.

Your Strategy

Your Solution

Solution on p. S24

OBJECTIVE B   **Histograms**

A **histogram** is a bar graph that represents the data in a frequency distribution. The width of a bar represents each class, and the height of the bar corresponds to the frequency of the class.

A survey of 105 households is conducted, and the number of kilowatt-hours (kWh) of electricity used by each household in a 1-month period is recorded in the frequency distribution shown at the left below. The histogram for the frequency distribution is shown in Figure 10.1.

Classes (kWh)	Frequency
850–900	9
900–950	14
950–1,000	17
1,000–1,050	25
1,050–1,100	16
1,100–1,150	14
1,150–1,200	10

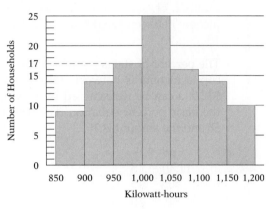

Figure 10.1

> **Take Note**
>
> Note that the upper class limit of one class is the lower class limit of the next class. In this case, a class contains the range of data from the lower class limit up to, but not including, the upper class limit. Therefore, if a data value for the distribution at the left were 900, that data value would be included in the 900–950 class.

From the frequency distribution or the histogram, we can see that 17 households used between 950 kWh and 1,000 kWh during the 1-month period.

EXAMPLE 2

Use the histogram in Figure 10.1 to find the number of households that used 950 kWh of electricity or less during the month.

Strategy

To find the number of households:

▶ Read the histogram to find the number of households whose use was between 850 and 900 kWh and the number whose use was between 900 and 950 kWh.

▶ Add the two numbers.

Solution

Number between 850 and 900 kWh: 9
Number between 900 and 950 kWh: 14

9 + 14 = 23

Twenty-three households used 950 kWh of electricity or less during the month.

YOU TRY IT 2

Use the histogram in Figure 10.1 to find the number of households that used 1,100 kWh of electricity or more during the month.

Your Strategy

Your Solution

Solution on pp. S24–S25

OBJECTIVE C Frequency polygons

A **frequency polygon** is a graph that displays information in a manner similar to a histogram. A dot is placed above the center of each class interval at a height corresponding to that class's frequency. The dots are then connected to form a broken-line graph. The center of a class interval is called the **class midpoint.**

The per capita incomes for the 50 states in a recent year are recorded in the frequency polygon in Figure 10.2. The number of states with a per capita income between $28,000 and $32,000 is 15.

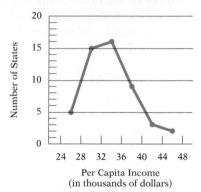

The percent of states for which the per capita income is between $28,000 and $32,000 can be determined by solving the basic percent equation. The base is 50 and the amount is 15.

$$pB = A$$
$$p(50) = 15$$
$$\frac{50p}{50} = \frac{15}{50}$$
$$p = 0.3$$

Figure 10.2

Source: Bureau of Economic Analysis

30% of the states had a per capita income between $28,000 and $32,000.

EXAMPLE 3

Use Figure 10.2 to find the number of states for which the per capita income was $40,000 or above.

Strategy

To find the number of states:

▶ Read the frequency polygon to find the number of states with a per capita income between $40,000 and $44,000 and the number of states with a per capita income between $44,000 and $48,000.
▶ Add the numbers.

Solution

Number with per capita income between $40,000 and $44,000: 3

Number with per capita income between $44,000 and $48,000: 2

3 + 2 = 5

The per capita income was $40,000 or above in 5 states.

YOU TRY IT 3

Use Figure 10.2 to find the ratio of the number of states with a per capita income between $24,000 and $28,000 to the number with a per capita income between $28,000 and $32,000.

Your Strategy

Your Solution

Solution on p. S25

10.1 Exercises

OBJECTIVE A **Frequency distributions**

1.  In your own words, describe a frequency distribution.

2. 📝 Explain what *class* and *frequency* indicate in a frequency distribution.

3. Use the following data for parts (a) to (e).

    58 82 67 55 59 77 98 76 82 88 91 75 75

    a. The range of the data is

    largest data value − smallest data value = _____ − _____ = _____

    b. For these data, the class width for a frequency distribution that has 5
    classes is

    $$\text{class width} = \frac{\text{range}}{\text{number of classes}} = \frac{\boxed{}}{\boxed{}} = 8.6 \approx \underline{}$$

    c. The lower class limit of the first class of the frequency distribution is the
    smallest data value, _____. The upper class limit of the first class is

    lower class limit + class width = _____ + _____ = _____

    d. The lower class limit of the second class of the frequency distribution
    is the next whole number greater than the upper class limit of the first
    class. The lower class limit of the second class is _____. The upper
    class limit of the second class is

    lower class limit of the second class + class width =
    _____ + _____ = _____

    e. The number of data values in the second class of the frequency distri-
    bution is _____.

Education   Use the table below for Exercises 4 to 11.

Tuition per Term at 40 Universities (in hundreds of dollars)

85	87	95	48	41	91	88	92
71	74	63	51	70	87	84	95
72	94	61	52	88	49	55	60
77	53	89	91	45	96	49	58
83	36	39	32	36	59	95	67

4. What is the range of the data in the tuition table?

5. Make a frequency distribution for the tuition table. Use 8 classes.

6. Which class has the greatest frequency?

7. How many tuitions are between $7,700 and $8,500?

8. How many tuitions are between $5,000 and $5,800?

9. How many tuitions are less than or equal to $6,700?

10. What percent of the tuitions are between $9,500 and $10,300?

11. What percent of the tuitions are between $3,200 and $4,000?

The Hotel Industry   Use the table below for Exercises 12 to 22.

Corporate Room Rate for 50 Hotels

60	87	77	117	114	82	91	65	69	63
106	71	74	86	106	78	101	100	107	109
57	106	103	100	95	68	99	112	107	77
64	68	99	112	107	76	116	100	82	86
81	98	92	78	95	89	91	102	115	127

12. Make a frequency distribution for the hotel room rates. Use 7 classes.

13. How many hotels charge a corporate room rate that is between $79 and $89 per night?

14. How many hotels charge a corporate room rate that is between $57 and $67 per night?

15. How many hotels charge a corporate room rate that is between $112 and $133 per night?

16. How many hotels charge a corporate room rate that is less than or equal to $100?

17. What percent of the hotels charge a corporate room rate that is between $101 and $111 per night?

18. What percent of the hotels charge a corporate room rate that is between $90 and $100 per night?

19. What percent of the hotels charge a corporate room rate that is greater than or equal to $101 per night?

20. What percent of the hotels charge a corporate room rate that is less than or equal to $78 per night?

21. What is the ratio of the number of hotels whose room rates are between $79 and $89 to those whose room rates are between $90 and $100?

22.  Suppose the hotel that charges a corporate room rate of $60 per night raises the rate. Give an example of a higher rate that will not change the frequency distribution you made in Exercise 12. Give an example of a higher rate that will change the frequency distribution.

OBJECTIVE B **Histograms**

Customer Credit   A total of 50 monthly credit account balances were recorded. A histogram of these data is shown at the right.

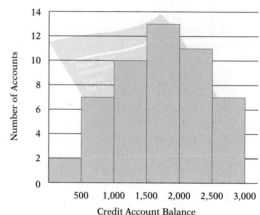

23. To find the number of account balances that were between $2,500 and $3,000, use the scale on the vertical axis to determine the height of the last bar. The height is _____ , so _____ account balances were between $2,500 and $3,000.

24. How many account balances were between $1,500 and $2,000?

25. How many account balances were less than $2,000?

26. To find the percent of the account balances that were between $2,500 and $3,000, use the basic percent equation.

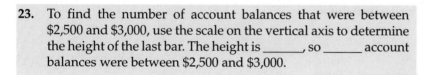

a. Because the total number of accounts is 50, replace the base B with 50. Because the number of account balances that were between $2,500 and $3,000 is 7 (see Exercise 23), replace A with 7.

$$pB = A$$
$$p(\underline{\quad}) = \underline{\quad}$$

b. Solve for p by dividing both sides of the equation by _____.

$$\frac{50p}{50} = \frac{7}{50}$$
$$p = \underline{\quad}$$

c. Write the decimal as a percent: _____% of the account balances were between $2,500 and $3,000.

27. What percent of the account balances were between $2,000 and $2,500?

28. What percent of the account balances were greater than $1,500?

Marathons   The times, in minutes, for 100 runners in a marathon were recorded. A histogram of these data is shown at the right.

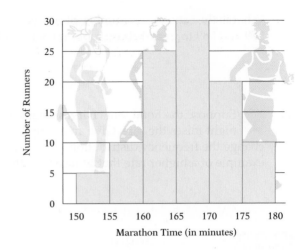

29. What is the ratio of the number of runners with times that were between 150 min and 155 min to those with times between 175 min and 180 min?

30. What is the ratio of the number of runners with times that were between 165 min and 170 min to those with times between 155 min and 160 min?

31. What percent of the runners had times greater than 165 min?

32. What percent of the runners had times less than 170 min?

33. ![icon] Which two consecutive classes have the greatest difference in frequency?

34. ![icon] Suppose you redraw the histogram with only three classes (150–160, 160–170, and 170–180). What are the heights of the three bars?

Cost of Living   A total of 40 apartment complexes were surveyed to find the monthly rent for a one-bedroom apartment. A histogram of these data is shown at the right.

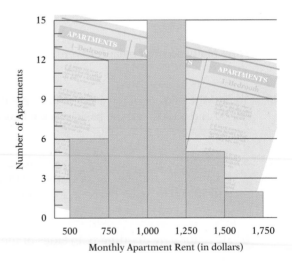

35. What percent of the apartments had rents between $1,250 and $1,500?

36. What percent of the apartments had rents between $500 and $750?

37. What percent of the apartments had rents greater than $1,000?

38. What percent of the apartments had rents less than $1,250?

OBJECTIVE C Frequency polygons

State Board Exams   The scores of 50 nurses taking a state board exam were recorded. A frequency distribution of these scores is shown at the right.

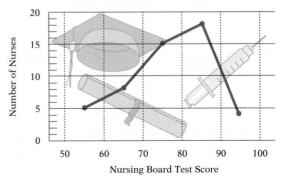

39. The highest dot on the frequency polygon represents the number of test scores between _____ and _____.

40. To find the number of test scores that were between 60 and 70, use the scale on the vertical axis to determine the height of the dot that is second from the left. The height is _____, so _____ test scores were between 60 and 70.

41. How many nurses had scores that were greater than 80?

42. How many nurses had scores that were less than 70?

43. What percent of the nurses had scores between 70 and 90?

44. What percent of the nurses had scores greater than 70?

45. Create the frequency distribution on which the frequency polygon of state board exam scores was based.

46. Create a possible data set for the state board exam scores frequency polygon.

Emergency Calls The response times for 75 emergency 911 calls for a small city were recorded. A frequency distribution of these times is shown at the right.

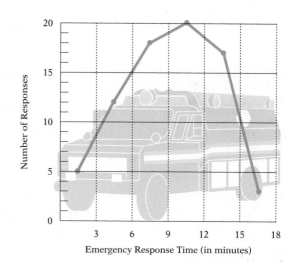
Emergency Response Time (in minutes)

47. What is the ratio of the number of response times between 6 min and 9 min to the number of response times between 15 min and 18 min?

48. What is the ratio of the number of response times of 0 min to 3 min to the total number of recorded response times?

49. What percent of the response times are greater than 9 min? Round to the nearest tenth of a percent.

CRITICAL THINKING

50. Toss two dice 100 times. Record the sum of the dots on the upward faces. Make a histogram showing the number of times the sum was 2, 3, . . . , 12.

51. How are a frequency table and a histogram alike? How are they different?

52. The frequency table at the right contains data from a survey of the type of vehicle a prospective buyer would consider. Explain why these data could be shown in a bar graph but not in a histogram.

Body Style	Frequency
Sedan	25
Convertible	16
SUV	9
Sports car	20
Truck	23

10.2 Statistical Measures

OBJECTIVE A The mean, median, and mode of a distribution

The average annual rainfall in Mobile, Alabama, is 67 in. The average annual snowfall in Syracuse, New York, is 111 in. The average daily low temperature in January in Bismarck, North Dakota, is −4°F. Each of these statements uses one number to describe an entire collection of numbers. Such a number is called an *average*. In statistics, there are various ways to calculate an average. Three of the most common—*mean, median,* and *mode*—are discussed here.

An automotive engineer tests the miles-per-gallon ratings of 15 cars and records the results as follows:

Miles-per-Gallon Ratings of 15 Cars

25   22   21   27   25   35   29   31   25   26   21   39   34   32   28

The **mean** of the data is the sum of the measurements divided by the number of measurements. The symbol for the mean is $\bar{x}$.

Formula for the Mean

$$\text{Mean} = \bar{x} = \frac{\text{sum of all data values}}{\text{number of data values}}$$

To find the mean for the data above, add the numbers and then divide by 15.

$$\bar{x} = \frac{25 + 22 + 21 + 27 + 25 + 35 + 29 + 31 + 25 + 26 + 21 + 39 + 34 + 32 + 28}{15}$$

$$= \frac{420}{15} = 28$$

The mean number of miles per gallon for the 15 cars tested was 28 mi/gal.

The mean is one of the most frequently computed averages. It is the one that is commonly used to calculate a student's performance in a class.

The scores for a history student on 5 tests were 78, 82, 91, 87, and 93. What was the mean score for this student?

To find the mean, add the numbers. Then divide by 5.

$$\bar{x} = \frac{78 + 82 + 91 + 87 + 93}{5}$$

$$= \frac{431}{5} = 86.2$$

The mean score for the history student was 86.2.

The **median** of a data set is the number that separates the data into two equal parts when the numbers are arranged from smallest to largest (or largest to smallest). There are always an equal number of values above the median and below the median.

To find the median of a set of numbers, first arrange the numbers from smallest to largest. The median is the number in the middle. The data for the miles-per-gallon ratings given on the previous page are arranged from smallest to largest below.

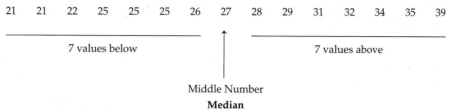

The median is 27.

> **Take Note**
>
> Half the data values are less than 27 and half the data values are greater than 27. The median indicates the center, or middle, of the set of data.

If the data contain an *even* number of values, the median is the sum of the two middle numbers, divided by 2.

The selling prices of the last six homes sold by a real estate agent were $175,000, $150,000, $250,000, $130,000, $245,000, and $190,000. Find the median selling price of these homes.

Arrange the numbers from smallest to largest. Because there are an even number of values, the median is the sum of the two middle numbers, divided by two.

130,000   150,000   175,000   190,000   245,000   250,000

middle 2 numbers

$$\text{Median} = \frac{175,000 + 190,000}{2} = 182,500$$

The median selling price of a home was $182,500.

> **Take Note**
>
> If the data contain an even number of values, the median is the mean of the two middle values.

The **mode** of a set of numbers is the value that occurs most frequently. If a set of numbers has no number that occurs more than once, then the data have no mode.

Here again are the data for the gasoline mileage ratings of cars.

Miles-per-Gallon Ratings of 15 Cars

25   22   21   27   25   35   29   31   25   26   21   39   34   32   28

25 is the number that occurs most frequently. The mode is 25.

> **Point of Interest**
>
> A set of data can have no mode, one mode, or several modes. The data
>
> 3, 7, 7, 12, 14, 15, 15, 19
>
> have two modes, 7 and 15. A set of data with two modes is called *bimodal*.

The gasoline mileage ratings data show that the mean, median, and mode of a set of numbers do not have to be the same value. For the data on rating the miles per gallon for 15 cars,

Mean = 28   Median = 27   Mode = 25

Although any of the averages can be used when the data collected consist of numbers, the mean and median are not appropriate for *qualitative* data. Examples of qualitative data are recording a person's favorite color or recording a person's preference from among classical, hard rock, jazz, rap, and country western music. It does not make sense to say that the *average* favorite color is red or the *average* musical choice is jazz. The mode is used to indicate the most frequently chosen color or musical category. The **modal response** is the category that receives the greatest number of responses.

A survey asked people to state whether they strongly disagree, disagree, have no opinion, agree, or strongly agree with the state governor's position on increasing taxes for health care. What was the modal response for these data?

Strongly disagree   57
Disagree   68
No opinion   12
Agree   45
Strongly agree   58

Because a response of "disagree" was recorded most frequently, the modal response was "disagree."

EXAMPLE 1

Twenty students were asked the number of units in which they were currently enrolled. The responses were

| 15 | 12 | 13 | 15 | 17 | 18 | 13 | 20 | 9 | 16 |

| 14 | 10 | 15 | 12 | 17 | 16 | 6 | 14 | 15 | 12 |

Find the mean and median number of units taken by these students.

Strategy

To find the mean number of units taken by the 20 students:
▶ Determine the sum of the numbers.
▶ Divide the sum by 20.

To find the median number of units taken by the 20 students:
▶ Arrange the numbers from smallest to largest.
▶ Because there is an even number of values, the median is the sum of the two middle numbers, divided by 2.

Solution

The sum of the numbers is 279.

$$\bar{x} = \frac{279}{20} = 13.95$$

The mean is 13.95 units.

| 6 | 9 | 10 | 12 | 12 | 12 | 13 | 13 | 14 | 14 |

| 15 | 15 | 15 | 15 | 16 | 16 | 17 | 17 | 18 | 20 |

$$\text{Median} = \frac{14 + 15}{2} = 14.5$$

The median is 14.5 units.

YOU TRY IT 1

The amounts spent by the last 10 customers at a fast-food restaurant were

| 4.32 | 6.21 | 5.45 | 5.90 | 5.58 | 4.45 | 5.05 | 6.00 | 3.59 | 4.75 |

Find the mean and median amount spent on lunch by these customers.

Your Strategy

Your Solution

Solution on p. S25

EXAMPLE 2

A bowler has scores of 165, 172, 168, and 185 for four games. What score must the bowler achieve on the next game so that the mean for the five games is 174?

Strategy

To find the score, use the formula for the mean, letting n be the score on the fifth game.

Solution

$$174 = \frac{165 + 172 + 168 + 185 + n}{5}$$

$$174 = \frac{690 + n}{5}$$

$$870 = 690 + n \quad \blacktriangleright \text{ Multiply each side by 5.}$$

$$180 = n$$

The score on the fifth game must be 180.

YOU TRY IT 2

You have scores of 82, 91, 79, and 83 on four exams. What score must you receive on the fifth exam to have a mean of 84 for the five exams?

Your Strategy

Your Solution

Solution on p. S25

OBJECTIVE B **Box-and-whiskers plots**

The purpose of calculating a mean or median is to obtain one number that describes a group of measurements. That one number alone, however, may not adequately represent the data. A **box-and-whiskers plot** is a graph that gives a more comprehensive picture of the data. A box-and-whiskers plot shows five numbers: the smallest value, the *first quartile*, the median, the *third quartile*, and the largest value. The **first quartile,** symbolized by Q_1, is the number below which one-quarter of the data lie. The **third quartile,** symbolized by Q_3, is the number above which one-quarter of the data lie.

Find the first quartile Q_1 and the third quartile Q_3 for the prices of 15 half-gallon cartons of ice cream.

| 3.26 | 4.71 | 4.18 | 4.45 | 5.49 | 3.18 | 3.86 | 3.58 | 4.29 | 5.44 | 4.83 | 4.56 | 4.36 | 2.39 | 2.66 |

To find the quartiles, first arrange the data from the smallest value to the largest value. Then find the median.

| 2.39 | 2.66 | 3.18 | 3.26 | 3.58 | 3.86 | 4.18 | <u>4.29</u> | 4.36 | 4.45 | 4.56 | 4.71 | 4.83 | 5.44 | 5.49 |

The median is 4.29.

Now separate the data into two groups: those values below the median and those values above the median.

Values Less Than the Median

| 2.39 | 2.66 | 3.18 | 3.26 | 3.58 | 3.86 | 4.18 |

$\uparrow$
Q_1

Values Greater Than the Median

| 4.36 | 4.45 | 4.56 | 4.71 | 4.83 | 5.44 | 5.49 |

$\uparrow$
Q_3

The first quartile Q_1 is the median of the lower half of the data: $Q_1 = 3.26$.
The third quartile Q_3 is the median of the upper half of the data: $Q_3 = 4.71$.

> **Take Note**
>
> To draw this box-and-whiskers plot, think of a number line that includes the five values listed. With this in mind, mark off the five values. Draw a box that spans the distance from Q_1 to Q_3. Draw a vertical line the height of the box at the median, Q_2.

Participants in Software Training

30	45	54	24	48	38	43
38	46	53	62	64	40	35

The **interquartile range** is the difference between Q_3 and Q_1.

$$\text{Interquartile range} = Q_3 - Q_1 = 4.71 - 3.26 = 1.45$$

A box-and-whiskers plot shows the data in the interquartile range as a box. The box-and-whiskers plot for the data on the cost of ice cream is shown below.

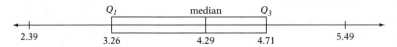

Note that the box-and-whiskers plot labels five values: the smallest, 2.39; the first quartile Q_1, 3.26; the median, 4.29; the third quartile Q_3, 4.71; and the largest value, 5.49.

For Example 3 and You Try It 3, use the data in the table at the left, which gives the number of people registered for a software training program.

EXAMPLE 3

Find Q_1 and Q_3 for the data in the software training table.

Strategy

To find Q_1 and Q_3, arrange the data from smallest to largest. Find the median. Then find Q_1, the median of the lower half of the data, and Q_3, the median of the upper half of the data.

Solution

24	30	35	**38**	38	40	43
45	46	48	**53**	54	62	64

$$\text{Median} = \frac{43 + 45}{2} = 44$$

$Q_1 = 38$ The median of the top row of data.
$Q_3 = 53$ The median of the bottom row of data.

YOU TRY IT 3

Draw the box-and-whiskers plot for the data in the software training table.

Your Strategy

Your Solution

Solution on p. S25

OBJECTIVE C **The standard deviation of a distribution**

Consider two students, each of whom has taken five exams.

Student A

84	86	83	85	87

$$\bar{x} = \frac{84 + 86 + 83 + 85 + 87}{5} = \frac{425}{5} = 85$$

The mean for Student A is 85.

Student B

90	75	94	68	98

$$\bar{x} = \frac{90 + 75 + 94 + 68 + 98}{5} = \frac{425}{5} = 85$$

The mean for Student B is 85.

For each of these students, the mean (average) for the 5 tests is 85. However, Student A has a more consistent record of scores than Student B. One way to measure the consistency, or "clustering," of data near the mean is the **standard deviation.**

To calculate the standard deviation:

1. Sum the squares of the differences between each value of data and the mean.
2. Divide the result in Step 1 by the number of items in the set of data.
3. Take the square root of the result in Step 2.

Here is the calculation for Student A. The symbol for standard deviation is the Greek letter *sigma*, denoted by σ.

Step 1

x	$(x - \bar{x})$	$(x - \bar{x})^2$
84	$(84 - 85)$	$(-1)^2 = 1$
86	$(86 - 85)$	$1^2 = 1$
83	$(83 - 85)$	$(-2)^2 = 4$
85	$(85 - 85)$	$0^2 = 0$
87	$(87 - 85)$	$2^2 = 4$
		Total $= 10$

Step 2 $\frac{10}{5} = 2$

Step 3 $\sigma = \sqrt{2} \approx 1.414$

The standard deviation for Student A's scores is approximately 1.414.

> **Take Note**
>
> The standard deviation of Student B's scores is greater than the standard deviation of Student A's scores, and the range of Student B's scores ($98 - 68 = 30$) is greater than the range of Student A's scores ($87 - 83 = 4$).

Following a similar procedure for Student B, we find that the standard deviation for Student B's scores is approximately 11.524. Because the standard deviation of Student B's scores is greater than that of Student A's scores ($11.524 > 1.414$), Student B's scores are not as consistent as those of Student A.

In this text, standard deviations are rounded to the nearest thousandth.

EXAMPLE 4

The weights in pounds of the five-man front line of a college football team are 210, 245, 220, 230, and 225. Find the standard deviation of the weights.

Strategy

To calculate the standard deviation:

▸ Find the mean of the weights.
▸ Use the procedure for calculating standard deviation.

Solution

$\bar{x} = \dfrac{210 + 245 + 220 + 230 + 225}{5} = 226$

Step 1

x	$(x - \bar{x})^2$
210	$(210 - 226)^2 = 256$
245	$(245 - 226)^2 = 361$
220	$(220 - 226)^2 = 36$
230	$(230 - 226)^2 = 16$
225	$(225 - 226)^2 = 1$
	Total $= 670$

Step 2 $\frac{670}{5} = 134$

Step 3 $\sigma = \sqrt{134} \approx 11.576$

The standard deviation of the weights is approximately 11.576 lb.

YOU TRY IT 4

The numbers of miles a runner recorded for the last six days of running were 5, 7, 3, 6, 9, and 6. Find the standard deviation of the miles run.

Your Strategy

Your Solution

Solution on p. S25

10.2 Exercises

OBJECTIVE A   **The mean, median, and mode of a distribution**

1.   Find the mean of the data set 62, 75, 87, 54, 70, 36.

   a. Use the formula for the mean.

   $$\bar{x} = \frac{\text{sum of all data values}}{\text{number of data values}} = \frac{\boxed{} + \boxed{} + \boxed{} + \boxed{} + \boxed{} + \boxed{}}{\boxed{}}$$

   b. Find the sum of the numbers in the numerator and then divide by 6.

   $$x = \frac{\boxed{}}{6} = \underline{}$$

   c. The mean of the set of data is _____.

2.   Find the median of the set of data in Exercise 1.

   a. Arrange the numbers from smallest to largest.

   _____ _____ _____ _____ _____ _____

   b. The number of values in the set of data is _____. Because there are an even number of data values, the median is the sum of the two middle numbers, _____ and _____ , divided by _____.

   $$\text{Median} = \frac{\boxed{} + \boxed{}}{2} = \underline{}$$

   c. The mean of the set of data is _____.

3.   *Business*   The number of big-screen televisions sold each month for one year was recorded by an electronics store. The results were 15, 12, 20, 20, 19, 17, 22, 24, 17, 20, 15, and 27. Calculate the mean and the median number of televisions sold per month.

4.   *The Airline Industry*   The numbers of seats occupied on a jet for 16 transatlantic flights were recorded. The numbers were 309, 422, 389, 412, 401, 352, 367, 319, 410, 391, 330, 408, 399, 387, 411, and 398. Calculate the mean and the median number of occupied seats.

5.   *Sports*   The times, in seconds, for a 100-meter dash at a college track meet were 10.45, 10.23, 10.57, 11.01, 10.26, 10.90, 10.74, 10.64, 10.52, and 10.78. Calculate the mean and median times for the 100-meter dash.

6.   *Consumerism*   A consumer research group purchased identical items in 8 grocery stores. The costs for the purchased items were $45.89, $52.12, $41.43, $40.67, $48.73, $42.45, $47.81, and $45.82. Calculate the mean and median cost of the purchased items.

7.   *Education*   Your scores on six history tests were 78, 92, 95, 77, 94, and 88. If an "average" score of 90 receives an A for the course, which average, the mean or the median, would you prefer the instructor use?

8. *Sports*   The numbers of yards gained by a college running back were recorded for 6 games. The numbers were 98, 105, 120, 90, 111, and 104. How many yards must this running back gain in the next game so that the average for the seven games is 100 yd?

9. *Sports*   The numbers of unforced errors a tennis player made in four sets of tennis were recorded. The numbers were 15, 22, 24, and 18. How many unforced errors did this player make in the fifth set so that the mean number of unforced errors for the five sets was 20?

10. *Sports*   The last five golf scores for a player were 78, 82, 75, 77, and 79. What score on the next round of golf will give the player a mean score of 78 for all six rounds?

11. *Business*   A survey by an ice cream store asked people to name their favorite ice cream from five flavors. The responses were mint chocolate chip, 34; pralines and cream, 27; German chocolate cake, 44; chocolate raspberry swirl, 34; and rocky road, 42. What was the modal response?

12. *Politics*   A newspaper survey asked people to rate the performance of the city's mayor. The responses were very unsatisfactory, 230; unsatisfactory, 403; satisfactory, 1,237; very satisfactory, 403. What was the modal response for this survey?

13. *Business*   The patrons of a restaurant were asked to rate the quality of the food. The responses were bad, 8; good, 21; very good, 43; excellent, 21. What was the modal response for this survey?

14. A set of data has a mean of 16, a median of 15, and a mode of 14. Which of these numbers must be a value in the data set?

15. True or false? The mean of 10 data values is m. If one of the data values is increased by 10, then the mean of the data values is $m + 10$.

16. True or false? A set of data has an even number of data values. If the median of the data is one of the numbers in the set of data, then the two middle numbers are equal.

OBJECTIVE B   Box-and-whiskers plots

17. **a.** Arrange the data set shown at the right from smallest to largest. Then circle the median of the data.

8	13	6	9	7	15
21	12	18	13	10	

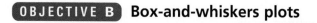

   b. The median of the lower half of the data is $Q_1 =$ _____.

   c. The median of the upper half of the data is $Q_3 =$ _____.

18. 🌑 *U.S. Presidents*   The box-and-whiskers plot at the right shows the distribution of the ages of the presidents of the United States at the time of their inauguration.

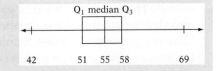

a. The youngest age in the set of data is _____ years.

b. The oldest age in the set of data is _____ years.

c. The median age at inauguration is _____ years.

d. The range of the data set is _____ − _____ = _____.

e. The interquartile range of the data set is $Q_1 - Q_3 =$ _____ − _____ = _____.

19. *Compensation*   The hourly wages for entry-level positions at various firms were recorded by a labor research firm. The results were as follows.

Starting Hourly Wages for 16 Companies

8.09	11.50	7.46	7.70	9.85	9.03	11.40	9.31
10.35	7.45	7.35	8.64	9.02	8.12	10.05	8.94

Find the first quartile and the third quartile, and draw a box-and-whiskers plot of the data.

20. *Health*   The cholesterol levels for 14 adults are recorded in the table below. Find the first quartile and the third quartile, and draw a box-and-whiskers plot of the data.

Cholesterol Levels for 14 Adults

375	185	254	221	183	251	258
292	214	172	233	208	198	211

21. *Fuel Efficiency*   The gasoline consumption of 19 cars was tested and the results recorded in the following table. Find the first quartile and the third quartile, and draw a box-and-whiskers plot of the data.

Miles per Gallon for 19 Cars

33	21	30	32	20	31	25	20	16	24
22	31	30	28	26	19	21	17	26	

22. *Education*   The ages of the accountants who passed the certified public accountant (CPA) exam at one test center are recorded in the table below. Find the first quartile and the third quartile, and draw a box-and-whiskers plot of the data.

Ages of Accountants Passing the CPA Exam

24	42	35	26	24	37	27	26	28
34	43	46	29	34	25	30	28	

23. *Manufacturing* The times for new employees to learn how to assemble a toy are recorded in the table below. Find the first quartile and the third quartile, and draw a box-and-whiskers plot of the data.

Times to Train Employees (in hours)

4.3	3.1	5.3	8.0	2.6	3.5	4.9	4.3
6.2	6.8	5.4	6.0	5.1	4.8	5.3	6.7

24. *Manufacturing* A manufacturer of light bulbs tested the life of 20 light bulbs. The results are recorded in the table below. Find the first quartile and the third quartile, and draw a box-and-whiskers plot of the data.

Life of 20 Light Bulbs (in hours)

1,010	1,235	1,200	998	1,400	789	986	905	1,050	1,100
1,180	1,020	1,381	992	1,106	1,298	1,268	1,309	1,390	890

OBJECTIVE C **The standard deviation of a distribution**

25. Find the standard deviation of the following set of data: 9, 13, 15, 8, 10.

a. Find the mean of the data.

$$\bar{x} = \frac{\boxed{} + \boxed{} + \boxed{} + \boxed{}}{\boxed{}}$$

$$= \frac{\boxed{}}{\boxed{}} = \boxed{}$$

b. Subtract the mean from each data value and then square the difference. Record your work in the table below.

x	$x - \bar{x}$	$(x - \bar{x})^2$
9	$9 - 11 = -2$	$(-2)^2 = \underline{}$
13	$\underline{} - 11 = \underline{}$	$\underline{} = \underline{}$
15	$\underline{} = \underline{}$	$\underline{} = \underline{}$
8	$\underline{} = \underline{}$	$\underline{} = \underline{}$
10	$\underline{} = \underline{}$	$\underline{} = \underline{}$

c. Find the sum of the squares.   Total = _____

d. Divide the sum of the squares by the number of data values: $\dfrac{\boxed{}}{\boxed{}} = \underline{}$.

e. The standard deviation of the data is the square root of the result in part (d).

$$\sigma = \sqrt{\boxed{}} \approx \underline{}$$

26. *The Airline Industry*   An airline recorded the times for a ground crew to unload the baggage from an airplane. The recorded times, in minutes, were 12, 18, 20, 14, and 16. Find the standard deviation of these times.

27. *Health*   The weights in ounces of newborn infants were recorded by a hospital. The weights were 96, 105, 84, 90, 102, and 99. Find the standard deviation of the weights.

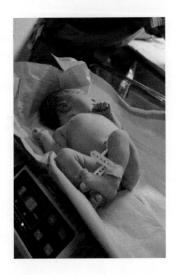

28. *Business*   The numbers of rooms occupied in a hotel on six consecutive days were 234, 321, 222, 246, 312, and 396. Find the standard deviation of the number of rooms occupied.

29. *Coin Tosses*   Seven coins were tossed 100 times. The numbers of heads recorded were 56, 63, 49, 50, 48, 53, and 52. Find the standard deviation of the number of heads.

30. *Meteorology*   The high temperatures for eleven consecutive days at a desert resort were 95°, 98°, 98°, 104°, 97°, 100°, 96°, 97°, 108°, 93°, and 104°. For the same days, the high temperatures in Antarctica were 27°, 28°, 28°, 30°, 28°, 27°, 30°, 25°, 24°, 26°, and 21°. Which location has the greater standard deviation of high temperatures?

31. *Sports*   The scores for five college basketball games were 56, 68, 60, 72, and 64. The scores for five professional basketball games were 106, 118, 110, 122, and 114. Which scores have the greater standard deviation?

32. On four tests, one student received scores of 85, 92, 86, and 89. A second student received scores of 90, 97, 91, and 94 (exactly 5 points more on each test). Are the means of the scores of the two students the same? If not, what is the relationship between the means of the two students' scores? Are the standard deviations of the scores of the two students the same? If not, what is the relationship between the standard deviations of the two students' scores?

CRITICAL THINKING

33. Determine whether the statement is always true, sometimes true, or never true.
   a. If there is an odd number of values in a set of data, the median is one of the numbers in the set.
   b. If there is an even number of values in a set of data, the median is not one of the numbers in the set.

34. A company is negotiating with its employees regarding salary raises. One proposal would add $1,500 a year to each employee's salary. The second proposal would give each employee a 4% raise. Explain how these proposals would affect the current mean and standard deviation of salaries for the company.

10.3 Introduction to Probability

OBJECTIVE A   The probability of simple events

A weather forecaster estimates that there is a 75% chance of rain. A state lottery director claims that there is a $\frac{1}{9}$ chance of winning a prize offered by the lottery. Each of these statements involves uncertainty to some extent. The degree of uncertainty is called **probability.** For the statements above, the probability of rain is 75% and the probability of winning a prize in the lottery is $\frac{1}{9}$.

A probability is determined from an **experiment,** which is any activity that has an observable outcome. Examples of experiments are

> Tossing a coin and observing whether it lands heads or tails
> Interviewing voters to determine their preference for a political candidate
> Recording the percent change in the price of a stock

All the possible outcomes of an experiment are called the **sample space** of the experiment. The outcomes of an experiment are listed between braces and frequently designated by S.

> For each experiment, list all the possible outcomes.
>
> 1. A number cube, which has the numbers from 1 to 6 written on its sides, is rolled once.
>   Any of the numbers from 1 to 6 could show on the top of the cube. $S = \{1, 2, 3, 4, 5, 6\}$
> 2. A fair coin is tossed once.
>   A fair coin is one for which heads and tails have an equal chance of being tossed. $S = \{H, T\}$, where H represents heads and T represents tails.
> 3. The spinner at the right is spun once.
>   Assuming that the spinner does not come to rest on a line, the arrow could come to rest in any one of the four sectors. $S = \{1, 2, 3, 4\}$

An **event** is one or more outcomes of an experiment. Events are denoted by capital letters. Consider the experiment of rolling the number cube given above. Some possible events are:

> The number is even. $E = \{2, 4, 6\}$
> The number is a prime number. $P = \{2, 3, 5\}$
> The number is less than 10. $T = \{1, 2, 3, 4, 5, 6\}$. Note that in this case, the event is the entire sample space.
> The number is greater than 20. This event is impossible for the given sample space. The impossible event is symbolized by $\varnothing$.

When discussing experiments and events, it is convenient to refer to the *favorable outcomes* of an experiment. These are the outcomes of an experiment that satisfy the requirements of the particular event. For instance, consider the experiment of rolling a fair die once. The sample space is $\{1, 2, 3, 4, 5, 6\}$, and one possible event E would be rolling a number that is divisible by 3. The outcomes of the experiment that are favorable to E are 3 and 6, and $E = \{3, 6\}$.

Point of Interest

It was dice playing that led Antoine Gombaud, Chevalier de Mere, to ask Blaise Pascal, a French mathematician, to figure out the probability of throwing two 6's. Pascal and Pierre Fermat solved the problem, and their explorations led to the birth of probability theory.

Probability Formula

The probability of an event E, written $P(E)$, is the ratio of the number of favorable outcomes of an experiment to the total number of possible outcomes of the experiment.

$$P(E) = \frac{\text{number of favorable outcomes}}{\text{number of possible outcomes}}$$

The outcomes of the experiment of tossing a fair coin are *equally likely*. Any one of the outcomes is just as likely as another. If a fair coin is tossed once, the probability of a head or a tail is $\frac{1}{2}$. Each event, heads or tails, is equally likely.

The probability formula applies to experiments for which the outcomes are equally likely.

Not all experiments have equally likely outcomes. Consider an exhibition baseball game between a professional team and a college team. Although either team *could* win the game, the probability that the professional team will win is greater than that of the college team. The outcomes are not equally likely. For the experiments in this section, assume that the outcomes of an experiment are equally likely.

There are five choices, *a* through *e*, for each question on a multiple-choice test. By just guessing, what is the probability of choosing the correct answer for a certain question?

It is possible to select any of the letters *a*, *b*, *c*, *d*, or *e*.	There are 5 possible outcomes of the experiment.
The event E is the correct answer.	There is 1 favorable outcome, guessing the correct answer.
Use the probability formula.	$P(E) = \dfrac{\text{number of favorable outcomes}}{\text{number of possible outcomes}} = \dfrac{1}{5}$

The probability of guessing the correct answer is $\frac{1}{5}$.

Each of the letters in the word *Tennessee* is written on a card, and the cards are placed in a hat. If one card is drawn at random from the hat, what is the probability that the card has the letter *e* on it?

The phrase "at random" means that each card has an equal chance of being drawn.	There are 9 letters in *Tennessee*. Therefore, there are 9 possible outcomes of the experiment.
There are 4 cards with an *e* on them.	There are 4 favorable outcomes of the experiment, the 4 *e*'s.
Use the probability formula.	$P(E) = \dfrac{\text{number of favorable outcomes}}{\text{number of possible outcomes}} = \dfrac{4}{9}$

The probability is $\frac{4}{9}$.

Calculating the probability of an event requires counting the number of possible outcomes of an experiment and the number of outcomes that are favorable to the event. One way to do this is to list the outcomes of the experiment in some systematic way. Using a table is often very helpful.

Q_1	Q_2	Q_3
T	T	T
T	T	F
T	F	T
T	F	F
F	T	T
F	T	F
F	F	T
F	F	F

A professor writes three true/false questions for a test. If the professor randomly chooses which questions will have a true answer and which will have a false answer, what is the probability that the test will have 2 true questions and 1 false question?

The experiment S consists of choosing T or F for each of the 3 questions. The possible outcomes of the experiment are shown in the table at the right.

$S = \{$TTT, TTF, TFT, TFF, FTT, FTF, FFT, FFF$\}$

There are 8 outcomes for S.

The event E consists of 2 true questions and 1 false question.

$E = \{$TTF, TFT, FTT$\}$

There are 3 outcomes for E.

Use the probability formula.

$P(E) = \dfrac{3}{8}$

The probability of 2 true questions and 1 false question is $\dfrac{3}{8}$.

The probabilities that we have calculated so far are referred to as *mathematical* or *theoretical probabilities*. The calculations are based on theory—for example, that either side of a coin is equally likely to land face up or that each of the six sides of a fair die is equally likely to be face up. Not all probabilities arise from such assumptions.

Empirical probabilities are based on observations of certain events. For instance, a weather forecast of a 75% chance of rain is an empirical probability. From historical records kept by the weather bureau, when a similar weather pattern existed, rain occurred 75% of the time. It is theoretically impossible to predict the weather, and only observations of past weather patterns can be used to predict future weather conditions.

Empirical Probability Formula

The empirical probability of an event E is the ratio of the number of observations of E to the total number of observations.

$P(E) = \dfrac{\text{number of observations of } E}{\text{total number of observations}}$

Records of an insurance company show that of 2,549 claims for theft filed by policyholders, 927 were claims for more than $5,000. What is the empirical probability that the next claim for theft this company receives will be a claim for more than $5,000?

The empirical probability of E is the ratio of the number of claims for over $5,000 to the total number of claims.

$P(E) = \dfrac{927}{2,549} \approx 0.36$

The probability is approximately 0.36.

Two dice are rolled, one after the other. The sample space is shown below. There are 36 possible outcomes.

Point of Interest

Romans called a die that was marked on four faces a *talus*, which meant "anklebone." The anklebone was considered an ideal die because it is roughly a rectangular solid and it has no marrow, so loose ones from sheep were more likely to be lying around after the wolves had left their prey.

Possible Outcomes from Rolling Two Dice

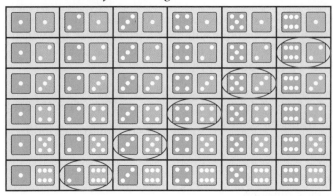

The outcomes that the sum of the numbers on the two dice is 8 are circled above. There are 5 possible outcomes that have a sum of 8. The probability that the sum of the numbers on the two dice is 8 is $P(E) = \frac{5}{36}$.

EXAMPLE 1

Two dice are rolled once. Calculate the probability that the sum of the numbers on the two dice is 7.

Strategy

To find the probability:

▶ Count the number of possible outcomes of the experiment.

▶ Count the outcomes of the experiment that are favorable to the event the sum is 7.

▶ Use the probability formula.

Solution

There are 36 possible outcomes.

There are 6 outcomes favorable for E: (1, 6), (2, 5), (3, 4), (4, 3), (5, 2), and (6, 1).

$$P(E) = \frac{6}{36} = \frac{1}{6}$$

The probability that the sum is 7 is $\frac{1}{6}$.

YOU TRY IT 1

Two dice are rolled once. Calculate the probability that the two numbers on the dice are equal.

Your Strategy

Your Solution

Solution on pp. S25–S26

EXAMPLE 2

A large box contains 25 red, 35 blue, and 40 white balls. If one ball is randomly selected from the box, what is the probability that it is blue? Write the answer as a percent.

YOU TRY IT 2

There are 8 covered circles on a "scratcher card" that is given to each customer at a fast-food restaurant. Under one of the circles is a symbol for a free soft drink. If the customer scratches off one circle, what is the probability that the soft drink symbol will be uncovered?

Strategy

To find the probability:

▶ Count the number of outcomes of the experiment.
▶ Count the number of outcomes of the experiment favorable to the event E that the ball is blue.
▶ Use the probability formula.

Your Strategy

Solution

There are 100 (25 + 35 + 40) balls in the box.

35 balls of the 100 are blue.

$$P(E) = \frac{35}{100} = 0.35 = 35\%$$

There is a 35% chance of selecting a blue ball.

Your Solution

Solution on p. S26

OBJECTIVE B   **The odds of an event**

Sometimes the chance of an event occurring is given in terms of *odds*. This concept is closely related to probability.

Odds in Favor of an Event

The **odds in favor** of an event is the ratio of the number of favorable outcomes of an experiment to the number of unfavorable outcomes.

$$\text{Odds in favor} = \frac{\text{number of favorable outcomes}}{\text{number of unfavorable outcomes}}$$

Odds Against an Event

The **odds against** an event is the ratio of the number of unfavorable outcomes of an experiment to the number of favorable outcomes.

$$\text{Odds against} = \frac{\text{number of unfavorable outcomes}}{\text{number of favorable outcomes}}$$

To find the odds in favor of a 4 when a die is rolled once, list the favorable outcomes and the unfavorable outcomes.

favorable outcomes: 4   unfavorable outcomes: 1, 2, 3, 5, 6

$$\text{Odds in favor of a 4} = \frac{\text{number of favorable outcomes}}{\text{number of unfavorable outcomes}} = \frac{1}{5}$$

Frequently the odds of an event are expressed as a ratio using the word *to*. For the last problem, the odds in favor of a 4 are 1 to 5.

It is possible to compute the probability of an event from the odds-in-favor fraction. The probability of an event is the ratio of the numerator to the sum of the numerator and denominator.

The odds in favor of winning a prize in a charity drawing are 1 to 19. What is the probability of winning a prize?

Write the ratio 1 to 19 as a fraction. $1 \text{ to } 19 = \dfrac{1}{19}$

The probability of winning a prize is the ratio of the numerator to the sum of the numerator and denominator. $\text{Probability} = \dfrac{1}{1 + 19} = \dfrac{1}{20}$

The probability of winning a prize is $\dfrac{1}{20}$.

EXAMPLE 3

In a horse race, the odds against a horse winning the race are posted as 9 to 2. What is the probability of the horse's winning the race?

Strategy

To calculate the probability of winning:

▶ Restate the odds against as odds in favor.
▶ Using the odds-in-favor fraction, the probability of winning is the ratio of the numerator to the sum of the numerator and denominator.

Solution

The odds against winning are 9 to 2. Therefore, the odds in favor of winning are 2 to 9.

$$\text{Probability of winning} = \frac{2}{2 + 9} = \frac{2}{11}$$

The probability of the horse winning the race is $\dfrac{2}{11}$.

YOU TRY IT 3

The odds in favor of contracting the flu during a flu epidemic are 2 to 13. Calculate the probability of getting the flu.

Your Strategy

Your Solution

Solution on p. S26

10.3 Exercises

OBJECTIVE A **The probability of simple events**

1. 🖊 Describe two situations in which probabilities are cited.

2. 🖊 Why can the probability of an event not be $\frac{5}{3}$?

Three face cards, a jack, a queen, and a king, are laid out in a row. Use this situation for Exercises 3 and 4.

3. **a.** Complete the following table to make a list of all the possible outcomes for this experiment. Use J for jack, Q for queen, and K for king.

Card 1	Card 2	Card 3
J	Q	K
J	K	
Q		

   b. The number of possible outcomes of this experiment is _____.
   c. The number of outcomes favorable to the event the first card is a queen is _____.

4. Use the results of Exercise 3 to find the probability that a random arrangement of the three cards will result in the event the first card is a queen.

$$P(\text{first card is a queen}) = \frac{\text{number of favorable outcomes}}{\text{number of possible outcomes}} = \frac{\boxed{}}{\boxed{}} = \underline{}$$

5. A coin is tossed 4 times. List all the possible outcomes of the experiment as a sample space. The table on page 629 shows an example of a systematic way of recording results for a similar problem.

6. Three cards—one red, one green, and one blue—are to be arranged in a stack. Using R for red, G for green, and B for blue, make a list of all the different stacks that can be formed. (Some computer monitors are called RGB monitors for the colors red, green, and blue.)

7. A tetrahedral die is one with four triangular sides. If two tetrahedral dice are rolled, list all the possible outcomes of the experiment as a sample space. (See the table for two dice, page 630, for assistance in listing the outcomes.)

8. A coin is tossed and then a die is rolled. List all the possible outcomes of the experiment as a sample space. (To get you started, (H, 1) is one of the possible outcomes.)

9. A coin is tossed four times. What is the probability that the outcomes of the tosses are exactly in the order HHTT? (See Exercise 5.)

10. A coin is tossed four times. What is the probability that the outcomes of the tosses are exactly in the order HTTH? (See Exercise 5.)

11. A coin is tossed four times. What is the probability that the outcomes of the tosses consist of two heads and two tails? (See Exercise 5.)

12. A coin is tossed four times. What is the probability that the outcomes of the tosses consist of one head and three tails? (See Exercise 5.)

13. If two dice are rolled, what is the probability that the sum of the dots on the upward faces is 5?

14. If two dice are rolled, what is the probability that the sum of the dots on the upward faces is 9?

15. If two dice are rolled, what is the probability that the sum of the dots on the upward faces is 15?

16. If two dice are rolled, what is the probability that the sum of the dots on the upward faces is less than 15?

17. If two dice are rolled, what is the probability that the sum of the dots on the upward faces is 2?

18. If two dice are rolled, what is the probability that the sum of the dots on the upward faces is 12?

19. A dodecahedral die has 12 sides. If the die is rolled once, what is the probability that the upward face shows an 11?

20. A dodecahedral die has 12 sides. If the die is rolled once, what is the probability that the upward face shows a 5?

21. If two tetrahedral dice are rolled (see Exercise 7), what is the probability that the sum on the upward faces is 4?

22. If two tetrahedral dice are rolled (see Exercise 7), what is the probability that the sum on the upward faces is 6?

23. A dodecahedral die has 12 sides. If the die is rolled once, what is the probability that the upward face shows a number divisible by 4?

24. A dodecahedral die has 12 sides. If the die is rolled once, what is the probability that the upward face shows a number that is a multiple of 3?

25. A survey of 95 people showed that 37 preferred a cash discount of 2% if an item was purchased using cash or a check. On the basis of this survey, what is the empirical probability that a person prefers a cash discount? Write the answer as a percent rounded to the nearest tenth of a percent.

26. A survey of 725 people showed that 587 had a group health insurance plan where they worked. Based on this survey, what is the empirical probability that an employee has a group health insurance plan? Write the answer as a percent rounded to the nearest tenth of a percent.

27. A signal light is green for 3 min, yellow for 15 s, and red for 2 min. If you drive up to this light, what is the probability that it will be green when you reach the intersection?

28. In a history class, a professor gave 4 A's, 8 B's, 22 C's, 10 D's, and 3 F's. If a single student's paper is chosen at random from this class, what is the probability that it received a B?

29. A television cable company surveyed some of its customers and asked them to rate the cable service as excellent, satisfactory, average, unsatisfactory, or poor. The results are recorded in the table at the right. What is the probability that a customer who was surveyed rated the service as satisfactory or excellent? Write the answer as a percent rounded to the nearest tenth of a percent.

Quality of Service	Number Who Voted
Excellent	98
Satisfactory	87
Average	129
Unsatisfactory	42
Poor	21

30. Using the television cable survey in Exercise 29, what is the probability that a customer who was surveyed rated the service as unsatisfactory or poor? Write the answer as a percent rounded to the nearest tenth of a percent.

31. Some people who cheat at gambling use dice that are loaded so that 7 occurs more frequently than expected. If these dice are used, are the probabilities of the outcomes equal? Why or why not?

32. If the spinner at the right is spun once, is each of the numbers 1 through 5 equally likely? Why or why not?

33. Use the situation described in Exercise 27. Suppose you decide to test your result empirically by recording the color of the light every time you come to the intersection during one week. During the week, you come to the intersection 21 times. Which number of green lights would confirm the result found in Exercise 27?

(i) 4   **(ii)** 7   **(iii)** 12   **(iv)** 14

34. Use the situation described in Exercise 28. What probability does the ratio $\frac{13}{47}$ represent?

35. Suppose the probability of an event occuring is p. Write an inequality that represents all possible values of p.

36. Suppose the probability of an event occuring is p. Write an expression for the probability of the event not occuring.

OBJECTIVE B   **The odds of an event**

37. Complete the equations using the words *favorable* and *unfavorable*.

a. Odds in favor of an event = $\dfrac{\text{number of} \rule{2cm}{0.4pt} \text{outcomes}}{\text{number of} \rule{2cm}{0.4pt} \text{outcomes}}$

b. Odds against an event = $\dfrac{\text{number of} \rule{2cm}{0.4pt} \text{outcomes}}{\text{number of} \rule{2cm}{0.4pt} \text{outcomes}}$

38. A single die is rolled once. Consider the event of rolling a number greater than 1.
a. List the favorable outcomes: _____
b. List the unfavorable outcomes: _____
c. The odds of rolling a number greater than 1 are

$\dfrac{\text{number of favorable outcomes}}{\text{number of unfavorable outcomes}} = \rule{1.5cm}{0.4pt}$, or _____ to _____

 For Exercises 39 and 40, use the following ratios.

(i) $\dfrac{a}{b}$ **(ii)** $\dfrac{b}{a}$ **(iii)** $\dfrac{a}{a+b}$ **(iv)** $\dfrac{b}{a+b}$

39. If the odds in favor of an event are a to b, which ratio represents the odds against the event?

40. If the odds in favor of an event are a to b, which ratio represents the probability of the event?

41. A coin is tossed once. What are the odds of its showing heads?

42. A coin is tossed twice. What are the odds of its showing tails both times?

43. The odds in favor of a candidate winning an election are 3 to 2. What is the probability of the candidate winning the election?

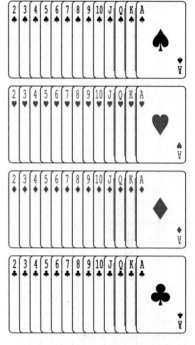

44. On a board game, the odds in favor of winning $5,000 are 3 to 7. What is the probability of winning the $5,000?

45. Two dice are rolled. What are the odds in favor of rolling a 7?

46. Two dice are rolled. What are the odds in favor of rolling a 12?

47. A single card is selected from a standard deck of playing cards. What are the odds against its being an ace?

48. A single card is selected from a standard deck of playing cards. What are the odds against its being a heart?

A Standard Deck of Playing Cards

49. At the beginning of the professional football season, one team was given 40 to 1 odds against its winning the Super Bowl. What is the probability of this team winning the Super Bowl?

50. At the beginning of the professional baseball season, one team was given 25 to 1 odds against its winning the World Series. What is the probability of this team winning the World Series?

51. A stock market analyst estimates that the odds in favor of a stock going up in value are 2 to 1. What is the probability of the stock's not going up in value?

52. The odds in favor of the occurrence of an event *A* are given as 5 to 2, and the odds against a second event *B* are given as 1 to 7. Which event, *A* or *B*, has the greater probability of occurring?

CRITICAL THINKING

53. Three line segments are randomly chosen from line segments whose lengths are 1 cm, 2 cm, 3 cm, 4 cm, and 5 cm. What is the probability that a triangle can be formed from the line segments?

Focus on **Problem Solving**

Applying Solutions to Other Problems

Problem solving in the previous chapters concentrated on solving specific problems. After a problem is solved, however, there is an important question to be asked: "Does the solution to this problem apply to other types of problems?"

To illustrate this extension of problem solving, we will consider *triangular numbers*, which were studied by ancient Greek mathematicians. The numbers 1, 3, 6, 10, 15, and 21 are the first six triangular numbers. What is the next triangular number?

To answer this question, note in the diagram below that a triangle can be formed using the number of dots that correspond to a triangular number.

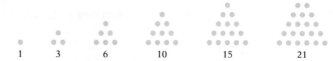

| 1 | 3 | 6 | 10 | 15 | 21 |

Observe that the number of dots in each row is one more than the number of dots in the row above. The total number of dots can be found by addition.

$$1 = 1 \qquad 1 + 2 = 3 \qquad 1 + 2 + 3 = 6 \qquad 1 + 2 + 3 + 4 = 10$$
$$1 + 2 + 3 + 4 + 5 = 15 \qquad 1 + 2 + 3 + 4 + 5 + 6 = 21$$

The pattern suggests that the next triangular number (the 7th one) is the sum of the first 7 natural numbers.

$$1 + 2 + 3 + 4 + 5 + 6 + 7 = 28$$

The 7th triangular number is 28. The diagram at the left shows the 7th triangular number.

Using the pattern for triangular numbers, the 10th triangular number is

$$1 + 2 + 3 + 4 + 5 + 6 + 7 + 8 + 9 + 10 = 55$$

Now consider a situation that may seem to be totally unrelated to triangular numbers. Suppose you are in charge of scheduling softball games for a league. There are seven teams in the league, and each team must play every other team once. How many games must be scheduled?

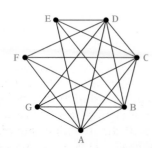

We can label the teams A, B, C, D, E, F, and G. (See the figure at the left.) A line between two teams indicates that the two teams play each other. Beginning with A, there are 6 lines for the 6 teams that A must play.

Now consider B. There are 6 teams that B must play, but the line between A and B has already been drawn, so there are only 5 remaining games to schedule for B. Now move on to C. The lines between C and A and between C and B have already been drawn, so there are 4 additional lines to be drawn to represent the teams that C will play. Moving on to D, the lines between D and A, D and B, and D and C have already been drawn, so there are 3 more lines to be drawn to represent the teams that D will play.

Note that each time we move from team to team, one fewer line needs to be drawn. When we reach F, there is only one line to be drawn, the one between F and G. The total number of lines drawn is $6 + 5 + 4 + 3 + 2 + 1 = 21$, the sixth triangular number. For a league with 7 teams, the number of games that must be scheduled so that each team plays every other team once is the 6th triangular number. If there were 10 teams in the league, the number of games that would have to be scheduled would be the 9th triangular number, which is 45.

A college chess team wants to schedule a match so that each of its 15 members plays each other member of the team once. How many matches must be scheduled?

Projects & Group Activities

Random Samples

Point of Interest

When a survey is taken to determine, for example, who is the most popular choice for a political office such as a governor or president, it would not be appropriate to survey only one ethnic group, or to survey only one religious group, or to survey only one political group. A survey done in that way would reflect only the views of that particular group of people. Instead, a *random sample* of people must be chosen. This sample would include people from different ethnic, religious, political, and income groups. The purpose of the random sample is to identify the popular choice of all the people by interviewing only a few people, the people in the random sample.

One way of choosing a random sample is to use a table of random numbers. An example of a portion of such a table is shown below.

Random Digits

40784	38916	12949
29798	57707	57392
42228	94940	10668
02218	89355	76117
15736	08506	29759
42658	32502	99698
98670	57794	64795
38266	30138	61250
68249	32459	41627
36910	85225	78541

In a random number table, each digit should occur with approximately the same frequency as any other digit.

1. Make a frequency table for the table of random digits given above. Do the digits occur with approximately the same frequency?

As an example of how to use a random number table, suppose there are 33 students in your class and you want to randomly select 6 students. Using the numbers 01, 02, 03, . . . , 31, 32, 33, assign each student a two-digit number. Starting with the first column of random numbers, move down the column, looking at only the first two digits. If the two digits are one of the numbers 01 through 33, write them down. If not, move to the next number. Continue in this way until six numbers have been selected.

For instance, the first two digits of 40784 are 40, which is not between 01 and 33. Therefore, move to the next number. The first two digits of 29798 are 29. Since 29 is between 01 and 33, write it down. Continuing in this way reveals that the random sample would be the students with numbers 29, 02, 15, 08, 32, and 30.

2. Once a random sample has been selected, it is possible to use this sample to estimate characteristics of the entire class. For example, find the mean height of the students in the random sample and compare it to the mean height of the entire class.

Random numbers also are used to *simulate* random events. In a simulation, it is not the actual activity that is performed, but instead one that is very similar.

For instance, suppose a coin is tossed repeatedly and the result, heads or tails, is recorded. The actual activity is tossing the coin. A simulation of tossing a coin uses a random number table. Starting with the first column of random numbers, move down the column, looking at the first digit. Associate an even digit with heads and an odd digit with tails. Using the first column of the table on the previous page, a simulation of the first 10 tosses of the coin would be H, H, H, H, T, H, T, T, H, and T.

Probabilities can be approximated by simulating an event. Consider the event of tossing two heads when a fair coin is tossed twice. We can simulate the event by associating a two-digit number with two even digits with two heads, one with two odd digits with two tails, and any other pair of digits with a head and tail or tail and head.

3. Using the 30 numbers in the table on the previous page, show that the probability of tossing two heads would be $\frac{7}{30} \approx 0.233$. Explain why the actual probability is 0.25.

4. Go to the library and find a table of random numbers. Use this table to simulate rolling a pair of dice. If you use the first two digits of a column, a valid roll consists of two digits that are both between 1 and 6, inclusive. Simulate the empirical probability that the sum on the upward faces of the dice is 7 by using 100 rolls of the dice.

Chapter 10 Summary

Key Words

Statistics is the study of collecting, organizing, and interpreting data. Data are collected from a **population,** which is the set of all observations of interest. A **frequency distribution** is one method of organizing data. A frequency distribution is constructed by dividing the data gathered from a population into **classes.** The **range** of a set of numerical data is the difference between the largest and smallest values. [10.1A, p. 607]

Examples

An Internet service provider (ISP) surveyed 1,000 of its subscribers to determine the time required for each subscriber to download a particular file. The results are summarized in the frequency distribution at the top of the next page. The distribution has 12 classes.

Download Time (in seconds)	Number of Subscribers
0–5	6
5–10	17
10–15	43
15–20	92
20–25	151
25–30	192
30–35	190
35–40	149
40–45	90
45–50	45
50–55	15
55–60	10

A **histogram** is a bar graph that represents the data in a frequency distribution. [10.1B, p. 609]

Below is a histogram for the frequency distribution given above.

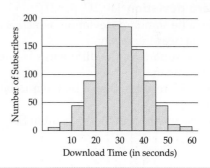

A **frequency polygon** is a graph that displays information in a manner similar to a histogram. A dot is placed above the center of each class interval at a height corresponding to that class's frequency. The dots are connected to form a broken-line graph. The center of a class interval is called the **class midpoint.** [10.1C, p. 610]

Below is a frequency polygon for the frequency distribution given above.

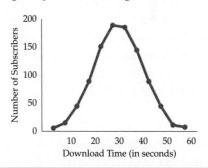

The **mean, median,** and **mode** are three types of averages used in statistics. The **median** of a data set is the number that separates the data into two equal parts when the data have been arranged from smallest to largest (or largest to smallest). The **mode** is the most frequently occurring data value. [10.2A, pp. 616–617]

Consider the following set of data:

24, 28, 33, 45, 45

The median is 33.
The mode is 45.

A **box-and-whiskers plot** is a graph that shows five numbers: the smallest value, the first quartile, the median, the third quartile, and the largest value. The **first quartile** Q_1 is the number below which one-fourth of the data lie. The **third quartile** Q_3 is the number above which one-fourth of the data lie. The box is placed around the values between the first quartile and the third quartile. The difference between Q_3 and Q_1 is the **interquartile range.** [10.2B, pp. 619–620]

The box-and-whiskers plot for a set of test scores is shown below.

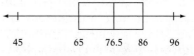

The interquartile range $= Q_3 - Q_1$
$= 86 - 65 = 21$

An **experiment** is an activity with an observable outcome. All the possible outcomes of an experiment are called the **sample space** of the experiment. An **event** is one or more outcomes of an experiment. [10.3A, p. 627]

Tossing a single die is an example of an experiment. The sample space for this experiment is the set of possible outcomes: $S = \{1, 2, 3, 4, 5, 6\}$. The event that the number landing face up is an odd number is represented by $E = \{1, 3, 5\}$.

Essential Rules and Procedures

Mean of a set of data [10.2A, p. 616]

$$\text{Mean} = \bar{x} = \frac{\text{sum of all data values}}{\text{number of data values}}$$

Consider the following set of data:

24, 28, 33, 45, 45

The mean is $\frac{24 + 28 + 33 + 45 + 45}{5} = 35$.

Standard deviation [10.2C, p. 621]

To determine standard deviation, which is a measure of the clustering of data near the mean:
1. Sum the squares of the differences between each value of data and the mean.
2. Divide the result in Step 1 by the number of items in the set of data.
3. Take the square root of the result in Step 2.

Consider the following set of data:

24, 28, 33, 45, 45

The mean is 35.

1.
x	$(x - \bar{x})$	$(x - \bar{x})^2$
24	$(24 - 35)$	$(-11)^2 = 121$
28	$(28 - 35)$	$(-7)^2 = 49$
33	$(33 - 35)$	$(-2)^2 = 4$
45	$(45 - 35)$	$(10)^2 = 100$
45	$(45 - 35)$	$(10)^2 = \underline{100}$
		Total $= 374$

2. $\frac{374}{5} = 74.8$

3. $\sigma = \sqrt{74.8} \approx 8.649$

Probability Formula [10.3A, p. 628]

$$P(E) = \frac{\text{number of favorable outcomes}}{\text{number of possible outcomes}}$$

A die is rolled. The probability of rolling a 2 or a 4 is

$$P(E) = \frac{2}{6} = \frac{1}{3}$$

Empirical Probability Formula [10.3A, p. 629]

$$P(E) = \frac{\text{number of observations of } E}{\text{total number of observations}}$$

A thumbtack is tossed 100 times. It lands point up 15 times and lands on its side 85 times. From this experiment, the empirical probability of "point up" is

$$P(\text{point up}) = \frac{15}{100} = \frac{3}{20}$$

Odds in Favor of an Event [10.3B, p. 631]

$$\text{Odds in favor} = \frac{\text{number of favorable outcomes}}{\text{number of unfavorable outcomes}}$$

A die is rolled.
The odds in favor of rolling a 2 or a 4:

$$\text{Odds in favor} = \frac{2}{4} = \frac{1}{2}$$

Odds Against an Event [10.3B, p. 631]

$$\text{Odds against} = \frac{\text{number of unfavorable outcomes}}{\text{number of favorable outcomes}}$$

The odds against rolling a 2 or a 4:

$$\text{Odds against} = \frac{4}{2} = \frac{2}{1}$$

Chapter 10 Review Exercises

Education   Use the data in the table below for Exercises 1 to 5.

Number of Students in 40 Mathematics Classes

30	45	54	24	48	12	38	31
15	36	37	27	40	35	55	32
42	14	21	18	29	25	16	42
44	41	28	32	27	24	30	24
21	35	27	32	39	41	35	48

1. Make a frequency distribution for these data using 6 classes.

2. Which class has the greatest frequency?

3. How many math classes have 35 or fewer students?

4. What percent of the math classes have 44 or more students?

5. What percent of the math classes have 27 or fewer students?

Meteorology   The high temperatures at a ski resort during a 125-day ski season are recorded in the figure at the right. Use this histogram for Exercises 6 and 7.

6. Find the number of days the high temperature was 45° or above.

7. How many days had a high temperature below 25°?

8. *Health*   A health clinic administered a test for cholesterol to 11 people. The results were 180, 220, 160, 230, 280, 200, 210, 250, 190, 230, and 210. Find the mean and median of these data.

9. *Health*   The weights, in pounds, of 10 babies born at a hospital were recorded as 6.3, 5.9, 8.1, 6.5, 7.2, 5.6, 8.9, 9.1, 6.9, and 7.2. Find the mean and median of these data.

10. *The Arts*   People leaving a new movie were asked to rate the movie as bad, good, very good, or excellent. The responses were bad, 28; good, 65; very good, 49; excellent, 28. What was the modal response for this survey?

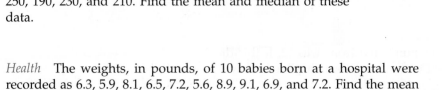

High Temperatures

Investments The frequency polygon in the figure at the right shows the numbers of shares of stock sold during particular hours on a stock exchange. Use the frequency polygon for Exercises 11 to 13.

11. How many shares of stock were sold between 7 A.M. and 10 A.M.?

12. Between which hours were less than 15 million shares sold?

13. What is the ratio of the number of shares of stock sold between 10 A.M. and 11 A.M. to the number that were sold between 11 A.M. and 12 P.M.?

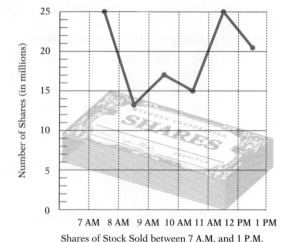

Shares of Stock Sold between 7 A.M. and 1 P.M.

14. *Sports* The numbers of points scored by a basketball team for 15 games were 89, 102, 134, 110, 121, 124, 111, 116, 99, 120, 105, 109, 110, 124, and 131. Find the first quartile, median, and third quartile. Draw a box-and-whiskers plot.

15. *Fuel Efficiency* A consumer research group tested the average miles per gallon for six cars. The results were 24, 28, 22, 35, 41, and 27. Find the standard deviation of the gasoline mileage ratings.

16. A charity raffle sells 2,500 raffle tickets for a big-screen television set. If you purchase 5 tickets, what is the probability that you will win the television?

17. A box contains 50 balls, of which 15 are red. If one ball is randomly selected from the box, what are the odds in favor of the ball's being red?

18. In professional jai alai, a gambler can wager on who will win the event. The odds against one of the players winning are given as 5 to 2. What is the probability of that player's winning?

19. A dodecahedral die has 12 sides numbered from 1 to 12. If this die is rolled once, what is the probability that a number divisible by 6 will be on the upward face?

20. One student is randomly selected from 3 first-year students, 4 sophomores, 5 juniors, and 2 seniors. What is the probability that the student is a junior?

Chapter 10 Test

1. *Communications*   The histogram below shows the cost for telephone service for 100 residences. For how many residences was the cost for telephone service $60 or more?

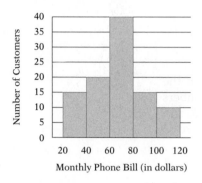

2. *Business*   The frequency polygon below shows the gross sales at a record store for 6 months. Find the total gross sales for January and February.

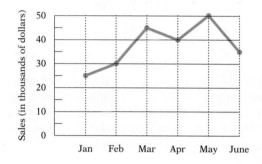

3. *Business*   The annual sales for the 50 restaurants in a restaurant chain are given in the frequency table at the right. What percent of the restaurants had annual sales between $750,000 and $1,000,000?

Annual Sales (in dollars)	Frequency
0–250,000	4
250,000–500,000	10
500,000–750,000	17
750,000–1,000,000	8
1,000,000–1,250,000	8
1,250,000–1,500,000	3

4. *Sports*   The bowling scores for eight people were 138, 125, 162, 144, 129, 168, 184, and 173. What was the mean score for these eight people?

5. *Emergency Calls*   The response times by an ambulance service to emergency calls were recorded by a public safety commission. The times (in minutes) were 17, 21, 11, 8, 22, 15, 11, 14, and 8. Determine the median response time for these calls.

6. *Education*   Recent college graduates were asked to rate the quality of their education. The responses were 47, excellent; 86, very good; 32, good; 20, poor. What was the modal response?

7. *Business*   The numbers of digital assistants sold by a store for the first 5 months of the year were 34, 28, 31, 36, and 38. How many digital assistants must be sold in the sixth month so that the mean number sold per month for the 6 months is 35?

8. *Manufacturing*   The average time, in minutes, it takes for a factory worker to assemble 14 different toys is given in the table. Determine the first quartile of the data.

10.5	21.0	17.3	11.2	9.3	6.5	8.6
9.8	20.3	19.6	9.8	10.5	11.9	18.5

9. *Business* The number of vacation days taken last year by each of the employees of a firm was recorded. The box-and-whiskers plot at the right represents the data. **a.** Determine the range of the data. **b.** What was the median number of vacation days taken?

10. *Sports* The scores of the 14 leaders in a college golf tournament are given in the table. Draw a box-and-whiskers plot of the data.

80	76	70	71	74	68	72
74	70	70	73	75	69	73

11. *Testing* An employee at a department of motor vehicles analyzed the written tests of the last 10 applicants for a driver's license. The numbers of incorrect answers for these applicants were 2, 0, 3, 1, 0, 4, 5, 1, 3, and 1. What is the standard deviation of the number of incorrect answers? Round to the nearest hundredth.

12. A coin is tossed and then a regular die is rolled. How many elements are in the sample space?

13. Three coins—a nickel, a dime, and a quarter—are stacked. List the elements in the sample space.

14. A cross-country flight has 14 passengers in first class, 32 passengers in business class, and 202 passengers in coach. If one passenger is selected at random, what is the probability that the person is in business class?

15. Three playing cards—an ace, a king, and a queen—are randomly arranged and stacked. What is the probability that the ace is on top of the stack?

16. A quiz contains three true/false questions. If a student attempts to answer the questions by just guessing, what is the probability that the student will answer all three questions correctly?

17. A package of flower seeds contains 15 seeds for red flowers, 20 seeds for white flowers, and 10 seeds for pink flowers. If one seed is selected at random, what is the probability that it is not a seed for a red flower?

18. The odds of winning a prize in a lottery are given as 1 to 12. What is the probability of winning a prize?

19. The probability of rolling a sum of 9 on two standard dice is $\frac{1}{9}$. What are the odds in favor of rolling a sum of 9 on these dice?

20. A dodecahedral die has 12 sides. If this die is tossed once, what is the probability that the number on the upward face is less than 6?

Cumulative Review Exercises

1. Simplify: $\sqrt{200}$

2. Solve: $7p - 2(3p - 1) = 5p + 6$

3. Evaluate $3a^2b - 4ab^2$ for $a = -1$ and $b = 2$.

4. Simplify: $-2[2 - 4(3x - 1) + 2(3x - 1)]$

5. Solve: $-\dfrac{2}{3}y - 5 = 7$

6. Simplify: $-\dfrac{4}{5}\left[\dfrac{3}{4} - \dfrac{7}{8} - \left(\dfrac{2}{3}\right)^2\right]$

7. Graph $y = \dfrac{4}{3}x - 3$.

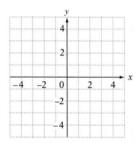

8. Graph $y = \dfrac{1}{3}x$.

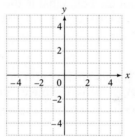

9. Subtract: $(7y^2 + 5y - 8) - (4y^2 - 3y + 1)$

10. Simplify: $(4a^2b)^3$

11. $16\dfrac{2}{3}\%$ of what number is 24?

12. Solve: $\dfrac{9}{8} = \dfrac{3}{n}$

13. Write 87,600,000,000 in scientific notation.

14. Find the perimeter of the square shown below.

12 in.

15. Multiply: $(5c^2d^4)(-3cd^6)$

16. Convert 40 km to meters.

17. What is the measure of $\angle n$ in the figure below?

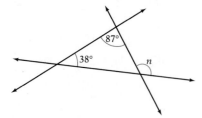

18. Find the area of the parallelogram shown below.

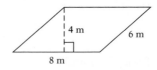

Solve.

19. *Simple Interest*   Find the simple interest on a 3-month loan of $25,000 at an annual interest rate of 7.5%.

20. *Probability*   A box contains 12 white, 15 blue, and 9 red balls. If one ball is randomly chosen from the box, what is the probability that the ball is not white?

21. *Education*   The scores for six students on an achievement test were 24, 38, 22, 34, 37, and 31. What were the mean and median scores for these students?

22. *Elections*   The results of a recent city election showed that 55,000 people voted out of a possible 230,000 registered voters. What percent of the registered voters did not vote in the election? Round to the nearest tenth of a percent.

23. *Measurement*   Eratosthenes (circa 300 B.C.), a Greek mathematician, calculated that an angle of 7.5° at Earth's center cuts an arc of 1 600 km on Earth's surface. Using this information, what is the approximate circumference of Earth?

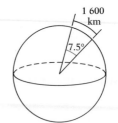

24. *Meteorology*   The annual rainfall totals, in inches, for a certain region for the last five years were 12, 16, 20, 18, and 14. Find the standard deviation of these rainfall totals.

25. *Compensation*   A chef's helper received a 10% hourly wage increase to $19.80 per hour. What was the chef's helper's hourly wage before the increase?

Final Examination

1. Estimate the sum of 672, 843, 509, and 417.

2. Simplify: $18 + 3(6 - 4)^2 \div 2$

3. Simplify: $-8 - (-13) - 10 + 7$

4. Evaluate $|a - b| - 3bc^3$ for $a = -2$, $b = 4$, and $c = -1$.

5. What is $5\frac{3}{8}$ minus $2\frac{11}{16}$?

6. Find the quotient of $\frac{7}{9}$ and $\frac{5}{6}$.

7. Simplify: $\dfrac{\frac{3}{4} - \frac{1}{2}}{\frac{5}{8} + \frac{1}{2}}$

8. Place the correct symbol, $<$ or $>$, between the two numbers.

   $\frac{5}{16}$ 0.313

9. Evaluate $-10qr$ for $q = -8.1$ and $r = -9.5$.

10. Divide and round to the nearest hundredth: $-15.32 \div 4.67$

11. Is -0.5 a solution of the equation $-90y = 45$?

12. Simplify: $\sqrt{162}$

13. Graph $x \geq -4$.

14. Simplify: $-\frac{5}{6}(-12t)$

15. Simplify: $2(x - 3y) - 4(x + 2y)$

16. Subtract: $(5z^3 + 2z^2 - 1) - (4z^3 + 6z - 8)$

17. Multiply: $(4x^2)(2x^5y)$

18. Multiply: $2a^2b^2(5a^2 - 3ab + 4b^2)$

19. Multiply: $(3x - 2)(5x + 3)$

20. Simplify: $(3x^2y)^4$

21. Evaluate: 4^{-3}

22. Simplify: $\dfrac{m^5n^8}{m^3n^4}$

23. Solve: $2 - \dfrac{4}{3}y = 10$

24. Solve: $6z + 8 = 5 - 3z$

25. Solve: $8 + 2(6c - 7) = 4$

26. Convert 2.48 m to centimeters.

27. Convert 2.6 mi to feet.

28. Solve: $\dfrac{n + 2}{8} = \dfrac{5}{12}$

29. Given that $\ell_1 \| \ell_2$, find the measures of angles a and b.

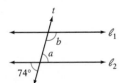

30. Find the unknown side of the triangle. Round to the nearest tenth.

31. Find the perimeter of the rectangle.

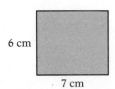

32. Find the volume of the rectangular solid.

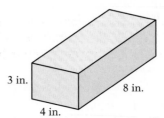

33. Graph $y = -2x + 3$.

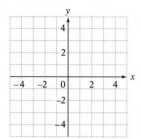

34. Graph $y = \frac{3}{5}x - 4$.

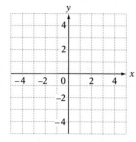

35. *Physics* Find the ground speed of an airplane traveling into a 22-mph wind with an air speed of 386 mph. Use the formula $g = a - h$, where g is the ground speed, a is the air speed, and h is the speed of the headwind.

36. *Manufacturing* A factory worker can inspect a product in $1\frac{1}{2}$ min. How many products can the worker inspect during an 8-hour day?

37. *Chemistry* The boiling point of bromine is 58.78°C. The melting point of bromine is −7.2°C. Find the difference between the boiling point and the melting point of bromine.

38. *Physics* One light-year, which is the distance that light travels through empty space in one year, is approximately 5,880,000,000,000 mi. Write this number in scientific notation.

39. *Physics* Two children are sitting on a seesaw that is 10 ft long. One child weighs 50 lb and the second child weighs 75 lb. How far from the 50-pound child should the fulcrum be placed so that the seesaw balances? Use the formula $F_1x = F_2(d - x)$.

40. *Consumerism* The fee charged by a ticketing agency for a concert is $10.50 plus $52.50 for each ticket purchased. If your total charge for tickets is $325.50, how many tickets are you purchasing?

41. *Taxes* The property tax on a $250,000 house is $3,750. At this rate, what is the property tax on a home appraised at $314,000?

42. 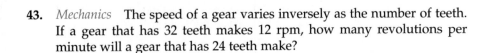 *Geography* The figure at the right represents the land area of the states in the United States. What percent of the states have a land area of 75,000 mi² or more?

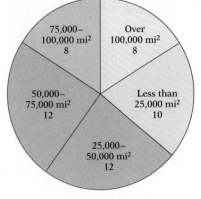

43. *Mechanics* The speed of a gear varies inversely as the number of teeth. If a gear that has 32 teeth makes 12 rpm, how many revolutions per minute will a gear that has 24 teeth make?

Land Area of the States in the United States

44. *Consumerism* A customer purchased a car for $32,500 and paid a sales tax of 5.5% of the cost. Find the total cost of the car including sales tax.

45. *Economics* Due to a recession, the number of housing starts in a community decreased from 124 to 96. What percent decrease does this represent? Round to the nearest tenth of a percent.

46. *Business* A necklace with a regular price of $245 is on sale for 35% off the regular price. Find the sale price.

47. *Finances* Find the simple interest on a 9-month loan of $25,000 at an annual interest rate of 8.6%.

48. *Labor Force* The numbers of hours per week that 80 twelfth-grade students spend at paid jobs are given in the figure at the right. What percent of the students work more than 15 h per week?

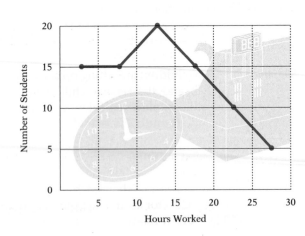

49. *Insurance* You requested rates for term life insurance from five different insurance companies. The annual premiums were $297, $425, $362, $281, and $309. Calculate the mean and median annual premiums for these five insurance companies.

50. *Probability* If two dice are tossed, what is the probability that the sum of the dots on the upward faces is divisible by 3?

▶ Subtract the fraction that named glazed, filled, or frosted from 1, the entire group surveyed.

Solution $\dfrac{2}{5} + \dfrac{8}{25} + \dfrac{3}{20} = \dfrac{40}{100} + \dfrac{32}{100} + \dfrac{15}{100}$

$$= \dfrac{87}{100}$$

$$1 - \dfrac{87}{100} = \dfrac{100}{100} - \dfrac{87}{100} = \dfrac{13}{100}$$

$\dfrac{13}{100}$ of the respondents did not name glazed, filled, or frosted as their favorite type of doughnut.

Section 3.5 *(pages 207–210)*

You Try It 1 $\quad -\dfrac{1}{5} = z - \dfrac{5}{6}$

$$-\dfrac{1}{5} + \dfrac{5}{6} = z - \dfrac{5}{6} + \dfrac{5}{6}$$

$$-\dfrac{6}{30} + \dfrac{25}{30} = z$$

$$\dfrac{-6 + 25}{30} = z$$

$$\dfrac{19}{30} = z$$

The solution is $\dfrac{19}{30}$.

You Try It 2 $\quad 26 = 4x$

$$\dfrac{26}{4} = \dfrac{4x}{4}$$

$$\dfrac{13}{2} = x$$

$$6\dfrac{1}{2} = x$$

The solution is $6\dfrac{1}{2}$.

You Try It 3 The unknown number: x

Negative five-sixths	is equal to	ten-thirds of a number

$$-\dfrac{5}{6} = \dfrac{10}{3}x$$

$$\dfrac{3}{10}\left(-\dfrac{5}{6}\right) = \dfrac{3}{10} \cdot \dfrac{10}{3}x$$

$$-\dfrac{3 \cdot 5}{10 \cdot 6} = x$$

$$-\dfrac{1}{4} = x$$

The number is $-\dfrac{1}{4}$.

You Try It 4

Strategy To find the total number of software products sold in January, write and solve an equation using s to represent the number of software products sold in January.

Solution

The number of computer software games sold in January	was	three-fifths of all the software products sold

$$450 = \dfrac{3}{5}s$$

$$\dfrac{5}{3} \cdot 450 = \dfrac{5}{3} \cdot \dfrac{3}{5}s$$

$$750 = s$$

BAL Software sold a total of 750 software products in January.

You Try It 5

Strategy To find the total number of points scored, replace A by 73 and N by 5 in the given formula and solve for T.

Solution $\quad A = \dfrac{T}{N}$

$$73 = \dfrac{T}{5}$$

$$5 \cdot 73 = 5 \cdot \dfrac{T}{5}$$

$$365 = T$$

The total number of points scored was 365.

Section 3.6 *(pages 214–219)*

You Try It 1 $\quad \left(\dfrac{2}{9}\right)^2 \cdot (-3)^4 = \dfrac{2}{9} \cdot \dfrac{2}{9} \cdot (-3)(-3)(-3)(-3)$

$$= \dfrac{2}{9} \cdot \dfrac{2}{9} \cdot 3 \cdot 3 \cdot 3 \cdot 3$$

$$= \dfrac{2}{9} \cdot \dfrac{2}{9} \cdot \dfrac{3}{1} \cdot \dfrac{3}{1} \cdot \dfrac{3}{1} \cdot \dfrac{3}{1}$$

$$= \dfrac{2 \cdot 2 \cdot 3 \cdot 3 \cdot 3 \cdot 3}{9 \cdot 9 \cdot 1 \cdot 1 \cdot 1 \cdot 1} = 4$$

You Try It 2 $\quad x^4 y^3$

$$\left(2\dfrac{1}{3}\right)^4 \cdot \left(\dfrac{3}{7}\right)^3 = \left(\dfrac{7}{3}\right)^4 \cdot \left(\dfrac{3}{7}\right)^3$$

$$= \dfrac{7}{3} \cdot \dfrac{7}{3} \cdot \dfrac{7}{3} \cdot \dfrac{7}{3} \cdot \dfrac{3}{7} \cdot \dfrac{3}{7} \cdot \dfrac{3}{7}$$

$$= \dfrac{7 \cdot 7 \cdot 7 \cdot 7 \cdot 3 \cdot 3 \cdot 3}{3 \cdot 3 \cdot 3 \cdot 3 \cdot 7 \cdot 7 \cdot 7} = \dfrac{7}{3} = 2\dfrac{1}{3}$$

You Try It 3
$$\frac{2y - 3}{y} = -2$$

$$\frac{2\left(-\dfrac{1}{2}\right) - 3}{-\dfrac{1}{2}} \;\bigg|\; -2 \qquad \blacktriangleright \text{ Replace } y \text{ with } -\dfrac{1}{2}.$$

$$\frac{-1 - 3}{-\dfrac{1}{2}} \;\bigg|\; -2 \qquad \blacktriangleright \text{ Simplify the complex fraction.}$$

$$\frac{-4}{-\dfrac{1}{2}} \;\bigg|\; -2$$

$$-4(-2) \;\bigg|\; -2$$

$$8 \neq -2$$

No, $-\dfrac{1}{2}$ is not a solution of the equation.

You Try It 4 $\dfrac{x}{y - z}$

$$\frac{2\dfrac{4}{9}}{3 - 1\dfrac{1}{3}} = \frac{\dfrac{22}{9}}{\dfrac{5}{3}} = \frac{22}{9} \div \frac{5}{3} = \frac{22}{9} \cdot \frac{3}{5}$$

$$= \frac{22}{15} = 1\frac{7}{15}$$

You Try It 5 $\left(-\dfrac{1}{2}\right)^3 \cdot \dfrac{7 - 3}{4 - 9} + \dfrac{4}{5}$

$$= \left(-\frac{1}{2}\right)^3 \cdot \frac{4}{-5} + \frac{4}{5} \qquad \blacktriangleright \text{ Simplify } \frac{7 - 3}{4 - 9}.$$

$$= -\frac{1}{8} \cdot \frac{4}{-5} + \frac{4}{5} \qquad \blacktriangleright \text{ Simplify } \left(-\frac{1}{2}\right)^3.$$

$$= \frac{1}{10} + \frac{4}{5} = \frac{1}{10} + \frac{8}{10} = \frac{9}{10}$$

Solutions to Chapter 4   *You Try Its*

Section 4.1 *(pages 237–242)*

You Try It 1   The digit 4 is in the thousandths place.

You Try It 2 $\dfrac{501}{1,000} = 0.501$
(five hundred one thousandths)

You Try It 3 $0.67 = \dfrac{67}{100}$ (sixty-seven hundredths)

You Try It 4   fifty-five and six thousand eighty-three ten-thousandths

You Try It 5   806.00491
↑ hundred-thousandths place

You Try It 6   0.065 = 0.0650
0.0650 < 0.0802
0.065 < 0.0802

You Try It 7   3.03, 0.33, 0.30, 3.30, 0.03
0.03, 0.30, 0.33, 3.03, 3.30
0.03, 0.3, 0.33, 3.03, 3.3

You Try It 8

3.675849

Given place value

4 < 5

3.675849 rounded to the nearest ten-thousandth is 3.6758.

You Try It 9

48.907

Given place value

0 < 5

48.907 rounded to the nearest tenth is 48.9.

You Try It 10

31.8652

Given place value

8 > 5

31.8652 rounded to the nearest whole number is 32.

You Try It 11

Strategy   To determine who had more home runs for every 100 times at bat, compare the numbers 7.03 and 7.09.

Solution   7.09 > 7.03

Ralph Kiner had more home runs for every 100 times at bat.

You Try It 12

Strategy   To determine the average annual precipitation to the nearest inch, round the number 2.65 to the nearest whole number.

Solution   2.65 rounded to the nearest whole number is 3.

To the nearest inch, the average annual precipitation in Yuma is 3 in.

Section 4.2 *(pages 247–251)*

You Try It 1
$$\begin{array}{r} \overset{1\ \ 1}{8.64} \\ 52.7 \\ +\ 0.39105 \\ \hline 61.73105 \end{array}$$

You Try It 2 $4.002 - 9.378 = 4.002 + (-9.378)$
$$= -5.376$$

You Try It 3
$$\begin{array}{r} \overset{4\ \ \ 9\ 10}{2\cancel{5}.\cancel{0}\cancel{0}} \\ -\ 4.91 \\ \hline 20.09 \end{array} \qquad \text{Check:} \qquad \begin{array}{r} 4.91 \\ +20.09 \\ \hline 25.00 \end{array}$$

You Try It 4
$6.514 \longrightarrow 7$
$8.903 \longrightarrow 9$
$2.275 \longrightarrow + 2$
$\overline{18}$

You Try It 5 $x + y + z$
$-7.84 + (-3.05) + 2.19$
$= -10.89 + 2.19$
$= -8.7$

You Try It 6 $-m + 16.9 = 40.7$

$\begin{array}{c|c} -(-23.8) + 16.9 & 40.7 \\ 23.8 + 16.9 & 40.7 \\ 40.7 & = 40.7 \end{array}$ ▶ Replace m
with -23.8.

Yes, -23.8 is a solution of the equation.

You Try It 7

Strategy To make the comparison:

▶ Find the number of hearing-impaired individuals under the age of 55 by adding the numbers who are aged 0–17 (1.37 million), aged 18–34 (2.77 million), aged 35–44 (4.07 million), and aged 45–54 (4.48 million).

▶ Find the number of hearing-impaired individuals who are 55 or older by adding the numbers who are aged 55–64 (4.31 million), aged 65–74 (5.41 million), aged 75 or older (3.80 million).

▶ Compare the two sums.

Solution $1.37 + 2.77 + 4.07 + 4.48 = 12.69$
$4.31 + 5.41 + 3.8 = 13.52$
$12.69 < 13.52$
The number of hearing-impaired individuals under the age of 55 is less than the number of hearing impaired who are 55 or older.

Section 4.3 *(pages 258–268)*

You Try It 1
$\begin{array}{r} 0.000081 \\ \times \quad 0.025 \\ \hline 405 \\ 162 \\ \hline 0.000002025 \end{array}$ ← 6 decimal places
← 3 decimal places

← 9 decimal places

You Try It 2
$6.407 \longrightarrow 6$
$0.959 \longrightarrow \times 1$
$\overline{6}$

You Try It 3 Move the decimal point 4 places to the right.
$1.756 \cdot 10^4 = 17,560$

You Try It 4 $(-0.7)(-5.8) = 4.06$

You Try It 5 $25xy$
$25(-0.8)(0.6) = -20(0.6) = -12$

You Try It 6
$\begin{array}{r} 48.2 \\ 6.53.\overline{)314.74.6} \\ -261\,2 \\ \hline 53\,54 \\ -52\,24 \\ \hline 1\,30\,6 \\ -1\,30\,6 \\ \hline 0 \end{array}$ Move the decimal point 2 places to the right.

You Try It 7 $62.7 \longrightarrow 60$
$3.45 \longrightarrow 3$
$60 \div 3 = 20$

You Try It 8
$\begin{array}{r} 6.0391 \approx 6.039 \\ 86\overline{)519.3700} \\ -516 \\ \hline 3\,3 \\ -\,0 \\ \hline 3\,37 \\ -2\,58 \\ \hline 790 \\ -774 \\ \hline 160 \\ -\,86 \\ \hline 74 \end{array}$

You Try It 9 Move the decimal point 2 places to the left.
$63.7 \div 100 = 0.637$

You Try It 10 The quotient is negative.
$-25.7 \div 0.31 \approx -82.9$

You Try It 11 $\dfrac{x}{y}$
$\dfrac{-40.6}{-0.7} = 58$

You Try It 12 $-2 = \dfrac{d}{-0.6}$

$\begin{array}{c|c} -2 & \dfrac{-1.2}{-0.6} \end{array}$ ▶ Replace d
with -1.2.
$-2 \neq 2$

No, -1.2 is not a solution of the equation.

You Try It 13 $\begin{array}{r} 0.8 \\ 5\overline{)4.0} \end{array}$ $\dfrac{4}{5} = 0.8$

You Try It 14 $\begin{array}{r} 0.8333 \\ 6\overline{)5.0000} \end{array}$ $1\dfrac{5}{6} = 1.8\overline{3}$

You Try It 15 $6.2 = 6\dfrac{2}{10} = 6\dfrac{1}{5}$

You Try It 16 $\dfrac{7}{12} \approx 0.5833$
$0.5880 > 0.5833$
$0.588 > \dfrac{7}{12}$

You Try It 17

Strategy   To find the change you receive:
- ▶ Multiply the number of stamps (12) by the cost of each stamp (41¢) to find the total cost of the stamps.
- ▶ Convert the total cost of the stamps to dollars and cents.
- ▶ Subtract the total cost of the stamps from $10.

Solution   12(41) = 492   The stamps cost 492¢.
492¢ = $4.92   The stamps cost $4.92.
10.00 − 4.92 = $5.08
You receive $5.08 in change.

You Try It 18

Strategy   To find the insurance premium due, replace B by 276.25 and F by 1.8 in the given formula and solve for P.

Solution $P = BF$
$P = 276.25(1.8)$
$P = 497.25$

The insurance premium due is $497.25.

Section 4.4 *(pages 279–280)*

You Try It 1

$$a - 1.23 = -6$$
$$a - 1.23 + 1.23 = -6 + 1.23$$
$$a = -4.77$$

The solution is −4.77.

You Try It 2

$$-2.13 = -0.71c$$
$$\frac{-2.13}{-0.71} = \frac{-0.71c}{-0.71}$$
$$3 = c$$

The solution is 3.

You Try It 3

Strategy   To find the assets, replace N by 24.3 and L by 17.9 in the given formula and solve for A.

Solution
$$N = A - L$$
$$24.3 = A - 17.9$$
$$24.3 + 17.9 = A - 17.9 + 17.9$$
$$42.2 = A$$

The assets of the business are $42.2 billion.

You Try It 4

Strategy   To find the markup, write and solve an equation using M to represent the amount of the markup.

Solution

The selling price	is	the sum of the amount paid by the store and the amount of the markup

$$295.50 = 223.75 + M$$
$$295.50 - 223.75 = 223.75 - 223.75 + M$$
$$71.75 = M$$

The markup is $71.75.

Section 4.5 *(pages 283–288)*

You Try It 1   Since $12^2 = 144$, $-\sqrt{144} = -12$.

You Try It 2   Since $\left(\frac{9}{10}\right)^2 = \frac{81}{100}$, $\sqrt{\frac{81}{100}} = \frac{9}{10}$.

You Try It 3 $4\sqrt{16} - \sqrt{9} = 4 \cdot 4 - 3$
$= 16 - 3 = 13$

You Try It 4 $5\sqrt{a + b}$
$5\sqrt{17 + 19} = 5\sqrt{36}$
$= 5 \cdot 6$
$= 30$

You Try It 5 $5\sqrt{23} \approx 23.9792$ ▶ Use a calculator.

You Try It 6   57 is between the perfect squares 49 and 64.
$\sqrt{49} = 7$ and $\sqrt{64} = 8$
$7 < \sqrt{57} < 8$

You Try It 7 $9^2 = 81$; 81 is too big.
$8^2 = 64$; 64 is not a factor of 80.
$7^2 = 49$; 49 is not a factor of 80.
$6^2 = 36$; 36 is not a factor of 80.
$5^2 = 25$; 25 is not a factor of 80.
$4^2 = 16$; 16 is a factor of 80. ($80 = 16 \cdot 5$)
$\sqrt{80} = \sqrt{16 \cdot 5} = \sqrt{16} \cdot \sqrt{5}$
$= 4 \cdot \sqrt{5} = 4\sqrt{5}$

You Try It 8

Strategy   To find the range, replace h by 6 in the given formula and solve for R.

Solution $R = 1.4\sqrt{h}$
$R = 1.4\sqrt{6}$
$R \approx 3.43$ ▶ Use a calculator.

The range of the periscope is 3.43 mi.

Section 4.6 *(pages 294–299)*

You Try It 1

You Try It 2

You Try It 3

You Try It 4
a. $x \geq 4$
$-1 \geq 4$ False
b. $x \geq 4$
$0 \geq 4$ False
c. $x \geq 4$
$4 \geq 4$ True
d. $x \geq 4$
$\sqrt{26} \geq 4$ True

The numbers 4 and $\sqrt{26}$ make the inequality true.

You Try It 5   All real numbers greater than −7 make the inequality $x > -7$ true.

You Try It 6

You Try It 7

Strategy
▶ To write the inequality, let s represent the speeds at which a motorist is ticketed. Motorists are ticketed at speeds greater than 55.

▶ To determine whether a motorist traveling at 58 mph will be ticketed, replace s in the inequality by 58. If the inequality is true, the motorist will be ticketed. If the inequality is false, the motorist will not be ticketed.

Solution
$s > 55$
$58 > 55$ True

Yes, a motorist traveling at 58 mph will be ticketed.

Solutions to Chapter 5   *You Try Its*

Section 5.1 *(pages 317–322)*

You Try It 1 $-6(-3p) = [-6(-3)]p = 18p$

You Try It 2 $(-2m)(-8n) = [(-2)(-8)](m \cdot n)$
$= 16mn$

You Try It 3 $(-12)(-d) = (-12)(-1d)$
$= [(-12)(-1)]d$
$= 12d$

You Try It 4 $6n + 9 + (-6n) = 6n + (-6n) + 9$
$= [6n + (-6n)] + 9$
$= 0 + 9$
$= 9$

You Try It 5 $-7(2k - 5) = -7(2k) - (-7)(5)$
$= -14k + 35$

You Try It 6 $-4(x - 2y) = -4(x) - (-4)(2y)$
$= -4x + 8y$

You Try It 7 $3(-2v + 3w - 7) = 3(-2v) + 3(3w) - 3(7)$
$= -6v + 9w - 21$

You Try It 8 $-4(2x - 7y - z)$
$= -4(2x) - (-4)(7y) - (-4)(z)$
$= -8x + 28y + 4z$

You Try It 9 $-(c - 9d + 1) = -c + 9d - 1$

Section 5.2 *(pages 327–330)*

You Try It 1 $\dfrac{x}{5} + \dfrac{2x}{5} = \dfrac{x + 2x}{5} = \dfrac{1x + 2x}{5} = \dfrac{(1 + 2)x}{5} = \dfrac{3x}{5}$

You Try It 2 $12a^2 - 8a + 3 - 16a^2 + 8a$
$= 12a^2 - 16a^2 - 8a + 8a + 3$
$= -4a^2 + 0a + 3$
$= -4a^2 + 3$

You Try It 3 $-7x^2 + 4xy + 8x^2 - 12xy$
$= -7x^2 + 8x^2 + 4xy - 12xy$
$= x^2 - 8xy$

You Try It 4 $-2r + 7s - 12 - 8r + s + 8$
$= -2r - 8r + 7s + s - 12 + 8$
$= -10r + 8s - 4$

You Try It 5 $8x^2y - 15xy^2 + 12xy^2 - 7x^2y$
$= 8x^2y - 7x^2y - 15xy^2 + 12xy^2$
$= x^2y - 3xy^2$

You Try It 6 $6 - 4(2x - y) + 3(x - 4y)$
$= 6 - 8x + 4y + 3x - 12y$
$= -5x - 8y + 6$

You Try It 7 $8c - 4(3c - 8) - 5(c + 4)$
$= 8c - 12c + 32 - 5c - 20$
$= -9c + 12$

You Try It 8 $6p + 5[3(2 - 3p) - 2(5 - 4p)]$
$= 6p + 5[6 - 9p - 10 + 8p]$
$= 6p + 5[-p - 4]$
$= 6p - 5p - 20$
$= p - 20$

Section 5.3 *(pages 335–338)*

You Try It 1
$(-4x^3 + 2x^2 - 8) + (4x^3 + 6x^2 - 7x + 5)$
$= (-4x^3 + 4x^3) + (2x^2 + 6x^2) - 7x + (-8 + 5)$
$= 8x^2 - 7x - 3$

You Try It 2 $\begin{array}{r} 6x^3 \phantom{{}+ 2x} + 2x + 8 \\ -9x^3 + 2x^2 - 12x - 8 \\ \hline -3x^3 + 2x^2 - 10x \phantom{{}+ 8} \end{array}$

You Try It 3 $(6a^4 - 5a^2 + 7) + (8a^4 + 3a^2 - 1)$
$= (6a^4 + 8a^4) + (-5a^2 + 3a^2) + (7 - 1)$
$= 14a^4 - 2a^2 + 6$

You Try It 4
The opposite of $3w^3 - 4w^2 + 2w - 1$ is
$-3w^3 + 4w^2 - 2w + 1$.

Add the opposite of $3w^3 - 4w^2 + 2w - 1$ to the first polynomial.
$(-4w^3 + 8w - 8) - (3w^3 - 4w^2 + 2w - 1)$
$= (-4w^3 + 8w - 8) + (-3w^3 + 4w^2 - 2w + 1)$
$= (-4w^3 - 3w^3) + 4w^2 + (8w - 2w) + (-8 + 1)$
$= -7w^3 + 4w^2 + 6w - 7$

You Try It 5
Add the opposite of $4y^2 - 6y - 9$ to $13y^3 - 6y - 7$.
$\begin{array}{r} 13y^3 \phantom{{}- 4y^2} - 6y - 7 \\ -4y^2 + 6y + 9 \\ \hline 13y^3 - 4y^2 \phantom{{}- 6y} + 2 \end{array}$

You Try It 6 $(-6n^4 + 5n^2 - 10) - (4n^2 + 2)$
$= (-6n^4 + 5n^2 - 10) + (-4n^2 - 2)$
$= -6n^4 + n^2 - 12$

You Try It 7 $(5y^2 - y) + (7y^2 + 4) = (5y^2 + 7y^2) - y + 4$
$$= 12y^2 - y + 4$$

The distance from Dover to Farley is $(12y^2 - y + 4)$ miles.

Section 5.4 *(pages 344–347)*

You Try It 1 $(-7a^4)(4a^2) = [-7(4)](a^4 \cdot a^2)$
$$= -28a^{4+2}$$
$$= -28a^6$$

You Try It 2 $(8m^3n)(-3n^5) = [8(-3)](m^3)(n \cdot n^5)$
$$= -24m^3n^{1+5}$$
$$= -24m^3n^6$$

You Try It 3 $(12p^4q^3)(-3p^5q^2) = [12(-3)](p^4 \cdot p^5)(q^3 \cdot q^2)$
$$= -36p^{4+5}q^{3+2}$$
$$= -36p^9q^5$$

You Try It 4 $(-y^4)^5 = [(-1)y^4]^5$
$$= (-1)^{1 \cdot 5}y^{4 \cdot 5}$$
$$= (-1)^5 y^{20}$$
$$= -1y^{20}$$
$$= -y^{20}$$

You Try It 5 $(-3a^4bc^2)^3 = (-3)^{1 \cdot 3}a^{4 \cdot 3}b^{1 \cdot 3}c^{2 \cdot 3}$
$$= (-3)^3 a^{12}b^3c^6 = -27a^{12}b^3c^6$$

Section 5.5 *(pages 350–351)*

You Try It 1 $-3a(-6a + 5b) = (-3a)(-6a) + (-3a)(5b)$
$$= 18a^2 - 15ab$$

You Try It 2 $3mn^2(2m^2 - 3mn - 1)$
$$= (3mn^2)(2m^2) - (3mn^2)(3mn) - (3mn^2)1$$
$$= 6m^3n^2 - 9m^2n^3 - 3mn^2$$

You Try It 3 $(3c + 7)(3c - 7)$
$$= (3c)(3c) + (3c)(-7) + 7(3c) + (7)(-7)$$
$$= 9c^2 - 21c + 21c - 49$$
$$= 9c^2 - 49$$

Section 5.6 *(pages 354–357)*

You Try It 1 **a.** $\dfrac{1}{d^{-6}} = d^6$

b. $\dfrac{n^6}{n^{11}} = n^{6-11} = n^{-5} = \dfrac{1}{n^5}$

You Try It 2 **a.** $4^{-2} = \dfrac{1}{4^2} = \dfrac{1}{16}$

b. $-8x^0 = -8(1) = -8$

You Try It 3 $0.000000961 = 9.61 \times 10^{-7}$

You Try It 4 $7.329 \times 10^6 = 7,329,000$

Section 5.7 *(pages 361–364)*

You Try It 1 <u>twice</u> x <u>divided by</u> the <u>difference</u> between x and 7

$$\dfrac{2x}{x - 7}$$

You Try It 2 the <u>product</u> of negative three and the <u>square</u> of d

$$-3d^2$$

You Try It 3 The smaller number is x.
The larger number is $16 - x$.
the <u>difference</u> between the larger number and <u>twice</u> the smaller number
$$(16 - x) - 2x$$
$$-3x + 16$$

You Try It 4 Let the unknown number be x.
the <u>difference</u> between fourteen and the <u>sum</u> of x and seven
$$14 - (x + 7)$$
$$14 - x - 7$$
$$-x + 7$$

You Try It 5 the pounds of caramel: c
the pounds of milk chocolate: $c + 3$

Solutions to Chapter 6   *You Try Its*

Section 6.1 *(pages 381–386)*

You Try It 1
$$7 + y = 12$$
$$7 - 7 + y = 12 - 7 \quad \blacktriangleright \text{Subtract 7.}$$
$$y = 5$$

The solution is 5.

You Try It 2
$$b - \frac{3}{8} = \frac{1}{2}$$
$$b - \frac{3}{8} + \frac{3}{8} = \frac{1}{2} + \frac{3}{8} \quad \blacktriangleright \text{Add } \frac{3}{8}.$$
$$b = \frac{4}{8} + \frac{3}{8}$$
$$b = \frac{7}{8}$$

The solution is $\dfrac{7}{8}$.

You Try It 3
$$-5r + 3 + 6r = 1$$
$$r + 3 = 1 \quad \blacktriangleright \text{Combine like terms.}$$
$$r + 3 - 3 = 1 - 3 \quad \blacktriangleright \text{Subtract 3.}$$
$$r = -2$$

The solution is -2.

You Try It 4

$$-60 = 5d$$
$$\frac{-60}{5} = \frac{5d}{5} \quad \blacktriangleright \text{ Divide by 5.}$$
$$-12 = d$$

The solution is -12.

You Try It 5

$$10 = \frac{-2x}{5}$$
$$\left(-\frac{5}{2}\right)10 = \left(-\frac{5}{2}\right)\left(-\frac{2}{5}x\right) \quad \blacktriangleright -\frac{2x}{5} = -\frac{2}{5}x$$
$$-25 = x$$

The solution is -25.

You Try It 6

$$\frac{1}{3}x - \frac{5}{6}x = 4$$
$$\frac{2}{6}x - \frac{5}{6}x = 4 \qquad \blacktriangleright \text{ Combine like terms.}$$
$$-\frac{1}{2}x = 4 \qquad \blacktriangleright -\frac{3}{6} = -\frac{1}{2}$$
$$-2\left(-\frac{1}{2}x\right) = -2(4) \quad \blacktriangleright \text{ Multiply by } -2.$$
$$x = -8$$

Check:

$$\frac{1}{3}x - \frac{5}{6}x = 4$$

$$\begin{array}{c|c} \frac{1}{3}(-8) - \frac{5}{6}(-8) & 4 \\ \hline -\frac{8}{3} - \left(-\frac{20}{3}\right) & 4 \\ -\frac{8}{3} + \frac{20}{3} & 4 \\ 4 & = 4 \end{array}$$

-8 checks as the solution.

The solution is -8.

Section 6.2 *(pages 389–390)*

You Try It 1

$$5v + 3 - 9v = 9$$
$$-4v + 3 = 9 \qquad \blacktriangleright \text{ Combine like terms.}$$
$$-4v + 3 - 3 = 9 - 3 \qquad \blacktriangleright \text{ Subtract 3.}$$
$$-4v = 6$$
$$\frac{-4v}{-4} = \frac{6}{-4} \qquad \blacktriangleright \text{ Divide by } -4.$$
$$v = -\frac{3}{2}$$

The solution is $-\frac{3}{2}$.

You Try It 2

Strategy   To find the pressure, replace P by its value and solve for D. $P = 45$.

Solution

$$P = 15 + \frac{1}{2}D$$
$$45 = 15 + \frac{1}{2}D$$

$$45 - 15 = 15 - 15 + \frac{1}{2}D$$
$$30 = \frac{1}{2}D$$
$$2(30) = 2\left(\frac{1}{2}D\right)$$
$$60 = D$$

When the pressure is 45 pounds per square inch, the depth is 60 ft.

Section 6.3 *(pages 395–398)*

You Try It 1

$$r - 7 = 5 - 3r$$
$$r + 3r - 7 = 5 - 3r + 3r \quad \blacktriangleright \text{ Add } 3r.$$
$$4r - 7 = 5$$
$$4r - 7 + 7 = 5 + 7 \qquad \blacktriangleright \text{ Add 7.}$$
$$4r = 12$$
$$\frac{4r}{4} = \frac{12}{4} \qquad \blacktriangleright \text{ Divide by 4.}$$
$$r = 3$$

The solution is 3.

You Try It 2

$$4a - 2 + 5a = 2a - 2 + 3a \quad \blacktriangleright \text{ Combine}$$
$$9a - 2 = 5a - 2 \qquad\qquad \text{ like terms.}$$
$$9a - 5a - 2 = 5a - 5a - 2 \quad \blacktriangleright \text{ Subtract } 5a.$$
$$4a - 2 = -2$$
$$4a - 2 + 2 = -2 + 2 \qquad \blacktriangleright \text{ Add 2.}$$
$$4a = 0$$
$$\frac{4a}{4} = \frac{0}{4} \qquad\qquad \blacktriangleright \text{ Divide by 4.}$$
$$a = 0$$

The solution is 0.

You Try It 3

$$6 - 5(3y + 2) = 26$$
$$6 - 15y - 10 = 26 \qquad \blacktriangleright \text{ Distributive Property}$$
$$-15y - 4 = 26 \qquad \blacktriangleright \text{ Combine like terms.}$$
$$-15y - 4 + 4 = 26 + 4 \quad \blacktriangleright \text{ Add 4.}$$
$$-15y = 30$$
$$\frac{-15y}{-15} = \frac{30}{-15} \qquad \blacktriangleright \text{ Divide by } -15.$$
$$y = -2$$

The solution is -2.

You Try It 4

$$2w - 7(3w + 1) = 5(5 - 3w)$$
$$2w - 21w - 7 = 25 - 15w \qquad\qquad \blacktriangleright \text{ Distributive}$$
$$-19w - 7 = 25 - 15w \qquad\qquad\qquad \text{Property}$$
$$-19w + 15w - 7 = 25 - 15w + 15w \quad \blacktriangleright \text{ Add } 15w.$$
$$-4w - 7 = 25$$
$$-4w - 7 + 7 = 25 + 7 \qquad \blacktriangleright \text{ Add 7.}$$
$$-4w = 32$$
$$\frac{-4w}{-4} = \frac{32}{-4} \qquad\qquad \blacktriangleright \text{ Divide by } -4.$$
$$w = -8$$

The solution is -8.

You Try It 5

Strategy   To find the location of the fulcrum when the system balances, replace the variables F_1, F_2, and d in the lever system equation by the given values and solve for x. $F_1 = 45$, $F_2 = 80$, $d = 25$.

Solution

$$F_1 x = F_2(d - x)$$
$$45x = 80(25 - x)$$
$$45x = 2000 - 80x$$
$$45x + 80x = 2000 - 80x + 80x$$
$$125x = 2000$$
$$\frac{125x}{125} = \frac{2000}{125}$$
$$x = 16$$

The fulcrum is 16 ft from the 45-pound force.

Section 6.4 *(pages 403–406)*

You Try It 1

The unknown number: x

Six more than one-half a number	is	the total of the number and nine

$$\frac{1}{2}x + 6 = x + 9$$
$$\frac{1}{2}x - x + 6 = x - x + 9$$
$$-\frac{1}{2}x + 6 = 9$$
$$-\frac{1}{2}x + 6 - 6 = 9 - 6$$
$$-\frac{1}{2}x = 3$$
$$(-2)\left(-\frac{1}{2}x\right) = (-2)3$$
$$x = -6$$

-6 checks as the solution. The solution is -6.

You Try It 2

The unknown number: x

Seven less than a number	is equal to	five more than three times the number

$$x - 7 = 3x + 5$$
$$x - 3x - 7 = 3x - 3x + 5$$
$$-2x - 7 = 5$$
$$-2x - 7 + 7 = 5 + 7$$
$$-2x = 12$$
$$\frac{-2x}{-2} = \frac{12}{-2}$$
$$x = -6$$

-6 checks as the solution.
The solution is -6.

You Try It 3

The smaller number: n
The larger number: $14 - n$

One more than three times the smaller number	equals	the sum of the larger number and three

$$3n + 1 = (14 - n) + 3$$
$$3n + 1 = 17 - n$$
$$3n + n + 1 = 17 - n + n$$
$$4n + 1 = 17$$
$$4n + 1 - 1 = 17 - 1$$
$$4n = 16$$
$$\frac{4n}{4} = \frac{16}{4}$$
$$n = 4$$

$14 - n = 14 - 4 = 10$

These numbers check as solutions.

The smaller number is 4.
The larger number is 10.

You Try It 4

Strategy   To find the number of pages in the Internal Revenue Service's tax code in 2001, write and solve an equation using n to represent the number of pages in the tax code in 2001.

Solution

66,498	is	20,836 pages more than in 2001

$$66,498 = n + 20,836$$
$$66,498 - 20,836 = n + 20,836 - 20,836$$
$$45,662 = n$$

In 2001, the Internal Revenue Service's tax code was 45,662 pages long.

You Try It 5

Strategy   To find the number of tickets that you are purchasing, write and solve an equation using x to represent the number of tickets purchased.

Solution

$9.50 plus $57.50 for each ticket	equals	$527

$$9.50 + 57.50x = 527$$
$$9.50 - 9.50 + 57.50x = 527 - 9.50$$
$$57.50x = 517.50$$
$$\frac{57.50x}{57.50} = \frac{517.50}{57.50}$$
$$x = 9$$

You are purchasing 9 tickets.

You Try It 6

Strategy   To find the number of days, write and solve an equation using d to represent the number of days.

Solution

10 + 5.50 per day	equals	43

$$10 + 5.50d = 43$$
$$10 - 10 + 5.50d = 43 - 10$$

$$5.50d = 33$$
$$\frac{5.50d}{5.50} = \frac{33}{5.50}$$
$$d = 6$$

The customer posted the ad for 6 days.

You Try It 7

Strategy   To find the length of each piece, write and solve an equation using x to represent the length of the shorter piece and $18 - x$ to represent the length of the longer piece.

Solution

1 ft more than twice the shorter piece	is	2 ft less than the longer piece

$$2x + 1 = (18 - x) - 2$$
$$2x + 1 = 16 - x$$
$$2x + x + 1 = 16 - x + x$$
$$3x + 1 = 16$$
$$3x + 1 - 1 = 16 - 1$$
$$3x = 15$$
$$\frac{3x}{3} = \frac{15}{3}$$
$$x = 5$$

$$18 - x = 18 - 5 = 13$$

The length of the shorter piece is 5 ft and the length of the longer piece is 13 ft.

Section 6.5 *(pages 411–414)*

You Try It 1

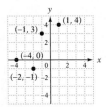

You Try It 2

$A(4, 2)$
$B(-3, 4)$
$C(-3, 0)$
$D(0, 0)$

You Try It 3

Strategy   Graph the ordered pairs on a rectangular coordinate system, where the horizontal axis represents the number of yards gained and the vertical axis represents the number of points scored.

Solution

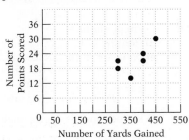

You Try It 4

Locate 20 in. on the x-axis. Follow the vertical line from 20 to a point plotted in the diagram. Follow a horizontal line from that point to the y-axis. Read the number where that line intersects the y-axis.

The ordered pair is (20, 37), which indicates that when the space between seats was 20 in., the evacuation time was 37 s.

Section 6.6 *(pages 420–423)*

You Try It 1

Replace x by -2 and y by 4.

$$y = -\frac{1}{2}x + 3$$

4	$\left(-\frac{1}{2}\right)(-2) + 3$
4	$1 + 3$
$4 = 4$	

Yes, $(-2, 4)$ is a solution of the equation $y = -\frac{1}{2}x + 3$.

You Try It 2

$y = 2x - 3$
$y = 2(0) - 3$ ▶ Replace x by 0.
$y = 0 - 3$
$y = -3$

The ordered-pair solution is $(0, -3)$.

You Try It 3

$y = 3x + 1$

x	y
0	1
-1	-2
1	4

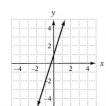

You Try It 4

Locate 5 on the y-axis. Follow the horizontal line from 5 to a point plotted on the graph. Follow a vertical line from that point to the x-axis. Read the number where that line intersects the x-axis.

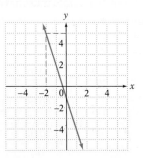

The x value is -2.

Solutions to Chapter 7   *You Try Its*

Section 7.1 *(pages 441–444)*

You Try It 1   The gram is the basic unit for measuring mass.

The amount of protein in a glass of milk is measured in grams.

You Try It 2   **a.** 1 295 m = 1.295 km
b. 7 543 g = 7.543 kg
c. 6.3 L = 6 300 ml
d. 2 kl = 2 000 L

You Try It 3

Strategy   To find the number of grams of cholesterol in one dozen eggs:

▶ Multiply the amount of cholesterol in one egg (274 mg) by number of eggs (12). This will be the amount of cholesterol in milligrams.

▶ Convert milligrams to grams.

Solution   274(12) = 3 288

3 288 mg = 3.288 g

One dozen eggs contain 3.288 g of cholesterol.

Section 7.2 *(pages 449–450)*

You Try It 1 $\dfrac{12}{20} = \dfrac{3}{5}$

12:20 = 3:5
12 to 20 = 3 to 5

You Try It 2 $\dfrac{20 \text{ bags}}{8 \text{ acres}} = \dfrac{5 \text{ bags}}{2 \text{ acres}}$

You Try It 3 $\dfrac{\$8.96}{3.5 \text{ lb}}$

8.96 ÷ 3.5 = 2.56

The unit rate is \$2.56/lb.

Section 7.3 *(pages 453–458)*

You Try It 1   The equivalence is 1 ft = 12 in.

The conversion rate must have feet in the numerator and inches in the denominator:

$\dfrac{1 \text{ ft}}{12 \text{ in.}}$

$40 \text{ in.} = 40 \text{ in.} \cdot 1 = \dfrac{40 \text{ in.}}{1} \cdot \dfrac{1 \text{ ft}}{12 \text{ in.}}$

$= \dfrac{40 \text{ in.} \cdot 1 \text{ ft}}{12 \text{ in.}}$

$= \dfrac{40 \text{ ft}}{12} = 3\dfrac{1}{3} \text{ ft}$

You Try It 2   The equivalence is 1 yd = 36 in.

The conversion rate must have inches in the numerator and yards in the denominator:

$\dfrac{36 \text{ in.}}{1 \text{ yd}}$

$2\dfrac{1}{2} \text{ yd} = 2\dfrac{1}{2} \text{ yd} \cdot 1 = \dfrac{5}{2} \text{ yd} \cdot \dfrac{36 \text{ in.}}{1 \text{ yd}}$

$= \dfrac{5 \text{ yd}}{2} \cdot \dfrac{36 \text{ in.}}{1 \text{ yd}}$

$= \dfrac{5 \text{ yd} \cdot 36 \text{ in.}}{2 \cdot 1 \text{ yd}}$

$= \dfrac{5 \cdot 36 \text{ in.}}{2}$

$= 90 \text{ in.}$

You Try It 3   We need to convert 1 mi to feet and feet to yards. The equivalences are 1 mi = 5,280 ft and 1 yd = 3 ft.

Choose the conversion rates so that we can divide by the unit "miles" and by the unit "feet."

$1 \text{ mi} = 1 \text{ mi} \cdot 1 \cdot 1$

$= \dfrac{1 \text{ mi}}{1} \cdot \dfrac{5,280 \text{ ft}}{1 \text{ mi}} \cdot \dfrac{1 \text{ yd}}{3 \text{ ft}}$

$= \dfrac{1 \text{ mi} \cdot 5,280 \text{ ft} \cdot 1 \text{ yd}}{1 \cdot 1 \text{ mi} \cdot 3 \text{ ft}}$

$= \dfrac{5,280 \text{ yd}}{3} = 1,760 \text{ yd}$

You Try It 4

Strategy   To find the number of gallons:

▶ Use the formula $V = LWH$ to find the volume in cubic inches.

▶ Use the conversion factor $\dfrac{1 \text{ gal}}{231 \text{ in}^3}$ to convert cubic inches to gallons.

Solution $V = LWH = 36 \text{ in.} \cdot 24 \text{ in.} \cdot 16 \text{ in.} = 13{,}824 \text{ in}^3$

$13{,}824 \text{ in}^3 = \dfrac{13{,}824 \text{ in}^3}{1} \cdot \dfrac{1 \text{ gal}}{231 \text{ in}^3} \approx 59.8 \text{ gal}$

The fishtank holds 59.8 gal of water.

You Try It 5 $45 \text{ cm} = \dfrac{45 \text{ cm}}{1}$

$= \dfrac{45 \text{ cm}}{1} \cdot \dfrac{1 \text{ in.}}{2.54 \text{ cm}}$

$= 17.72 \text{ in.}$

You Try It 6 $75 \text{ km per hour} = \dfrac{75 \text{ km}}{1 \text{ h}}$

$\approx \dfrac{75 \text{ km}}{1 \text{ h}} \cdot \dfrac{1 \text{ mi}}{1.61 \text{ km}}$

$\approx \dfrac{46.58 \text{ mi}}{\text{h}}$

75 km per hour is approximately 46.58 mph.

You Try It 7 $3.59 \text{ per gallon} = \dfrac{\$3.59}{\text{gal}}$

$$\approx \dfrac{\$3.59}{\text{gal}} \cdot \dfrac{1 \text{ gal}}{3.79 \text{ L}}$$

$$\approx \$.95 \text{ per liter}$$

The price is approximately $.95 per liter.

You Try It 8 $2.65 \text{ per liter} = \dfrac{\$2.65}{1 \text{ L}}$

$$\approx \dfrac{\$2.65}{1 \text{ L}} \cdot \dfrac{3.79 \text{ L}}{1 \text{ gal}}$$

$$\approx \$10.04 \text{ per gal}$$

The price is approximately $10.04 per gallon.

Section 7.4 *(pages 464–467)*

You Try It 1 $\dfrac{50}{3} \diagdown\diagup \dfrac{250}{12} \longrightarrow 3 \cdot 250 = 750$
$\longrightarrow 50 \cdot 12 = 600$

$$750 \neq 600$$

The proportion is not true.

You Try It 2 $\dfrac{7}{12} = \dfrac{42}{x}$

$$12 \cdot 42 = 7 \cdot x$$
$$504 = 7x$$
$$72 = x$$

You Try It 3 $\dfrac{5}{n} = \dfrac{3}{322}$

$$n \cdot 3 = 5 \cdot 322$$
$$3n = 1{,}610$$
$$\dfrac{3n}{3} = \dfrac{1{,}610}{3}$$
$$n \approx 536.67$$

You Try It 4 $\dfrac{4}{5} = \dfrac{3}{x - 3}$

$$5 \cdot 3 = 4(x - 3)$$
$$15 = 4x - 12$$
$$27 = 4x$$
$$6.75 = x$$

You Try It 5

Strategy To find the number of gallons, write and solve a proportion using n to represent the number of gallons needed to travel 832 mi.

Solution $\dfrac{396 \text{ mi}}{11 \text{ gal}} = \dfrac{832 \text{ mi}}{n \text{ gal}}$

$$11 \cdot 832 = 396 \cdot n$$
$$9{,}152 = 396n$$
$$23.1 \approx n$$

To travel 832 mi, approximately 23.1 gal of gas are needed.

You Try It 6

Strategy

To find the number of defective transmissions, write and solve a proportion using n to represent the number of defective transmissions in 120,000 cars.

Solution

$$\dfrac{15 \text{ defective transmissions}}{1{,}200 \text{ cars}} = \dfrac{n \text{ defective transmissions}}{120{,}000 \text{ cars}}$$
$$1{,}200 \cdot n = 15 \cdot 120{,}000$$
$$1{,}200n = 1{,}800{,}000$$
$$n = 1{,}500$$

1,500 defective transmissions would be found in 120,000 cars.

Section 7.5 *(pages 472–475)*

You Try It 1

Strategy To find the constant of variation, substitute 120 for y and 8 for x in the direct variation equation $y = kx$ and solve for k.

Solution
$$y = kx$$
$$120 = k \cdot 8 \qquad \blacktriangleright\ y = 120 \text{ when } x = 8.$$
$$15 = k \qquad \blacktriangleright\ \text{Solve for } k.$$

The constant of variation is 15.

You Try It 2

Strategy To find S when $R = 200$:

▶ Write the basic direct variation equation, replace the variables by the given values, and solve for k.

▶ Write the direct variation equation, replacing k by its value. Substitute 200 for R and solve for S.

Solution
$$S = kR$$
$$8 = k \cdot 30 \qquad \blacktriangleright\ S = 8 \text{ when } R = 30.$$
$$\dfrac{8}{30} = k \qquad \blacktriangleright\ \text{Solve for } k.$$
$$\dfrac{4}{15} = k$$
$$S = \dfrac{4}{15}R \qquad \blacktriangleright\ k = \dfrac{4}{15}$$
$$S = \dfrac{4}{15}(200) \qquad \blacktriangleright\ R = 200$$
$$S = \dfrac{160}{3} \approx 53.3 \qquad \blacktriangleright\ \text{Solve for } S.$$

The value of S is approximately 53.3 when $R = 200$.

You Try It 3

Strategy To find the distance:

▶ Write the basic direct variation equation, replace the variables by the given values, and solve for k.

▶ Write the direct variation equation, replacing k by its value. Substitute 9 for the time and solve for the distance.

Solution
$$d = kt^2$$
$$64 = k \cdot 2^2 \qquad \blacktriangleright\ d = 64 \text{ when } t = 2.$$
$$64 = k \cdot 4 \qquad \blacktriangleright\ \text{Solve for } k.$$
$$16 = k$$
$$d = 16t^2$$
$$d = 16 \cdot 9^2 \qquad \blacktriangleright\ t = 9$$
$$d = 16 \cdot 81 \qquad \blacktriangleright\ \text{Solve for } d.$$
$$d = 1{,}296$$

The object will fall 1,296 ft.

You Try It 4

Strategy   To find the resistance:
- ▶ Write the basic inverse variation equation, replace the variables by the given values, and solve for k.
- ▶ Write the inverse variation equation, replacing k by its value. Substitute 0.02 for the diameter and solve for the resistance.

Solution

$$R = \frac{k}{d^2}$$

$$0.5 = \frac{k}{(0.01)^2} \qquad ▶ \; R = 0.5 \text{ when } d = 0.01.$$

$$0.5 = \frac{k}{0.0001} \qquad ▶ \; \text{Solve for } k.$$

$$0.00005 = k$$

$$R = \frac{0.00005}{d^2}$$

$$R = \frac{0.00005}{(0.02)^2} \qquad ▶ \; d = 0.02$$

$$R = \frac{0.00005}{0.0004} \qquad ▶ \; \text{Solve for } R.$$

$$R = 0.125$$

The resistance is 0.125 ohm.

Solutions to Chapter 8 *You Try Its*

Section 8.1 *(pages 493–494)*

You Try It 1 $110\% = 110\left(\frac{1}{100}\right) = \left(\frac{110}{100}\right) = 1\frac{1}{10}$
$110\% = 110(0.01) = 1.10$

You Try It 2 $16\frac{3}{8}\% = 16\frac{3}{8}\left(\frac{1}{100}\right) = \frac{131}{8}\left(\frac{1}{100}\right)$
$= \frac{131}{800}$

You Try It 3 $0.8\% = 0.8(0.01) = 0.008$

You Try It 4 $0.038 = 0.038(100\%) = 3.8\%$

You Try It 5 $\frac{9}{7} = \frac{9}{7}(100\%) = \frac{900}{7}\% = 128\frac{4}{7}\%$

You Try It 6 $1\frac{5}{9} = \frac{14}{9} = \frac{14}{9}(100\%) = \frac{1,400}{9}\% \approx 155.6\%$

Section 8.2 *(pages 498–503)*

You Try It 1

Strategy   To find the amount, solve the basic percent equation. Percent $= 33\frac{1}{3}\% = \frac{1}{3}$,
base $= 45$, amount $= n$

Solution   Percent · base = amount

$$\frac{1}{3}(45) = n$$

$$15 = n$$

15 is $33\frac{1}{3}\%$ of 45.

You Try It 2

Strategy   To find the percent, solve the basic percent equation. Percent $= n$, base $= 40$, amount $= 25$

Solution   Percent · base = amount
$$n \cdot 40 = 25$$
$$\frac{40n}{40} = \frac{25}{40} = 0.625$$
$$n = 62.5\%$$

25 is 62.5% of 40.

You Try It 3

Strategy   To find the base, solve the basic percent equation. Percent $= 16\frac{2}{3}\% = \frac{1}{6}$, base $= n$, amount $= 15$

Solution   Percent · base = amount
$$\frac{1}{6} \cdot n = 15$$
$$6 \cdot \frac{1}{6}n = 15 \cdot 6$$
$$n = 90$$

$16\frac{2}{3}\%$ of 90 is 15.

You Try It 4   Percent $= 25$, base $= n$, amount $= 8$
$$\frac{25}{100} = \frac{8}{n}$$
$$25 \cdot n = 100 \cdot 8$$
$$25n = 800$$
$$\frac{25n}{25} = \frac{800}{25}$$
$$n = 32$$

8 is 25% of 32.

You Try It 5   Percent $= 0.74$, base $= 1{,}200$, amount $= n$
$$\frac{0.74}{100} = \frac{n}{1{,}200}$$
$$100 \cdot n = 0.74 \cdot 1{,}200$$
$$100n = 888$$
$$\frac{100n}{100} = \frac{888}{100}$$
$$n = 8.88$$

0.74% of 1,200 is 8.88.

You Try It 6   Percent $= n$, base $= 180$, amount $= 54$
$$\frac{n}{100} = \frac{54}{180}$$
$$n \cdot 180 = 100 \cdot 54$$
$$180n = 5{,}400$$
$$\frac{180n}{180} = \frac{5{,}400}{180}$$
$$n = 30$$

30% of 180 is 54.

You Try It 7

Strategy   To find the percent, use the basic percent equation. Percent = n, base = 4,330, amount = 649.50

Solution   Percent · base = amount

$$n \cdot 4{,}330 = 649.50$$
$$\frac{4{,}330n}{4{,}330} = \frac{649.50}{4{,}330}$$
$$n = 0.15$$

15% of the instructor's salary is deducted for income tax.

You Try It 8

Strategy   To find the number, solve the basic percent equation.

Percent = 19% = 0.19, base = 2.4 million, amount = n

Solution   Percent · base = amount

$$0.19 \cdot 2.4 = n$$
$$0.456 = n$$

0.456 million = 456,000

There are approximately 456,000 female surfers in this country.

You Try It 9

Strategy   To find the increase in the hourly wage:
- ▶ Find last year's wage. Solve the basic percent equation.
  Percent = 115% = 1.15, base = n, amount = 30.13
- ▶ Subtract last year's wage from this year's wage.

Solution   Percent · base = amount

$$1.15 \cdot n = 30.13$$
$$\frac{1.15n}{1.15} = \frac{30.13}{1.15}$$
$$n = 26.20$$
$$30.13 - 26.20 = 3.93$$

The increase in the hourly wage was $3.93.

Section 8.3 *(pages 509–510)*

You Try It 1

Strategy   To find the percent increase in mileage:
- ▶ Find the amount of increase in mileage.
- ▶ Solve the basic percent equation.
  Percent = n, base = 17.5, amount = amount of increase

Solution $18.2 - 17.5 = 0.7$

Percent · base = amount

$$n \cdot 17.5 = 0.7$$
$$\frac{17.5n}{17.5} = \frac{0.7}{17.5}$$
$$n = 0.04$$

The percent increase in mileage is 4%.

You Try It 2

Strategy   To find the value of the car:
- ▶ Solve the basic percent equation to find the amount of decrease in value.
  Percent = 24% = 0.24, base = 47,000, amount = n
- ▶ Subtract the amount of decrease from the cost.

Solution   Percent · base = amount

$$0.24 \cdot 47{,}000 = n$$
$$11{,}280 = n$$
$$47{,}000 - 11{,}280 = 35{,}720$$

The value of the car is $35,720.

Section 8.4 *(pages 514–517)*

You Try It 1

Strategy   To find the markup, solve the formula $M = r \cdot C$ for M.
$r = 45\% = 0.45$, $C = 650$

Solution $M = r \cdot C$
$$M = 0.45 \cdot 650$$
$$M = 292.50$$

The markup is $292.50.

You Try It 2

Strategy   To find the markup rate:
- ▶ Solve the formula $M = P - C$ for M.
  $P = 1{,}450$, $C = 950$
- ▶ Solve the formula $M = r \cdot C$ for r.

Solution $M = P - C$
$$M = 1{,}450 - 950$$
$$M = 500$$

$$M = r \cdot C$$
$$500 = r \cdot 950$$
$$\frac{500}{950} = \frac{950r}{950}$$
$$0.526 \approx r$$

The markup rate is 52.6%.

You Try It 3

Strategy   To find the selling price:
- ▶ Find the markup by solving the equation $M = r \cdot C$ for M. $r = 42\% = 0.42$, $C = 82.50$
- ▶ Solve the formula $P = C + M$ for P.
  $C = 82.50$, M = the markup

Solution $M = r \cdot C$
$$M = 0.42 \cdot 82.50$$
$$M = 34.65$$

$$P = C + M$$
$$P = 82.50 + 34.65$$
$$P = 117.15$$

The selling price is $117.15.

You Try It 4

Strategy   To find the discount rate:

▶ Find the discount by solving the formula $D = R - S$ for D. $R = 325$, $S = 253.50$

▶ Solve the formula $D = r \cdot R$ for r. D = the discount, $R = 325$

Solution $D = R - S$
$D = 325 - 253.50$
$D = 71.50$

$D = r \cdot R$
$71.50 = r \cdot 325$
$\dfrac{71.50}{325} = \dfrac{325r}{325}$
$0.22 = r$

The discount rate is 22%.

You Try It 5

Strategy   To find the sale price, solve the formula $S = (1 - r)R$ for S. $r = 25\% = 0.25$, $R = 312$

Solution $S = (1 - r)R$
$S = (1 - 0.25)312$
$S = (0.75)312$
$S = 234$

The sale price is $234.

You Try It 6

Strategy   To find the regular price, solve the formula $S = (1 - r)R$ for R. $S = 1{,}495$, $r = 35\% = 0.35$

Solution $S = (1 - r)R$
$1{,}495 = (1 - 0.35)R$
$1{,}495 = 0.65R$
$\dfrac{1{,}495}{0.65} = \dfrac{0.65R}{0.65}$
$2{,}300 = R$

The regular price is $2,300.

Section 8.5 *(pages 521–522)*

You Try It 1

Strategy   To calculate the maturity value:

▶ Find the simple interest due on the loan by solving the simple interest formula for I.

$t = \dfrac{8}{12}$, $P = 12{,}500$, $r = 9.5\% = 0.095$

▶ Use the formula for the maturity value of a simple interest loan, $M = P + I$.

Solution $I = Prt$

$I = 12{,}500(0.095)\left(\dfrac{8}{12}\right)$

$I \approx 791.67$

$M = P + I$
$M = 12{,}500 + 791.67$
$M = 13{,}291.67$

The total amount due on the loan is $13,291.67.

Solutions to Chapter 9   *You Try Its*

Section 9.1 *(pages 537–546)*

You Try It 1 $QR + RS + ST = QT$
$24 + RS + 17 = 62$ ▶ $QR = 24$, $ST = 17$, $QT = 62$
$41 + RS = 62$ ▶ Add 24 and 17.
$RS = 21$ ▶ Subtract 41 from each side.

$RS = 21$ cm

You Try It 2 $AC = AB + BC$

$AC = \dfrac{1}{4}(BC) + BC$ ▶ AB is one-fourth BC.

$AC = \dfrac{1}{4}(16) + 16$ ▶ $BC = 16$

$AC = 4 + 16$
$AC = 20$
$AC = 20$ ft

You Try It 3

Strategy   Supplementary angles are two angles whose sum is 180°. To find the supplement, let x represent the supplement of a 129° angle. Write an equation and solve for x.

Solution $x + 129° = 180°$
$x = 51°$

The supplement of a 129° angle is a 51° angle.

You Try It 4

Strategy   To find the measure of $\angle a$, write an equation using the fact that the sum of the measure of $\angle a$ and 68° is 118°. Solve for $\angle a$.

Solution $\angle a + 68° = 118°$
$\angle a = 50°$

The measure of $\angle a$ is 50°.

You Try It 5

Strategy   The angles labeled are adjacent angles of intersecting lines and are therefore supplementary angles. To find x, write an equation and solve for x.

Solution $(x + 16°) + 3x = 180°$
$4x + 16° = 180°$
$4x = 164°$
$x = 41°$

You Try It 6

Strategy $3x = y$ because corresponding angles have the same measure. $y + (x + 40°) = 180°$ because adjacent angles of intersecting lines are supplementary angles. Substitute $3x$ for y and solve for x.

Solution $3x + (x + 40°) = 180°$
$4x + 40° = 180°$
$4x = 140°$
$x = 35°$

You Try It 7

Strategy
- ▶ To find the measure of angle b, use the fact that $\angle b$ and $\angle x$ are supplementary angles.
- ▶ To find the measure of angle c, use the fact that the sum of the interior angles of a triangle is $180°$.
- ▶ To find the measure of angle y, use the fact that $\angle c$ and $\angle y$ are vertical angles.

Solution
$$\angle b + \angle x = 180°$$
$$\angle b + 100° = 180°$$
$$\angle b = 80°$$

$$\angle a + \angle b + \angle c = 180°$$
$$45° + 80° + \angle c = 180°$$
$$125° + \angle c = 180°$$
$$\angle c = 55°$$

$$\angle y = \angle c = 55°$$

You Try It 8

Strategy
To find the measure of the third angle, use the fact that the measure of a right angle is $90°$ and the fact that the sum of the measures of the interior angles of a triangle is $180°$. Write an equation using x to represent the measure of the third angle. Solve the equation for x.

Solution
$$x + 90° + 34° = 180°$$
$$x + 124° = 180°$$
$$x = 56°$$

The measure of the third angle is $56°$.

Section 9.2 *(pages 553–562)*

You Try It 1

Strategy
To find the perimeter, use the formula for the perimeter of a square. Substitute 60 for s and solve for P.

Solution
$$P = 4s$$
$$P = 4(60)$$
$$P = 240$$

The perimeter of the infield is 240 ft.

You Try It 2

Strategy
To find the perimeter, use the formula for the perimeter of a rectangle. Substitute 11 for L and $8\frac{1}{2}$ for W and solve for P.

Solution
$$P = 2L + 2W$$
$$P = 2(11) + 2\left(8\frac{1}{2}\right)$$
$$P = 2(11) + 2\left(\frac{17}{2}\right)$$
$$P = 22 + 17$$
$$P = 39$$

The perimeter of a standard piece of typing paper is 39 in.

You Try It 3

Strategy
To find the circumference, use the circumference formula that involves the diameter. Leave the answer in terms of π.

Solution
$$C = \pi d$$
$$C = \pi(9)$$
$$C = 9\pi$$

The circumference is 9π in.

You Try It 4

Strategy
To find the number of rolls of wallpaper to be purchased:
- ▶ Use the formula for the area of a rectangle to find the area of one wall.
- ▶ Multiply the area of one wall by the number of walls to be covered (2).
- ▶ Divide the area of wall to be covered by the area one roll of wallpaper will cover (30).

Solution
$$A = LW$$
$$A = 12 \cdot 8 = 96 \quad \text{The area of one wall is 96 ft}^2.$$
$$2(96) = 192 \quad \text{The area of the two walls is 192 ft}^2.$$
$$192 \div 30 = 6.4$$

Because a portion of a seventh roll is needed, 7 rolls of wallpaper should be purchased.

You Try It 5

Strategy
To find the area, use the formula for the area of a circle. An approximation is asked for; use the π key on a calculator. $r = 11$

Solution
$$A = \pi r^2$$
$$A = \pi(11)^2$$
$$A = 121\pi$$
$$A \approx 380.13$$

The area is approximately 380.13 cm^2.

Section 9.3 *(pages 571–576)*

You Try It 1

Strategy
To find the measure of the other leg, use the Pythagorean Theorem. $a = 2, c = 6$

Solution
$$a^2 + b^2 = c^2$$
$$2^2 + b^2 = 6^2$$
$$4 + b^2 = 36$$
$$b^2 = 32$$
$$b = \sqrt{32} \quad \blacktriangleright \text{ The Principal Square}$$
$$b \approx 5.66 \qquad \text{Root Property}$$

The measure of the other leg is approximately 5.66 m.

You Try It 2

Strategy
To find FG, write a proportion using the fact that, in similar triangles, the ratio of corresponding sides equals the ratio of corresponding heights. Solve the proportion for FG.

Solution

$$\frac{AC}{DF} = \frac{CH}{FG}$$

$$\frac{10}{15} = \frac{7}{FG}$$

$$10(FG) = 15(7)$$
$$10(FG) = 105$$
$$FG = 10.5$$

The height FG of triangle DEF is 10.5 m.

You Try It 3

Strategy   To determine whether the triangles are congruent, determine whether one of the rules for congruence is satisfied.

Solution $PR = MN$, $QR = MO$, and $\angle QRP = \angle OMN$. Two sides and the included angle of one triangle equal two sides and the included angle of the other triangle.

The triangles are congruent by the SAS Rule.

Section 9.4 *(pages 582–587)*

You Try It 1

Strategy   To find the volume, use the formula for the volume of a cube. $s = 2.5$

Solution $V = s^3$
$V = (2.5)^3 = 15.625$

The volume of the cube is 15.625 m^3.

You Try It 2

Strategy   To find the volume:
▶ Find the radius of the base of the cylinder. $d = 8$
▶ Use the formula for the volume of a cylinder. Leave the answer in terms of π.

Solution $r = \frac{1}{2}d = \frac{1}{2}(8) = 4$

$V = \pi r^2 h = \pi(4)^2(22) = \pi(16)(22) = 352\pi$

The volume of the cylinder is 352π ft^3.

You Try It 3

Strategy   To find the surface area of the cylinder:
▶ Find the radius of the base of the cylinder. $d = 6$
▶ Use the formula for the surface area of a cylinder. An approximation is asked for; use the π key on a calculator.

Solution $r = \frac{1}{2}d = \frac{1}{2}(6) = 3$

$SA = 2\pi r^2 + 2\pi rh$
$SA = 2\pi(3)^2 + 2\pi(3)(8)$
$= 2\pi(9) + 2\pi(3)(8)$
$= 18\pi + 48\pi$
$= 66\pi$
≈ 207.35

The surface area of the cylinder is approximately 207.35 ft^2.

You Try It 4

Strategy   To find which solid has the larger surface area:
▶ Use the formula for the surface area of a cube to find the surface area of the cube. $s = 10$
▶ Find the radius of the sphere. $d = 8$
▶ Use the formula for the surface area of a sphere to find the surface area of the sphere. Because this number is to be compared to another number, use the π key on a calculator to approximate the surface area.
▶ Compare the two numbers.

Solution $SA = 6s^2$
$SA = 6(10)^2 = 6(100) = 600$

The surface area of the cube is 600 cm^2.

$r = \frac{1}{2}d = \frac{1}{2}(8) = 4$

$SA = 4\pi r^2$
$SA = 4\pi(4)^2 = 4\pi(16) = 64\pi \approx 201.06$

The surface area of the sphere is 201.06 cm^2.

$600 > 201.06$

The cube has a larger surface area than the sphere.

Solutions to Chapter 10   *You Try Its*

Section 10.1 *(pages 607–610)*

You Try It 1

Strategy   To make the frequency distribution:
▶ Find the range.
▶ Divide the range by 8, the number of classes. The quotient is the class width.
▶ Tabulate the data for each class.

Solution   Range $= 998 - 118 = 880$

Class width $= \dfrac{880}{8} = 110$

Dollar Amount of Insurance Claims

Classes	Tally	Frequency
118–228	///	3
229–339	////	4
340–450	/////	5
451–561	////////	8
562–672	//////	6
673–783	///////	7
784–894	//////////	10
895–1,005	///////	7

You Try It 2

Strategy   To find the number:
▶ Read the histogram to find the number of households using between 1,100 and 1,150 kWh and the number using between 1,150 and 1,200 kWh.
▶ Add the two numbers.

Solution   Number between 1,100 and 1,150 kWh: 14
Number between 1,150 and 1,200 kWh: 10

$14 + 10 = 24$

24 households used 1,100 kWh of electricity or more during the month.

You Try It 3

Strategy   To find the ratio:

▶ Read the frequency polygon to find the number of states with a per capita income between $24,000 and $28,000 and between $28,000 and $32,000.

▶ Write the ratio of the number of states with a per capita income between $24,000 and $28,000 to the number of states with a per capita income between $28,000 and $32,000.

Solution   Number of states with a per capita income between $24,000 and $28,000: 5

Number of states with a per capita income between $28,000 and $32,000: 15

$$\frac{\text{Income between \$24,000 and \$28,000}}{\text{Income between \$28,000 and \$32,000}}$$

$$= \frac{5}{15} = \frac{1}{3}$$

The ratio is $\frac{1}{3}$ or 1 to 3.

Section 10.2 *(pages 616–621)*

You Try It 1

Strategy   To calculate the mean amount spent:

▶ Calculate the sum of the amounts spent by the customers.

▶ Divide the sum by the number of customers.

To calculate the median amount spent by the customers:

▶ Arrange the numbers from smallest to largest.

▶ Because there is an even number of values, the median is the sum of the two middle numbers, divided by 2.

Solution   The sum of the numbers is 51.30.

$$\bar{x} = \frac{51.30}{10} = 5.13$$

The mean amount spent by the customers was $5.13.

Arrange the numbers from smallest to largest.

3.59   4.32   4.45   4.75   5.05
5.45   5.58   5.90   6.00   6.21

$$\text{Median} = \frac{5.05 + 5.45}{2} = 5.25$$

The median is $5.25.

You Try It 2

Strategy   To find the score, use the formula for the mean, letting n be the score on the fifth exam.

Solution $$84 = \frac{82 + 91 + 79 + 83 + n}{5}$$

$$84 = \frac{335 + n}{5}$$

$$5 \cdot 84 = 5\left(\frac{335 + n}{5}\right)$$

$$420 = 335 + n$$

$$85 = n$$

The score on the fifth test must be 85.

You Try It 3

Strategy   To draw the box-and-whiskers plot:

▶ Use the value of the first quartile, the median, and the third quartile from Example 3.

▶ Determine the smallest and largest data values.

▶ Draw the box-and-whiskers plot.

Solution $Q_1 = 38$, median $= 44$, $Q_3 = 53$

Smallest value: 24
Largest value: 64

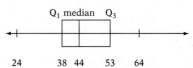

You Try It 4

Strategy   To calculate the standard deviation:

▶ Find the mean of the number of miles run.

▶ Use the procedure for calculating the standard deviation.

Solution

$$\bar{x} = \frac{5 + 7 + 3 + 6 + 9 + 6}{6} = 6$$

Step 1

x	$(x - \bar{x})^2$
5	$(5 - 6)^2 = 1$
7	$(7 - 6)^2 = 1$
3	$(3 - 6)^2 = 9$
6	$(6 - 6)^2 = 0$
9	$(9 - 6)^2 = 9$
6	$(6 - 6)^2 = 0$
	Total $= 20$

Step 2 $\frac{20}{6} = \frac{10}{3}$

Step 3 $\sigma = \sqrt{\frac{10}{3}} \approx 1.826$

The standard deviation is 1.826 mi.

Section 10.3 *(pages 627–632)*

You Try It 1

Strategy   To calculate the probability:

▶ Count the number of possible outcomes of the experiment.

> ▶ Count the outcomes of the experiment that are favorable to the event the two numbers are the same.

> ▶ Use the probability formula.

Solution   There are 36 possible outcomes.
There are 6 favorable outcomes: (1, 1), (2, 2), (3, 3), (4, 4), (5, 5), and (6, 6).

$$P(E) = \frac{6}{36} = \frac{1}{6}$$

The probability that the two numbers are equal is $\frac{1}{6}$.

You Try It 2

Strategy   To calculate the probability:

> ▶ Count the number of possible outcomes of the experiment.

> ▶ Count the number of outcomes of the experiment that are favorable to the event E that the soft drink symbol is uncovered.

> ▶ Use the probability formula.

Solution   There are 8 possible outcomes of the experiment.
There is 1 favorable outcome for E, uncovering the soft drink symbol.

$$P(E) = \frac{1}{8}$$

The probability is $\frac{1}{8}$ that the soft drink symbol will be uncovered.

You Try It 3

Strategy

To calculate the probability:

▶ Write the odds in favor of contracting the flu as a fraction.

▶ The probability of contracting the flu is the numerator of the odds-in-favor fraction over the sum of the numerator and the denominator.

Solution

Odds in favor of contracting the flu $= \dfrac{2}{13}$

Probability of contracting the flu $= \dfrac{2}{2 + 13} = \dfrac{2}{15}$

The probability of contracting the flu is $\dfrac{2}{15}$.

Answers to Chapter 1 *Exercises*

Prep Test *(page 2)*

1. 8   **2.** 1 2 3 4 5 6 7 8 9 10   **3.** a and D; b and E; c and A; d and B; e and F; f and C   **4.** 0   **5.** fifty

1.1 Exercises *(pages 13–18)*

1. is greater than
3. (number line with point at 2, 0–12)   **5.** (number line with point at 9, 0–12)   **7.** (number line with point at 8, 0–12)
9. 5   **11.** 5   **13.** 0   **15.** 27 < 39   **17.** 0 < 52   **19.** 273 > 194   **21.** 2,761 < 3,857   **23.** 4,610 > 4,061
25. 8,005 < 8,050   **27.** Yes   **29.** 11, 14, 16, 21, 32   **31.** 13, 48, 72, 84, 93   **33.** 26, 49, 77, 90, 106   **35.** 204, 399, 662, 736, 981
37. 307, 370, 377, 3,077, 3,700   **39.** comma; three   **41.** five hundred eight   **43.** six hundred thirty-five   **45.** four thousand
seven hundred ninety   **47.** fifty-three thousand six hundred fourteen   **49.** two hundred forty-six thousand fifty-three
51. three million eight hundred forty-two thousand nine hundred five   **53.** 496   **55.** 53,340   **57.** 502,140   **59.** 9,706
61. 5,012,907   **63.** 8,005,010   **65.** 7,000 + 200 + 40 + 5   **67.** 500,000 + 30,000 + 2,000 + 700 + 90 + 1
69. 5,000 + 60 + 4   **71.** 20,000 + 300 + 90 + 7   **73.** 400,000 + 2,000 + 700 + 8   **75.** 8,000,000 + 300 + 10 + 6
77. tens   **79.** greater   **81.** 7,110   **83.** 5,000   **85.** 28,600   **87.** 7,000   **89.** 94,000   **91.** 630,000   **93.** 350,000
95. sometimes true   **97.** never true   **99.** blue   **101.** Billy Hamilton   **103.** *Fiddler on the Roof*   **105.** two tablespoons of
peanut butter   **107.** St. Louis to San Diego   **109.** Neptune   **111.** 300,000 km/s   **113.** 160,000 acres   **115a.** 1985
b. decrease   **117.** 999; 10,000

1.2 Exercises *(pages 33–40)*

1a. 12; 2; 1   **b.** 1; 2; 6; 9   **c.** 92   **3.** 1,383,659   **5.** 6,043   **7.** 12,548   **9.** 199,556   **11.** 327,473   **13.** 168,574
15. 7,947   **17.** 99,637   **19.** 1,872 students   **21.** 15,040; 15,000   **23.** 1,388,917; 1,400,000   **25.** 1,998; 2,000
27. 329,801; 307,000   **29.** 1,272   **31.** 12,150   **33.** 89,900   **35.** 1,572   **37.** 14,591   **39.** 56,010
41. The Commutative Property of Addition   **43.** The Associative Property of Addition   **45.** The Addition Property of Zero
47. 28   **49.** 4   **51.** 15   **53.** The Commutative Property of Addition   **55.** Yes   **57.** No   **59.** Yes   **61a.** 4; 12
b. 8; 12; 4   **c.** 2; 4; 2   **63.** 416   **65.** 188   **67.** 464   **69.** 208   **71.** 3,557   **73.** 2,836   **75.** 1,437   **77.** 20,148
79. 1,618   **81.** 7,378   **83.** 17,548   **85.** 15 ft   **87.** 2,136; 2,000   **89.** 38,283; 40,000   **91.** 31,195; 35,000
93. 125,665; 100,000   **95.** 13   **97.** 643   **99.** 355   **101.** 5,211   **103.** 766   **105.** 18,231   **107.** Yes   **109.** No
111. Yes   **113.** addition   **115.** 210   **117.** 901   **119.** 370 calories   **121.** 78 m   **123.** 43 in.   **125.** 43 more orbits
127. 10,818 seats   **129.** $1,645   **131.** 20,000 mi   **133.** January to February; 24 cars   **135.** $13,275   **137.** $261,000
139. 350 mph   **141a.** 9,571 drivers   **b.** 4,211 drivers   **143.** More people are driving at or below the posted speed limit.
145. 11   **147a.** always true   **b.** always true

1.3 Exercises *(pages 59–66)*

3a. 4; 12; 2; 1   **b.** 2; 6; 6; 1; 7   **c.** 72   **5.** 2,492   **7.** 18,040   **9.** 420,329   **11.** 54,372   **13.** 388,832   **15.** 3,324,048
17. 2,400   **19.** 1,400   **21.** fg **23.** 1,244,653; 1,200,000   **25.** 1,138,134; 1,200,000   **27.** 46,935; 42,000   **29.** 6,491,166;
6,300,000   **31.** 14,880   **33.** 3,255   **35.** 1,800   **37.** 3,082   **39.** Answers may vary. One possible answer is 5 and 20.
41. The Multiplication Property of One   **43.** The Commutative Property of Multiplication   **45.** 30   **47.** 0   **49.** Yes
51. No   **53.** Yes   **55a.** 3; 4   **b.** 3; 3; 3; 3; 81   **57.** $2^3 \cdot 7^5$ **59.** $2^2 \cdot 3^3 \cdot 5^4$ **61.** c^2 **63.** x^3y^3 **65.** 32   **67.** 1,000,000
69. 200   **71.** 9,000   **73.** 0   **75.** 540   **77.** 144   **79.** 512   **81.** a^4 **83.** 24   **85.** 320   **87.** 225   **89a.** 6 · 2; 12
b. 2^6; 64   **91.** 495; 6   **93.** 307   **95.** 309 r4   **97.** 2,550   **99.** 21 r9   **101.** 147 r38   **103.** 200 r8   **105.** 404 r34
107. 16 r97   **109.** 907   **111.** 881 r1   **113.** $\dfrac{c}{d}$ **115.** 776; 800   **117.** 5,129; 5,000   **119.** 493 r37; 500   **121.** 1,516; 1,500
123. 48   **125.** undefined   **127.** 9,800   **129.** Yes   **131.** No   **133.** factors; 1; 3; 5; 15   **135.** 1, 2, 5, 10   **137.** 1, 2, 3, 4, 6, 12
139. 1, 2, 4, 8   **141.** 1, 13   **143.** 1, 2, 3, 6, 9, 18   **145.** 1, 5, 25   **147.** 1, 2, 4, 7, 8, 14, 28, 56   **149.** 1, 2, 4, 7, 14, 28   **151.** 1, 2, 3,
4, 6, 8, 12, 16, 24, 48   **153.** 1, 2, 3, 6, 9, 18, 27, 54   **155.** 2^4 **157.** $2^2 \cdot 3$ **159.** 3 · 5   **161.** $2^3 \cdot 5$ **163.** prime
165. 5 · 13   **167.** $2^2 \cdot 7$ **169.** 2 · 3 · 7   **171.** 3 · 17   **173.** 2 · 23   **175.** division   **177.** 460 calories   **179.** 4,325
gal   **181a.** 78 m   **b.** 360 m²   **183.** 81 ft²   **185.** $16,000   **187.** $6,840   **189.** 9 h   **191.** $21   **193.** an approximation
195. 222

1.4 Exercises *(pages 71–72)*

1. 5; 15   **3.** 14   **5.** 25   **7.** 5   **9.** 13   **11.** 7   **13.** 9   **15.** 8   **17.** 1   **19.** 0   **21.** 76   **23.** $n + 8 = 13$ **25.** 24
27. 6   **29.** 12   **31.** 12 in.   **33.** 190 mi   **35.** 36 payments   **37.** 5 h

A2 CHAPTER 2

1.5 Exercises (pages 75–76)

1. to multiply 7 times 2   **3.** 4   **5.** 29   **7.** 13   **9.** 19   **11.** 11   **13.** 6   **15.** 61   **17.** 54   **19.** 19
21. 24   **23.** 186   **25.** 39   **27.** 18   **29.** 14   **31.** 14   **33.** 2   **35.** 57   **37.** 8   **39.** 68   **41.** 16
43. $12 + (9 - 5) \cdot 3 > 11 + (8 + 4) \div 6$ **45.** $5 + 7 \cdot (3 - 1)$ **47.** $5 + (7 \cdot 3) - 1$ **49.** 97

Chapter 1 Review Exercises (pages 83–84)

Note: The numbers in brackets following the answers in the Chapter Review are a reference to the objective that corresponds with that problem. For example, the reference [1.2A] stands for Section 1.2, Objective A. This notation will be used for all Chapter Reviews, Chapter Tests, and Cumulative Reviews throughout the text.

1. ⊢⊢⊢⊢⊢⊢⊢⊢●⊢⊢⊢⊢→ [1.1A]   **2.** 10,000 [1.3B]   **3.** 2,583 [1.2B]   **4.** $3^2 \cdot 5^4$ [1.3B]   **5.** 1,389 [1.2A]
  0 1 2 3 4 5 6 7 8 9 10 11 12
6. 38,700 [1.1C]   **7.** 247 > 163 [1.1A]   **8.** 32,509 [1.1B]   **9.** 700 [1.3A]   **10.** 2,607 [1.3C]   **11.** 4,048 [1.2B]
12. 1,500 [1.2A]   **13.** 1, 2, 5, 10, 25, 50 [1.3D]   **14.** Yes [1.2B]   **15.** 18 [1.5A]   **16.** The Commutative Property of Addition [1.2A]   **17.** four million nine hundred twenty-seven thousand thirty-six [1.1B]   **18.** 675 [1.3B]
19a. 16 times more   **b.** 61 times more [1.3E]   **20.** 67 r70 [1.3C]   **21.** 2,636 [1.3A]   **22.** 137 [1.2B]
23. $2 \cdot 3^2 \cdot 5$ [1.3D]   **24.** 80 [1.3C]   **25.** 1 [1.3A]   **26.** 9 [1.4A]   **27.** 932 [1.2A]   **28.** 432 [1.3A]
29. 56 [1.5A]   **30.** Kareem Abdul-Jabbar [1.1A]   **31.** $182,000 [1.3E]   **32a.** 74 m   **b.** 300 m² [1.2C, 1.3E]
33a. 1960s   **b.** 4,792,000 students [1.2C]   **34.** 42 mi [1.3E]   **35.** $449 [1.2C]

Chapter 1 Test (pages 85–86)

1. 329,700 [1.3A]   **2.** 16,000 [1.3B]   **3.** 4,029 [1.2B]   **4.** $x^4 y^3$ [1.3B]   **5.** Yes [1.2A]   **6.** 3,000 [1.1C]
7. 7,177 < 7,717 [1.1A]   **8.** 8,490 [1.1B]   **9.** three hundred eighty-two thousand nine hundred four [1.1B]
10. 2,000 [1.2A]   **11.** 11,008 [1.3A]   **12.** 2,400,000 [1.3A]   **13.** 1, 2, 4, 23, 46, 92 [1.3D]   **14.** $2^4 \cdot 3 \cdot 5$ [1.3D]
15. 30,866 [1.2B]   **16.** The Commutative Property of Addition [1.2A]   **17.** 897 [1.3C]   **18.** 26 [1.5A]   **19.** $284 [1.2B]
20. 51 [1.4A]   **21.** 44 [1.4A]   **22.** 56 [1.5A]   **23.** 7 [1.2A]   **24.** 78 [1.4B]   **25.** 720 [1.3E]   **26.** $556 [1.2C]
27a. 96 cm   **b.** 576 cm² [1.3E]   **28.** $4,456 [1.2C]   **29a.** 2004 to 2005   **b.** 117,749 hybrid cars [1.2C]   **30.** $960 [1.3E]
31. $11 [1.3E]

Answers to Chapter 2 *Exercises*

Prep Test (page 88)

1. 54 > 45 [1.1A]   **2.** 4 units [1.1A]   **3.** 15,847 [1.2A]   **4.** 3,779 [1.2B]   **5.** 26,432 [1.3A]   **6.** 6 [1.3B]   **7.** 13 [1.4A]
8. 5 [1.4A]   **9.** $172 [1.2C]   **10.** 31 [1.5A]

2.1 Exercises (pages 95–100)

1a. left   **b.** right   **3.** ←⊢●⊢⊢⊢⊢⊢⊢⊢⊢⊢⊢→   **5.** ←●⊢⊢⊢⊢⊢⊢⊢⊢⊢⊢⊢→
       −6 −5 −4 −3 −2 −1 0 1 2 3 4 5 6   −6 −5 −4 −3 −2 −1 0 1 2 3 4 5 6
7. ←⊢⊢⊢⊢⊢⊢⊢⊢⊢⊢●⊢→   **9.** ←⊢⊢●⊢⊢⊢⊢⊢⊢⊢⊢⊢→   **11.** 1   **13.** −1   **15.** 3
    −6 −5 −4 −3 −2 −1 0 1 2 3 4 5 6   −6 −5 −4 −3 −2 −1 0 1 2 3 4 5 6
17. A is −4. C is −2.   **19.** A is −7. D is −4.   **21.** −2 > −5   **23.** 3 > −7   **25.** −42 < 27   **27.** 53 > −46
29. −51 < −20   **31.** −131 < 101   **33.** −7, −2, 0, 3   **35.** −5, −3, 1, 4   **37.** −4, 0, 5, 9   **39.** −10, −7, −5, 4, 12
41a. never true   **b.** sometimes true   **c.** sometimes true   **d.** always true   **43.** minus; negative   **45.** −45   **47.** 88
49. −n   **51.** d   **53.** the opposite of negative thirteen   **55.** the opposite of negative p   **57.** five plus negative ten
59. negative fourteen minus negative three   **61.** negative thirteen minus eight   **63.** m plus negative n   **65.** 7
67. −46   **69.** 73   **71.** z   **73.** −p   **75.** negative   **77a.** −6   **b.** 6   **c.** 6   **79.** 4   **81.** 9   **83.** 11
85. 12   **87.** 23   **89.** −27   **91.** 25   **93.** −41   **95.** −93   **97.** 10   **99.** 8   **101.** 6   **103.** |−12| > |8|
105. |6| < |13| **107.** |−1| < |−17| **109.** |x| = |−x| **111.** −|6|, −(4), |−7|, −(−9)   **113.** −9, −|−7|, −(5), |4|
115. −|10|, −|−8|, −(−2), −(−3), |5| **117.** 11, −11   **119.** −6, −5, −4, −3, −2, −1, 0, 1, 2, 3, 4, 5, 6   **121.** Answers will vary. For example, a = −5 and b = 2.   **123.** −73; −73   **125.** −9°F   **127.** −35°F   **129.** −30°F with a 5-mph wind
131a. −27¢   **b.** −40¢   **133.** Yes; 2008   **135.** Stock B   **137.** third quarter   **139a.** −2 and 6   **b.** −2 and 8

2.2 Exercises (pages 111–116)

1. the same; negative   **3.** −11   **5.** −5   **7.** 8   **9.** −4   **11.** −2   **13.** −9   **15.** 1   **17.** −15   **19.** 0
21. −21   **23.** −14   **25.** 19   **27.** −5   **29.** −30   **31.** 9   **33.** −12   **35.** −28   **37.** −13   **39.** −18   **41.** 11
43. 1   **45.** x + (−7)   **47a.** −$132,800,000,000   **b.** −$126,600,000,000   **c.** −$284,300,000,000   **49.** 5   **51.** −2

53. −11   **55.** −17   **57.** The Addition Property of Zero   **59.** The Associative Property of Addition   **61.** 0   **63.** 18
65. No   **67.** Yes   **69.** No   **71.** sometimes true   **73.** always true   **75.** 4; (−3)   **77.** −3   **79.** −13
81. 7   **83.** 0   **85.** −17   **87.** −3   **89.** 12   **91.** 27   **93.** −106   **95.** −67   **97.** −6   **99.** −15   **101.** 82°C
103. −9   **105.** 11   **107.** 0   **109.** −138   **111.** 26   **113.** 13   **115.** −8   **117.** 5   **119.** 2   **121.** −6
123. 12   **125.** −3   **127.** 18   **129.** Yes   **131.** No   **133.** Yes   **135.** sometimes true   **137.** subtract; from; 5,642
139. Asia   **141.** add; 12; to; −5   **143.** 86°   **145.** 36°   **147.** −3   **149.** 19   **151.** Answers will vary. Possible
answers include −1 and −6, −2 and −5, −3 and −4.

2.3 Exercises *(pages 123–128)*

3. −24   **5.** 6   **7.** 18   **9.** −20   **11.** −16   **13.** 25   **15.** 0   **17.** 42   **19.** −128   **21.** 208   **23.** −243
25. −115   **27.** 238   **29.** −96   **31.** −210   **33.** −224   **35.** −40   **37.** 180   **39.** −qr   **41a.** −$2,368,000
b. −$1,540,000   **c.** −$8,216,000   **43.** The Multiplication Property of One   **45.** The Associative Property of Multiplication
47. −6   **49.** 1   **51.** −24   **53.** −60   **55.** 357   **57.** −56   **59.** −1,600   **61.** No   **63.** No   **65.** Yes
67. positive   **69.** different; negative   **71.** −6   **73.** 8   **75.** −49   **77.** 8   **79.** −11   **81.** 14   **83.** 13

85. 1   **87.** 26   **89.** 23   **91.** −110   **93.** 111   **95.** $\dfrac{-9}{x}$ **97.** −$368 million   **99.** −9   **101.** 9   **103.** −6

105. 6   **107.** Yes   **109.** No   **111.** Yes   **113.** $\dfrac{a}{b}$ **115.** $-\dfrac{a}{b}$ **117.** sum; 5   **119.** −3   **121.** −4°   **123.** −63°F

125. 135, −405, 1,215   **127.** −192, −768, −3,072   **129a.** 81   **b.** −17

2.4 Exercises *(pages 133–134)*

1. 3; −10   **3.** 15   **5.** 11   **7.** −7   **9.** −16   **11.** −8   **13.** 0   **15.** −2   **17.** 20   **19.** −5   **21.** −2   **23.** 5
25. 10   **27.** −20   **29.** 0   **31.** −6 = n + 12   **33.** 25   **35.** −15   **37.** −8   **39.** −$421,067 million   **41.** 3°C
43. $480   **45.** $57 million   **47.** 28

2.5 Exercises *(pages 137–138)*

1. division   **3.** −3   **5.** −6   **7.** −5   **9.** −12   **11.** −3   **13.** 19   **15.** 2   **17.** 1   **19.** 14   **21.** 42   **23.** −13
25. −20   **27.** 32   **29.** 30   **31.** −27   **33.** 27   **35.** 2   **37.** 8   **39.** 1   **41.** 15   **43.** 32   **45.** 1   **47.** 1
49. 28   **51.** (6 − 12) ÷ 2 · 3 − 5²   **53.** 6 − 12 ÷ (2 · 3) − 5²   **55.** −4

Chapter 2 Review Exercises *(pages 143–144)*

1. eight minus negative one [2.1B]   **2.** −36 [2.1C]   **3.** 200 [2.3A]   **4.** −9 [2.3B]   **5.** −14 [2.2A]   **6.** 13 [2.1B]
7. [2.1A]   **8.** 4 [2.4A]   **9.** 17 [2.3B]   **10.** −210 [2.3B]   **11.** −2 [2.2B]
12. −18 [2.3A]   **13.** −1 [2.2A]   **14.** −72 [2.3A]   **15.** −4 [2.5A]   **16.** −2 [2.2B]   **17.** 11 strokes [2.2B]   **18.** 13 [2.2B]
19. The Commutative Property of Multiplication [2.3A]   **20.** Yes [2.2B]   **21.** 14 [2.2B]   **22.** 0 [2.3B]   **23.** −60 [2.3A]
24. −12 [2.2A]   **25.** 5 [2.5A]   **26.** −8 > −10 [2.1A]   **27.** 21 [2.2A]   **28.** 27 [2.1C]   **29.** −8 [2.4B]   **30.** −12°C [2.1D]
31. −238°C [2.3C]   **32.** −3°C [2.2C]   **33.** 12 [2.2C]

Chapter 2 Test *(pages 145–146)*

1. negative three plus negative five [2.1B]   **2.** −34 [2.1C]   **3.** 18 [2.2B]   **4.** −20 [2.2A]   **5.** 24 [2.3A]
6. The Commutative Property of Addition [2.2A]   **7.** 12 [2.3B]   **8.** 2 [2.2A]   **9.** 16 > −19 [2.1A]   **10.** −2 [2.2B]
11. −3 [2.2B]   **12.** 49 [2.1B]   **13.** −250 [2.3A]   **14.** −|5|, −(3), |−9|, −(−11) [2.1C]   **15.** No [2.2B]   **16.** −3 [2.1A]
17. 10 strokes [2.2B]   **18.** 0 [2.3B]   **19.** 19 [2.5A]   **20.** −25 [2.1B]   **21.** 16 [2.4A]   **22.** −11 [2.2B]   **23.** 24 [2.3B]
24. 10 [2.5A]   **25.** −7 [2.3B]   **26.** 60 [2.3A]   **27.** −11 [2.4A]   **28.** −10 [2.2B]   **29.** 5°C [2.2C]   **30.** −64°F [2.3C]
31. −5°C [2.2C]   **32.** 16 units [2.2C]   **33.** $24 million [2.4B]

Cumulative Review Exercises *(pages 147–148)*

1. 5 [2.2B]   **2.** 12,000 [1.3A]   **3.** 3,209 [1.3C]   **4.** 2 [1.5A]   **5.** −82 [2.1C]   **6.** 309,480 [1.1B]   **7.** 2,400 [1.3A]
8. 21 [2.3B]   **9.** −11 [2.2B]   **10.** −40 [2.2A]   **11.** 1, 2, 4, 11, 22, 44 [1.3D]   **12.** 1,936 [1.3B]   **13.** 630,000 [1.1C]
14. 1,300 [1.2A]   **15.** 9 [2.2B]   **16.** −2,500 [2.3A]   **17.** 3 · 23 [1.3D]   **18.** −16 [2.4A]   **19.** −32 [2.5A]
20. −4 [2.3B]   **21.** −3 [2.3B]   **22.** −62 < 26 [2.1A]   **23.** 126 [2.3A]   **24.** −9 [2.4A]   **25.** 2⁵ · 7² [1.3B]
26. 47 [1.5A]   **27.** 10,062 [1.2A]   **28.** −26 [2.2B]   **29.** 5,000 [1.2B]   **30.** 2,025 [1.3B]   **31.** 1,722,685 mi² [1.2C]
32. 76 years old [1.2C]   **33.** $14,200 [1.2C]   **34.** $92,250 [1.3E]   **35.** −5°C [2.2C]   **36a.** 168°F   **b.** Alaska [2.2C]
37. $24,900 [1.2C]   **38.** −8 [2.2C]

Answers to Chapter 3   *Exercises*

Prep Test *(page 150)*

1. 20 [1.3A]   **2.** 120 [1.3B]   **3.** 9 [1.3A]   **4.** −2 [2.2A]   **5.** −13 [2.2B]   **6.** 2 r3 [1.3C]   **7.** 24 [1.3C]
8. 4 [1.3C]   **9.** 59 [1.5A]   **10.** 7 [1.2A]   **11.** 44 < 48 [1.1A]

3.1 Exercises *(pages 155–157)*

1a. 6, 12, 18, 24, 30, 36   **b.** 9, 18, 27, 36, 45   **c.** 18   **3.** 8   **5.** 14   **7.** 30   **9.** 45   **11.** 48   **13.** 20   **15.** 42
17. 72   **19.** 120   **21.** 30   **23.** 24   **25.** 180   **27.** 90   **29.** 78   **31.** True   **33a.** 2, 4, 8   **b.** 2, 4, 7, 14, 28
c. 4   **35.** 3   **37.** 6   **39.** 14   **41.** 16   **43.** 1   **45.** 4   **47.** 4   **49.** 6   **51.** 12   **53.** 15   **55.** 2   **57.** 3
59. 21   **61.** 12   **63a.** For example, 6 and 8. The GCF is 2 and the LCM is 24.   **b.** i   **65.** LCM   **67.** GCF
69. every 6 min   **71.** 25 copies   **73.** 12:20 P.M.; 12:20 P.M.

3.2 Exercises *(pages 166–171)*

1a. 9; 4   **b.** improper   **c.** divided by; 9; 4   **3.** $\frac{4}{5}$ **5.** $\frac{1}{4}$ **7.** $\frac{4}{3}$; $1\frac{1}{3}$ **9.** $\frac{13}{5}$; $2\frac{3}{5}$ **11.** $3\frac{1}{4}$ **13.** 4

15. $2\frac{7}{10}$ **17.** 7   **19.** $1\frac{8}{9}$ **21.** $2\frac{2}{5}$ **23.** 18   **25.** $2\frac{2}{15}$ **27.** 1   **29.** $9\frac{1}{3}$ **31.** $\frac{9}{4}$ **33.** $\frac{11}{2}$

35. $\frac{14}{5}$ **37.** $\frac{47}{6}$ **39.** $\frac{7}{1}$ **41.** $\frac{33}{4}$ **43.** $\frac{31}{3}$ **45.** $\frac{55}{12}$ **47.** $\frac{8}{1}$ **49.** $\frac{64}{5}$ **51.** $a > b$ **53.** 4; 4; 4; 20

55. $\frac{6}{12}$ **57.** $\frac{9}{24}$ **59.** $\frac{6}{51}$ **61.** $\frac{24}{32}$ **63.** $\frac{108}{18}$ **65.** $\frac{30}{90}$ **67.** $\frac{14}{21}$ **69.** $\frac{42}{49}$ **71.** $\frac{8}{18}$ **73.** $\frac{28}{4}$ **75.** $\frac{1}{4}$

77. $\frac{3}{4}$ **79.** $\frac{1}{6}$ **81.** $\frac{8}{33}$ **83.** 0   **85.** $\frac{7}{6}$ **87.** 1   **89.** $\frac{3}{5}$ **91.** $\frac{4}{15}$ **93.** $\frac{3}{5}$ **95.** $\frac{2m}{3}$ **97.** $\frac{y}{2}$

99. $\frac{2a}{3}$ **101.** c **103.** $6k$ **105.** ii   **107.** (iii); One example is $\frac{5}{6}$ and $\frac{15}{18}$.   **109.** $\frac{3}{8} < \frac{2}{5}$ **111.** $\frac{3}{4} < \frac{7}{9}$

113. $\frac{2}{3} > \frac{7}{11}$ **115.** $\frac{17}{24} > \frac{11}{16}$ **117.** $\frac{7}{15} > \frac{5}{12}$ **119.** $\frac{5}{9} > \frac{11}{21}$ **121.** $\frac{7}{12} < \frac{13}{18}$ **123.** $\frac{4}{5} > \frac{7}{9}$ **125.** $\frac{9}{16} > \frac{5}{9}$

127. $\frac{5}{8} < \frac{13}{20}$ **129.** Examples will vary. One possible example is $\frac{4}{5}$ and $\frac{3}{7}$.   **131a.** 5; 3; 6; 14   **b.** 5; 14   **133.** $\frac{3}{8}$

135. $\frac{1}{3}$ **137.** location   **139a.** $\frac{1}{4}$ **b.** $\frac{1}{13}$ **141.** Yes   **143.** Yes; yes; yes   **145a.** $\frac{3}{a}$ **b.** $\frac{6}{b}$ and $\frac{7}{b}$

3.3 Exercises *(pages 182–189)*

1a. can   **b.** product; product   **3.** $\frac{3}{5}$ **5.** $-\frac{11}{14}$ **7.** $\frac{4}{5}$ **9.** 0   **11.** $\frac{1}{10}$ **13.** $-\frac{3}{8}$ **15.** $\frac{63}{xy}$ **17.** $-\frac{yz}{30}$

19. $\frac{1}{9}$ **21.** $-\frac{7}{30}$ **23.** $\frac{3}{16}$ **25.** 1   **27.** 6   **29.** $-7\frac{1}{2}$ **31.** $-3\frac{11}{15}$ **33.** 0   **35.** $\frac{1}{2}$ **37.** 19

39. $-2\frac{1}{3}$ **41.** 1   **43.** $7\frac{7}{9}$ **45.** −30   **47.** 42   **49.** $5\frac{1}{2}$ **51.** $\frac{7}{10}$ **53.** $-\frac{1}{12}$ **55.** $-\frac{1}{21}$ **57.** $1\frac{4}{5}$

59. $4\frac{1}{2}$ **61.** $13,000   **63.** $-\frac{7}{48}$ **65.** $3\frac{1}{2}$ **67.** $-17\frac{1}{2}$ **69.** −8   **71.** $\frac{1}{5}$ **73.** $\frac{1}{6}$ **75.** $-3\frac{2}{3}$

77. Yes   **79.** No   **81.** No   **83.** less than   **85.** $\frac{3}{7}$ **87.** $1\frac{11}{14}$ **89.** −1   **91.** 0   **93.** $-\frac{2}{3}$ **95.** $\frac{5}{6}$

97. 0   **99.** 8   **101.** $-\frac{1}{8}$ **103.** undefined   **105.** $-\frac{8}{9}$ **107.** $\frac{1}{6}$ **109.** $-\frac{32}{xy}$ **111.** $\frac{bd}{30}$ **113.** $5\frac{1}{3}$

115. −8   **117.** $-\frac{6}{7}$ **119.** $\frac{1}{2}$ **121.** $5\frac{2}{7}$ **123.** −12   **125.** $1\frac{29}{31}$ **127.** $1\frac{1}{5}$ **129.** $-1\frac{1}{24}$ **131.** $\frac{7}{26}$

133. $-\frac{10}{11}$ **135.** $\frac{1}{12}$ **137.** undefined   **139.** −48   **141.** $\frac{4}{29}$ **143.** $-1\frac{3}{5}$ **145.** 2   **147.** greater than

149. division   **151.** 32 servings   **153.** 30 min   **155.** $16\frac{1}{2}$ ft; 198 in.   **157.** 234 h   **159.** 30 houses

161. 14 in. by 7 in. by $1\frac{3}{4}$ in.   **163.** 96 m²   **165.** 2 bags   **167.** $3\frac{1}{2}$ mph   **169.** 1,250 mi

3.4 Exercises (pages 199–206)

1. denominators   **3.** $\frac{9}{11}$ **5.** 1   **7.** $1\frac{2}{3}$ **9.** $1\frac{1}{6}$ **11.** $\frac{16}{b}$ **13.** $\frac{9}{c}$ **15.** $\frac{11}{x}$ **17.** $\frac{11}{12}$ **19.** $\frac{11}{12}$

21. $1\frac{7}{12}$ **23.** $2\frac{2}{15}$ **25.** $-\frac{1}{12}$ **27.** $-\frac{1}{3}$ **29.** $\frac{11}{24}$ **31.** $\frac{1}{12}$ **33.** $15\frac{2}{3}$ **35.** $5\frac{2}{3}$ **37.** $15\frac{1}{20}$ **39.** $10\frac{7}{36}$

41. $7\frac{5}{12}$ **43.** $-\frac{7}{18}$ **45.** $\frac{3}{4}$ **47.** $-1\frac{1}{2}$ **49.** $6\frac{5}{24}$ **51.** $1\frac{2}{5}$ **53.** $-\frac{1}{12}$ **55.** $1\frac{13}{18}$ **57.** $-\frac{19}{24}$ **59.** $1\frac{5}{24}$

61. $11\frac{2}{3}$ **63.** $14\frac{3}{4}$ **65.** Yes   **67.** Yes   **69.** $\frac{31}{50}$ **71.** iii   **73a.** $\frac{1}{2}; -\frac{3}{7}$ **b.** $\frac{1}{2}; \frac{3}{7}$ **75.** $\frac{1}{6}$ **77.** $\frac{1}{6}$

79. $\frac{5}{d}$ **81.** $-\frac{5}{n}$ **83.** $\frac{1}{14}$ **85.** $\frac{1}{2}$ **87.** $\frac{1}{4}$ **89.** $-\frac{7}{8}$ **91.** $-1\frac{1}{10}$ **93.** $\frac{1}{4}$ **95.** $\frac{13}{36}$ **97.** $2\frac{1}{3}$

99. $6\frac{3}{4}$ **101.** $1\frac{1}{12}$ **103.** $3\frac{3}{8}$ **105.** $5\frac{1}{9}$ **107.** $2\frac{3}{4}$ **109.** $1\frac{17}{24}$ **111.** $4\frac{19}{24}$ **113.** $1\frac{7}{10}$ **115.** $-1\frac{13}{36}$

117. $-\frac{5}{24}$ **119.** $6\frac{5}{12}$ **121.** $\frac{1}{3}$ **123.** $-1\frac{1}{3}$ **125.** $\frac{1}{12}$ **127.** $\frac{1}{6}$ **129.** $1\frac{1}{9}$ **131.** $4\frac{2}{5}$ **133.** $2\frac{2}{9}$

135. $4\frac{11}{12}$ **137.** No   **139.** Yes   **141.** Examples will vary. One example is $\frac{1}{2} - \frac{3}{4} = -\frac{1}{4}$.   **143.** not possible

145. subtraction   **147.** $1\frac{3}{4}$ acres   **149.** $7\frac{3}{4}$ h   **151.** $6\frac{3}{4}$ lb   **153a.** 5 meals   **b.** $\frac{23}{100}$ **c.** $\frac{49}{100}$; less than half

155. $29\frac{1}{2}$ ft   **157.** $\frac{3}{32}$ in.   **159.** $\frac{17}{20}$ **161.** Answers will vary.

3.5 Exercises (pages 211–213)

1. 8   **3.** $\frac{1}{5}$ **5.** 36   **7.** -12 **9.** 25   **11.** -12 **13.** $\frac{1}{2}$ **15.** $\frac{7}{12}$ **17.** $\frac{3}{4}$ **19.** $-\frac{4}{5}$ **21.** $\frac{2}{3}$

23. $1\frac{1}{2}$ **25.** $-1\frac{1}{3}$ **27.** $-\frac{4}{9}$ **29.** iii   **31.** $\frac{n}{6} = \frac{2}{9}$ **33.** $\frac{5}{6}$ **35.** $1\frac{1}{2}$ **37.** -3 **39.** $-\frac{2}{9}$

41. 298,000 people   **43.** 25 qt   **45.** $4,500   **47.** 532 mi   **49.** a

3.6 Exercises (pages 220–223)

1. three   **3.** $\frac{9}{16}$ **5.** $-\frac{1}{216}$ **7.** $5\frac{1}{16}$ **9.** $\frac{5}{128}$ **11.** $\frac{4}{45}$ **13.** $-\frac{1}{10}$ **15.** $1\frac{1}{7}$ **17.** $-\frac{27}{49}$ **19.** $\frac{16}{81}$

21. $\frac{25}{144}$ **23.** $\frac{2}{3}$ **25.** True   **27.** $\frac{3}{4}$ **29.** $-\frac{8}{9}$ **31.** $\frac{1}{6}$ **33.** 6   **35.** $\frac{18}{35}$ **37.** $-\frac{1}{2}$ **39.** $1\frac{7}{25}$ **41.** $3\frac{3}{11}$

43. 17   **45.** $-\frac{4}{5}$ **47.** 1   **49.** No   **51.** 1   **53.** 0   **55.** addition, division, subtraction   **57.** $\frac{2}{5}$ **59.** $\frac{5}{36}$

61. $\frac{11}{32}$ **63.** 1   **65.** 4   **67.** 0   **69.** $\frac{3}{10}$ **71.** $1\frac{1}{9}$ **73.** $1\frac{15}{16}$ **75.** $\frac{1}{2}$ **77.** 1   **79.** No   **81.** $0; \frac{9}{16}$

Chapter 3 Review Exercises (pages 229–230)

1. $9\frac{1}{2}$ [3.2A]   **2.** $2\frac{5}{6}$ [3.4B]   **3.** $1\frac{1}{2}$ [3.3B]   **4.** -1 [3.3A]   **5.** 2 [3.3B]   **6.** $2\frac{2}{3}$ [3.3A]   **7.** $2\frac{11}{12}$ [3.6B]

8. $\frac{3}{5} > \frac{7}{15}$ [3.2C]   **9.** 150 [3.1A]   **10.** $11\frac{13}{30}$ [3.4A]   **11.** $3\frac{1}{3}$ [3.3A]   **12.** $\frac{10}{7}; 1\frac{3}{7}$ [3.2A]   **13.** $\frac{7}{8} > \frac{17}{20}$ [3.2C]

14. $\frac{3}{5}$ [3.6B]   **15.** $\frac{32}{72}$ [3.2B]   **16.** $-\frac{1}{3}$ [3.6A]   **17.** $\frac{2}{7}$ [3.6C]   **18.** 21 [3.1B]   **19.** $\frac{33}{14}$ [3.2A]   **20.** $\frac{3}{8}$ [3.4A]

21. $-\frac{5}{6}$ [3.3B]   **22.** $1\frac{3}{40}$ [3.6C]   **23.** -14 [3.3A]   **24.** $\frac{1}{18}$ [3.4B]   **25.** $1\frac{17}{24}$ [3.4B]   **26.** $2\frac{1}{4}$ [3.6A]   **27.** $9\frac{1}{12}$ [3.4A]

28. $\frac{2}{7}$ [3.2B] **29.** $4\frac{7}{10}$ [3.4B] **30.** $-\frac{13}{18}$ [3.5A] **31.** $\frac{2}{3}$ [3.2D] **32.** $68\frac{1}{6}$ yd [3.4C] **33.** $6\frac{1}{4}$ lb [3.4C]

34. 192 units [3.3C] **35.** $150 [3.3C] **36.** 496 ft/s [3.3C]

Chapter 3 Test *(pages 231–232)*

1. $2\frac{4}{7}$ [3.2A] **2.** $3\frac{11}{12}$ [3.4B] **3.** $22\frac{1}{2}$ [3.3A] **4.** $\frac{7}{12}$ [3.3A] **5.** 90 [3.1A] **6.** $\frac{13}{24}$ [3.4A] **7.** $2\frac{11}{32}$ [3.6A]

8. $\frac{19}{5}$ [3.2A] **9.** $\frac{7}{9}$ [3.3B] **10.** 2 [3.6C] **11.** 7 [3.6B] **12.** 18 [3.1B] **13.** $\frac{1}{6}$ [3.4B] **14.** $\frac{4}{5}$ [3.2B]

15. $2\frac{17}{24}$ [3.4A] **16.** $\frac{5}{6} > \frac{11}{15}$ [3.2C] **17.** $3\frac{16}{25}$ [3.6C] **18.** $\frac{5}{6}$ [3.6B] **19.** $\frac{1}{4}$ [3.6B] **20.** $-1\frac{1}{2}$ [3.3B]

21. $-\frac{1}{2}$ [3.5A] **22.** No [3.4A] **23.** $2\frac{1}{11}$ [3.3A] **24.** $\frac{1}{2}$ [3.5A] **25.** $\frac{12}{28}$ [3.2B] **26.** $\frac{5}{6}$ [3.5B]

27. $\frac{5}{12}$ [3.2D] **28.** $10\frac{5}{24}$ lb [3.4C] **29.** $17\frac{1}{2}$ lb [3.3C] **30.** 120 in² [3.3C] **31.** 10 h [3.4C] **32.** 80 units [3.3C]

33. $5,100 [3.3C]

Cumulative Review Exercises *(pages 233–234)*

1. 39 [1.5A] **2.** $3\frac{1}{2}$ [3.3A] **3.** $8\frac{11}{18}$ [3.4A] **4.** −15 [2.2B] **5.** 36 [3.1B] **6.** 16 [3.3A] **7.** $-1\frac{1}{9}$ [3.3B]

8. $-\frac{4}{15}$ [3.4B] **9.** 9 [3.6B] **10.** $\frac{7}{11} < \frac{4}{5}$ [3.2C] **11.** $-1\frac{22}{27}$ [3.3B] **12.** $\frac{1}{15}$ [3.3A] **13.** 2 [3.3A]

14. $7\frac{1}{28}$ [3.4B] **15.** $\frac{23}{24}$ [3.4B] **16.** $1\frac{7}{12}$ [3.6C] **17.** $1\frac{5}{8}$ [3.4B] **18.** $6\frac{3}{16}$ [3.4A] **19.** −4 [2.4A] **20.** $4\frac{5}{9}$ [3.2A]

21. $\frac{1}{7}$ [3.4B] **22.** $\frac{3}{28}$ [3.6A] **23.** −21 [2.5A] **24.** 11,272 [1.2A] **25.** 48 [1.5A] **26.** $-\frac{11}{20}$ [3.5A]

27. 20,000 [1.2B] **28.** −13 [2.2B] **29.** $\frac{31}{4}$ [3.2A] **30.** $2^2 \cdot 5 \cdot 7$ [1.3D] **31.** 40 calories [1.3E]

32. 1,740,000 people [1.2C] **33.** 9 years [3.5B] **34.** 66 ft [3.3C] **35.** $4\frac{1}{8}$ mi [3.3C] **36.** $22\frac{3}{8}$ lb/in² [3.3C]

Answers to Chapter 4 *Exercises*

Prep Test *(page 236)*

1. $\frac{3}{10}$ [3.2A] **2.** 36,900 [1.1C] **3.** four thousand seven hundred ninety-one [1.1B] **4.** 6,842 [1.1B]

5. [2.1A] **6.** 9,320 [2.2A] **7.** 3,168 [2.2B] **8.** 76,804 [2.3A] **9.** 278 r18 [1.3C] **10.** 64 [1.3B]

4.1 Exercises *(pages 243–246)*

1. hundredths; thousandths; hundred-thousandths; millionths **3.** and; thousandths **5.** thousandths **7.** ten-thousandths

9. hundredths **11.** 0.3 **13.** 0.21 **15.** 0.461 **17.** 0.093 **19.** $\frac{1}{10}$ **21.** $\frac{47}{100}$ **23.** $\frac{289}{1,000}$ **25.** $\frac{9}{100}$

27. thirty-seven hundredths **29.** nine and four tenths **31.** fifty-three ten-thousandths **33.** forty-five thousandths
35. twenty-six and four hundredths **37.** 3.0806 **39.** 407.03 **41.** 246.024 **43.** 73.02684 **45.** >; > **47.** 0.7 > 0.56
49. 3.605 > 3.065 **51.** 9.004 < 9.04 **53.** 9.31 > 9.031 **55.** 4.6 < 40.6 **57.** 0.07046 > 0.07036 **59.** 0.609, 0.66, 0.696,
0.699 **61.** 1.237, 1.327, 1.372, 1.732 **63.** 21.78, 21.805, 21.87, 21.875 **65.** 0.62 > 0.062 **67.** 3; 2; 2, 7, and 4 **69.** 5.4
71. 30.0 **73.** 413.60 **75.** 6.062 **77.** 97 **79.** 5,440 **81.** 0.0236 **83.** 0.18 oz **85.** Walter Payton
87a. $20.00 **b.** $35.00 **c.** $30.00 **d.** $16.99 **e.** $35.00 **f.** $20.00 **g.** $25.00 **89a.** 1.500 **b.** 0.908
c. 60.07 **d.** 0.0032

b. The Inverse Property of Multiplication **c.** The Multiplication Property of One **11.** $(x + 4) + y$ **13.** $\dfrac{1}{5}$ **15.** 0

17. $-\dfrac{3}{2}$ **19a.** Associative **b.** -24 **21.** $12x$ **23.** $-15x$ **25.** $21t$ **27.** $-21p$ **29.** $12q$ **31.** $2x$ **33.** $-15w$
35. x **37.** $6x^2$ **39.** $-27x^2$ **41.** x^2 **43.** x **45.** c **47.** a **49.** $12w$ **51.** $16vw$ **53.** $-28bc$ **55.** 0 **57.** 0
59. 9 **61.** 7 **63.** -15 **65.** $-5y$ **67.** $13b$ **69.** negative **71.** positive **73.** $10z + 4$ **75.** $12y + 30z$
77. $21x - 27$ **79.** $-2x + 7$ **81.** $4x + 9$ **83.** $-5y - 15$ **85.** $-12x + 18$ **87.** $-20n + 40$ **89.** $48z - 24$
91. $24p + 42$ **93.** $10a + 15b + 5$ **95.** $12x - 4y - 4$ **97.** $36m - 9n + 18$ **99.** $12v - 18w - 42$ **101.** $20x + 4$
103. $20a - 25b + 5c$ **105.** $-18p + 12r + 54$ **107.** $-5a + 9b - 7$ **109.** $-11p + 2q + r$ **111.** ii

5.2 Exercises (pages 331–334)

1. $-2x; -9; 5x^2; -2x; -9$ **3.** $-7y^2, -2y, \underline{6}$ **5.** $8n^2, \underline{-1}$ **7.** $\boxed{6x^2 y}, \boxed{7xy^2}$ **9.** $\boxed{-2n^2}, \boxed{5n}$ **11.** $16a$ **13.** $27x$ **15.** $3z$
17. $8x$ **19.** $-7z$ **21.** $-6w$ **23.** 0 **25.** s **27.** $\dfrac{4n}{5}$ **29.** $\dfrac{x}{2}$ **31.** $\dfrac{4y}{7}$ **33.** $\dfrac{2c}{3}$ **35.** $6x - 3y$ **37.** $2r + 13p$
39. $-3w + 2v$ **41.** $-9p + 11$ **43.** $2p$ **45.** 6 **47.** $13y^2 + 1$ **49.** $12w^2 - 16$ **51.** $-14w$ **53.** $5a^2b + 8ab^2$
55. 5 **57.** $5y^2 - 3y + 10$ **59.** (ii) and (iii) **61.** 12; 3; 5; 10 **63.** $7x + 2$ **65.** $3n + 3$ **67.** $4a + 4$ **69.** $4a + 1$
71. $8x + 42$ **73.** $-12x + 28$ **75.** $-18m - 52$ **77.** $20c + 23$ **79.** $8a + 5b$ **81.** $15z - 12$ **83.** -19
85. $-13x - 2y$ **87.** $-2v + 13$ **89.** $-5c - 6$ **91.** $2a + 21$ **93.** $11n - 26$ **95.** $-9x + 6$ **97.** $-3r - 24$ **99.** i

5.3 Exercises (pages 339–343)

1. Yes **3.** No **5.** Yes **7.** No **9.** Yes **11.** No **13.** 3 **15.** 1 **17.** binomial **19.** monomial
21. $3x^3 + 8x^2 - 2x - 6$ **23.** $5a^3 - 3a^2 + 2a + 1$ **25.** $-b^2 + 4$ **27a.** $-2y^3; -6y; 4$ **b.** $3; -4; -3$ **c.** $3y^3 - 4y - 3$
29. $12m^2 + m - 4$ **31.** $-4x^2 - 23x - 2$ **33.** $11p^3 + p^2 - 14p + 4$ **35.** $4x^3 - 15x^2 + 14x - 21$ **37.** $-6y^3 - 2$
39. $2y^2 - 2$ **41.** $11k^2 + 2k - 18$ **43.** $17x^3 - 5$ **45.** $16b^3 + 14b^2 + 1$ **47.** $8p^3 + 9p^2 - 6p - 7$ **49.** $-6a^3 - 17$
51. $3d^4 - 10d^2 - 3$ **53.** (i) and (iii) **55.** opposite **57.** $-8x^3 - 5x^2 + 3x + 6$ **59.** $9a^3 - a^2 + 2a - 9$
61. $4y^2 - 10y - 1$ **63.** $6w^3 + 3w^2 + 9w - 19$ **65.** $-2t^3 + 16t + 5$ **67.** $8p^3 - 9p^2 + 14p + 12$ **69.** $-6v^3 + 9v^2 + v - 3$
71. $5m^2 - 2m - 11$ **73.** $3b^2 - 15b - 18$ **75.** $10y^3 - 6y^2 - 10y - 20$ **77.** $-3a^3 + a + 24$ **79.** $-2m^3 + m^2 + 7m - 6$
81. $-10q^3 + 16q$ **83.** $12x^4 + 11x^2 - 17$ **85.** True **87.** $(10x^2 + 2x + 2)$ km **89.** $(3y + 11)$ ft
91. $-0.2n^2 + 480n; 75n + 4000$ **93.** $(-0.6n^2 + 200n - 4000)$ dollars **95.** $(-n^2 + 700n - 1500)$ dollars

5.4 Exercises (pages 348–349)

1. $6; 3; x^9$ **3.** a^9 **5.** x^{16} **7.** n^6 **9.** z^8 **11.** a^8b^3 **13.** $-m^9n^3$ **15.** $10x^7$ **17.** $8x^3y^6$ **19.** $-12m^7$
21. $-14v^3w$ **23.** $-2ab^5c^5$ **25.** $24a^3b^5c^2$ **27.** $40r^3t^7v$ **29.** $-27m^2n^4p^3$ **31.** $24x^7$ **33.** $6a^6b^5$ **35.** No
37. No **39.** $3; 4; x^{12}$ **41.** p^{15} **43.** b^8 **45.** p^{28} **47.** c^{28} **49.** $9x^2$ **51.** $x^{12}y^{18}$ **53.** $r^{12}t^4$ **55.** y^4
57. $8x^{12}$ **59.** $-8a^6$ **61.** $9x^4y^2$ **63.** $8a^9b^3c^6$ **65.** $m^4n^{20}p^{12}$ **67.** No **69.** Yes **71.** $(49y^{10})$ cm^2

5.5 Exercises (pages 352–353)

1. Distributive; $-3y; -3y; -3y; -3y^2; 21y$ **3.** $x^3 - 3x^2 - 4x$ **5.** $8a^3 + 12a^2 - 24a$ **7.** $-6a^3 - 18a^2 + 14a$
9. $4m^4 - 9m^3$ **11.** $10x^5 - 12x^4y + 4x^3y^2$ **13.** $-6r^7 + 12r^6 + 36r^5$ **15.** $12a^4 + 24a^3 - 28a^2$ **17.** $-6n^2 + 8n^5 + 10n^7$
19. $3a^3b^2 - 4a^2b^3 + ab^4$ **21.** $-4x^7y^5 + 5x^5y^4 + 7x^3y^3$ **23.** $6r^2t^3 - 6r^3t^4 - 6r^5t^6$ **25.** $4x^2; -3x; 8x; -6; 4x^2 + 5x - 6$
27. $y^2 + 12y + 27$ **29.** $x^2 + 11x + 30$ **31.** $a^2 - 11a + 24$ **33.** $10z^2 + 9z + 2$ **35.** $40c^2 - 11c - 21$
37. $10v^2 - 11v + 3$ **39.** $35t^2 + 18t - 8$ **41.** $24x^2 - x - 10$ **43.** $25r^2 - 4$ **45.** positive
47. $(2x^2 - 9x - 18)$ mi^2

5.6 Exercises (pages 358–360)

1. $1; 1; 1$ **3.** 1 **5.** -1 **7.** $\dfrac{1}{9}$ **9.** $\dfrac{1}{8}$ **11.** $\dfrac{1}{x^5}$ **13.** $\dfrac{1}{w^8}$ **15.** $\dfrac{1}{y}$ **17.** a^5 **19.** b^3 **21.** a^6 **23.** q^4

25. mn^2 **27.** t^2u^3 **29.** $\dfrac{1}{x^5}$ **31.** $\dfrac{1}{b^4}$ **33.** No **35.** No **37.** $1; 10; 10$ **39.** $9;$ right; -9 **41.** 2.37×10^6

43. 4.5×10^{-4} **45.** 3.09×10^5 **47.** 6.01×10^{-7} **49.** 5.7×10^{10} **51.** 1.7×10^{-8} **53.** 710,000 **55.** 0.000043
57. 671,000,000 **59.** 0.00000713 **61.** 5,000,000,000,000 **63.** 0.00801 **65.** No. 84.3 is not a number between 1 and 10.
67. No. 2.5 is not an integer. **69.** 1×10^{21} **71.** 4×10^{-10} **73.** 1.2×10^{10} **75.** 2.45×10^9 **77.** $m \times 10^8 > n \times 10^6$
79a. $>$ **b.** $>$ **c.** $>$ **d.** $<$

5.7 Exercises (pages 365–368)

1. sum of; times **3.** difference between; divided by **5.** $t + 3$ **7.** q^4 **9.** $-2 + z$ **11.** $2q + 5$ **13.** $8d - 7$

15. $6c - 12$ **17.** $2(3 + w)$ **19.** $4(2r - 5)$ **21.** $\dfrac{v}{v - 4}$ **23.** $4t^2$ **25.** $m^2 + m^3$ **27.** $(31 - s) + 5$ **29.** iii

31. $\dfrac{1}{5}; \dfrac{3}{5}; \dfrac{4}{5}n$ **33.** $x - (x + 12); -12$ **35.** $\dfrac{2}{3}x - \dfrac{3}{8}x; \dfrac{7}{24}x$ **37.** $2(7x + 6); 14x + 12$ **39.** $11x + 3x; 14x$

41. $9(x + 7); 9x + 63$ **43.** $(x + 5) + 7; x + 12$ **45.** $7(x - 4); 7x - 28$ **47.** $10x - 3x; 7x$ **49.** $x + (7x - 8); 8x - 8$

51. $5 + 2(x + 15); 2x + 35$ **53.** $14 - (x + 13); -x + 1$ **55.** $8x(2); 16x$ **57.** $(x + 9) + (x - 3); 2x + 6$

59. $5(9 - y); -5y + 45$ **61.** $(17 - m) - 9; -m + 8$ **63.** $6W$ **65.** Let d be the distance from Earth to the moon; $390d$

67. Let G be the number of genes in the roundworm genome; $G + 11,000$ **69.** Let A be the amount of cashews in the

mixture; $3A$ **71.** Let p be the original price; $\dfrac{3}{4}p$ **73.** Let L be the length of the longer piece; $3 - L$

75. Let L be the length of the shorter piece; $12 - L$ **77.** $2x$

Chapter 5 Review Exercises *(pages 373–374)*

1. $6z^2 - 6z$ [5.2A] **2.** $-18z - 2$ [5.1B] **3.** $10z^2 - z - 15$ [5.3A] **4.** $-8m^5n^2$ [5.4A] **5.** $\dfrac{1}{243}$ [5.6A] **6.** $-\dfrac{3}{7}$ [5.1A]

7. x [5.1A] **8.** $-4s + 43t$ [5.2B] **9.** $15x^3y^7$ [5.4A] **10.** $21a^2 - 10a - 24$ [5.5B] **11.** $-3b^3 + 4b - 18$ [5.3B]

12. $32z^{20}$ [5.4B] **13.** $6w$ [5.1A] **14.** $-15x^3yz^3 + 30xy^2z^4 - 5x^4y^5z^2$ [5.5A] **15.** $-\dfrac{4}{9}$ [5.1A] **16.** $-12c + 32$ [5.1B]

17. $-2m + 16$ [5.2A] **18.** $-12a^5b^{15}$ [5.4A] **19.** The Distributive Property [5.1B] **20.** p^6q^9 [5.4B] **21.** $\dfrac{1}{a^7}$ [5.6A]

22. 3.97×10^{-5} [5.6B] **23.** The Commutative Property of Addition [5.1A] **24.** $3y^3 + 8y^2 + 8y - 19$ [5.3A]

25. $12c - 4d$ [5.2B] **26.** $14m - 42$ [5.1B] **27.** x^2y^4 [5.6A] **28.** $-5a^2 + 3a + 9$ [5.2A] **29.** $12p^2 - 15p - 63$ [5.5B]

30. $-8a^5b + 10a^3b^3 - 6a^2b^5$ [5.5A] **31.** $3x - 4y$ [5.2A] **32.** $-a + 13b$ [5.2B] **33.** $\dfrac{1}{c^5}$ [5.6A] **34.** $6x^3 - 10x$ [5.3B]

35. $240,000$ [5.6B] **36.** $(3b^2 + 3)$ ft [5.3C] **37.** $\dfrac{4x}{7} - 9$ [5.7A] **38.** $3x + (x - 7); 4x - 7$ [5.7B] **39.** 6.023×10^{23} [5.6B]

40. Let p be the number of pounds of mocha java beans; $30 - p$ [5.7C]

Chapter 5 Test *(pages 375–376)*

1. $-r$ [5.1A] **2.** $-15y + 21$ [5.1B] **3.** $3y + 3$ [5.2A] **4.** $-4x^2 + 5z$ [5.2A] **5.** $-3a - 6b + 18$ [5.2A] **6.** $\dfrac{4}{5}$ [5.1A]

7. $4x + 13y$ [5.2B] **8.** $-10a + b + 9$ [5.2B] **9.** 7.9×10^{-7} [5.6B] **10.** $4,900,000$ [5.6B] **11.** $6x^2 - 5x + 5$ [5.3A]

12. v^8w^{20} [5.4B] **13.** $27m^6n^9$ [5.4B] **14.** $-10v^5z^3$ [5.4A] **15.** $6p^2 - p - 40$ [5.5B] **16.** $-8m^3n^5 + 4m^5n^2 - 6m^2n^6$ [5.5A]

17. $4w$ [5.1A] **18.** xy^3 [5.6A] **19.** $\dfrac{1}{a^5}$ [5.6A] **20.** The Associative Property of Multiplication [5.1A]

21. $-3a^3 + a^2 - 10$ [5.3B] **22.** c^6 [5.6A] **23.** The Distributive Property [5.1B] **24.** 0 [5.1A] **25.** $9x^2 - 49y^2$ [5.5B]

26. $\dfrac{4}{7}$ [5.1A] **27.** 7.2×10^8 [5.6B] **28.** $12a^2 - 18a - 12$ [5.5B] **29.** $23a - 12b$ [5.2B] **30.** $\dfrac{m^2}{n^3}$ [5.6A]

31. $3x + 5$ [5.7A] **32.** $x + (x - 6); 2x - 6$ [5.7B] **33.** Let s be the number of cups of sugar in the batter; $s + 3$. [5.7C]

Cumulative Review Exercises *(pages 377–378)*

1. -12.4 [4.3B] **2.** $14v - 2$ [5.2A] **3.** $6x^2 + 2x - 20$ [5.5B] **4.** $\dfrac{3}{8}$ [3.4B] **5.** 24 [4.5A]

6. [number line from -6 to 6 with closed point/ray] [4.6A] **7.** x^7 [5.6A] **8.** -9 [2.4A] **9.** 8.4×10^{-7} [5.6B]

10. $9x^2 - 2x - 4$ [5.3A] **11.** -35 [4.5A] **12.** $\dfrac{11}{20}$ [3.6B] **13.** $-12a^7b^9$ [5.4A] **14.** $\dfrac{1}{x^2}$ [5.6A] **15.** $\dfrac{2}{5}$ [3.6A]

16. $-48p$ [5.1A] **17.** 200 [4.2A] **18.** $-12a^3b^3 - 15a^2b^3 + 6a^2b^4$ [5.5A] **19.** $18x$ [5.2B] **20.** -7 [2.3B] **21.** $\dfrac{9}{16}$ [4.3C]

22. 4 [2.5A] **23.** $10\sqrt{3}$ [4.5B] **24.** $5y^2 - 2y - 5$ [5.3B] **25.** 1 [3.3A] **26.** -1 [5.6A] **27.** $32a^{20}b^{15}$ [5.4B]

28. 90 [2.5A] **29.** $2\dfrac{2}{5}$ [3.3A] **30.** 0.0000623 [5.6B] **31.** $\dfrac{10}{x - 9}$ [5.7A] **32.** $2(x + 4) - 2; 2x + 6$ [5.7B]

33. 30.78 in. [4.2B] **34.** 657 lb [4.2B] **35.** Let d be the distance from Earth to the sun; $30d$ [5.7C] **36.** $\$3,075$ [4.3D]

Answers to Chapter 6 *Exercises*

Prep Test *(page 380)*

1. -4 [2.2B] **2.** 1 [3.3A] **3.** -10 [3.3A] **4.** 1 [2.3B] **5.** $7y$ [5.2A] **6.** -9 [5.2A] **7.** -5 [2.5A] **8.** 13 [2.5A]

6.1 Exercises *(pages 387–388)*

3. 6 **5.** 9 **7.** 17 **9.** 1 **11.** -7 **13.** -9 **15.** 5 **17.** 9 **19.** 0 **21.** -8 **23.** -6 **25.** 12 **27.** 6 **29.** -12 **31.** $\frac{2}{7}$ **33.** $\frac{1}{2}$ **35.** ii **39.** 3 **41.** -3 **43.** -8 **45.** 7 **47.** 0 **49.** 6 **51.** -4 **53.** 4 **55.** $\frac{5}{2}$ **57.** $-\frac{7}{2}$ **59.** 6 **61.** -28 **63.** 10 **65.** $\frac{7}{10}$ **67.** -5 **69.** -7 **71.** ii **73.** $x = \frac{b}{a}$; no, $a \neq 0$

6.2 Exercises *(pages 391–394)*

3. 4; 7 **5.** 2 **7.** 10 **9.** 2 **11.** -1 **13.** -2 **15.** -4 **17.** 3 **19.** 9 **21.** 4 **23.** $\frac{3}{5}$ **25.** $\frac{7}{2}$ **27.** $\frac{3}{4}$ **29.** $\frac{3}{2}$ **31.** $\frac{7}{4}$ **33.** $-\frac{5}{2}$ **35.** $\frac{9}{2}$ **37.** 0 **39.** $-\frac{7}{6}$ **41.** $\frac{1}{3}$ **43.** 1 **45.** 10 **47.** 28 **49.** -45 **51.** -12 **53.** 10 **55.** $\frac{24}{7}$ **57.** $\frac{5}{6}$ **59.** $\frac{7}{5}$ **61.** $\frac{9}{4}$ **63.** 1.1 **65.** 5.8 **67.** 4 **69.** 1 **71.** $\frac{11}{3}$ **73.** negative **75a.** C; V; t **b.** 76,000; 76,000; 76,000 **c.** $-7,700$ **d.** $-5,500$ **e.** 1.4 **f.** years; $68,300 **77.** 2.8 years **79.** $14,450.87 **81.** 1952 **83.** 168 ft **85.** 41,493

6.3 Exercises *(pages 399–402)*

1a. $2x$ **b.** 3 **3.** 3 **5.** 3 **7.** -1 **9.** -4 **11.** 5 **13.** 2 **15.** 1 **17.** $\frac{7}{6}$ **19.** 3 **21.** $\frac{8}{5}$ **23.** $\frac{7}{2}$ **25.** $\frac{7}{2}$ **27.** 3 **29.** -1 **31.** 0 **33.** negative **35.** $3x - 18$ **37.** iv **39.** $\frac{5}{6}$ **41.** -3 **43.** 3 **45.** 2 **47.** 2 **49.** 2 **51.** 3 **53.** $\frac{1}{2}$ **55.** 8 **57.** -2 **59.** $-\frac{6}{5}$ **61.** 60; 84; 12; x **63.** 6 ft **65.** 34.6 lb **67.** 325 units **69.** 400 units **71.** 108

6.4 Exercises *(pages 407–410)*

1. $3x = -30$ **3.** x; 12; 3 **5.** $7 + x = 40$; 33 **7.** $x + 4 = -2$; -6 **9.** $x + 12 = 20$; 8 **11.** $\frac{3}{5}x = -30$; -50 **13.** $3x + 4 = 13$; 3 **15.** $9x - 6 = 12$; 2 **17.** $5x - 17 = 3$; 4 **19.** $40 = 7x - 9$; 7 **21.** $2(x - 25) = 3x$; -50 **23.** $3x = 2(20 - x)$; 8 and 12 **25.** $2x = (21 - x) + 3$; 8 and 13 **27.** (ii) and (iv) **29.** $42 = 2x$; 21 million passengers **31.** $6.3 = x + 3.7$; $2.6 billion **33.** $3,400,000 = 20x$; 170,000 U.S. military personnel **35.** $58 = 2x - 2$; 30 million tons **37.** $18 + 1.50n = 34.50$; 11 min **39.** $2x = (12 - x) - 3$; 3 ft and 9 ft **41.** $2x = (7,000 - x) - 1,000$; $5,000 **43.** $x + (x + 1) + (x + 1 + 2) = 10$; 2 lb of Colombian coffee, 3 lb of French Roast, and 5 lb of Java **45.** identity **47.** conditional; $y = -\frac{5}{16}$ **49.** contradiction

6.5 Exercises *(pages 415–419)*

1. right; down **5.** I **7.** II **9.** y-axis **11a.** The abscissa is positive and the ordinate is positive. **b.** The abscissa is negative and the ordinate is negative.

13. 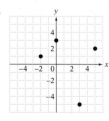 **15.** 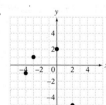 **17.** 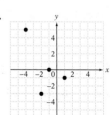 **19.**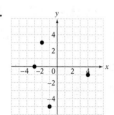

21.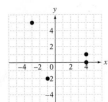

23. $A(0, 2)$, $B(-4, -1)$, $C(2, 0)$, $D(1, -3)$ **25.** $A(0, 4)$, $B(-4, 3)$, $C(-2, 0)$, $D(2, -3)$

27. $A(3, 5)$, $B(1, -4)$, $C(-3, -5)$, $D(-5, 0)$ **29.** $A(1, -4)$, $B(-3, -6)$, $C(-2, 0)$, $D(3, 5)$ **31a.** 2; -4 **b.** 1; -3

33a. 4; -3 **b.** -2; 2 **35.** II   **37.** $(5, 250)$; $(10, 250)$

39. 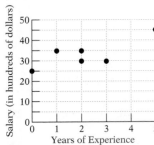   **41.**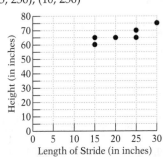

43. The point $(10, 1)$ will move up to $(10, 2)$. The graph will have the same number of points.   **45.** 200 s   **47.** The point will be graphed with an x-coordinate of $1,200$ and a y-coordinate equal to the record time for the $1,200$-meter race. The graph will have an additional point.   **49a.** 35 mpg   **b.** 30 mpg   **51.** 4 units   **53.** 2 units   **55.** 5 units

6.6 Exercises (pages 424–429)

1. No   **3.** Yes   **5.** No   **7.** x; y **9.** Yes   **11.** No   **13.** No   **15.** Yes   **17.** No   **19.** $(3, 7)$ **21.** $(6, 3)$

23. $(0, 1)$ **25.** $(-5, 0)$ **27.** iv   **29.** Yes   **31.** No   **33a.** -1; -11; $(-1, -11)$ **b.** 0; -5; $(0, -5)$ **c.** 1; 1; $(1, 1)$

35. 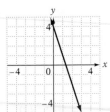   **37.** 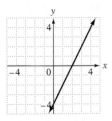   **39.** 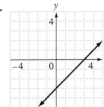   **41.**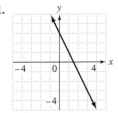

43. 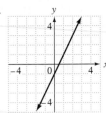   **45.** 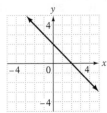   **47.** 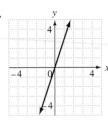   **49.**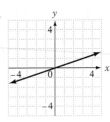

51. 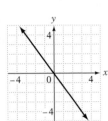   **53.** 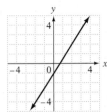   **55.** 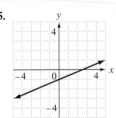   **57.**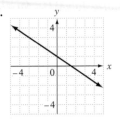

59. 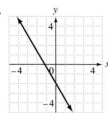 **61.** 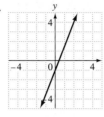 **63.** 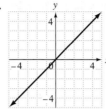 **65.** 4 **67.** −1 **69.** 2

71. 2 **73.** 1 **75.** −3 **77.** same sign **79.** (0, 1)

Chapter 6 Review Exercises *(pages 433–434)*

1. −3 [6.1A] **2.** 3 [6.1B] **3.** $\frac{3}{2}$ [6.2A] **4.** −7 [6.1A] **5.** −24 [6.1B] **6.** $-\frac{15}{32}$ [6.1B] **7.** 2 [6.2A]

8. $\frac{17}{3}$ [6.3B] **9.** $\frac{1}{2}$ [6.3A] **10.** −2 [6.3A] **11.** 3 [6.3B] **12.** −4 [6.1B] **13.** $-\frac{10}{9}$ [6.3B] **14.** 4 [6.3B]

15. −4 [6.2A] **16.** Yes [6.6A] **17.** 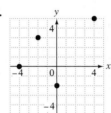 [6.5A] **18.** 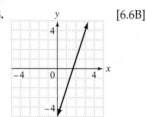 [6.6B]

19. 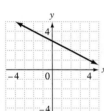 [6.6B] **20.** (2, −1) [6.6A] **21.** $7 - 5x = 37$; −6 [6.4A] **22.** 16 in. [6.4B] **23.** 7 h [6.4B]

24. 55 m [6.4B] **25.** 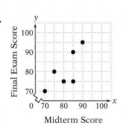 [6.5B] **26.** 12.5 lb [6.3C] **27.** 147 amplifiers [6.2B]

Chapter 6 Test *(pages 435–436)*

1. −5 [6.1A] **2.** −10 [6.1B] **3.** −3 [6.2A] **4.** 3 [6.2A] **5.** 4 [6.3A] **6.** 3 [6.3A] **7.** 1 [6.3B] **8.** $\frac{15}{4}$ [6.3B]

9. $-\frac{7}{24}$ [6.2A] **10.** $\frac{7}{3}$ [6.3A] **11.** (−3, 1) [6.5A] **12.** 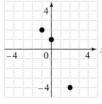 [6.5A]

13. 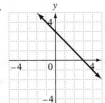 [6.6B] **14.** 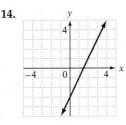 [6.6B] **15.** 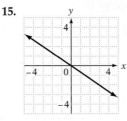 [6.6B]

16. 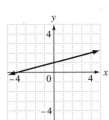 [6.6B] **17.** 3 [6.3B] **18.** −6 [6.2A] **19.** (6, −2) [6.6A] **20.** $4 + \frac{1}{3}n = 9$; 15 [6.4A]

21. $8 + 2x = -4$; −6 [6.4A] **22.** 5 and 12 [6.4A] **23.** [6.5B] **24.** 5 h [6.4B]

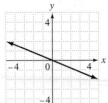

25. 100 ft [6.2B]

Cumulative Review Exercises *(pages 437–438)*

1. 18 [2.3A] **2.** −12p + 21 [5.1B] **3.** 0 [3.6C] **4.** −18 [6.1B] **5.** 8 [2.5A] **6.** −60 [2.5A] **7.** 11 [4.5A]
8. $4\sqrt{3}$ [4.5B] **9.** 2v + 7 [5.2B] **10.** 12m [5.1A] **11.** No [2.3A] **12.** 1 [6.3A] **13.** −10z + 12 [5.2B]
14. $\frac{5}{4}$ [3.6C] **15.** $\frac{3}{2}$ [6.2A] **16.** $32m^{10}n^{25}$ [5.4B] **17.** $-6a^5 - 9a^4b + 12a^3b^2$ [5.5A] **18.** $6x^2 - 7x - 3$ [5.5B]
19. $\frac{1}{16}$ [5.6A] **20.** x^6 [5.6A] **21.** $15x^8y^3$ [5.4A] **22.** $\frac{31}{8}$ [6.3B] **23.** [6.6B]

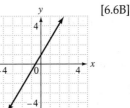

24. [6.6B] **25.** 0.000000035 [5.6B] **26.** 5(n + 2); 5n + 10 [5.7B] **27.** 1.5 s [6.2B]

28. Let w be the number of wolves in the world; 1,000w [5.7C] **29.** $1,192.5 million [4.2B] **30.** $576.50 [6.4B]
31. 1,929 stories [4.3D] **32.** $4,000 and $8,000 [6.4B]

Answers to Chapter 7 *Exercises*

Prep Test *(page 440)*

1. $\frac{4}{5}$ [3.2B] **2.** 24.8 [4.3C] **3.** 4 [3.3A] **4.** 10 [3.3A] **5.** 46 [3.3A] **6.** 238 [1.3C] **7.** 37,320 [4.3A]
8. 0.04107 [4.3B] **9.** 5.125 [4.2A] **10.** 5.96 [4.2A] **11.** 0.13 [4.3A] **12.** 56.35 [3.3A, 4.3A] **13.** 0.5 [3.3A, 4.3A]
14. 3.75 [4.3B]

7.1 Exercises *(pages 445–448)*

3a. row 2: 10^9; row 3: 1 000 000; row 4: k, 10^3; row 5: 10^2; row 6: 10; row 7: 0.1; row 8: c, 0.01; row 9: m, $\dfrac{1}{10^3}$; row 10: 0.000 001;

row 11: 0.000 000 001; row 12: $\dfrac{1}{10^{12}}$ **b.** The exponent on 10 indicates the number of places to move the decimal point. For the prefixes tera-, giga-, mega-, kilo-, hecto-, and deca-, move the decimal point to the right. For the other prefixes shown, move the decimal point to the left.   **5.** kilogram   **7.** milliliter   **9.** kiloliter   **11.** centimeter   **13.** kilogram   **15.** liter
17. gram   **19.** meter or centimeter   **21.** kilogram   **23.** milliliter   **25.** kilogram   **27.** kilometer   **29.** 910 mm
31. 1.856 kg   **33.** 7.285 L   **35.** 8 000 mm   **37.** 0.034 g   **39.** 29.7 ml   **41.** 7.530 km   **43.** 9 200 g   **45.** 36 L
47. 2 350 m   **49.** 83 mg   **51.** 0.716 m   **53.** 3.206 kl   **55.** 99 cm   **57.** 60.5 cm   **59.** cm   **61.** cg   **63.** cm
65. centimeters; divide   **67.** 3 g   **69.** 24 L   **71.** 16 servings   **73a.** 0.186 kg   **b.** 0.42 g   **75.** 10 L   **77.** 500 s
79. \$186.50   **81.** 2 720 ml or 2.720 L

7.2 Exercises *(pages 451–452)*

1. are not; are   **3.** $\dfrac{2}{3}$, 2:3, 2 to 3   **5.** $\dfrac{3}{8}$, 3:8, 3 to 8   **7.** $\dfrac{9}{2}$, 9:2, 9 to 2   **9.** $\dfrac{1}{2}$, 1:2, 1 to 2   **11.** $\dfrac{9}{13}$ **13.** $\dfrac{7\text{ children}}{5\text{ families}}$

15. $\dfrac{80\text{ mi}}{3\text{ h}}$ **17.** $\dfrac{\$9}{1\text{ h}}$ **19.** \$11.50/h   **21.** 55.4 mph   **23.** \$4.24/lb   **25.** $\dfrac{1}{56}$
27. Australia: 6.8 people/mi^2; India: 863.2 people/mi^2; U.S.: 80.8 people/mi^2 **29.** False

7.3 Exercises *(pages 459–463)*

1. denominator; numerator   **3.** $5\dfrac{1}{3}$ ft   **5.** $2\dfrac{5}{8}$ lb   **7.** $1\dfrac{1}{2}$ mi   **9.** $\dfrac{1}{4}$ ton   **11.** $2\dfrac{1}{2}$ gal   **13.** 20 fl oz   **15.** 11,880 ft

17. $\dfrac{5}{8}$ ft   **19.** $7\dfrac{1}{2}$ c   **21.** $3\dfrac{3}{4}$ qt   **23.** $\dfrac{86{,}400\text{ s}}{1\text{ day}}$ **25a.** $\dfrac{1\text{ c}}{8\text{ fl oz}}$; $\dfrac{1\text{ qt}}{4\text{ c}}$ **b.** $\dfrac{\$3.81}{1\text{ qt}}$ **27.** 1,103,760,000 s   **29.** $12\dfrac{1}{2}$ gal

31. $7\dfrac{1}{2}$ gal   **33.** 25 qt   **35.** 750 mph   **37.** \$65,340   **39.** $\dfrac{28.35\text{ g}}{1\text{ oz}}$ **41.** 65.91 kg   **43.** 8.48 c   **45.** 54.20 L
47. 189.2 lb   **49.** 1.38 in.   **51.** 6.33 gal   **53.** 18.29 m/s   **55.** \$2.18   **57.** 49.69 mph   **59.** \$2.98   **61.** 40 068.07 km
63. $\dfrac{39.36\text{ in.}}{1\text{ m}}$ **65.** Answers will vary.

7.4 Exercises *(pages 468–471)*

1. proportion; *a*; *d*; *b*; *c*; equal   **3.** Not true   **5.** True   **7.** True   **9.** Not true   **11.** 10   **13.** 2.4   **15.** 4.5
17. 17.14   **19.** 25.6   **21.** 20.83   **23.** 4.35   **25.** 10.97   **27.** 1.15   **29.** 38.73   **31.** 0.5   **33.** 10

35. 0.43   **37.** −1.6   **39.** 6.25   **41.** 32   **43.** 6.2   **45.** 5.8   **47.** Answers will vary.
a. One example is $\dfrac{2}{9} = \dfrac{4}{18}$.   **b.** One example is $\dfrac{2}{4} = \dfrac{4}{8}$.   **49.** $\dfrac{n\text{ mi}}{30\text{ min}}$ **51.** 0.0052 in.   **53.** 24 robes   **55.** \$6,720
57. 406 mi   **59.** 438 lights   **61.** 60 weeks   **63.** 26.92 mi   **65.** \$2,250   **67.** 20 mi   **69.** 1.6 lb   **71.** No

7.5 Exercises *(pages 476–479)*

3a. 1.8; 1.5   **b.** divide; 1.5; 1.2   **c.** 1.2   **d.** 3.5; 3.5; 4.2   **e.** 4.2   **5.** $\dfrac{1}{6}$ **7.** 6   **9.** 28   **11.** 10   **13.** 162
15. \$307.50   **17.** 5.4 lb/in^2 **19.** 287.3 ft   **21.** 3 amps   **25.** 40   **27.** 24   **29.** 10   **31.** 10   **33.** 10 ft
35. $1.\overline{6}$ ohms   **37.** 2,160 computers   **39.** 51.2 lumens   **41.** 416 items   **43.** indirect variation
45a. *y* is 4 times larger.   **b.** *y* is 8 times larger.

Chapter 7 Review Exercises *(pages 485–486)*

1. 1 250 m [7.1A]   **2.** 450 mg [7.1A]   **3.** $\dfrac{1}{1}$, 1:1, 1 to 1 [7.2A]   **4.** $\dfrac{2\text{ roof supports}}{1\text{ ft}}$ [7.2A]   **5.** \$15.70/h [7.2A]

6. $\dfrac{8}{15}$ [7.2A]   **7.** $2\dfrac{2}{3}$ yd [7.3A]   **8.** $4\dfrac{1}{2}$ lb [7.3A]   **9.** $4\dfrac{1}{2}$ c [7.3A]   **10.** 6,600 ft [7.3A]   **11.** 1.6 [7.4A]

12. $\dfrac{5\text{ lb}}{4\text{ trees}}$ [7.2A]   **13.** 57 mph [7.2A]   **14.** 6.86 [7.4A]   **15.** 943.40 ml [7.3C]   **16.** 8.84 m [7.3C]   **17.** 328 ft [7.3C]

18. 4.62 lb [7.3C]   **19.** 48.3 km/h [7.3C]   **20.** 46.58 mph [7.3C]   **21.** $\dfrac{3\text{ c}}{4\text{ pt}}$ [7.2A]   **22.** $\dfrac{1}{3}$ [7.5A]   **23.** 28,800 [7.5A]

24. 0.04 [7.5B] **25.** 181 cm [7.1A] **26.** $\frac{2}{5}$ [7.2A] **27.** 12 oz [7.3B] **28.** $12,000 [7.4B] **29.** 11.2 in. [7.5A]

30. 2.75 lb [7.4B] **31.** 127.6 ft/s [7.3B] **32.** 1.25 ft³ [7.5B] **33.** $64,000 [7.4B]

Chapter 7 Test *(pages 487–488)*

1. 46.50 m [7.1A] **2.** 4 100 ml [7.1A] **3.** $\frac{1}{8}$, 1:8, 1 to 8 [7.2A] **4.** $\frac{1\text{ oz}}{4\text{ cookies}}$ [7.2A] **5.** 0.6 mi/min [7.2A]

6. $\frac{2}{1}$ [7.2A] **7.** 5,200 lb [7.3A] **8.** 20 fl oz [7.3A] **9.** 52 oz [7.3A] **10.** 102 in. [7.3A] **11.** 0.75 [7.4A]

12. 2 ft/s [7.2A] **13.** 34.4 fl oz [7.3A] **14.** 126 ft [7.3A] **15.** 476.67 ft²/h [7.2A] **16.** 3.56 [7.4A] **17.** 340.2 g [7.3C]
18. 166.77 m [7.3C] **19.** 1,090 yd [7.3C] **20.** 4.18 lb [7.3C] **21.** 56.35 km/h [7.3C] **22.** 37.27 mph [7.3C]

23. 20 [7.5B] **24.** 75 [7.5A] **25.** 80 [7.5B] **26.** $\frac{33}{38}$ [7.2A] **27.** $3,136 [7.4B] **28.** 243,750 voters [7.4B]

29. $114 [7.3B] **30.** 50 ft [7.4B] **31.** 76.27 ft/s [7.3B] **32.** 292.5 ft [7.5A] **33.** 100 rpm [7.5B]

Cumulative Review Exercises *(pages 489–490)*

1. 57 [3.6C] **2.** 4.8 qt [7.3A] **3.** $3\frac{31}{36}$ [3.4B] **4.** $1\frac{2}{3}$ [3.6C] **5.** −114 [2.3B] **6.** 22 [2.5A] **7.** 4 [6.2A]

8. 3 [6.3B] **9.** [number line] [4.6A] **10.** [number line] [4.6B] **11.** 21 [2.5A] **12.** $-\frac{3}{8}$ [3.6A]

13. 13 [4.5A] **14.** $8a - 3$ [5.2B] **15.** $-20a^5b^4$ [5.4A] **16.** $-4y^2 - 3y$ [5.2A] **17.** (−1, −5) [6.6A] **18.** $\frac{3}{10}$ [7.2A]

19. $3,885/month [7.2A] **20.** $.78 [7.3C] **21.** 32 [7.4A] **22.** $\frac{10}{11}$ [3.6B] **23.** −10 [4.5A] **24.** −2 [6.3B]

25. $28.67 [4.3D] **26.** 12 [6.4A] **27.** $4x - 3(x + 2)$; $x - 6$ [5.7B] **28.** 64 mi [1.2C] **29.** $265.48 [4.2B] **30.** $\frac{4}{15}$ [3.4C]

31. 20,854 votes [7.4B] **32.** 35 mi [7.2A] **33.** 3,750 rpm [6.4B]

Answers to Chapter 8 *Exercises*

Prep Test *(page 492)*

1. $\frac{19}{100}$ [3.3A] **2.** 0.23 [4.3A] **3.** 47 [4.3A] **4.** 2,850 [4.3A] **5.** 4,000 [4.3B] **6.** 32 [3.3B] **7.** 62.5 [3.3A, 4.3C]

8. $66\frac{2}{3}$ [3.2A] **9.** 1.75 [4.3B]

8.1 Exercises *(pages 495–497)*

3. 0.01; 0.01; 0.53 **5.** $\frac{1}{100}$; $\frac{1}{100}$; $\frac{4}{5}$ **7.** $\frac{1}{20}$, 0.05 **9.** $\frac{3}{10}$, 0.30 **11.** $\frac{5}{2}$, 2.50 **13.** $\frac{7}{25}$, 0.28 **15.** $\frac{7}{20}$, 0.35

17. $\frac{29}{100}$, 0.29 **19.** $\frac{1}{9}$ **21.** $\frac{3}{8}$ **23.** $\frac{2}{3}$ **25.** $\frac{1}{15}$ **27.** $\frac{1}{200}$ **29.** $\frac{1}{16}$ **31.** 0.073 **33.** 0.158 **35.** 0.003

37. 1.212 **39.** 0.6214 **41.** 0.0825 **43.** less than **45.** $\frac{6}{25}$ **47.** 100%; 100%; 125% **49.** 100%; 100%; 700; 140%

51. 37% **53.** 2% **55.** 12.5% **57.** 136% **59.** 96% **61.** 7% **63.** 83% **65.** 33.3% **67.** 44.4% **69.** 45%

71. 250% **73.** 16.7% **75.** 68% **77.** $56\frac{1}{4}$% **79.** $262\frac{1}{2}$% **81.** $283\frac{1}{3}$% **83.** $23\frac{1}{3}$% **85.** $22\frac{2}{9}$%

87. greater than

8.2 Exercises *(pages 504–508)*

1. 0.12; n; 68 **3.** 0.08; 450; n **5.** 8 **7.** 0.075 **9.** $16\frac{2}{3}$% **11.** 37.5% **13.** 100 **15.** 1,200 **17.** 51.895

19. 13 **21.** 2.7% **23.** 400% **25.** 7.5 **27.** $x > y$ **29.** 36; n; 25 **31.** 65 **33.** 25% **35.** 75 **37.** 12.5%
39. 400 **41.** 19.5 **43.** 14.8% **45.** 62.62 **47.** 5 **49.** 45 **51.** 300% **53.** ii **55.** True **57.** n; 200; 24

59. 12%   **61.** 76.5%; more than   **63.** 6.15 million workers   **65.** 31.6%   **67.** 50 million students   **69.** 20%
71. 7,944 computer boards   **73.** 12,151 more faculty members   **75.** No   **77.** less than

8.3 Exercises *(pages 511–513)*

1. original; new; 18; 15; 3   **3.** 12.5%   **5.** 115.4%   **7.** 50%   **9.** 69.0%   **11.** new; original; 48; 38; 10   **13.** 15.6%
15. 37.5%   **17.** $15,330   **19.** 53.7%   **21.** 93.3%   **23a.** '01 and '02; '01 and '03; '02 and '03
b. '03 and '04; '03 and '05; '03 and '06; '04 and '05; '04 and '06; '05 and '06   **25.** $1,890; no; 37%

8.4 Exercises *(pages 518–520)*

1a. pays for a product   **b.** an amount of money; a percent   **c.** selling price; cost; cost   **3.** $60.50   **5.** 60%   **7.** 44.4%
9. $2,187.50   **11.** $82.25   **13.** True   **15.** True   **17a.** markdown; discount   **b.** an amount of money; a percent
c. regular price; sale price; regular price   **19.** $110   **21.** 23.2%   **23.** 38%   **25.** $1,396.50   **27.** $25.20   **29.** $300
31. $123.08

8.5 Exercises *(pages 523–524)*

1. Row 1: 5,000, 0.06, $\frac{1}{12}$, $25; Row 2: 5,000, 0.06, $\frac{2}{12}$, $50; Row 3: 5,000, 0.06, $\frac{3}{12}$, $75; Row 4: 5,000, 0.06, $\frac{4}{12}$, $100; Row 5: 5,000, 0.06,

$\frac{5}{12}$, $125   **3.** Multiply the interest due by 7.   **5.** $146.40   **7.** $3,166.67   **9.** $16   **11.** $168.75   **13.** $27,050
15. $5,517.50   **17.** 7.5%   **19.** 5.84%

Chapter 8 Review Exercises *(pages 529–530)*

1. $\frac{8}{25}$ [8.1A]   **2.** 0.22 [8.1A]   **3.** $\frac{1}{4}$, 0.25 [8.1A]   **4.** $\frac{17}{500}$ [8.1A]   **5.** 17.5% [8.1B]   **6.** 128.6% [8.1B]   **7.** 280% [8.1B]
8. 21 [8.2A/8.2B]   **9.** 500% [8.2A/8.2B]   **10.** 66.7% [8.2A/8.2B]   **11.** 30 [8.2A/8.2B]   **12.** 15.3 [8.2A/8.2B]
13. 562.5 [8.2A/8.2B]   **14.** 5.625% [8.2A/8.2B]   **15.** 0.014732 [8.2A/8.2B]   **16.** 9.6 [8.2A/8.2B]   **17.** 34.0% [8.2C]
18. $8,400 [8.2C]   **19.** 3,952 telephones [8.2C]   **20.** 36.5% [8.2C]   **21.** $165,000 [8.2C]   **22.** 1,620 seats [8.2C]
23. 196 tickets [8.2C]   **24.** 22.7% [8.2C]   **25.** $11.34 [8.3A]   **26.** 25% [8.3B]   **27.** $19,610 [8.4A]   **28.** 65% [8.4B]
29. $56 [8.4B]   **30.** $390 [8.4B]   **31.** $31.81 [8.5A]   **32.** 9.00% [8.5A]   **33.** $10,630 [8.5A]

Chapter 8 Test *(pages 531–532)*

1. 0.864 [8.1A]   **2.** 40% [8.1B]   **3.** 125% [8.1B]   **4.** $\frac{5}{6}$ [8.1A]   **5.** $\frac{8}{25}$ [8.1A]   **6.** 118% [8.1B]   **7.** 90 [8.2A/8.2B]
8. 49.64 [8.2A/8.2B]   **9.** 56.25% [8.2A/8.2B]   **10.** 200 [8.2A/8.2B]   **11.** 33 accidents [8.2C]   **12.** 82.2% [8.2C]
13. $80 [8.2C]   **14.** 139.4% [8.3A]   **15.** $16\frac{2}{3}$% [8.3A]   **16a.** $83\frac{1}{3}$%   **b.** 100%   **c.** 50% [8.3B]   **17.** 8% [8.3B]
18. $300 [8.2C]   **19.** $12.60 [8.4A]   **20.** 55.1% [8.4A]   **21.** $300 [8.4B]   **22.** 14% [8.4B]   **23.** $315 [8.5A]
24. $41,520.55 [8.5A]   **25.** 8.4% [8.5A]

Cumulative Review Exercises *(pages 533–534)*

1. 24.954 [4.2A]   **2.** 625 [1.3B]   **3.** 14.04269 [4.3A]   **4.** $4x^2 - 16x + 15$ [5.5B]   **5.** $1\frac{25}{62}$ [3.3B]

6. $6a^3b^3 - 8a^4b^4 + 2a^3b^4$ [5.5A]   **7.** 42 [8.2A/8.2B]   **8.** −3 [6.1A]   **9.** 100,500 [4.3A]   **10.** $\frac{23}{24}$ [3.4B]

11. $1\frac{23}{28}$ [3.6B]   **12.** $-12a^7b^5$ [5.4A]   **13.** [6.6B]   **14.** [6.6B]

15. $2\frac{4}{5}$ [3.3B]   **16.** 4 [2.2B]   **17.** -12 [6.1B]   **18.** 5 [6.3A]   **19.** 64.48 mph [7.2A]   **20.** 44.8 [7.4A]

21. 8.3% [8.2A/8.2B]   **22.** 67.2 [8.2A/8.2B]   **23.** 58 [2.5A]   **24.** 5 [6.3B]   **25.** $\frac{1}{10}$ [8.1A]   **26.** $129.60 [8.4B]

27. $42 [8.4A]   **28.** 117 games [7.4B]   **29.** $2\frac{1}{4}$ lb [3.4C]   **30.** 72 ft/s [4.5C]   **31.** 36.11 m/s [7.1A]

32. 36 h [6.4B]   **33.** 5 ohms [7.5B]

Answers to Chapter 9   *Exercises*

Prep Test *(page 536)*

1. 56 [1.5A]   **2.** 56.52 [4.3A]   **3.** 113.04 [4.3A]   **4.** 43 [6.1A]   **5.** 51 [6.1A]   **6.** 14.4 [7.4A]

9.1 Exercises *(pages 547–552)*

1. 40°; acute   **3.** 115°; obtuse   **5.** 90°; right   **7.** 12; 5; x; 4   **9.** 160°; 140°; 360°   **11.** 28°   **13.** 18°   **15.** 14 cm
17. 28 ft   **19.** 30 m   **21.** 86°   **23.** 30°   **25.** 36°   **27.** 71°   **29.** 127°   **31.** 116°   **33.** 20°   **35.** 20°   **37.** 20°
39. 141°   **41.** $90° - x$ **43.** $a; b$ **45.** $c; d;$ 180°   **47.** 106°   **49.** 11°   **51.** $\angle a = 38°, \angle b = 142°$ **53.** $\angle a = 47°,$
$\angle b = 133°$ **55.** 20°   **57.** False   **59.** True   **61a.** $\angle a, \angle b,$ and $\angle c$ **b.** $\angle y$ and $\angle z$ **c.** $\angle x$
63. $\angle x = 155°; \angle y = 70°$ **65.** $\angle a = 45°; \angle b = 135°$ **67.** 60°   **69.** 35°   **71.** True   **73.** 360°

9.2 Exercises *(pages 563–570)*

1. hexagon   **3.** pentagon   **5.** scalene   **7.** equilateral   **9.** obtuse   **11.** acute   **13.** 56 in.   **15.** 14 ft   **17.** 47 mi
19. 8π cm; 25.13 cm   **21.** 11π mi; 34.56 mi   **23.** 17π ft; 53.41 ft   **25.** 17.4 cm   **27.** 8 cm   **29.** 24 m   **31.** 17.5 in.
33. 48.8 cm   **35.** 1.5π in.   **37.** 226.19 cm   **39.** 60 ft   **41.** 44 ft   **43.** 120 ft   **45.** 10 in.   **47.** 12 in.   **49.** 2.55 cm
51. 13.19 ft   **53.** 50.27 ft   **55.** 39,935.93 km   **57.** A square whose side is 1 ft   **59.** 60 ft^2 **61.** 20.25 in^2 **63.** 546 ft^2
65. 16π cm^2; 50.27 cm^2 **67.** 30.25π mi^2; 95.03 mi^2 **69.** 72.25π ft^2; 226.98 ft^2 **71.** 156.25 cm^2 **73.** 570 in^2 **75.** 192 in^2
77. 13.5 ft^2 **79.** 330 cm^2 **81.** 25π in^2 **83.** $10,000\pi$ in^2 **85.** 126 ft^2 **87.** 7,500 yd^2 **89a.** LW **b.** 80; 16; 80
c. 16; 5   **d.** 5   **91.** 10 in.   **93.** 20 m   **95.** 2 qt   **97.** $74   **99.** 113.10 in^2 **101.** $638   **103.** 216 m^2
105a. No; area is measured in square units.   **b.** No; A cannot be a whole number.   **107.** 4 times

9.3 Exercises *(pages 577–581)*

1. the hypotenuse; a leg   **3.** 5 in.   **5.** 8.6 cm   **7.** 11.2 ft   **9.** 4.5 cm   **11.** 12.7 yd   **13.** iii   **15.** 7.4 m
17. 24.3 cm   **19.** $FD; BA$ **21.** $\frac{1}{2}$ **23.** $\frac{3}{4}$ **25.** 7.2 cm   **27.** 3.3 m   **29.** 12 m   **31.** 12 in.   **33.** 56.3 cm^2
35. True   **37.** 18 ft   **39.** 16 m   **41.** $\angle E$ **43.** Yes, SAS Rule   **45.** Yes, SSS Rule   **47.** Yes, ASA Rule
49. Yes, SAS Rule   **51.** No   **53.** No   **55.** Yes

9.4 Exercises *(pages 588–592)*

1a. cone   **b.** cube   **c.** sphere   **d.** cylinder   **3.** 840 in^3 **5.** 15 ft^3 **7.** 4.5π cm^3; 14.14 cm^3 **9.** 34 m^3
11. 15.625 in^3 **13.** 36π ft^3 **15.** 8,143.01 cm^3 **17.** 392.70 cm^3 **19.** 216 m^3 **21.** 2.5 ft   **23.** 4.00 in.
25. length: 5 in.; width: 5 in.   **27.** 75.40 m^3 **29.** iv   **31.** $s^2 + 2sl; l; s$ **33.** 94 m^2 **35.** 56 m^2 **37.** 96π in^2; 301.59 in^2
39. 184 ft^2 **41.** 69.36 m^2 **43.** 225π cm^2 **45.** 402.12 in^2 **47.** 6π ft^2 **49.** 297 in^2 **51.** 3 cm   **53.** 3,217 ft^2
55. 456 in^2 **57.** 22.53 cm^2 **59.** $SA = \pi r l$ **61a.** always true   **b.** never true   **c.** sometimes true

Chapter 9 Review Exercises *(pages 599–600)*

1. $\angle x = 22°, \angle y = 158°$ [9.1C]   **2.** 24 in. [9.3B]   **3.** 168 in^3 [9.4A]   **4.** 68° [9.1B]   **5.** Yes, by the SAS Rule [9.3C]
6. 125.66 m^2 [9.4B]   **7.** 44 cm [9.1A]   **8.** 19° [9.1A]   **9.** 32 in^2 [9.2B]   **10.** 96 cm^3 [9.4A]   **11.** 42 in. [9.2A]
12. $\angle a = 138°, \angle b = 42°$ [9.1B]   **13.** 220 ft^2 [9.4B]   **14.** 9.75 ft [9.3A]   **15.** 42.875 in^3 [9.4A]   **16.** 148° [9.1A]
17. 39 ft^3 [9.4A]   **18.** 95° [9.1C]   **19.** 8 cm [9.2B]   **20.** 288π mm^3 [9.4A]   **21.** 21.5 cm [9.2A]   **22.** 4 cans [9.4B]
23. 208 yd [9.2A]   **24.** 90.25 m^2 [9.2B]   **25.** 276 m^2 [9.2B]

Chapter 9 Test *(pages 601–602)*

1. 7.55 cm [9.3A]   **2.** congruent, SAS [9.3C]   **3.** 111 m^2 [9.2B]   **4.** 42 ft^2 [9.2B]   **5.** $\frac{784\pi}{3}$ cm^3 [9.4A]   **6.** 75 m^2 [9.4B]

7. 4,618.14 cm^3 [9.4A]   **8.** 159 in^2 [9.2B]   **9.** 20° [9.1B]   **10.** 75 m^2 [9.4B]   **11.** 34° [9.1B]   **12.** octagon [9.2A]
13. not necessarily congruent [9.3C]   **14.** 168 ft^3 [9.4A]   **15.** 8.06 m [9.3A]   **16.** 143° [9.1B]   **17.** 500π cm^2 [9.4B]

18. 61° [9.1C]   **19.** 6.67 ft [9.3B]   **20.** 4.27 ft [9.3B]   **21.** 20 m [9.2A]   **22.** 26 cm [9.2A]   **23.** 51.6 ft [9.3A]
24. 102° [9.1C]   **25.** 139° [9.1A]

Cumulative Review Exercises *(pages 603–604)*

1. 204 [8.2A/8.2B]   **2.** 1, 2, 3, 6, 13, 26, 39, 78 [1.3D]   **3.** $\frac{5}{6}$ [3.3B]   **4.** $7x^2 + 4x + 5$ [5.3A]   **5.** 12.8 [4.3B]

6. 2.9×10^{-5} [5.6B]   **7.** 131° [9.1B]   **8.** 26 cm [9.3A]   **9.** 56 in² [9.2B]   **10.** 40° [9.1B]   **11.** $-12x^5y^3$ [5.4A]

12. $\frac{1}{2}$ [6.3B]   **13.** 58° [9.1C]   **14.** 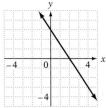 [4.6B]   **15.** $7x + 18$ [5.2B]   **16.** -2 [2.5A]   **17.** $\frac{4}{5}$ [3.6C]

18. 96.6 km/h [7.3C]   **19.** 5 [6.3A]   **20.** [6.6B]   **21.** 3.482 km [7.1A]   **22.** 37.5% [8.1B]

23. $1,313.01 [8.5A]   **24.** 23 gal [7.3B]   **25.** 87 min [6.4B]   **26.** $4.50 [7.4B]   **27.** 5.4% [8.3A]   **28.** 3 ft [9.4A]
29. 40 ft [6.2B]   **30.** 4,000 mi [4.5C]

Answers to Chapter 10 *Exercises*

Prep Test *(page 606)*

1. $\frac{1}{3}$ [3.2B]   **2.** 3.606 [4.5B]   **3.** 48.0% [8.2C]   **4.** Between 2009 and 2010; $5,318 [1.2C]   **5a.** $\frac{5}{3}$ [7.2A]   **b.** 1:1 [7.2A]

6a. 3.9, 3.9, 4.2, 4.5, 5.2, 5.5, 7.1 [4.1B]   **b.** 4.9 million [4.3D]   **7a.** 210,000 women [8.2C]   **b.** $\frac{3}{20}$ [8.1A]

10.1 Exercises *(pages 611–615)*

3a. 98; 55; 43   **b.** 43; 5; 9   **c.** 55; 55; 9; 64   **d.** 65; 65; 9; 74   **e.** 1   **5.** *Tuition at 40 Universities*

Classes	Tally	Frequency
32–40	////	4
41–49	/////	5
50–58	/////	5
59–67	/////	5
68–76	////	4
77–85	////	4
86–94	/////////	9
95–103	////	4

7. 4 tuitions   **9.** 19 tuitions   **11.** 10%   **13.** 7 hotels   **15.** 7 hotels   **17.** 20%   **19.** 34%   **21.** 7 to 11   **23.** 7; 7
25. 32 account balances   **27.** 22%   **29.** 1 to 2   **31.** 60%   **33.** 155–160 and 160–165   **35.** 12.5%   **37.** 55%
39. 80; 90   **41.** 22 nurses   **43.** 66%   **45.** *Nursing Board Test Scores*   **47.** 6 to 1   **49.** 53.3%

Classes (test score)	Frequency
50–60	5
60–70	8
70–80	15
80–90	18
90–100	4

10.2 Exercises *(pages 622–626)*

1a. 62; 75; 87; 54; 70; 36; 6   **b.** 384; 64   **c.** 64   **3.** mean: 19 televisions; median: 19.5 televisions   **5.** mean: 10.61 s;
median: 10.605 s   **7.** median   **9.** 21 unforced errors   **11.** German chocolate cake   **13.** very good   **15.** False
17a. 6; 7; 8; 9; 10; ⑫; 13; 13; 15; 18; 21   **b.** 8   **c.** 15   **19.** $Q_1 = 7.895$, $Q_3 = 9.95$

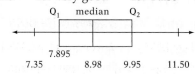

21. $Q_1 = 20$, $Q_3 = 30$

Q₁   median   Q₃ box plot

16   20   25   30   33

23. $Q_1 = 4.3$, $Q_3 = 6.1$

Q₁   median   Q₃ box plot

2.6   4.3   5.2   6.1   8.0

25a. 9; 13; 15; 8; 10; 5; 55; 5; 11   **b.** Row 1: $9 - 11 = -2$, $(-2)^2 = 4$; Row 2: $13 - 11 = 2$, $(2)^2 = 4$; Row 3: $15 - 11 = 4$, $(4)^2 = 16$; Row 4: $8 - 11 = -3$, $(-3)^2 = 9$; Row 5: $10 - 11 = -1$, $(-1)^2 = 1$ **c.** 34   **d.** 34; 5; 6.8
e. 6.8; 2.608   **27.** 7.141 oz   **29.** 4.781 heads   **31.** The standard deviations are the same.   **33a.** always true
b. sometimes true

10.3 Exercises *(pages 633–637)*

3a.

Card 1	Card 2	Card 3
J	Q	K
J	K	Q
Q	J	K
Q	K	J
K	J	Q
K	Q	J

b. 6   **c.** 2

5. HHHH, HHHT, HHTH, HTHH, THHH, HHTT, HTTH, TTHH, HTHT, THTH, THHT, HTTT, TTTH, TTHT, THTT, TTTT

7. (1, 1), (1, 2), (1, 3), (1, 4), (2, 1), (2, 2), (2, 3), (2, 4), (3, 1), (3, 2), (3, 3), (3, 4), (4, 1), (4, 2), (4, 3), (4, 4)   **9.** $\frac{1}{16}$ **11.** $\frac{3}{8}$

13. $\frac{1}{9}$ **15.** 0   **17.** $\frac{1}{36}$ **19.** $\frac{1}{12}$ **21.** $\frac{3}{16}$ **23.** $\frac{1}{4}$ **25.** 38.9%   **27.** $\frac{4}{7}$ **29.** 49.1%
31. No, because the dice are weighted so that some numbers occur more often than other numbers.   **33.** iii

35. $0 \le p \le 1$ **37a.** favorable; unfavorable   **b.** unfavorable; favorable   **39.** ii   **41.** 1 to 1   **43.** $\frac{3}{5}$

45. $\frac{1}{5}$ **47.** $\frac{12}{1}$ **49.** $\frac{1}{41}$ **51.** $\frac{1}{3}$ **53.** $\frac{3}{10}$

Chapter 10 Review Exercises *(pages 643–644)*

1.
Classes	Tally	Frequency
12–19	/////	5
20–27	/////////	9
28–35	///////////	11
36–43	/////////	9
44–51	////	4
52–59	//	2

[10.1A]   **2.** 28–35 [10.1A]   **3.** 25 classes [10.1A]   **4.** 15% [10.1A]

5. 35% [10.1A]   **6.** 25 days [10.1B]   **7.** 29 days [10.1B]   **8.** mean: $214.\overline{54}$; median: 210 [10.2A]   **9.** mean: 7.17 lb;

median: 7.05 lb [10.2A]   **10.** good [10.2A]   **11.** 55 million shares [10.1C]   **12.** 8 A.M.–9 A.M. [10.1C]   **13.** $\frac{3}{5}$ [10.1C]

14. $Q_1 = 105$, median = 111, $Q_3 = 124$

Q₁   median   Q₂ box plot

89   105   111   124   134

[10.2B]   **15.** 6.551 mpg [10.2C]

16. $\frac{1}{500}$ [10.3A]   **17.** $\frac{3}{7}$ [10.3B]   **18.** $\frac{2}{7}$ [10.3B]   **19.** $\frac{1}{6}$ [10.3A]   **20.** $\frac{5}{14}$ [10.3A]

Chapter 10 Test *(pages 645–646)*

1. 65 residences [10.1B]   **2.** $55,000 [10.1C]   **3.** 16% [10.1A]   **4.** 152.875 [10.2A]   **5.** 14 min [10.2A]   **6.** very good
[10.2A]   **7.** 43 digital assistants [10.2A]   **8.** 9.8 [10.2B]   **9a.** 22 days   **b.** 14 vacation days [10.2B]

10.

Q₁   median   Q₃   [10.2B]   **11.** 1.612 incorrect answers [10.2C]   **12.** 12 elements [10.3A]

68   70   72.5   74   80

13. (N, D, Q), (N, Q, D), (D, N, Q), (D, Q, N), (Q, N, D), (Q, D, N) [10.3A]   **14.** $\frac{4}{31}$ [10.3A]   **15.** $\frac{1}{3}$ [10.3A]

16. $\frac{1}{8}$ [10.3A]   **17.** $\frac{2}{3}$ [10.3A]   **18.** $\frac{1}{13}$ [10.3B]   **19.** 1 to 8 [10.3B]   **20.** $\frac{5}{12}$ [10.3A]

Cumulative Review Exercises *(pages 647–648)*

1. $10\sqrt{2}$ [4.5B]   **2.** -1 [6.3B]   **3.** 22 [2.5A]   **4.** $12x - 8$ [5.2B]   **5.** -18 [6.2A]   **6.** $\frac{41}{90}$ [3.6C]

7.   [6.6B]   **8.**   [6.6B]   **9.** $3y^2 + 8y - 9$ [5.3B]   **10.** $64a^6b^3$ [5.4B]

11. 144 [8.2A/B]   **12.** $\frac{8}{3}$ [7.4A]   **13.** 8.76×10^{10} [5.6B]   **14.** 48 in. [9.2A]   **15.** $-15c^3d^{10}$ [5.4A]

16. 40 000 m [7.1A]   **17.** 125° [9.1C]   **18.** 32 m² [9.2B]   **19.** $468.75 [8.5A]   **20.** $\frac{2}{3}$ [10.3A]

21. mean: 31; median: 32.5 [10.2A]   **22.** 76.1% [8.2C]   **23.** 76 800 km [7.4B]   **24.** 2.828 in. [10.2C]
25. $18.00 [8.3A]

Answers to Final Examination *(pages 649–652)*

1. 2,400 [1.2A]   **2.** 24 [1.5A]   **3.** 2 [2.2B]   **4.** 18 [2.5A]   **5.** $2\frac{11}{16}$ [3.4B]   **6.** $\frac{14}{15}$ [3.3B]   **7.** $\frac{2}{9}$ [3.6B]

8. $\frac{5}{16} < 0.313$ [4.3C]   **9.** -769.5 [4.3A]   **10.** -3.28 [4.3B]   **11.** Yes [4.3A]   **12.** $9\sqrt{2}$ [4.5B]

13.   [4.6B]   **14.** $10t$ [5.1A]   **15.** $-2x - 14y$ [5.2B]   **16.** $z^3 + 2z^2 - 6z + 7$ [5.3B]   **17.** $8x^7y$
-4 0

[5.4A]   **18.** $10a^4b^2 - 6a^3b^3 + 8a^2b^4$ [5.5A]   **19.** $15x^2 - x - 6$ [5.5B]   **20.** $81x^8y^4$ [5.4B]   **21.** $\frac{1}{64}$ [5.6A]   **22.** m^2n^4 [5.6A]

23. -6 [6.2A]   **24.** $-\frac{1}{3}$ [6.3A]   **25.** $\frac{5}{6}$ [6.3B]   **26.** 248 cm [7.1A]   **27.** 13,728 ft [7.3A]   **28.** $\frac{4}{3}$ [7.4A]

29. $\angle a = 74°$; $\angle b = 106°$ [9.1B]   **30.** 10.6 ft [9.3A]   **31.** 26 cm [9.2A]   **32.** 96 in³ [9.4A]
33.   [6.6B]   **34.**   [6.6B]   **35.** 364 mph [1.2C]   **36.** 320 products [3.3C]

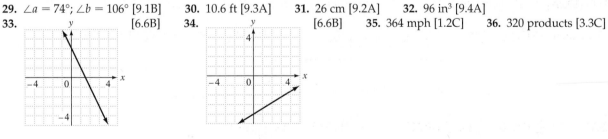

37. 65.98°C [4.2B]   **38.** 5.88×10^{12} [5.6B]   **39.** 6 ft [6.3C]   **40.** 6 tickets [6.4B]   **41.** $4,710 [7.4B]   **42.** 32% [8.2C]
43. 16 rpm [7.5B]   **44.** $34,287.50 [8.2C]   **45.** 22.6% [8.3B]   **46.** $159.25 [8.4B]   **47.** $1,612.50 [8.5A]
48. 37.5% [10.1C]   **49.** mean: $334.80; median: $309 [10.2A]   **50.** $\frac{1}{3}$ [10.3A]

Glossary

abscissa The first number in an ordered pair. It measures a horizontal distance and is also called the first coordinate. [6.5]

absolute value of a number The distance between zero and the number on the number line. [2.1]

acute angle An angle whose measure is between 0° and 90°. [9.1]

acute triangle A triangle that has three acute angles. [9.2]

addend In addition, one of the numbers added. [1.2]

addition The process of finding the total of two numbers. [1.2]

Addition Property of Zero Zero added to a number does not change the number. [1.2]

additive inverses Numbers that are the same distance from zero on the number line, but on opposite sides; also called opposites. [2.2; 5.1]

adjacent angles Two angles that share a common side. [9.1]

alternate exterior angles Two nonadjacent angles that are on opposite sides of the transversal and outside the parallel lines. [9.1]

alternate interior angles Two nonadjacent angles that are on opposite sides of the transversal and between the parallel lines. [9.1]

angle An angle is formed when two rays start at the same point; it is measured in degrees. [1.2; 9.1]

approximation An estimated value obtained by rounding an exact value. [1.1]

area A measure of the amount of surface in a region. [1.3; 7.3; 9.2]

Associative Property of Addition Numbers to be added can be grouped (with parentheses, for example) in any order; the sum will be the same. [1.2]

Associative Property of Multiplication Numbers to be multiplied can be grouped (with parentheses, for example) in any order; the product will be the same. [1.3]

average One number that describes an entire collection of numbers. [10.2]

axes The two number lines that form a rectangular coordinate system; also called coordinate axes. [6.5]

bar graph A graph that represents data by the height of the bars. [1.1]

base In exponential form, the factor that is multiplied the number of times shown by the exponent. [1.3]

base of a triangle The side that the triangle rests on. [3.3; 9.2]

basic percent equation Percent times base equals amount. [8.2]

binomial A polynomial of two terms. [5.3]

borrowing In subtraction, taking a unit from the next larger place value in the minuend and adding it to the number in the given place value in order to make that number larger than the number to be subtracted from it. [1.2]

box-and-whiskers plot A graph that shows the smallest value in a set of numbers, the first quartile, the median, the third quartile, and the greatest value. [10.2]

broken-line graph A graph that represents data by the position of the lines and shows trends and comparisons. [1.1]

capacity A measure of liquid substances. [7.1; 7.3]

carrying In addition, transferring a number to another column. [1.2]

center of a circle The point from which all points on the circle are equidistant. [9.2]

center of a sphere The point from which all points on the surface of the sphere are equidistant. [9.4]

circle A plane figure in which all points are the same distance from point O, which is called the center of the circle. [9.2]

circle graph A graph that represents data by the size of the sectors. [1.1]

circumference The distance around a circle. [9.2]

class frequency The number of occurrences of data in a class on a histogram; represented by the height of each bar. [10.1]

class midpoint The center of a class interval in a frequency polygon. [10.1]

class width Range of numbers represented by the width of a bar on a histogram. [10.1]

coefficient The number part of a variable term. [5.2]

combining like terms Using the Distributive Property to add the coefficients of like variable terms; adding like terms of a variable expression. [5.2]

common factor A number that is a factor of two or more numbers is a common factor of those numbers. [3.1]

common multiple A number that is a multiple of two or more numbers is a common multiple of those numbers. [3.1]

Commutative Property of Addition Two numbers can be added in either order; the sum will be the same. [1.2]

Commutative Property of Multiplication Two numbers can be multiplied in either order; the product will be the same. [1.3]

complementary angles Two angles whose sum is 90°. [9.1]

complex fraction A fraction whose numerator or denominator contains one or more fractions. [3.6]

composite number A number that has natural number factors besides 1 and itself. For instance, 18 is a composite number. [1.3]

congruent objects Objects that have the same shape and the same size. [9.3]

congruent triangles Triangles that have the same shape and the same size. [9.3]

constant of proportionality k in a variation equation; also called the constant of variation. [7.5]

constant of variation k in a variation equation; also called the constant of proportionality. [7.5]

constant term A term that includes no variable part; also called a constant. [5.2]

coordinate axes The two number lines that form a rectangular coordinate system; also simply called axes. [6.5]

coordinates of a point The numbers in an ordered pair that is associated with a point. [6.5]

corresponding angles Two angles that are on the same side of the transversal and are both acute angles or are both obtuse angles. [9.1]

cost The price that a business pays for a product. [8.4]

counting numbers The numbers 1, 2, 3, 4, 5, ... [1.1]

cross product In a proportion, the product of the numerator on the left side of the proportion times the denominator on the right, and the product of the denominator on the left side of the proportion times the numerator on the right. [7.4]

cube A rectangular solid in which all six faces are squares. [9.4]

cylinder A geometric solid in which the bases are circles and are perpendicular to the height. [9.4]

decimal A number written in decimal notation. [4.1]

decimal notation Notation in which a number consists of a whole number part, a decimal point, and a decimal part. [4.1]

decimal part In decimal notation, that part of the number that appears to the right of the decimal point. [4.1]

decimal point In decimal notation, the point that separates the whole number part from the decimal part. [4.1]

degree Unit used to measure angles; one complete revolution is 360°. [1.2; 9.1]

denominator The part of a fraction that appears below the fraction bar. [3.2]

descending order The terms of a polynomial in one variable arranged so that the exponents on the variable decrease from left to right. The polynomial $9x^5 - 2x^4 + 7x^3 + x^2 - 8x + 1$ is in descending order. [5.3]

diameter of a circle A line segment with endpoints on the circle and going through the center. [9.2]

diameter of a sphere A line segment with endpoints on the sphere and going through the center. [9.4]

difference In subtraction, the result of subtracting two numbers. [1.2]

direct variation A special function that can be expressed as the equation $y = kx$, where k is a constant called the constant of variation or the constant of proportionality. [7.5]

discount The difference between the regular price and the sale price. [8.4]

discount rate The percent of a product's regular price that is represented by the discount. [8.4]

dividend In division, the number into which the divisor is divided to yield the quotient. [1.3]

division The process of finding the quotient of two numbers. [1.3]

divisor In division, the number that is divided into the dividend to yield the quotient. [1.3]

double-bar graph A graph used to display data for purposes of comparison. [1.1]

empirical probability The ratio of the number of observations of an event to the total number of observations. [10.3]

equation A statement of the equality of two numerical or variable expressions. [1.2; 6.1]

equilateral triangle A triangle that has three sides of equal length; the three angles are also of equal measure. [9.2]

equivalent fractions Equal fractions with different denominators. [3.2]

estimate An approximation. [1.2]

evaluating a variable expression Replacing the variable or variables with numbers and then simplifying the resulting numerical expression. [1.2]

event One or more outcomes of an experiment. [10.3]

expanded form The number 46,208 can be written in expanded form as $40,000 + 6,000 + 200 + 8$. [1.1]

experiment Any activity that has an observable outcome. [10.3]

exponent In exponential form, the raised number that indicates how many times the base occurs in the multiplication. [1.3]

exponential form The expression 2^5 is in exponential form. [1.3]

exterior angle An angle adjacent to an interior angle in a triangle. [9.1]

extremes in a proportion The first and fourth terms in a proportion. [7.4]

factors In multiplication, the numbers that are multiplied. [1.3]

factors of a number The natural number factors of a number divide that number evenly. [1.3]

favorable outcomes The outcomes of an experiment that satisfy the requirements of a particular event. [10.3]

first-degree equation in one variable An equation that has only one variable, and each instance of the variable is to the first power. [6.1]

first quartile In a set of numbers, the number below which one-quarter of the data lie. [10.2]

FOIL A method of finding the product of two binomials; the letters stand for First, Outer, Inner, and Last. [5.5]

fraction The notation used to represent the number of equal parts of a whole. [3.2]

fraction bar The bar that separates the numerator of a fraction from the denominator. [3.2]

frequency distribution A method of organizing the data collected from a population by dividing the data into classes. [10.1]

frequency polygon A graph that displays information similarly to a histogram. A dot is placed above the center of each class interval at a height corresponding to that class's frequency. [10.1]

geometric solid A figure in space. [9.4]

gram The basic unit of mass in the metric system. [7.1]

graph a point in the plane To place a dot at the location given by the ordered pair; also called plotting a point in the plane. [6.5]

graph of an equation in two variables A graph of the ordered-pair solutions of the equation. [6.6]

graph of an ordered pair The dot drawn at the coordinates of the point in the plane. [6.5]

graph of a whole number A heavy dot placed directly above that number on the number line. [1.1]

greater than A number that appears to the right of a given number on the number line is greater than the given number. [1.1]

greatest common factor (GCF) The largest common factor of two or more numbers. [3.1]

height of a parallelogram The distance between parallel sides. [9.2]

height of a triangle A line segment perpendicular to the base from the opposite vertex. [3.3; 9.2]

histogram A bar graph in which the width of each bar corresponds to a range of numbers called a class interval. [10.1]

hypotenuse The side opposite the right angle in a right triangle. [9.3]

improper fraction A fraction greater than or equal to 1. [3.2]

inequality An expression that contains the symbol $<$, $>$, $\geq$ (is greater than or equal to), or $\leq$ (is less than or equal to). [1.1; 4.6]

integers The numbers $\ldots, -3, -2, -1, 0, 1, 2, 3, \ldots$. [2.1]

interest Money paid for the privilege of using someone else's money. [8.5]

interest rate The percent used to determine the amount of interest. [8.5]

interior angles The angles within the region enclosed by a triangle. [9.1]

interquartile range The difference between the third quartile and the first quartile. [10.2]

intersecting lines Lines that cross at a point in the plane. [1.2; 9.1]

inverse variation A function that can be expressed as the equation $y = \dfrac{k}{x}$, where k is a constant. [7.5]

inverting a fraction Interchanging the numerator and denominator. [3.3]

irrational number The decimal representation of an irrational number never repeats or terminates and can only be approximated. [4.6]

isosceles triangle A triangle that has two sides of equal length; the angles opposite the equal sides are of equal measure. [9.2] least common denominator (LCD) The least common multiple of denominators. [3.2; 3.4]

least common multiple (LCM) The smallest common multiple of two or more numbers. [3.1]

legs of a right triangle The two shortest sides of a right triangle. [9.3]

length A measure of distance. [7.1]

less than A number that appears to the left of a given number on the number line is less than the given number. [1.1]

like terms Terms of a variable expression that have the same variable part. [5.2]

line A line extends indefinitely in two directions in a plane; it has no width. [1.2; 9.1]

linear equation in two variables An equation of the form $y = mx + b$, where m is the coefficient of x and b is a constant. [6.6]

line segment Part of a line; it has two endpoints. [1.2; 9.1]

liter The basic unit of capacity in the metric system. [7.1]

markup The difference between selling price and cost. [8.4]

markup rate The percent of a product's cost that is represented by the markup. [8.4]

mass The amount of material in an object. On the surface of Earth, mass is the same as weight. [7.1]

maturity value of a loan The principal of a loan plus the interest owed on it. [8.5]

mean The sum of all values divided by the number of those values; also known as the average value. [10.2]

median The value that separates a list of values in such a way that there is the same number of values below the median as above it. [10.2]

meter The basic unit of length in the metric system. [7.1]

metric system A system of measurement based on the decimal system. [7.1]

minuend In subtraction, the number from which another number (the subtrahend) is subtracted. [1.2]

mixed number A number greater than 1 that has a whole number part and a fractional part. [3.2]

mode In a set of numbers, the value that occurs most frequently. [10.2]

monomial A number, a variable, or a product of numbers and variables; a polynomial of one term. [5.3]

multiples of a number The products of that number and the numbers 1, 2, 3, ... [3.1]

multiplication The process of finding the product of two numbers. [1.3]

Multiplication Property of One The product of a number and one is the number. [1.3]

Multiplication Property of Zero The product of a number and zero is zero. [1.3]

multiplicative inverse The reciprocal of a number. [5.1]

natural numbers The numbers 1, 2, 3, 4, 5, . . . [1.1]

negative integers The numbers . . . , −5, −4, −3, −2, −1. [2.1]

negative numbers Numbers less than zero. [2.1]

number line A line on which a number can be graphed; also called the real number line. [1.1; 4.6]

numerator The part of a fraction that appears above the fraction bar. [3.2]

numerical coefficient The number part of a variable term. When the numerical coefficient is 1 or −1, the 1 is usually not written. [5.2]

obtuse angle An angle whose measure is between 90° and 180°. [9.1]

obtuse triangle A triangle that has one obtuse angle. [9.2]

odds against an event The ratio of the number of unfavorable outcomes of an experiment to the number of favorable ones. [10.3]

odds in favor of an event The ratio of the number of favorable outcomes of an experiment to the number of unfavorable ones. [10.3]

opposite numbers Two numbers that are the same distance from zero on the number line, but on opposite sides. [2.1]

opposite of a polynomial The polynomial created when the sign of each term of the original polynomial is changed. [5.3]

Order of Operations Agreement A set of rules that tells us in what order to perform the operations that occur in a numerical expression. [1.5]

ordered pair Pair of numbers of the form (x, y) that can be used to identify a point in the plane determined by the axes of a rectangular coordinate system. [6.5]

ordinate The second number in an ordered pair. It measures a vertical distance and is also called the second coordinate. [6.5]

origin The point corresponding to 0 on the number line. Also the point of intersection of the two coordinate axes that form a rectangular coordinate system. [2.1; 6.5]

parallel lines Lines that never meet; the distance between them is always the same. [1.2; 9.1]

parallelogram A quadrilateral that has opposite sides equal and parallel. [9.2]

percent Parts per hundred. [8.1]

percent decrease A decrease of a quantity, expressed as a percent of its original value. [8.3]

percent increase An increase of a quantity, expressed as a percent of its original value. [8.3]

perfect square The square of an integer. [4.5]

perimeter The distance around a plane figure. [1.2; 9.2]

period In a number written in standard form, each group of digits separated from other digits by a comma or commas. [1.1]

perpendicular lines Intersecting lines that form right angles. [9.1]

pictograph A graph that uses symbols to represent information. [1.1]

place value The position of each digit in a number written in standard form determines that digit's place value. [1.1]

place-value chart A chart that indicates the place value of every digit in a number. [1.1]

plane A flat surface. [1.2; 6.5; 9.1]

plane figures Figures that lie totally in a plane. [1.2; 9.1]

plot a point in the plane To place a dot at the location given by the ordered pair; to graph a point in the plane. [6.5]

polygon A closed figure determined by three or more line segments that lie in a plane. [1.2; 9.2]

polynomial A variable expression in which the terms are monomials. [5.3]

population The set of all observations of interest. [10.1]

positive integers The numbers 1, 2, 3, 4, 5, . . . ; also called the natural numbers. [2.1]

positive numbers Numbers greater than zero. [2.1]

prime factorization The expression of a number as the product of its prime factors. [1.3]

prime number A number whose only natural number factors are 1 and itself. For instance, 13 is a prime number. [1.3]

principal The amount of money originally deposited or borrowed. [8.5]

probability A number from 0 to 1 that tells us how likely it is that a certain outcome of an experiment will happen. [10.3]

product In multiplication, the result of multiplying two numbers. [1.3]

proper fraction A fraction less than 1. [3.2]

proportion An equation that states the equality of two ratios or rates. [7.4]

Pythagorean Theorem The square of the hypotenuse of a right triangle is equal to the sum of the squares of the two legs. [9.3]

quadrant One of the four regions into which the two axes of a rectangular coordinate system divide the plane. [6.5]

quadrilateral A four-sided polygon. [1.2; 9.2]

quotient In division, the result of dividing the divisor into the dividend. [1.3]

radical The symbol $\sqrt{}$, which is used to indicate the positive square root of a number. [4.5]

radicand In a radical expression, the expression under the radical sign. [4.5]

radius of a circle A line segment going from the center to a point on the circle. [9.2]

radius of a sphere A line segment going from the center to a point on the sphere. [9.4]

range In a set of numbers, the difference between the largest and smallest values. [10.1]

rate A comparison of two quantities that have different units. [7.2]

ratio A comparison of two quantities that have the same units. [7.2]

rational number A number that can be written in the form a/b, where a and b are integers and b is not equal to zero. [4.6]

ray A ray starts at a point and extends indefinitely in one direction. [1.2; 9.1]

real numbers The rational numbers and the irrational numbers. [4.6]

reciprocal of a fraction The fraction with the numerator and denominator interchanged. [3.3]

rectangle A quadrilateral in which opposite sides are parallel, opposite sides are equal in length, and all four angles are right angles. [1.2; 9.2]

rectangular coordinate system System formed by two number lines, one horizontal and one vertical, that intersect at the zero point of each line. [6.5]

rectangular solid A solid in which all six faces are rectangles. [9.4]

regular polygon A polygon in which each side has the same length and each angle has the same measure. [9.2]

remainder In division, the quantity left over when it is not possible to separate objects or numbers into a whole number of equal groups. [1.3]

repeating decimal A decimal in which a block of one or more digits repeats forever. [4.2]

right angle A 90° angle. [1.2; 9.1]

right triangle A triangle that contains one right angle. [9.2; 9.3]

rounding Giving an approximate value of an exact number. [1.1]

sample space All the possible outcomes of an experiment. [10.3]

scalene triangle A triangle that has no sides of equal length; no two of its angles are of equal measure. [9.2]

scatter diagram A graph of collected data as points in a coordinate system. [6.5]

scientific notation Notation in which a number is expressed as a product of two factors, one a number between 1 and 10 and the other a power of 10. [5.6]

selling price The price for which a business sells a product to a customer. [8.4]

sides of an angle The rays that form the angle. [9.1]

sides of a polygon The line segments that form the polygon. [1.2]

similar objects Objects that have the same shape but not necessarily the same size. [9.3]

similar triangles Triangles that have the same shape but not necessarily the same size. [9.3]

simple interest Interest computed on the original principal. [8.4]

simplest form of a fraction A fraction is in simplest form when there are no common factors in the numerator and denominator. [3.2]

simplest form of a rate A rate is in simplest form when the numbers that make up the rate have no common factor. [7.2]

simplest form of a ratio A ratio is in simplest form when the two numbers do not have a common factor. [7.2]

solution of an equation A number that, when substituted for the variable, results in a true equation. [1.2; 6.1]

solution of an equation in two variables An ordered pair whose coordinates make the equation a true statement. [6.6]

solving an equation Finding a solution of the equation. [1.4; 6.1]

sphere A solid in which all points are the same distance from point O, which is called the center of the sphere. [9.4]

square A rectangle that has four equal sides. [1.3; 9.2]

square root A square root of a positive number x is a number a for which $a^2 = x$. [4.5; 9.3]

standard deviation A measure of the consistency, or "clustering," of data near the mean. [10.2]

standard form A whole number is in standard form when it is written using the digits 0, 1, 2, . . . , 9. An example is 46,208. [1.1]

statistics The branch of mathematics concerned with data, or numerical information. [10.1]

straight angle A 180° angle. [9.1]

subtraction The process of finding the difference between two numbers. [1.2]

subtrahend In subtraction, the number that is subtracted from another number (the minuend). [1.2]

sum In addition, the total of the numbers added. [1.2]

supplementary angles Two angles whose sum is 180°. [9.1]

surface area The total area on the surface of a solid. [9.4]

terminating decimal A decimal that has a finite number of digits after the decimal point, which means that it comes to an end and does not go on forever. [4.2]

terms of a variable expression The addends of the expression. [5.2]

theoretical probability A fraction with the number of favorable outcomes of an experiment in the numerator and the total number of possible outcomes of the experiment in the denominator. [10.3]

third quartile In a set of numbers, the number above which one-quarter of the data lie. [10.2]

transversal A line intersecting two other lines at two different points. [9.1]

triangle A three-sided polygon. [1.2; 9.1]

trinomial A polynomial of three terms. [5.3]

unit rate A rate in which the number in the denominator is 1. [7.2]

variable A letter used to stand for a quantity that is unknown or that can change. [1.2]

variable expression An expression that contains one or more variables. [1.2]

variable part In a variable term, the variable or variables and their exponents. [5.2]

variable term A term composed of a numerical coefficient and a variable part. [5.2]

vertex The common endpoint of two rays that form an angle. [9.1] vertical angles Two angles that are on opposite sides of the intersection of two lines. [9.1]

volume A measure of the amount of space inside a closed surface. [9.4]

weight A measure of how strongly Earth is pulling on an object. [7.1; 7.3]

whole number part In decimal notation, that part of the number that appears to the left of the decimal point. [4.1]

whole numbers The whole numbers are 0, 1, 2, 3, ... [1.1]

x-coordinate The abscissa in an xy-coordinate system. [6.5]

y-coordinate The ordinate in an xy-coordinate system. [6.5]

Index

Table of Geometric Formulas and Properties

PERIMETER

Triangle: $P = a + b + c$

Rectangle: $P = 2L + 2W$

Square: $P = 4s$

Circle: $C = \pi d$ or $C = 2\pi r$

AREA

Triangle: $A = \dfrac{1}{2} bh$

Rectangle: $A = LW$

Square: $A = s^2$

Circle: $A = \pi r^2$

Parallelogram: $A = bh$

Trapezoid: $A = \dfrac{1}{2} h(b_1 + b_2)$

VOLUME

Rectangular solid: $V = LWH$

Cube: $V = s^3$

Sphere: $V = \dfrac{4}{3} \pi r^3$

Right circular cylinder: $V = \pi r^2 h$

Right circular cone: $V = \dfrac{1}{3} \pi r^2 h$

Regular pyramid: $V = \dfrac{1}{3} s^2 h$

SURFACE AREA

Rectangular solid: $SA = 2LW + 2LH + 2WH$

Cube: $SA = 6s^2$

Sphere: $SA = 4\pi r^2$

Right circular cylinder: $SA = 2\pi r^2 + 2\pi rh$

Right circular cone: $SA = \pi r^2 + \pi rl$

Regular pyramid: $SA = s^2 + 2sl$

TRIANGLES

Sum of the measures of the interior angles = 180°

Sum of an interior and corresponding exterior angle = 180°

An **isosceles triangle** has two sides equal in length;
 the angles opposite the equal sides are of equal measure.

In an **equilateral triangle**, the three sides are of equal length;
 the three angles are of equal measure.

A **scalene triangle** has no two sides of equal length;
 no two angles are of equal measure.

An **acute triangle** has three acute angles.

An **obtuse triangle** has one obtuse angle.

A **right triangle** has a right angle.

In **similar triangles**:
 The ratios of corresponding sides are equal.
 The ratio of corresponding heights is equal to the ratio of corresponding sides.

Rules used to determine **congruent triangles**: SSS rule,
 SAS rule, ASA rule.

PYTHAGOREAN THEOREM

If a and b are the legs of a right triangle and c is the length of the hypotenuse, then $c^2 = a^2 + b^2$.